Proceedings of the International Petroleum and Petrochemical Technology Conference 2019

Jia'en Lin
Editor

Proceedings of the International Petroleum and Petrochemical Technology Conference 2019

Springer

Editor
Jia'en Lin
School of International Education
Xi'an Shiyou University
Xi'an, Shaanxi, China

ISBN 978-981-15-0859-2 ISBN 978-981-15-0860-8 (eBook)
https://doi.org/10.1007/978-981-15-0860-8

This Springer imprint is published by the registered company Springer Nature Singapore Pte Ltd.
The registered company address is: 152 Beach Road, #21-01/04 Gateway East, Singapore 189721, Singapore

Contents

Analysis of Initial Productivity Evaluation Method for Horizontal Well in Offshore Block Oilfield

Li-jun Zhang(✉), Wei Zheng, and Guo-jin Zhu

CNOOC Reasearch Institute Co., Ltd., Beijing, China
{zhanglj8, zhengwei8, zhugj}@cnooc.com.cn

Abstract. Due to the short testing time and unstable productivity of horizontal wells in offshore fault-block oilfields, it is very important to evaluate the stable productivity of horizontal wells and the comprehensive correction coefficient of productivity by using limited horizontal well testing data. In this paper, the unsteady flow stage of horizontal wells is decomposed, and the productivity equations of vertical radial flow, linear flow and pesudo radial flow stage are obtained respectively. On this basis, the test time correction coefficients under different formation properties and horizontal well length conditions are plotted, unsteady productivity changes in horizontal wells under different fault conditions are compared and analyzed, the productivity correction coefficients under different fault conditions are obtained. The distance between well spacing fault and horizontal well is generally 0.2–1.0, its productivity correction coefficient is 0.68–0.85. The reliability of the charts are verified by testing data and production performance data of an offshore fault block oilfield. The methods and charts can effectively guide the determination of productivity of horizontal well testing method in offshore fault block oilfield.

Keywords: Horizontal well · Productivity · Well testing · Fault · Correction coefficient

J. Lin (ed.), *Proceedings of the International Petroleum and Petrochemical Technology Conference 2019*, pp. 1–13, 2020.
https://doi.org/10.1007/978-981-15-0860-8_1

1 Introduction

Horizontal well placement has evolved from the science of drilling wells at 90° from the kickoff point to the art of utilizing technologies that place wells with maximum reservoir contact. And we also known many scholars have studied the productivity of horizontal well.

Merkulov deduced the analytical formula of horizontal well productivity for zonal and circular reservoirs, which is suitable for pseudo-radial and parallel flow [1]. Borisov summarizes the production principle and development process of horizontal wells, and puts forward the equation of steady productivity calculation for horizontal wells [2]. Giger obtained the productivity ratio equation of horizontal well and vertical well in homogeneous isotropic reservoir based on Borisov formula, and compared the productivity of horizontal well and vertical well [3]. Joshi simplified the three-dimensional seepage problem of horizontal wells into two-dimensional seepage problem in vertical and horizontal planes. The steady-state productivity equation of horizontal wells in homogeneous isotropic reservoirs was derived by using potential energy theory [4]. Renard et al. summarized the productivity equations of Joshi and Giger horizontal wells and introduced skin factor to modify the steady-state equation [5, 6].

The above is the main steady-state productivity equation. Mutalik assumed the shape coefficients and corresponding equivalent skin coefficients of horizontal wells at any position in a certain rectangular oil release zone. According to the productivity formula of fractured vertical wells, the productivity formula for predicting quasi-steady state of horizontal wells was given [7]. Babu set up a mathematical model for three-dimensional unsteady seepage flow in horizontal wells for arbitrary box type closed reservoirs [8]. Kuchuk et al. used an approximate infinite diversion method to derive the inflow performance equations of horizontal wells with constant pressure and impermeable top-bottom boundary conditions [9].

Frick et al. deduced an analytical formula for skin factor of horizontal wells, which can be directly added to the productivity formula of horizontal wells. The form of the equation is similar to the traditional productivity formula of vertical wells, and it is suitable for both unstable early stage and quasi-stable flow [10].

Although the steady-state model of horizontal wells is widely used in productivity prediction of horizontal wells, in fact, it is difficult for any reservoir to appear in a steady-state form when well testing. The unstable pressure formula is researched by many scholar.

Gringarten and Ramey introduce the point source function and Green function to solve unstable reservoir seepage problem, based on the theory of point source function, horizontal well testing technology has been developed [11]. Daviau first proposed a horizontal well test analysis model [12]. Goode considered reservoir anisotropy in the horizontal well test model [13]. Kuchuk proposed a horizontal well test model with different boundary between top and bottom of reservoir [14, 15]. Ozkan obtained a dynamic mathematical model of horizontal well with closed boundary, and solved the model in pull space [16].

In the process of horizontal well testing, unstable flow can be divided into three sections: vertical radial flow, linear flow and quasi-radial flow. Because of the short test time at sea, the productivity of quasi-radial flow is still not stable. In evaluating the stable productivity of horizontal well, we must introduce productivity correction coefficient.

2 Unsteady Productivity Equation of Horizontal Well

Suppose a horizontal well is drilled in the middle of the reservoir, the half length of the horizontal well is L, the thickness of the reservoir is h, and the horizontal permeability and vertical permeability are different. The horizontal well location is shown in Fig. 1.

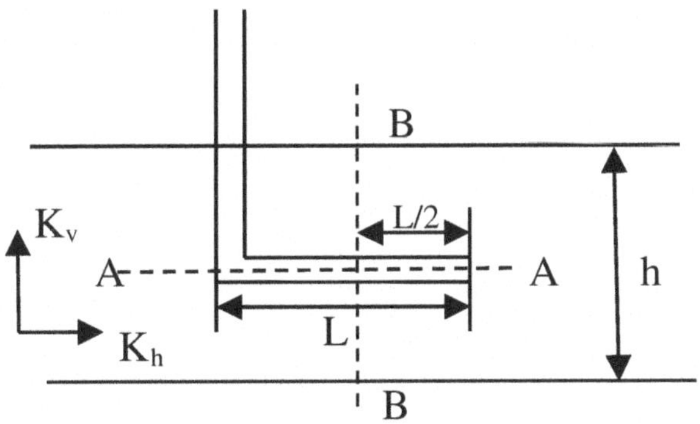

Fig. 1 Horizontal wells in central reservoir schematic diagram

Assuming that the upper and lower boundaries of the strata are closed and the horizontal direction is infinite, the seepage control equation of horizontal wells is:

$$K_h \frac{\partial^2 p}{\partial x^2} + K_h \frac{\partial^2 p}{\partial y^2} + K_v \frac{\partial^2 p}{\partial z^2} = \phi \mu C_t \frac{\partial p}{\partial t} \tag{1}$$

where $z^* = z\sqrt{K_h/K_v}$

The above formula can be turned into the standard equation:

$$\frac{\partial^2 p}{\partial x^2} + \frac{\partial^2 p}{\partial y^2} + \frac{\partial^2 p}{\partial z^{*2}} = \frac{1}{\eta} \frac{\partial p}{\partial t} \tag{2}$$

Initial and boundary conditions of reservoir:

$$p(x, y, z, t)|_{t=0} = p_{\mathrm{i}}$$

$$\left.\frac{\partial p}{\partial x}\right|_{x \to \infty} = 0 \left.\frac{\partial p}{\partial y}\right|_{y \to \infty} = 0 \left.\frac{\partial p}{\partial z}\right|_{z=0} = \left.\frac{\partial p}{\partial z}\right|_{z=h} = 0 \tag{3}$$

The above is a differential equation for unsteady seepage in horizontal wells with an upper and lower boundary in an infinite reservoir.

Using Newman product method and instantaneous source solutions under various conditions, the formation pressure distribution formulas of the above models are obtained.

$$\tilde{p}(x,y,z,t) = p_i - \frac{dv}{\phi C_t}\frac{1}{2}\left\{\text{erf}\left[\frac{L+(x-x_w)}{\sqrt{4\eta(t-\tau)}}\right] + \text{erf}\left[\frac{L-(x-x_w)}{\sqrt{4\eta(t-\tau)}}\right]\right\}$$
$$\frac{1}{\sqrt{4\eta(t-\tau)}}\exp\left[-\frac{(y-y_w)^2}{4\eta(t-\tau)}\right] \tag{4}$$
$$\frac{1}{h^*}\left\{1+2\sum_{n=1}^{\infty}\exp\left[-\frac{n^2\pi^2\eta_V(t-\tau)}{h^{*2}}\right]\cos\frac{m\pi z_w}{h}\cos\frac{m\pi z}{h}\right\}$$

where $h^* = h\sqrt{K_h/K_v}$

The following dimensionless quantities are defined:

$$h_D^* = h^*/L \; \eta_V = \frac{K_v}{K_h}\eta = \beta^2\eta \; t_D = \eta\frac{t}{L^2} \; p_D = \frac{2\pi K_h h^*[p_i - p(x,y,z,t)]}{q\mu B} \; L_D = \frac{L}{h^*}$$

So the dimensionless horizontal well pressure distribution can be written:

$$p_D(x_D,y_D,z_D,t_D) = \frac{\sqrt{\pi}}{4}\int_0^{t_D}\frac{1}{\sqrt{t}}\exp[-\frac{(y_D-y_{wD})^2}{4t}]$$
$$\{\text{erf}[\frac{1+(x_D-x_{wD})}{\sqrt{4t}}] + \text{erf}[\frac{1-(x_D-x_{wD})}{\sqrt{4t}}]\} \tag{5}$$
$$[1+2\sum_{n=1}^{\infty}\exp(-\frac{n^2\pi^2\beta^2 t}{h_D^{*2}})\cos n\pi z_{wD}\cos n\pi z_D]dt$$

The above formula is a dimensionless bottom hole pressure solution for horizontal wells without considering wellbore storage effect and skin effect.

when considering well storage The dimensionless pressure drop at the bottom of well is considered as follows:

$$p_{wDC} = \int_0^{t_D}[1 - C_D\frac{dp_{wDC}}{dt_D}]\frac{dp_D(t-\tau)}{d\tau}d\tau \tag{6}$$

where

$$C_D = \frac{C}{2\pi h^*\phi C_f L^2} \tag{7}$$

when considering skin effect, The dimensionless pressure drop at the bottom of well is considered as follows:

$$p_{\text{wDS}} = \frac{1}{L_D}(1 - C_D \frac{dp_{\text{wDS}}}{dt_D})S \tag{8}$$

When evaluating the initial productivity of horizontal oil wells, we should analyze the influence of different testing time and different types of boundary. Therefore, the following dimensionless productivity index is defined as follows:

$$J_D = \frac{q}{(p_i - p_{wf})} \cdot \frac{1.842 \times 10^{-3} \mu B}{K_h h \sqrt{\frac{K_h}{K_v}}} = \frac{1}{p_{\text{wDs}}} \tag{9}$$

Because of the length of the horizontal well, the location of the horizontal well in the reservoir, and the heterogeneity of the reservoir, the flow pattern of the horizontal well is different from the vicinity of the wellbore to the middle of the reservoir. In the shorter test time, the horizontal well can not reach stable flow or quasi-stable flow, so when using the horizontal well test data to evaluate productivity, the test time correction coefficient and boundary correction coefficient should also be defined.

The determination principle of test time correction factor: Taking upper and lower layers closed and infinite plane reservoirs as an example, dimensionless recovery index of horizontal wells gradually decreases and finally achievs stability with the extension of production time. The stable production index and the ratio of production index to different test time are defined as test time correction coefficient, as shown in formula 10.

$$\Delta J_t = \frac{J_{DTest}}{J_{Dstable}} \tag{10}$$

The determining principle of boundary correction factor:

Because the horizontal well has a certain length, the stable productivity varies with the distribution of oil wells and faults. The stable recovery index of infinite reservoir is taken as the base, and the ratio of stable recovery index with different fault distances is defined as the boundary correction coefficient, as shown in formula 11.

$$\Delta J_b = \frac{J_{Dbstable}}{J_{DIstable}} \tag{11}$$

2.1 Test Time Correction Coefficient Plates

The actual offshore oilfield DST test time is very short, generally in about 5–15 h. During the horizontal well testing, the vertical radial flow occurs earlier and lasts less than an hour. Therefore, in the DST test time range, the duration of linear flow and quasi-radial flow is long, and sometimes only the linear flow section may not appear quasi-radial flow section. According to the productivity equation of horizontal wells in different testing stages, the productivity curves of horizontal wells under different reservoir and well parameters are established, as shown in Figs. 2 and 3.

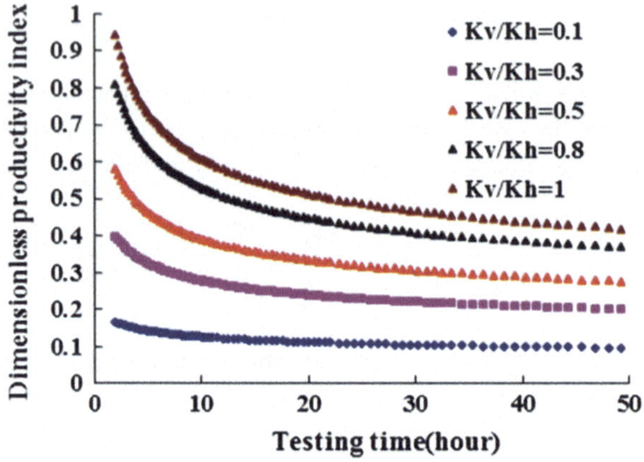

Fig. 2 Horizontal well dimensionless PI versus time curves in different Kv/Kh

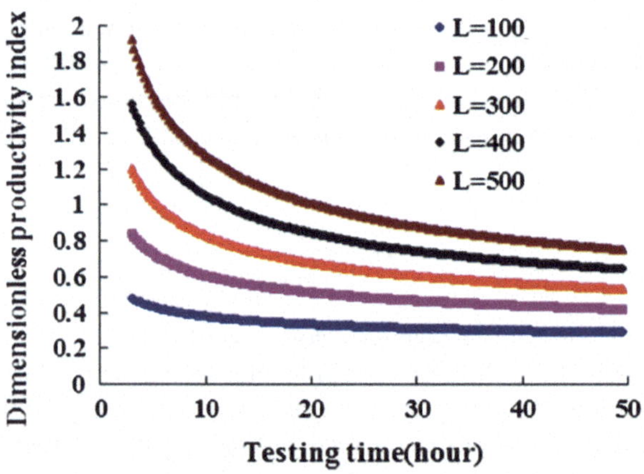

Fig. 3 Horizontal well dimensionless PI versus time curves in different well length

In the process of horizontal well testing, the pressure and flow sections are affected by the length of horizontal well and the anisotropy of oil field, so the length and anisotropy of different horizontal sections are determined separately when evaluating the correction of testing time for productivity. Figure 4 gives the time productivity correction factor for homogeneous and anisotropic horizontal wells.

The time productivity correction coefficients are different under different anisotropic conditions. When the length of horizontal wells is the same, the productivity is greatly affected by anisotropy, that is, the greater the ratio of vertical permeability to horizontal permeability, the smaller the correction coefficient of test time; under the same reservoir physical conditions, the longer the horizontal section, the greater the influence of time on productivity, the smaller the correction coefficient. The average test time is about 5–15 h, and the correction time is about 0.5–0.7.

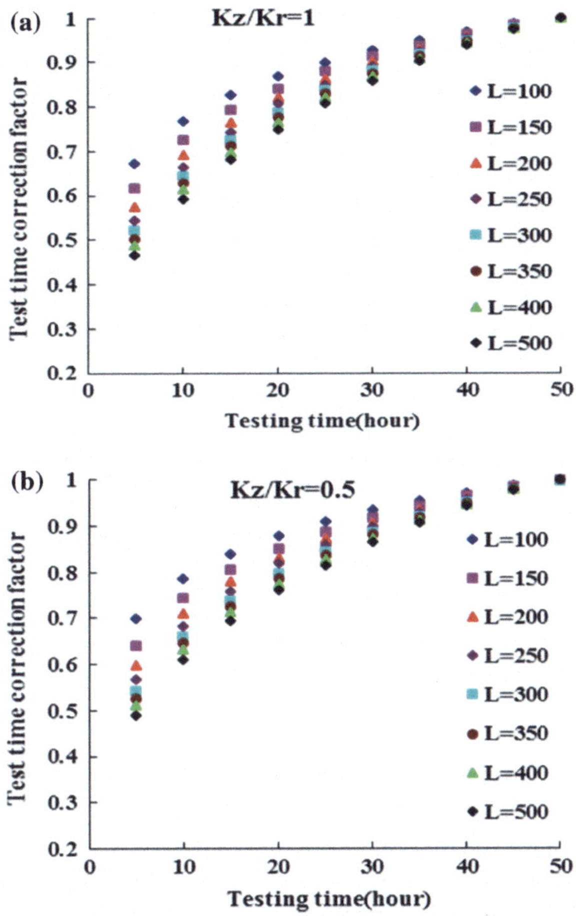

Fig. 4 **a** Test time correction coefficient in different well length (Kv/Kh = 1.0). **b** Test time correction coefficient in different well length (Kv/Kh = 0.5)

2.2 Boundary Correction Coefficient Plates

Taking a fault as an example, as shown in Fig. 5, a single fault has the following distribution relationship with a horizontal well, parallel to and perpendicular vertical to the fault. The productivity of horizontal well under these two conditions is analyzed.

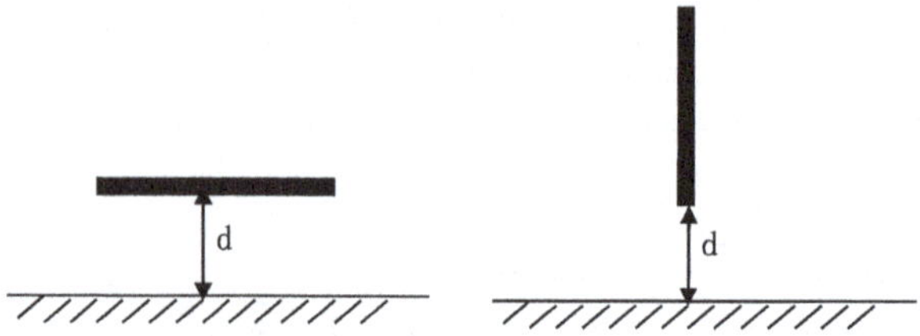

Fig. 5 Horizontal well models near faults schematic diagram

When the horizontal well is parallel to the fault, the fault response occurs earlier in well testing curves than the horizontal well is vertical to the fault. Sometimes the linear flow sections are covered when the horizontal well is perpendicular to the fault, the fault response time is later, and the pesudo radial flow occur when the linear flow ends. In both cases, the pressure derivatives eventually converge, but the pressure drop parallel to the fault is greater than that perpendicular to the fault, as shown in Fig. 6.

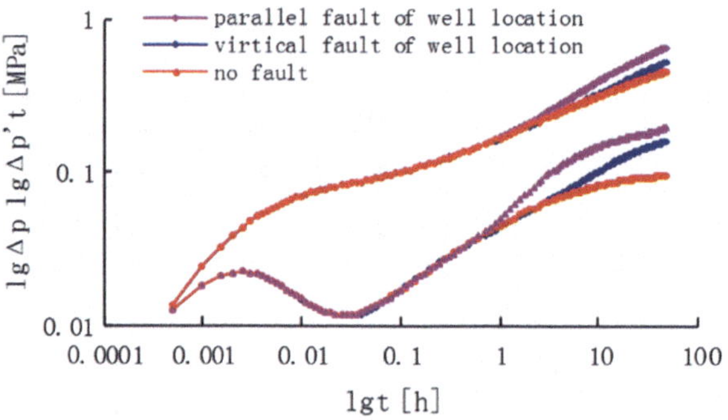

Fig. 6 Horizontal well testing type curves under different fault conditions

The curves of dimensionless PI versus time under two conditions are shown in Fig. 7. The vertical radial flow lasted for a short time, and the dimensionless production index changed the same. The dimensionless production index of horizontal wells parallel to faults decreases rapidly and early than the well perpendicular to faults. The dimensionless PI of horizontal wells is ultimately stable due to the influence of faults, but the pseudo stable flow productivity index of well parallel to the fault is smaller than that perpendicular to the fault.

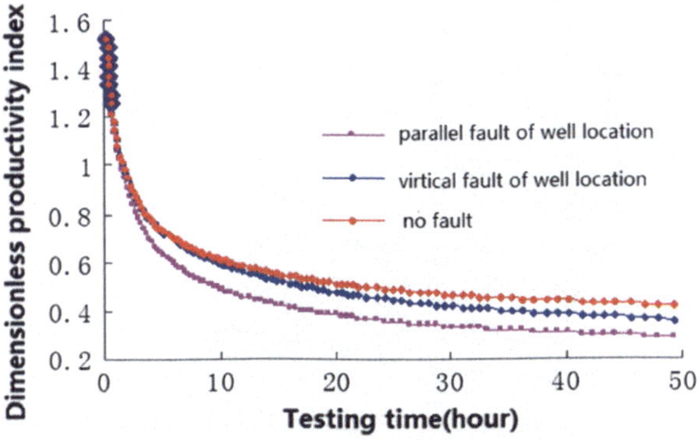

Fig. 7 Horizontal well dimensionless PI versus time curves under different fault distances

The smaller of distance horizontal well to fault, and the smaller the ratio of vertical permeability to horizontal permeability, the boundary correction coefficient is larger. The ratio of actual well spacing fault distance to horizontal well length is generally between 0.2 and 1.0, and the boundary correction coefficient is between 0.68 and 0.85. The horizontal well boundary correction coefficient plates are shown in Fig. 8.

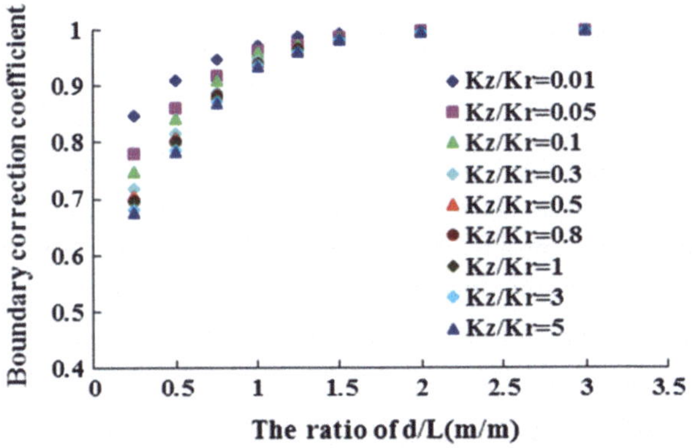

Fig. 8 Boundary correction coefficient plates in different Kv/Kh

2.3 Case Studies

A reservoir is a deep-water submarine fan channel deposit. The porosity is medium and the permeability is high in this reservoir. There are some faults in the sand layer. Well A9 is a horizontal well in this reservoir. The effective length of the horizontal well drilled into the reservoir is 260 m. The effective thickness of the reservoir is 8.5 m. Horizontal wells pass through faults. The distance between the two ends of the horizontal well is about 100–200 m. The horizontal well is completed with high quality screen pipe, Well A9 location map is shown in Fig. 9.

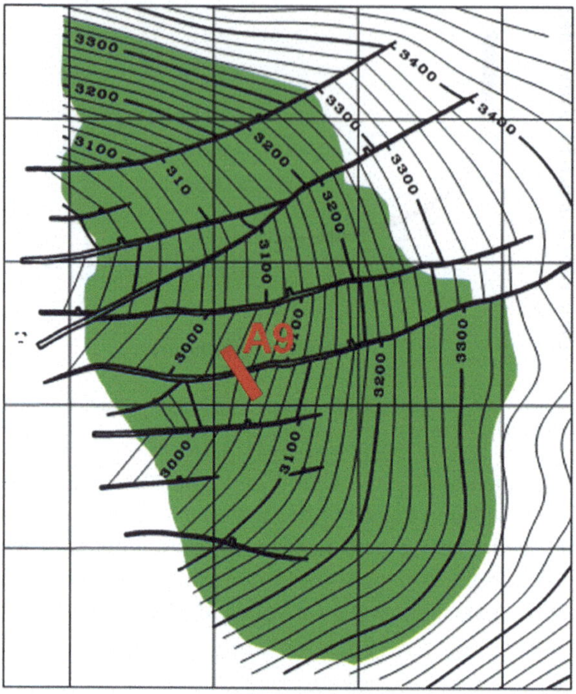

Fig. 9 Well A9 location map

The well test time is about 50 h, and the nozzle is replaced every four hours during well testing. Because the duration of each flow section is short, the flow is not completely stable or quasi-stable, so the test time and faults have a great influence on the productivity. The curve of productivity index with time in the test phase is shown in Fig. 10, the productivity index decreases with time.

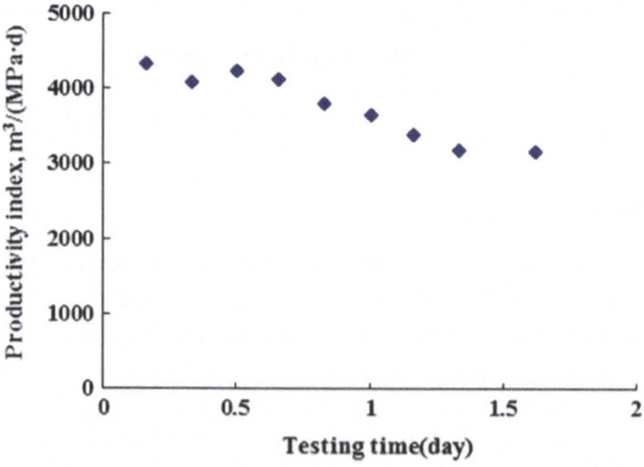

Fig. 10 A9 well blow off testing PI versus time curves

The vertical radial flow, linear flow and quasi-radial flow all occur in the log-log curve of horizontal well test, but the response of fault on the well test curve may coincide with the linear flow and can not be distinguished. The well test fitting curve is shown in Fig. 11.

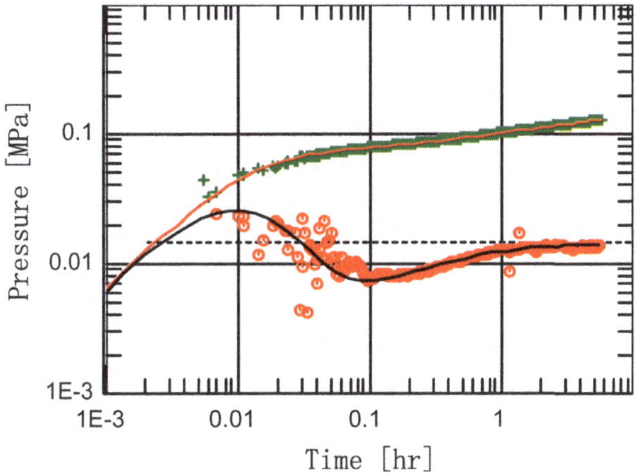

Fig. 11 A9 well testing fitting curves

Because of unstable flow in the well test stage, the testing time is about 1.5 days and the length of horizontal well is 260 m. According to previous theoretical analysis, the time correction coefficient is about 0.7–0.8, the ratio of horizontal well spacing fault distance to horizontal well length is about 0.3, so the boundary correction coefficient is 0.7. The composite correction coefficient is equal to the product of the time correction coefficient and the boundary correction coefficient, which is about 0.52.

We use the production data to verify the composite correction coefficient. The test productivity and production productivity are converted to the ideal production index. The ideal productivity index of the well test is 5150 $m^3/(MPa{\cdot}d)$, the well initial stable ideal productivity index is 2450 $m^3/(MPa{\cdot}d)$, so the actual composite correction coefficient is close to 0.47, which is close to the theoretical analysis results (Fig. 12).

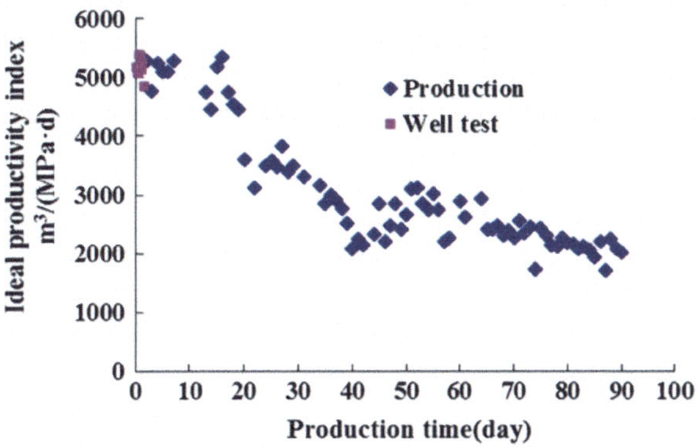

Fig. 12 A9 well early period production curves

3 Conclusion

Aiming at the problems of short test time and unstable production capacity of DST in offshore block oilfield evaluation well. The productivity equations of different flow sections of horizontal wells are established, and the time-dependent variation of dimensionless recovery index of horizontal wells under different conditions is analyzed.

(1) Horizontal well productivity test time correction coefficient chart and boundary correction coefficient chart are established.
(2) when the test time is 5–15 h, Test time correction coefficient is about 0.8.
(3) The ratio of actual well spacing fault distance to horizontal well length is generally between 0.2 and 1.0, and its productivity correction coefficient is between 0.68 and 0.85.
(4) Combining with the actual horizontal well test and the production performance data, the reliability of the theoretical chart is verified.

The establishment of the two kinds of chart has guiding significance for the future productivity evaluation of horizontal wells in fault-block oilfields.

References

1. Merkulov, V.P.: Le debit des puits devies et horizontaux. Neft. Khoz. **6**, 51–56 (1958)
2. Borisov, J.P.: Oil Production Using Horizontal and Multiple Deviation Wells. The R & D Library Translation, Oklahoma (1984)
3. Giger, F.M.: Horizontal wells production techniques in heterogenous reservoir. SPE 13710 (1985)
4. Joshi, S.D.: Augmentation of well production using slant and horizontal wells. SPE 15375 (1986)
5. Renard, G., Dupuy, J.M.: Formation damage effects on horizontal-well flow efficiency. SPE 19414 (1991)
6. Elgaghad, S.A., Osisanya, S.O., Tiab, D.: A simple productivity equation for horizontal wells based on drainage area concept. SPE 35713 (1996)
7. Mutalik, P.N., Godbole, S.P., Joshi, S.D.: Effect of drainage area shapes on the productivity of horizontal wells. SPE 18301 (1988)
8. Babu, D.K., Odeh Aziz, S.: Productivity of a horizontal well. SPE 18298 (1989)
9. Kuchuk, F.J., Goode, P.A., Wilkson, D.J., et al.: Pressure-transient behavior of horizontal wells with and without gas cap or aquifer. SPE 17413 (1991)
10. Frick, T.P., Economides, M.J.: Horizontal well damages characterization and removal. SPE 21795 (1991)
11. Gringarten, A.C., Ramey, H.J.: The use of source and green's functions in solving unsteady-flow problem in reservoir. pe 3818 (1973)
12. Davian,F., Mouronal, G., Boudarot, G., Curutchet, P.: Pressure analysis for horizontal wells. SPEFE, 716–724 (1988)
13. Goode, P.A.: Pressure drawdown and buildup analysis of horizontal wells in anisotropic media. SPE 14250 (1987)

14. Kuchuk, F.J., Goode, P.A.: A pressure transient analysis and inflow performance for horizontal wells. SPE 18300
15. Kuchuk, F.J.: Well testing and interpretation for horizontal wells. JPT. 36–40 (1995)
16. Ozkan, E., Raghavan, R., Joshi, S.D.: Horizontal well pressure analysis. SPE 20271

Unstable Polymer Flooding and Its Application in Offshore Heavy-Oil Field of China

Zhi-jie Wei[1,2(✉)], Xiao-dong Kang[1,2], Jian Zhang[1,2], Yu-yang Liu[1,2], and Yang Zeng[1,2]

[1] State Key Laboratory of Offshore Oil Exploitation, Beijing, China
{weizhj5, kangxd, zhangjian}@cnooc.com.cn, {liuyy52, zengyang}@cnooc.com.cn
[2] CNOOC Research Institute Co., Ltd., Beijing, China

Abstract. As the injection conformance reversion induced by single-slug continuous polymer flooding usually compromises EOR results, a novel and unstable discontinuous polymer injection pattern and the corresponding black-oil-based multi-component polymer model is proposed, which allows to conduct research on mechanisms and key injection parameter optimization in offshore heavy-oil reservoirs. In addition, its pilot test and results in J oilfield of Bohai Bay of China will be discussed. It shows that unstable polymer flooding pattern can improve compatibility between the injected liquid and reservoir heterogeneity, which makes the alternation law of flow resistance force of layers tend to be the same and contain conformance reversion, which make the injection conformance of heavy-oil reservoir turn to 'U' type from 'V' type, finally enhances liquid suction amount of the low-permeability zone and its remaining oil mobilization. Proper alternating injection cycle can best suppress conformance reversion and creates best EOR results. An alternating injection cycle optimization model is developed based on uniform design method. Good EOR results are shown in the pilot test, the central production well witness more than 7.3 percentage decrement in water cut and more than 70% increment in oil production rate. These can provide technical guidance for sustainable and efficient development of heavy-oil reservoir.

Keywords: Polymer flooding · Conformance reversion · Unstable injection pattern · Enhanced oil recovery · Vertical sweep

J. Lin (ed.), *Proceedings of the International Petroleum and Petrochemical Technology Conference 2019*, pp. 14–25, 2020.
https://doi.org/10.1007/978-981-15-0860-8_2

1 Introduction

Polymer flooding is proved as a successful EOR technology in offshore oilfields of Bohai Bay, China [1–3]. However, some production wells witness adverse performance, such as early water cut rebound, high produced polymer concentration and fast growth, which compromise EOR results of polymer flooding. It is found that it is subject to the conformance reversion induced by the conventional single-slug continuous polymer injection pattern [4, 5], which indicates that the relative suction amount of low-permeability zone firstly increases and then decreases during polymer flooding. After that, the vertical sweep is continuously weaken, which leads to low-efficient circulation of polymer solution in high permeable zone [6]. To effectively contain conformance reversion and further improve EOR results of polymer flooding, a novel and unstable injection pattern is proposed, i.e., polymer is not injected by the conventional single-slug continuous way with the same chemical agents, but by an alternating way with different concentration or different chemicals.

A black-oil-based multi-component polymer model is developed and the corresponding numerical solver is programmed with the fully implicit method in this study. Then, simulation cases are run to explore the mechanisms and parameter optimization for offshore heavy-oil reservoirs. In addition, its pilot test have been implemented in J oilfield in Bohai Bay and its results will be discussed.

2 Mathematical Model

Unstable polymer flooding probably involve multi component polymer, i.e., many types of polymer. Similar with the conventional single polymer, injection of many types of polymer can also increase viscosity of displacing phase and decrease the effective permeability of displaced phase, both can significantly improve the adverse mobility ratio for heavy-oil reservoir, which favor the volumetric sweep and finally enhance oil recovery [7, 8]. But, as for unstable injection with multi component polymer, different polymers within water phase can interact with each other, and also interact with rock during transportation in the porous media, these create more complicated physicochemical phenomena, such as competitive adsorption, viscosity enhancement, inaccessible pore volume and residual resistance factor, and etc. Polymer used in chemical flooding is water soluble and its impact on oil phase can be ignored [9, 10]. By considering all of these displacement mechanisms and physicochemical phenomena, a black-oil-based multi-component polymer mathematical model is developed as:

$$\begin{aligned}
&\text{oil: } \nabla\cdot\left(\frac{Kk_{ro}}{B_o\mu_o}\nabla\Phi_o\right)+q_o=\frac{\partial}{\partial t}\left(\frac{\phi S_o}{B_o}\right)\\
&\text{gas: } \nabla\cdot\left(\frac{Kk_{rg}}{B_g\mu_g}\nabla\Phi_g\right)+\nabla\cdot\left(\frac{Kk_{ro}}{B_o\mu_o}R_s\nabla\Phi_o\right)+q_g+R_sQ_o=\frac{\partial}{\partial t}\left(\frac{\phi S_g}{B_g}+R_s\frac{\phi S_o}{B_o}\right)\\
&\text{water: } \nabla\cdot\left(\frac{Kk_{rw}}{B_w\mu_pR_K}\nabla\Phi_w\right)+q_w=\frac{\partial}{\partial t}\left(\frac{\phi S_w}{B_w}\right)\\
&\text{polymer } i\cdot\nabla\cdot\left(\frac{Kk_{rw}c_{p,i}}{B_w\mu_pR_K}\nabla\Phi_w\right)+q_wc_{p,i}=\frac{\partial}{\partial t}\left(\frac{(1-I_{PV,i})\phi S_wc_{p,i}}{B_w}\right)\\
&\qquad+\frac{\partial}{\partial t}[\rho_rc_{p,i}^a(1-\phi)]\quad i=1,2,\ldots,m
\end{aligned} \tag{1}$$

where K is permeability, mD; ϕ is porosity; k_r is relative permeability; B is liquid compressibility, MPa^{-1}; μ is viscosity, mPa·s; Φ is potential, Pa; g is gravity, m/s^2; Z is depth, m; q is source or sink term, m^3/day; ρ_r is density ratio of rock to water; S is saturation; R_s is dissolved gas-oil ratio, m^3/m^3; R_K is general residual resistance factor; $c_{p,i}$ is concentration of polymer i in water, mg/L; $c_{p,i}^a$ is adsorption amount of polymer i, µg/g; $I_{PV,i}$ is inaccessible pore volume of polymer i; m is total polymer number; o, w, g, r indicates oil, water, gas and rock, respectively; i indicates polymer number; ∇ indicates gradient; $\nabla\cdot$ indicates divergence.

From Eq. (1), the conservative equations of multi-component polymer is as the same as that of single polymer for oil and gas, but different water and polymer. The difference are PVT properties and water-rock interaction parameters, the former mainly consist of viscosity of multi-component polymer solution, the latter consist of multi-component polymer adsorption, general inaccessible pore volume and etc.

2.1 Multi-component Polymer Viscosity Equation

Multi component polymers within water phase significantly affect its viscosity. The water viscosity depends on the concentration of each polymer. The general viscosity multiplier, defined as the viscosity ratio of polymer solution to water under original state, is introduced, that is:

$$\mu_p=R_\mu\cdot\mu_w \tag{2}$$

where R_μ is general viscosity multiplier, $R_{\mu,i}$ is viscosity multiplier of polymer i.

According to the lab data and history matching results, the general viscosity multiplier R_μ is the logarithmic weighted sum of viscosity multiplier of each polymer:

$$ln(R_\mu)=\sum_{i=1}^{m}w_i\cdot ln[R_{\mu,i}(c_{p,i})]\rightarrow R_\mu=\prod_{i=1}^{m}R_{\mu,i}^{wi} \tag{3}$$

where w_i is weight. The introduction of w_i help more accurate describe viscosity of multi-component polymer mixture.

2.2 Multi-component Polymer Adsorption Equation

The adsorption of single component polymer is commonly described by Langmuir type curve, but no such curves have been reported to our best knowledge. With the fact that rock has different adsorption capacity to each polymer, and competitive relationship exists for polymers to the limited adsorption space of rock, which is similar to multi-component gas adsorption to coal rock [11], an extended Langmuir type curves is applied to describe multi-component polymer adsorption, and each polymer's adsorption amount can be expressed as:

$$c_{p,i}^{a} = c_{p,i}^{\max} \frac{b_i c_{p,i}}{1 + \sum_{i=1}^{m} b_i c_{p,i}} \tag{4}$$

where $c_{p,i}^{\max}$ is the maximum adsorption amount of polymer i, μg/g; b_i is Langmuir adsorption constant of polymer i, L/mg.

The black-oil-based multi-component polymer model for unstable discontinuous polymer flooding is developed. The governing equations are mass conservative equations (Eq. 1), and the primary variables consist of water pressure p_w, water saturation S_w, oil saturation S_o, polymer concentration $c_{p,i}$. The full implicit finite volume method is utilized to discrete Eq. (1) and the corresponding simulation code is programmed with linear solver called CPR (constrained pressure residual) is released in this study. In the following, illustration cases are designed and run to explore the EOR mechanisms and parameter optimization of unstable polymer flooding in offshore heavy-oil reservoirs.

3 EOR Mechanisms

Based on the reservoir properties of J oilfield of Bohai Bay, a conception simulation model is designed: (1) dual layered reservoir with determinant well pattern, permeability, porosity, PVT properties, and polymer physicochemical parameters and etc. are all adopted from the real reservoir data; (2) displacing process consist of first water flooding, and then polymer flooding for 0.67 pore volume, and finally chasing water flooding, with injection rate 0.045 PV/a and voidage replacement ratio of 1.0; (3) two injection pattern is modeled during polymer flooding, one is continuous single polymer injection with medium molecular weight and mass concentration of 1500 mg/L, the other is unstable discontinuous polymer injection pattern with 2000 mg/L high molecular weight polymer first alternated by 1000 mg/L medium molecular weight polymer, the cyclic can vary from one to eight. The total mass of chemical agent is the same.

To explore the EOR mechanisms of unstable polymer flooding in heavy-oil reservoirs, two simulation cases are conducted: the continuous single-slug polymer injection and alternating injection with one cyclic, and the corresponding suction profiles are shown in Fig. 1. As for continuous injection, the suction amount of low permeability layer firstly decrease quickly and go stable in water flooding stage; and increase linearly when polymer injection starts, and then turns to decrease later. Generally, the suction profile display V type curve. Alternating injection significantly alter the profile and make it tend to U type. During high molecular polymer injection, the suction of low permeability layer increase more quickly in the beginning and then gradually goes stable, while in the sequent medium molecular polymer injection, it decreases smoothly. Unstable polymer injection can suppress conformance reversion, which enhances liquid suction amount of the low-permeability layer by 8.5% and creates bigger and longer water cut reduction funnel, the maximum water cut reduction can reaches 14.3 percentage from 9.5 percentage, the ultimate recovery factor can further reaches 11.1% from 9.2%.

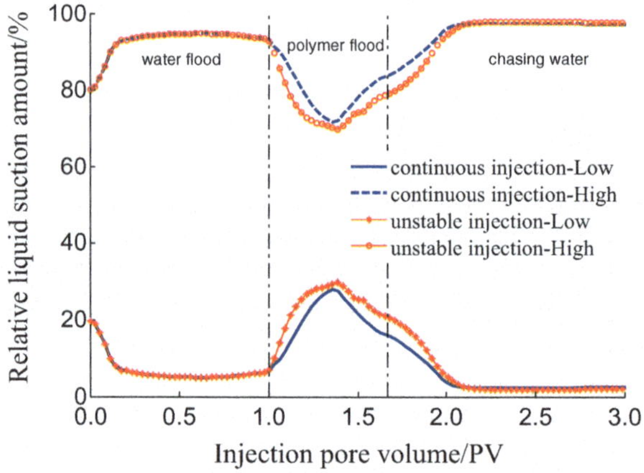

Fig. 1 Liquid suction profile comparison between continuous and unstable injection pattern

According to Darcy's law, the relative suction of low permeability zone and its derivative are:

$$\begin{cases} f_1 = \frac{q_1}{q_1 + q_2} = \frac{\frac{K_1 H_1}{R_1}\nabla p}{\frac{K_1 H_1}{R_1}\nabla p + \frac{K_2 H_2}{R_2}\nabla p} = \frac{R_2}{MR_1 + R_2} \\ f_1' = (\frac{R_2}{MR_1 + R_2})' = \frac{MR_1 R_2}{(MR_1 + R_2)^2}[\frac{(R_2)'}{R_2} - \frac{(R_1)'}{R_1}] \\ M = \frac{K_2 H_2}{K_1 H_1} \end{cases} \quad (5)$$

where f is relative liquid suction amount; R is flow resistance force; H is net thickness, m; M is formation coefficient ratio of high permeability zone to low permeability zone; subscript 1 and 2 indicates low permeability zone, high permeability zone. Equation (5) shows that the relative suction depend on flow resistance force of each layer, and its trend depends on the variation ratio of flow resistance force. There are three scenarios:

when $\frac{(R_2)'}{R_2} > \frac{(R_1)'}{R_1} \rightarrow f_1' > 0$, then the relative sunction of low permeability zone increase

when $\frac{(R_2)'}{R_2} = \frac{(R_1)'}{R_1} \rightarrow f_1' = 0$, then conformance reversion point

when $\frac{(R_2)'}{R_2} < \frac{(R_1)'}{R_1} \rightarrow f_1' < 0$, then the relative sunction of low permeability zone decrease

The flow resistance force of each layer and its variation ratio are shown in Fig. 2a, b. At the early stage of polymer flooding, high molecular and high concentration polymer is preferentially injected into the high permeability layer, which leads to fast and dramatically increment of resistance force for high permeability layer but slow and moderate for low permeability layer (see red circled line), thus $(R_2)'/R_2 > (R_1)'/R_1$ and suction amount of low permeability layer increases; as more polymer solution is injected into low permeability layer, its growth ratio increase and opposite situation happen for high permeability layer, which reduce the difference on resistance force variation and slow the rise velocity of water suction, conformance reversion takes place when $(R_2)'/R_2 = (R_1)'/R_1$. When medium molecular low concentration polymer starts, growth ratio of force decreases fast for high permeability layer, but relative moderate for low permeability layer (see Fig. 2b), which results in that $(R_1)'/R_1 > (R_2)'/R_2$ and suction amount of low permeability layer decreases.

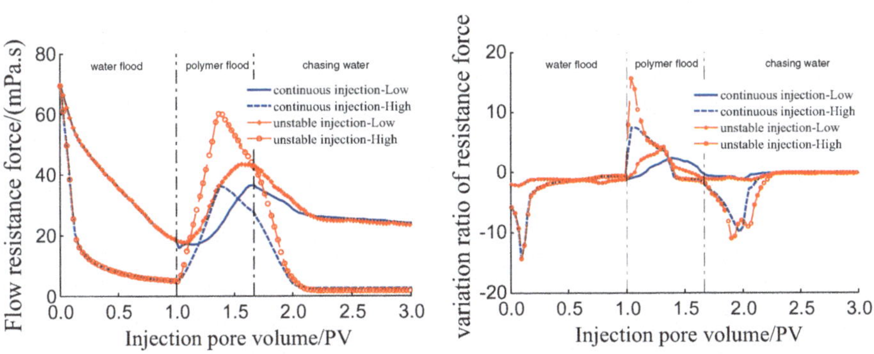

(a) flow resistance force (b) variation ratio of resistance force

Fig. 2 Flow resistance force and its variation ratio of layers with continuous and unstable injection pattern

4 Key Alternating Parameter Optimization

Alternating cycle is an important parameter for unstable polymer injection. 9 scenarios are designed to investigate its impact on injection conformance behavior and EOR results, these are single slug continuous polymer flooding and unstable polymer flooding with alternating cycle from 1 to 8, same polymer usage is applied. Injection conformance behavior is shown in Fig. 3, the suction profile of low permeability layer firstly turns from V type to U type as cycle number increases, plateau appear and its duration extended, which make liquid suction amount increases, and the best conformance is obtained at 3 cycles; after that it gradually returns to V type and its suction amount falls back. The increased recovery factor relative to water flooding is shown in Fig. 4, unstable flooding with different alternating cycles can increase the recovery factor by 2.0–3.1 percentage, and the best EOR result is achieved at 3 cycles and meanwhile the suction profile of low permeability layer reaches the best state, which indicates that its liquid suction can increased by 10.5%. Proper alternating cycle can best suppress conformance reversion and results in best EOR results.

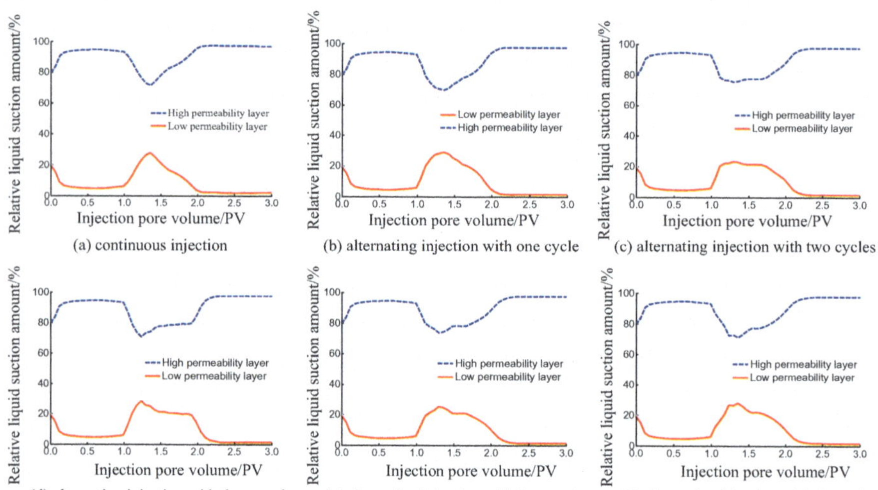

(a) continuous injection (b) alternating injection with one cycle (c) alternating injection with two cycles (d) alternating injection with three cycles (e) alternating injection with four cycles (f) alternating injection with six cycles

Fig. 3 Conformance behavior for unstable injection pattern with different cycles

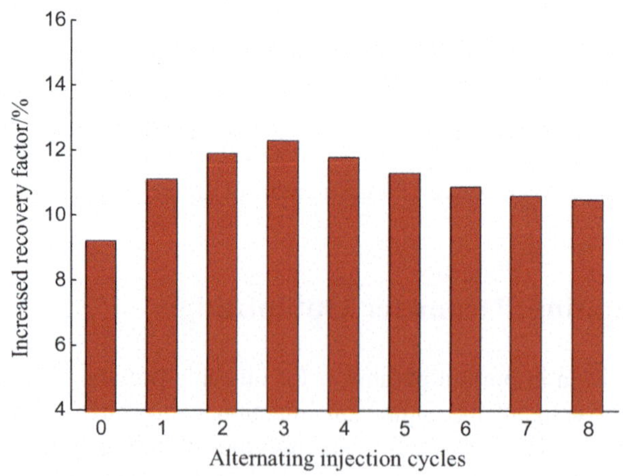

Fig. 4 Increased recovery factor for unstable injection pattern with different alternating cycles

4.1 Impact Factor Investigation

Impact of permeability contrast and thickness portion of low permeability layer on best alternating cycles are investigated. Thickness portion of low permeability layer is defined as the ratio of its thickness to the gross thickness.

Increased recovery factor for different alternating cycles under various permeability contrast is shown in Table 1, the best alternating cycle is 2, 4 and 5 respectively when the permeability contrast is 3, 5, and 7. The cycles needed to achieve the best EOR result increases as permeability contrast increase from 1 to 8.

Table 1 Increased recovery factor for different alternating cycles under various permeability contrast

Cyclic number	Permeability contrast		
	3	5	7
0	9.9	8.9	7.3
1	11.7	10.9	9.7
2	12.7	11.6	10.5
3	12.5	12.1	11.0
4	12.2	12.3	11.3
5	11.8	11.7	11.5
6	11.4	11.2	10.5
7	11.1	10.7	9.8
8	10.9	10.3	9.3

Increased recovery factor for different alternating cycles under various thickness portion of low permeability layer is also shown in Table 2, the best alternating cycle is 6, 4 and 3 respectively when the thickness portion is 33%, 40%, and 50%. The cycles needed to achieve the best EOR result decreases as thickness portion of low permeability layer increase from 33 to 50%. As a result, permeability contrast and thickness portion of low permeability layer are important factors.

Table 2 Increased recovery factor for different alternating cycles under low permeability layer thickness portion

Cyclic number	Low permeability layer thickness portion/%		
	33	40	50
0	10.4	9.9	9.2
1	12.2	12.1	11.1
2	12.7	12.8	11.9
3	13.1	13.4	12.3
4	13.5	13.7	11.8
5	13.8	13.2	11.3
6	14.0	12.7	10.9
7	13.4	12.4	10.6
8	13.0	12.2	10.5

4.2 Alternating Injection Cycle Optimization Model

An alternating injection cycle optimization model is developed in this subsection to provide practical technical guidance for unstable polymer flooding plan. Uniform design method is applied to arrange the simulation cases with varying permeability contrast and thickness portion of low permeability layer; then run simulations and obtain the best alternating cycle for each cases; finally summarize the simulation data and multiple linear regressions are run to establish the optimal alternating injection cycle model. Uniform design table $U_{12}^{*}(12^{10})$ with the first and fifth column is used [12]. With consideration of the variation range of permeability contrast and low permeability layer thickness portion of J oilfield, 12 simulation scenarios are conducted and the corresponding best alternating cycles are obtained, shown in Table 3. Multiple linear regressions are applied to establish the optimal alternating injection cycle model:

Table 3 Alternating injection cycle experiment design and results

No	Permeability contrast	Low permeability layer thickness portion/%	Optimal cycles
1	2	60	0
2	3	50	2
3	4	30	7
4	5	20	10
5	6	70	2
6	7	50	5
7	2	40	3
8	3	20	8
9	4	70	0
10	5	60	2
11	6	40	6
12	7	30	9

$$y = 11.548 + 0.838K_{\text{ratio}} - 31.834H_{\text{ratio}} + 16.428H_{\text{ratio}}^2 - 0.148K_{\text{ratio}} \cdot H_{\text{ratio}} \tag{6}$$

where y is the best alternating cycles; K_{ratio} is permeability contrast; H_{ratio} is thickness portion of low permeability layer. The coefficient of determination R^2 is 0.983, P value is 0.0003, which indicates that the overall effect of linear regression is good. Significance probability of each variables is less than 0.05. Optimal alternating cycle regression result and input data are shown in Fig. 5.

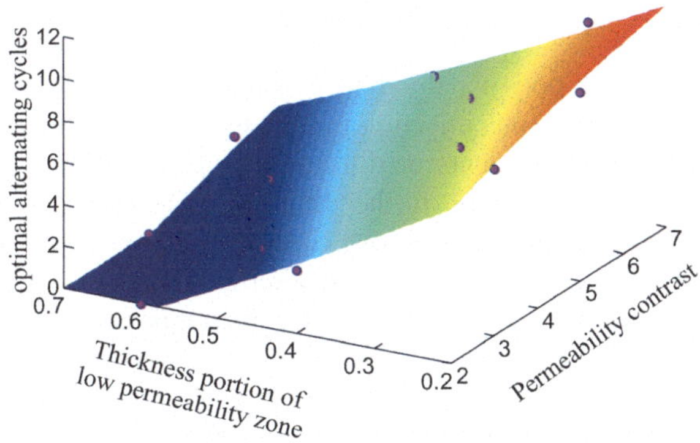

Fig. 5 Optimal alternating cycle regression result and input data

5 Pilot Test

To further validate the novel and unstable polymer injection pattern, a pilot test in J oilfield of Bohai Bay is designed and implemented. The mathematical model and key alternating parameter optimization model mentioned in the above sections are applied to design pilot plan. The target well group screened consists of 5 injectors and 12 producers, where 3 injectors with single polymer but altering concentration and the other 2 injectors with gel alternated by water. Since March of 2017, good EOR results are observed. The daily oil rate of 9 producers can increase by more than 10% and 4 producers among them can even increase by more than 30%. As for the central wells, water cut of C1 can decrease by 11.7% from 84.5 to 72.8%, and its oil rate can increase by 71% from 17.6 to 30.0 m^3/day, see Fig. 6a; water cut of C2 can decrease 7.3% from 96.1 to 88.8%, and its oil rate can increase by 260% from 12.3 to 32.1 m^3/day, see Fig. 6b. Generally speaking, the novel and unstable polymer flooding creates additional oil recovery of 16 thousand m^3 with input-output ratio of 1:6, compared with conventional single slug continuous polymer flooding.

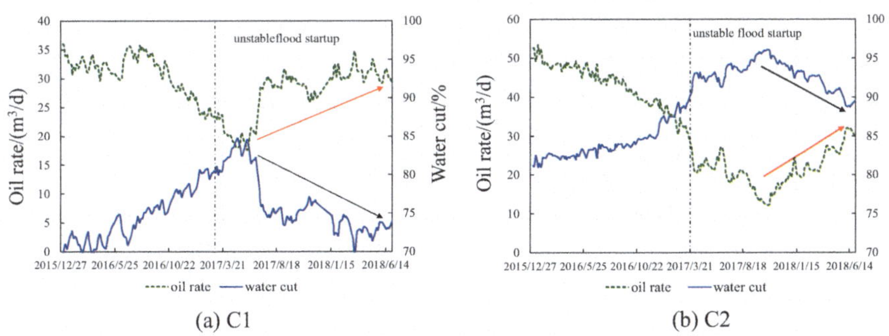

Fig. 6 Production curves of central wells of pilot test

6 Conclusions

(1) A novel and unstable polymer flooding is proposed, which changes the conventional single-slug continuous injection pattern into discontinuous injection pattern with multi-component polymers or different polymer concentrations. It can contain conformance reversion and further improve EOR results.
(2) Unstable polymer flooding can improve compatibility between injected liquid and reservoir heterogeneity, which makes the alternation law of flow resistance force of layers tend to be the same and make the injection conformance of heavy-oil reservoir turn to 'U' type from 'V' type, finally enhances liquid suction amount of the low-permeability zone and its remaining oil mobilization.
(3) A black-oil-based multi-component polymer model is developed and its numerical solver is programmed with the fully implicit method in this study.
(4) Proper alternating injection cycle can best suppress conformance reversion and results in best EOR results. The optimal alternating cycle increases as permeability contrast increases, but decreases as the relative thickness of low permeability layer increases. An alternating cycle optimization model is developed based on uniform design.
(5) Good EOR results are shown in the pilot test of J oilfield of Bohai Bay, the central producer witness more than 7.3 percentage decrement in water cut and more than 70% increment in oil production rate.

Acknowledgements. This work is supported by the China National Science and Technology Major Project Program (Number 2016ZX05025-003) and CNOOC Science and Technology Major Project Program (Number CNOOC-KJ135ZDXM36TJ02ZY).

References

1. Zhou, S., Han, M., Xiang, W., et al.: Application of EOR technology by means of polymer flooding in Bohai oilfields. China Offshore Oil Gas **18**(6), 386–389 (2006)
2. Sun, H.: Practice and understanding on tertiary recovery in Shengli oilfield. Petrol. Exploration Develop. **33**(3), 262–266 (2006)
3. Wang, D., Cheng, J., Wu, J., et al.: Application of polymer flooding technology in Daqing oilfield. Acta Petrol. Sinica **26**(1), 74–78 (2005)
4. Cao, R., Han, P., Hou, W., et al.: Section inversion rules and inversion mechanism of polymer flooding. Acta Petrol. Sinica **30**(2), 267–270 (2009)
5. Cao, R.: Laboratory research on section reversion mechanism of polymer flooding. Petrol. Geol. Recovery Efficiency **16**(4), 71–73 (2009)
6. Cao, R., Wang, X., Han, P., et al.: Multi-slug alternative injection method and field application of polymer flooding. Petrol. Geol. Recovery Efficiency **19**(3), 71–75 (2012)
7. Xiong, J., Liu, J., Liu, X., et al.: Two-phase seepage theory in unsaturated rock and soil based on Buckley-Levererett equation. J. Liaoning Tech. Univ. **12**(2), 213–215 (2007)
8. Buckley, S., Levertt, M.: Mechanism of fluid displacement in sands. AIME **46**, 107–116 (1942)
9. Deniz, M., Birol, D.: Analytical solution of nonisothermal Buckley-Leverett flow including tracers. SPE Reservoir Eval. Eng. **11**(3), 65–74 (2008)

10. Courant, R., Friedrichs, K., Lewy, H.: On the partial difference equations of mathematical physics. IBM J. Res. Dev. **11**(2), 215–234 (1967)
11. Wei, Z., Zhang, D.: A fully coupled multiphase multicomponent flow and geomechanics model for enhanced coalbed-methane recovery and CO_2 storage. SPE Journal **18**(3), 1–20 (2013)
12. Liu, W.: Design of Experiments. Tsinghua University Press, Beijing (2005)

Study on Deconvolution Well Test Interpretation Model for Low-Permeability Oil Wells in Offshore Reservoir

Qing Feng[1(✉)], Xiaonan Li[1], Tao Wang[1], Zijun Huang[1], Yu Li[2], Yingying Wang[3], Mingqiang Wei[4], Weixing Yang[1], and Xuxing Wang[1]

[1] Oilfield Production Optimization Institution, China Oilfield Services Limited, 300450 Tianjin, China
493748023@qq.com

[2] CNOOC EnerTech-Drilling & Production Co.,, 300452 Tianjin, China

[3] Exploration and Development, Research Institute of Huabei Oilfield Company, 062550 Renqiu, China

[4] State Key Laboratory of Oil and Gas Reservoir Geology and Exploitation, Southwest Petroleum University, 610500 Chengdu, Sichuan, China

Abstract. In order to avoid reservoir damage and productivity impact caused by shut-in, some offshore low-permeability oil wells have a short shut-in test time and slow pressure recovery, which leads to unapparent or missing pressure characteristics in well test interpretation. In order to improve the accuracy of well test interpretation results, obtain more reservoir information, and solve the problem of missing radial flow characteristics and boundary features on conventional well test pressure and pressure derivative curves, one convolution and deconvolution well test model is established based on the dynamic variation law of oil wells' bottom-hole flow pressure with variable production. Duhamel principle is used to obtain the equivalent constant flow pressure response for the entire production period, and the error caused by incomplete production or pressure history is reduced, which is verified by oil field examples. The result shows that the deconvolution well test interpretation model has the advantages that the conventional interpretation method does not have. By integrating the production and pressure information in the whole test process, it can eliminate

This paper was prepared for presentation at the 2019 International Petroleum and Petrochemical Technology Conference in Beijing, China, 27–29, March, 2019.

J. Lin (ed.), *Proceedings of the International Petroleum and Petrochemical Technology Conference 2019*, pp. 26–44, 2020.
https://doi.org/10.1007/978-981-15-0860-8_3

the interference of wellbore storage effects on radial flow, enlarge the detection range of pressure fluctuation, update interpretation results of conventional well test and ensure consistent geological understanding with real reservoir information. The deconvolution model can also explain the distance between oil well and edge-bottom water invasion position. This study can interpret reservoir information that cannot be reflected by conventional well test, providing a new theoretical interpretation method for offshore well test interpretation with large production changes and short shut-in time. It is of great significance to provide important reservoir parameters and test well boundary types for reservoir development.

Keywords: Low-permeability reservoir · Well test interpretation · Convolution and deconvolution · Pressure characteristics · Reservoir information

Nomenclature

φ_i	Reservoir porosity, fraction;
P_{wf}	Bottom-hole flow pressure, MPa;
P_i	Original reservoir pressure, MPa;
q	Flow rate, m^3/d;
B	Fluid volume coefficient, dimensionless;
μ	Fluid viscosity, mPa·S;
h	Reservoir thickness, m;
K	Reservoir effective permeability, mD;
r_w	Wellbore radius, m;
C_t	Reservoir comprehensive compression factor, MPa^{-1};
t_p	Production time, d;
Δt	Shut-in pressure recovery time, d;
P_{ws}	Pressure recovery during shut-in, MPa;
$E_i(-x)$	Power integral function, $E_i(-x) = -\int_x^{\infty} \frac{e^{-y}}{y} dy$;
ΔP_v	Equivalent pressure response per unit, MPa;
q_{sc}	Fluid rate under the standard condition, m^3/d;
S	Reservoir skin factor, dimensionless;
L	The number of history rate nodes;
N	The number of pressure response nodes;
$x_{qi}(i = 1, \ldots L)$	The fluid rate of the time node σ_i;
K	The number of stress tested values.
S_i	The error limit of corresponding constraints, for pressure constraints, $S_i = 0.07$ kPa; for the smoothness constraint, $S_i = 0.05$ kPa.

1 Introduction

In the process of unstable well test interpretation, how to obtain a complete formation flow pattern and reducing multi-solution is a key step, which is directly related to whether the interpretation results truly reflect the reservoir characteristics [1–3]. Due to the influences of complex offshore climate and high test cost, effective shut-in pressure measurement time of production wells is relatively short (6–20 h). At the same time, the slow pressure recovery and low permeability in media leads to the loss of typical seepage characteristics of well test interpretation curve, which makes it difficult for interpretation workers to accurately explain reliable and real reservoir information, and the multi-solution probability of interpretation results is enhanced [4–6].

In establishing well test interpretation model or developing well test interpretation chart, the rate is usually set as a man-made constant value, but the actual production of the tested well is always changing with reservoir fluids' output. If when rate changes' range is small accurate interpretation results may be obtained [7, 8]. Pressure changes caused by production history are generally practiced through the superposition principle of pressure drop. However, for long-term large-scale production changes, the bottom-hole flow pressure is more sensitive to the yield. As a result, the error of the superposition processing accuracy increases, and the pressure response curve is messy and difficult to explain [9, 10], which brings difficulties to reservoir cognition and evaluation.

However, the deconvolution method may take entire pressure history and production history of wells into consideration [11, 12], and uses integral transformation to obtain the equivalent pressure of the whole production stage under each unit. The reservoir and boundary information of the whole production process of the tested wells is obtained by numerical matching of the equivalent pressure in the log-log diagnosis chart, which helps the interpretation workers to select a reasonable interpretation model. Therefore, based on the varying law of bottom-hole pressure in the seepage of oil wells in reservoir with infinite boundary, this paper establishes the convolution and deconvolution pressure recovery model for variable production and shut-in. Combined with the pressure and production data of the whole production history, the equivalent pressure under each unit is obtained by deconvolution algorithm. And then the double logarithmic diagnostic chart of the equivalent pressure can be established, reliable reservoir parameters and test boundary types are obtained by deconvolution well test interpretation.

Through the application research of the deconvolution well test model, it is found that it can make up for the shortcomings of conventional interpretation methods, expand the detection range of pressure data information, and reflect more reliable reservoir information, which eliminate the effects of simplified production and superposition calculation. Therefore, the problems about shut-in pressure recovery of short test time and difficulty in accurate interpretation of the outer boundary of offshore reservoir can be solved.

2 Model Establishment

Both convolution and deconvolution are integral transformation mathematical methods, which are mainly applied to pressure changes correction during variable production and early data correction for pressure recovery [13–15]. However, due to the unit ability of the complex calculation process, the application of deconvolution method has been limited in well test interpretation. In this paper, combined with the analytical solution of the dynamic pressure response in reservoirs with infinite boundary, convolution and deconvolution models considering the whole history of production are established.

2.1 Establishment of Convolution Model

Based on the principle of mass conservation and Darcy's law of fluids flow, the well test model with infinite reservoir boundary is established. The model assumptions are as follows: (1) The reservoir is a single-porosity media, ignoring the influence of fluids' gravity and capillary force; (2) At the beginning, formation pressure is original that amounts to value P_{i}, and the seepage accords with isothermal Darcy law; (3) Including the influence of wellbore storage and skin effects; (4) Fluids and rock are slightly compressible, whose compressibility coefficient is constant.

Pressure drop and shut-in pressure recovery curve of production wells is shown in Fig. 1.

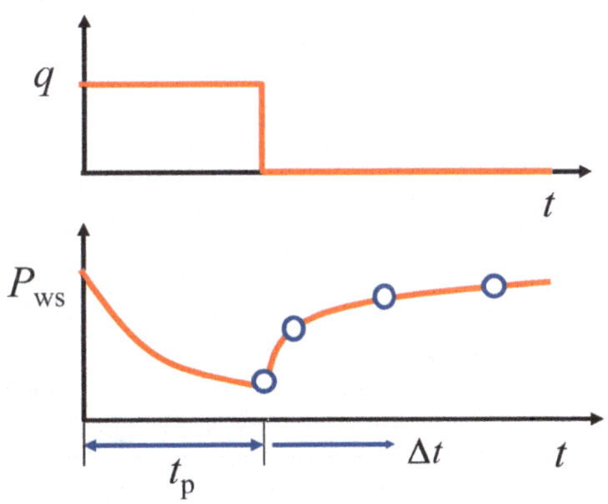

Fig. 1 Schematic diagram of oil wells' production and pressure along with time

If an oil well in the reservoir with infinite boundary is in the production stage or shut-in stage, the dynamic change of bottom hole flowing pressure [16, 17] is described as follows:

$$p_{wf}(t) = p_i - \frac{0.921\, q\mu B}{Kh}\left[-E_i\left(-\frac{\phi\mu C_t r_w^2}{0.0144Kt}\right)\right], 0<t<t_p \quad (1)$$

$$p_{ws}(t) = p_i + \frac{0.921\, q\mu\, B}{Kh}\left\{E_i\left[-\frac{\phi\mu C_t r_w^2}{0.0144K(t_p+\Delta t)}\right] + E_i\left(-\frac{\phi\mu C_t r_w^2}{0.0144K\Delta t}\right)\right\}, t_p < t < t_p + \Delta t \quad (2)$$

When $\frac{\phi\mu C_t r_w^2}{0.0144Kt} < 0.01$, Eqs. (1) and (2) can be simplified as:

$$p_{wf}(t) = p_i - \frac{2.12 q\mu\, B}{Kh}\left(\lg\frac{Kt}{\phi\mu C_t r_w^2} + 0.9077\right), 0 < t < t_p \quad (3)$$

$$p_{ws}(t) = p_i - \frac{2.12\, q\mu\, B}{Kh}\lg\frac{t_p+\Delta t}{\Delta t}, t_p < t < t_p + \Delta t \quad (4)$$

After being transformed by Eq. (1) it is obtained as follows:

$$\frac{Kh}{0.921\, q\mu\, B}(p_i - p_{wf}) = -E_i\left[-\frac{\phi\mu C_t r_w^2}{0.0144Kt}\right] \quad (5)$$

Set $\Delta p_v = \frac{Kh}{0.921\, q\mu\, B}(p_i - p_{wf})$, this leads to:

$$\Delta p_v'(t) = \frac{1}{t}\mathrm{e}^{-\frac{\phi\mu C_t r_w^2}{0.0144Kt}} \quad (6)$$

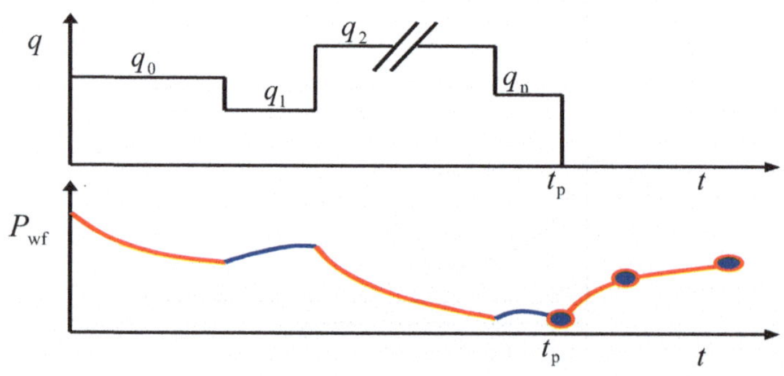

Fig. 2 Pressure drop and pressure recovery curve during variable production

If the production is variable as shown in Fig. 2, the production process is divided into n time-step ($q_i = q(t_i), i = 0, 1, 2\ldots n.$). According to the pressure drop superposition principle [18, 19], pressure model of variable yields can be written as follows:

$$p_{wf}(t) = p_i + \frac{0.921\ \mu\ B}{Kh}\sum_{j=0}^{n-2} q(\tau_{j,j+1})\left\{E_i\left[-\frac{\phi\mu C_t r_w^2}{0.0144K(t-\tau_j)}\right] - E_i\left[-\frac{\phi\mu C_t r_w^2}{0.0144K(t-\tau_{j+1})}\right]\right\} + \frac{0.921\ \mu\ B}{Kh} q(\tau_{n-1,n}) E_i\left[-\frac{\phi\mu C_t r_w^2}{0.0144K(t-\tau_{n-1})}\right] \tag{7}$$

Equation (7) can be simplified as:

$$p_{wf}(t) = p_i + \frac{0.921\ \mu\ B}{Kh}\sum_{j=0}^{n-2} q(\tau_{j,j+1})\left\{-\frac{d}{d\tau}E_i\left[-\frac{\phi\mu C_t r_w^2}{0.0144K(t-\tau)}\right]\right\}\Delta\tau + \frac{0.921\ \mu\ B}{Kh} q(\tau_{n-1,n}) E_i\left[-\frac{\phi\mu C_t r_w^2}{0.0144K(t-\tau_{n-1})}\right] \tag{8}$$

Set $\Delta\tau = \tau_{j+1} - \tau_j$, if when $n \to \infty$, $\max(\Delta\tau) \to 0$, Eq. (8) can be transformed furtherly as follows:

$$p_{wf}(t) = p_i + \frac{0.921\ \mu\ B}{Kh}\int_0^t q(\tau)\frac{d}{d\tau}\left\{-E_i\left[-\frac{\phi\mu C_t r_w^2}{0.0144K(t-\tau)}\right]\right\}d\tau \tag{9}$$

Based on Eq. (9) the equation is developed as follows:

$$p_i - p_{wf}(t) = \frac{0.921\ \mu\ B}{Kh}\int_0^t q(\tau)[\frac{1}{t-\tau}\mathrm{e}^{-\frac{\phi\mu C_t r_w^2}{0.0144K(t-\tau)}}]d\tau \tag{10}$$

Substituting Eqs. (6) into (10), the following can be obtained:

$$p_i - p_{wf}(t) = \frac{0.921\ \mu\ B}{Kh}\int_0^t q(\tau)\Delta p'_{u\,u}(t-\tau)d\tau \tag{11}$$

Equation (11) is the convolution form of bottom-hole pressure of production wells in reservoirs with infinite boundary.

Dimensionless definitions: $q_D = \frac{q(t)}{q_{sc}}; p_D = \frac{Kh[p_i - p(r,t)]}{0.921qB\mu}; t_D = \frac{0.0036Kt}{\phi\mu C_t r_w^2}; r_D = \frac{r}{r_w}$.

Dimensionless formula of Eq. (11) can be written as:

$$p_{wD}(t_D) = \int_0^{t_D} q_D(\tau)p'_v(t_D - \tau)d\tau \tag{12}$$

Based on the superposition method proposed by Umarov [20] and Dominic [21] and pressure drop caused by skin effects, Eq. (12) can be converted into

$$p_{wD}(t_D) = \int_0^{t_D} q_D(\tau)p'_v(t_D - \tau)d\tau + Sq_D(t_D) \tag{13}$$

Uniting Eqs. (3) and (13), the result can be obtained as follows:

$$\Delta p_{wf}(t) = m\int_0^{t} q'_D(\tau)[\log(t-\tau)]d\tau + mq_D(t)\bar{S} \tag{14}$$

where: $\bar{S} = -0.87S + \log(\frac{K}{\phi\mu C_t r_w^2}) - 2.1$, $m = \frac{2.12qB\mu}{Kh}$.

After being divided by q_D Eq. (14) can be written as follows:

$$\frac{\Delta p_{wf}(t)}{q_D} = \frac{m\int_0^t q'_D(\tau)[\log(t-\tau)]d\tau}{q_D} + b, b = m\bar{S} \tag{15}$$

Equation (15) is the convolution calculation expression of bottom-hole pressure during variable production, from which we can obtain one linear law with a slope of m and an intercept of b.

If pressure drop superposition principle is applied in Eq. (13) when the shut-in time is Δt, pressure recovery is obtained as follows:

$$p_{wsD}(t_D + \Delta t_D) = \int_0^{\Delta t_D} q_D(\tau)p'_v(\Delta t_D - \tau)d\tau + Sq_D(\Delta t_D) + p_D(t_D + \Delta t_D) - p_D(t_D) \tag{16}$$

With the assistance of Eqs. (4), (16) can be simplified as:

$$\Delta p = p_{ws} - p_i = m[\log\frac{t+\Delta t}{t} + \int_0^{\Delta t} q'_D(\tau)p_v(\Delta t - \tau)d\tau + \bar{S}(1 - q_D)] \tag{17}$$

After being divided by (1-q_D) Eq. (17) is transformed as:

$$\frac{\Delta p}{1-q_D} = m[\frac{\log(t+\Delta t)/t + \int_0^{\Delta t} q'_D(\tau)p_v(\Delta t - \tau)d\tau}{1-q_D} + \bar{S}] \tag{18}$$

In Eq. (18) linear law with a slope of m and an intercept of $\bar{S}$ can be obtained.

If the shut-in time Δt is greatly smaller than production time t, then Eq. (18) can be simplified as

$$\frac{\Delta p}{1 - q_D} = m[\frac{\int_0^{\Delta t} q'_D(\tau)p_v(\Delta t - \tau)d\tau}{1 - q_D} + \bar{S}] \tag{19}$$

Equation (19) can be used to calculate the convolution pressure when the shut-in pressure builds-up occurs.

The convolution curve of shut-in pressure recovery calculated by Eq. (19) is shown in Fig. 3.

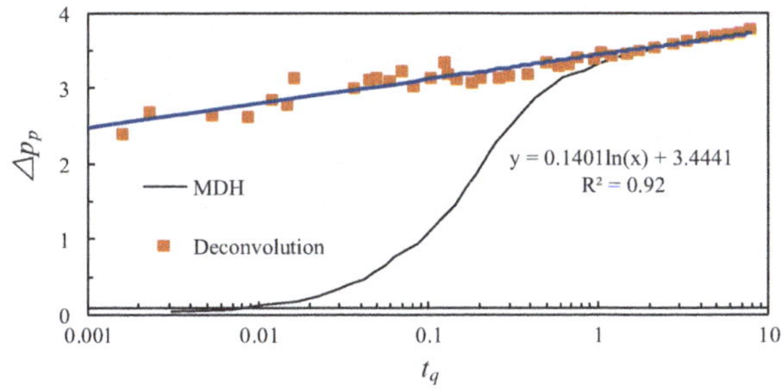

Fig. 3 Comparison of pressure recovery convolution curve and MDH curve

Where: $\Delta p_p = \frac{\Delta p}{1-q_D}$, $t_q = \frac{\int_0^{\Delta t} q'_D(\tau)p_v(\Delta t-\tau)d\tau}{1-q_D}$.

2.2 Establishment of Deconvolution Model

When q_D is constant, Eq. (13) is processed to obtain the deconvolution pressure equation with constant production as follows:

$$p_v(t_D) + S = \frac{\Delta p_{wf}(t_D)}{q_D(t_D)} \tag{20}$$

If the production is variable, Eq. (13) can be linearized as following:

$$p_v(t_{Dn+1/2}) + S = \frac{\Delta p_{wf}(t_{Dn+1}) - \sum_{i=0}^{n-1} p_D(t_{Di+1/2})[q_D(t_{Dn+1} - t_{Di}) - q_D(t_{Dn+1} - t_{Di+1})]}{q_D(t_{Dn+1} - t_{Dn})} \tag{21}$$

3 Model Solution

Based on the convolution property and integral variable substitution, Eq. (12) can be transformed as follows:

$$p_{wD}(t_D) = \int_0^{t_D} q_D(t_D - \tau)p'_v(\tau)d\tau \tag{22}$$

Set $\sigma = \ln \tau$, $0 < \tau < t_D$, then $\ln \frac{dp_v}{d\ln(\tau)} = \ln p'_v(\sigma), z(\sigma) = \ln p'_v(\sigma)$, then Eq. (22) can be converted into

$$p_{wD}(t_D) = \int_{-\infty}^{\ln t_D} q_D(t_D - e^{\sigma})e^{z(\sigma)}d\sigma \tag{23}$$

By the definition mentioned above: $p'_v(\sigma) = e^{z(\sigma)}$, integrating the equation leads to

$$p_v(\sigma) = \int_{-\infty}^{\ln t_D} e^{z(\sigma)}d\sigma \tag{24}$$

Presupposing that time variable σ is evenly split into N nodes, $\sigma = \ln t_i, i = 0, 1, 2\ldots N$, $\sigma_0 < \sigma_1 \ldots < \sigma_N = \ln t_D$, σ_0 and σ_N separately refers to the starting time and end time of test, σ_N.If when $\sigma_0 \to -\infty$, the deconvolution pressure of Eq. (24) can be written as follows:

$$p_v(\sigma_i) = p_v(\sigma_1) + \int_{\sigma_1}^{\sigma_i} e^{z(\sigma)}d\sigma = p_v(\sigma_1) + \sum_{j=1}^{i-1} \int_{\sigma_j}^{\sigma_{j+1}} e^{z(\sigma)}d\sigma \tag{25}$$

When $\sigma \in (\sigma_j, \sigma_{j+1}), j = 1 \sim N\text{–}1$, the linear interpolation of function $z(\sigma)$ leads to

$$\begin{gathered} z(\sigma) = a_j\sigma + b_j \\ a_j = \tfrac{z(\sigma_{j+1}) - z(\sigma_j)}{\sigma_{j+1} - \sigma_j}, b_j = z(\sigma_j) - a_j\sigma_j \end{gathered} \tag{26}$$

Equation (26) is used to optimize Eq. (25), then the following can be obtained:

$$p_v(\sigma) = p_v(\sigma_1) + \sum_{j=1}^{N-1} \frac{e^{Z(\sigma)_{j+1}} - e^{Z(\sigma)_j}}{a_j} d\sigma \tag{27}$$

Since $\sigma = \ln t$, Eq. (27) can be further simplified as

$$p_v(t) = p_v(t_1) + \sum_{j=1}^{N-1} \frac{e^{z(\sigma_{j+1})} - e^{z(\sigma_j)}}{z(\sigma_{j+1}) - z(\sigma_j)} \ln \frac{t_{j+1}}{t_j} \tag{28}$$

Set $x_{qi} = q_D(\sigma_i), i = 0, 1, 2\ldots L$, $q_D(\sigma_i)$ representing historical production rates, and then Eq. (12) can be converted into one cumulative mode:

$$\Delta p(t) = x_{q0} p_v(t) + \sum_{i=1}^{L} [x_{q_i} - x_{q_{i-1}}] p_v(t - t_{Di}) \tag{29}$$

Ω is designed to represent predicted pressure model in Eq. (29), the following can be obtained:

$$\Omega_i = p_i - x_{q0} p_v(t_i) - \sum_{j=1}^{i} [x_{q_j} - x_{q_{j-1}}] p_v(t_i - t_{Dj}), 1 < i < L \tag{30}$$

Set $x_{zi} = z(\sigma_i)$, i = 1–N, then all parameters are displayed by vector form as follows:

$$x = \{p_v(\sigma_1), x_{z1}, \ldots, x_{zN}, x_{q1}, x_{q2}, \ldots, x_{qL}\} \tag{31}$$

According to the research on the stability of deconvolution algorithm [22–24], many constraint types are contained such as pressure constraint, rate constraint and smoothness constraint. In the three types of constraints, variable w is used to represent their tested values as follows:

$$w = \{w_1, w_2, \ldots, w_K, w_{K+1}, w_{K+2}, \ldots, w_{K+L}, w_{K+L+1}, w_{K+L+2}, \ldots, w_{K+L+N-1}\} \tag{32}$$

In order to achieve the matching accuracy between the observed pressure and predicted pressure, the minimum difference solution is performed as follows:

$$\phi(x) = \min\left[\sum_{i=1}^{K+L+N-1} \left(\frac{\omega_i - \Omega_i^2}{s_i}\right)\right] \tag{33}$$

The second derivative of the predictive value under the smoothness constraint is calculated by the following formula

$$\Omega_v(\sigma_i) = \frac{x_{zi-1} + x_{zi+1} - 2x_{zi}}{\Delta\sigma_i^2} \tag{34}$$

Through minimizing the parameter x_{zi} in Eqs. (33) and (34), the objective function about K pressure, L rate history and N–1 curvature vectors can be written as follows:

$$\phi(x) = \min\left[\sum_{i=1}^{L} [q_i - x_{q_i}]^2 + \sum_{i=1}^{N-1} \left[\frac{x_{zi-1} + x_{zi+1} - 2x_{zi}}{\Delta\sigma_i^2}\right]^2 + \sum_{i=1}^{K} [p_i - \Omega_i]^2\right] \tag{35}$$

After obtaining the values of vector x_{zi} in Eq. (35), the deconvolution pressure response can be obtained as follows:

$$p_v(t_i) = p_v(t_1) + \sum_{j=1}^{i} \frac{e^{x_{zj+1}} - e^{x_{zj}}}{x_{zj+1} - x_{zj}} \ln \frac{t_{j+1}}{t_j}, 2 < i < N - 1 \tag{36}$$

4 Application Research

With the application of the deconvolution model tested wells' pressure recovery data are processed. Taking wells' entire production and shut-in history into consideration, the detection range of the pressure response is prolonged to obtain more reservoir information.

4.1 Identify the Boundary Characteristics of Composite Reservoir

According to the morphological characteristics of log-log pressure recovery curve in Fig. 4, the conventional well test interpretation shows one constant pressure boundary with drainage radius of 16 m. However, according to drilling fluid logging and well logging of adjacent wells, it can be proved that drainage boundary of well PX1 is greater than 16 m in both the plane and the longitudinal direction. So the conventional interpretation is obviously wrong.

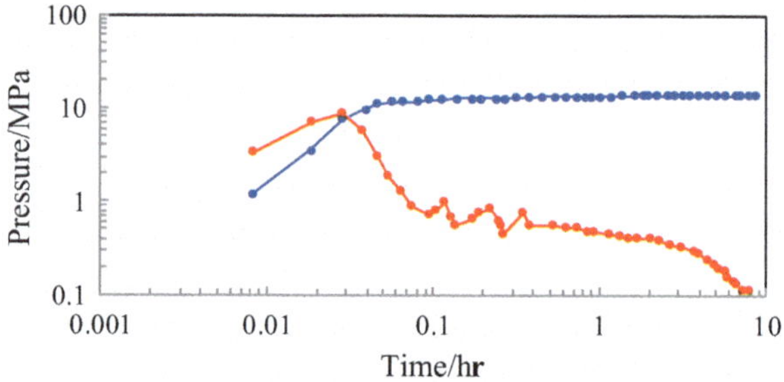

Fig. 4 Log-log pressure and pressure derivative plot of well PX1

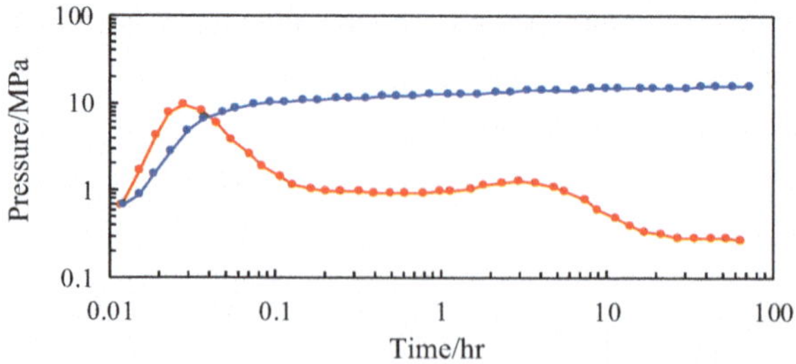

Fig. 5 Log-log pressure and pressure derivative plot of well PX1 by deconvolution

In Fig. 5, the deconvolution method presented in this paper is used to reinterpret test data of well PX1, whose result proves that the constant pressure boundary in Fig. 4 is not the real boundary. It truly reflects composite boundary exists in the reservoir, better physical properties in the outer zone leads to conventional misinterpretation as the constant pressure boundary.

4.2 Eliminate the Influences of Wellbore Storage Effects

Figure 6 shows that the horizontal radial flow segment with a value of 0.5 doesn't appear in log-log pressure derivative curve. Serious wellbore storage effects of well PX2 cover pressure characteristics of radial flow, which brings great difficulty to interpretation model selection in well test interpretation.

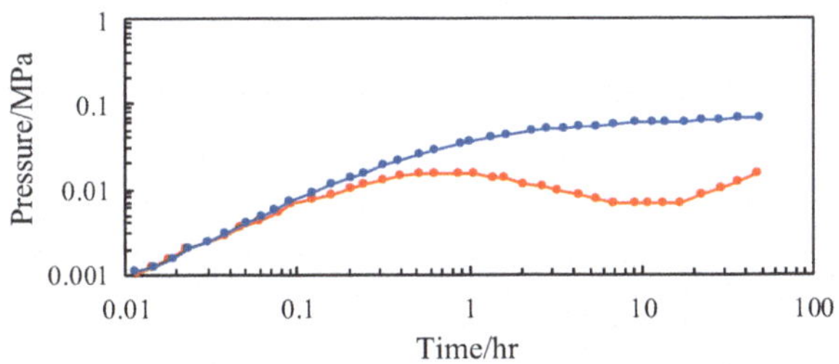

Fig. 6 Log-log pressure and pressure derivative plot of well PX2

In order to eliminate the influence of wellbore storage effects and recover the normal early flow state of well PX2, one-log periodic radial flow feature through the deconvolution operation is displayed in Fig. 7.

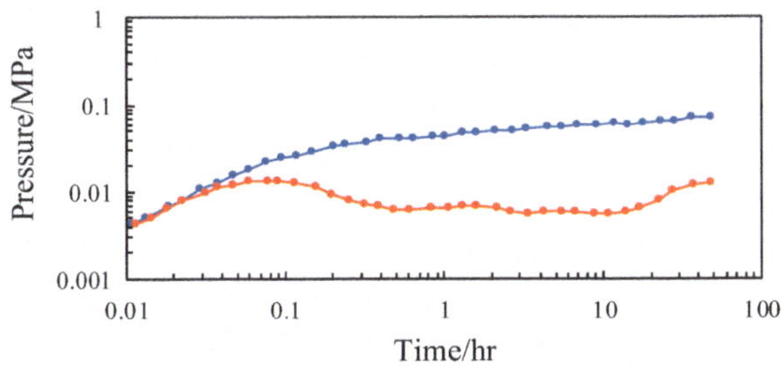

Fig. 7 Log-log pressure and pressure derivative plot of well PX2 by deconvolution

4.3 Identify the Closed Boundary Outside Parallel Faults

As can be seen from Fig. 8, a line segment with a slope of 1/2 appears in the later stage of the pressure derivative curve, which displays a typical feature of parallel faults, but the boundary features outside the parallel faults are not reflected. Meanwhile pressure history matching in Fig. 9 is poor. Measured data of well PX3 are largely greater than the theoretical values, and actual values have a large drop range, indicating that the current parallel faults do not completely reflect the seepage boundary of the reservoir.

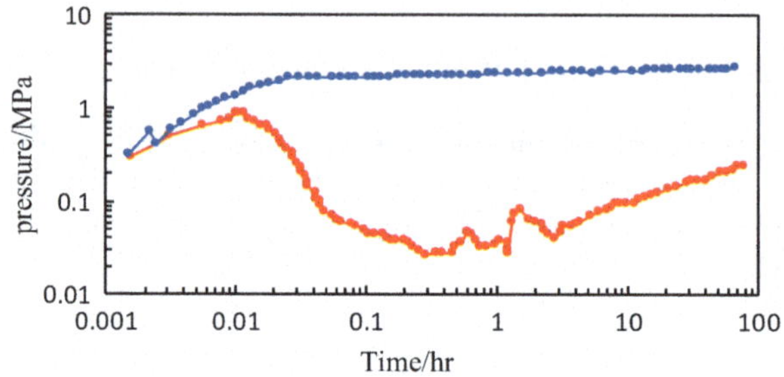

Fig. 8 Log-log pressure and pressure derivative plot of well PX3

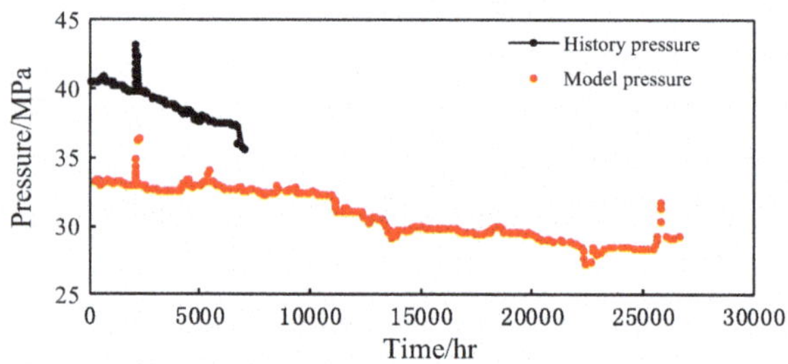

Fig. 9 Pressure history matching plot of well PX3

To obtain one more accurate boundary model, measured pressure data need to be processed by deconvolution. The deconvolution operation yields a pressure drop and pressure derivative curve as shown in Fig. 10.

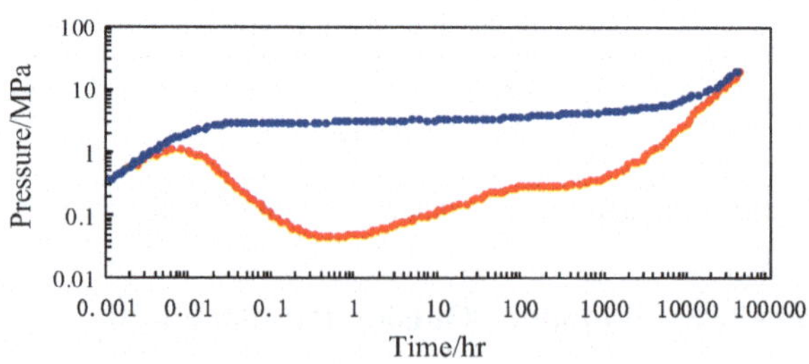

Fig. 10 Log-log pressure and pressure derivative plot of well PX3 by deconvolution

As can be seen from Fig. 10, in addition to the parallel fault reflection features with a slope of 1/2 in Fig. 9, there are also radial flow and pseudo steady flow stages. Well test interpretation model under the joint influence of parallel fault and closed boundary is adopted to match pressure history of well PX3. History matching results are shown in Fig. 11.

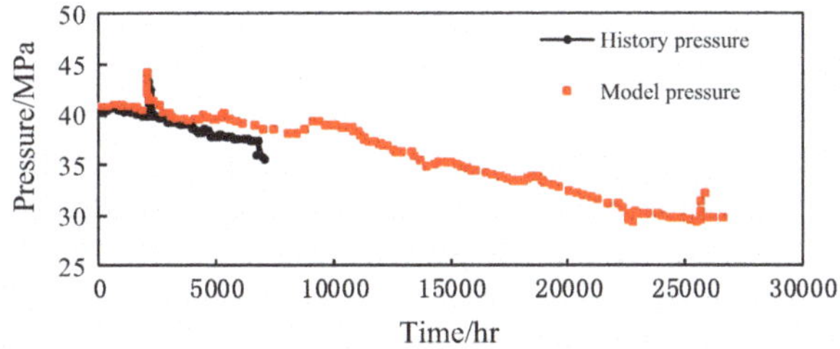

Fig. 11 Pressure history matching plot of well PX3 by deconvolution

As shown in Fig. 11, the pressure history is well matched, indicating that the boundary model interpreted after deconvolution accords with the real reservoir formation.

4.4 Identify the Distance to Edge-Bottom Aquifer

One PDG gauge for pressure was located in the downhole pipe string of well PX4 in Bohai oilfield. PDG gauge is affected by two short-time shut-in of well PX4. PDG gauge test data is interpreted to display the characteristics of radial composite reservoir in Fig. 12. However, reservoir geology and well logging results of well PX4 show that formation has strong edge-bottom water energy, water cut and formation pressure are maintained at a high level. With the assistance of the deconvolution interpretation method, two shut-in time (8.2, 15.6 h) of PDG gauge in well PX4 are extended to

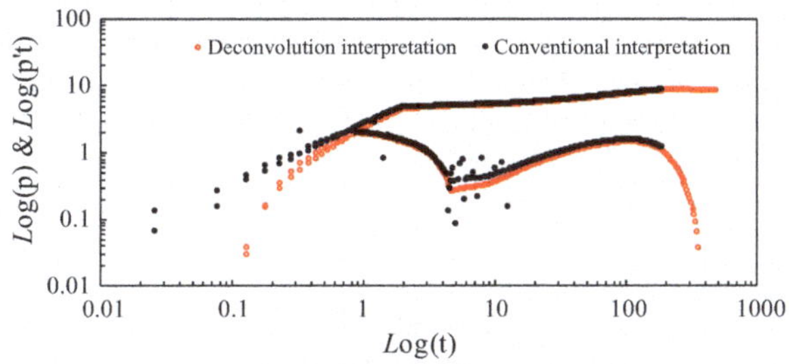

Fig. 12 Identification of edge-bottom water features by deconvolution in well PX4

47.5 h, and then pressure derivative generates obvious constant pressure boundary feature with the distance of 90 m in Fig. 12. The deconvolution interpretation brings about similar permeability and skin factor as conventional interpretation.

5 Case Study

Well JZ9-1-A1 from Bohai Oilfield is one production well with maximum hole deviation angle of 17°, whose perforation locates in layer Ed_2^L III with the thickness of 24.6 m. In August 3rd 2017 fracturing packing sand control was carried out in well JZ9-1-A1. In order to further master, the reservoir information of well JZ9-1-A1, shut-in formation test was taken with one working system of "three opening and three shut-in". But shut-in time were so short that formation pressure was still in the recovery. Three-time shut-in pressure history of well JZ9-1-A1 is shown in Fig. 13.

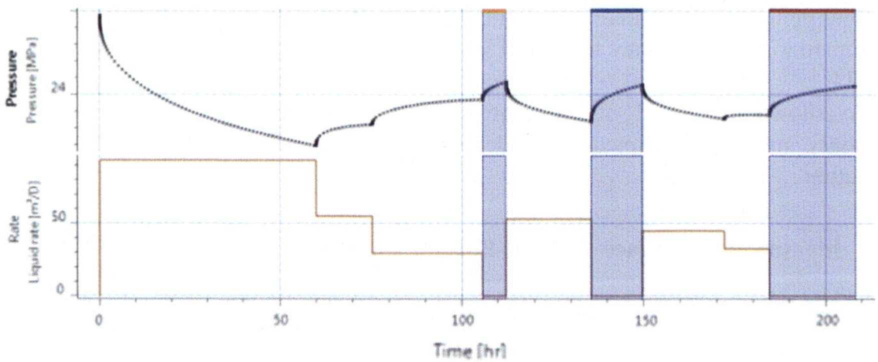

Fig. 13 Formation test yield and pressure history of well JZ9-1-A1

Test data of well JZ9-1-A1 in Fig. 13 are imported into the well test software Saphir 5.0, which generates pressure and pressure derivative curves as shown in Fig. 14.

The radial flow period shown in Fig. 14 is very short, and presents the linear flow feature generated by fracturing packing process. However, the radial flow position and boundary information of well JZ9-1-A1 cannot be determined by the conventional interpretation method, the finite conductivity fracture and radial composite reservoir model is not ideal for interpreting well JZ9-1-A1.

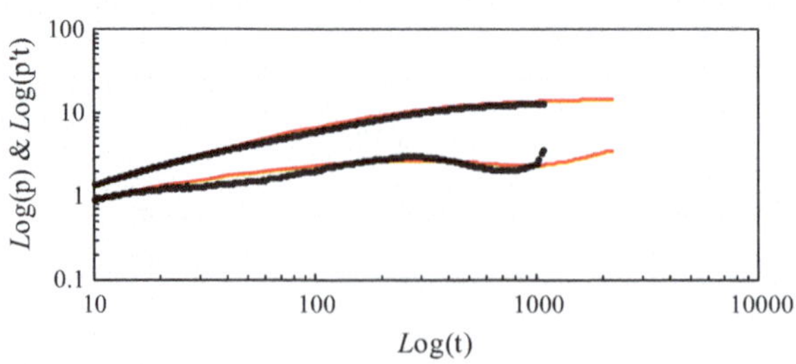

Fig. 14 Log-log fitting curve of shut-in pressure build-up data of well JZ9-1-A1

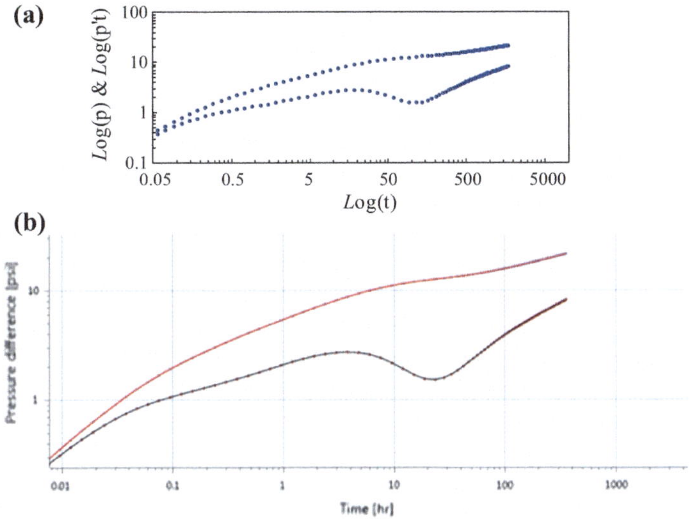

Fig. 15 **a** Log-log deconvolution plot of well JZ9-1-A1. **b** Log-log deconvolution fitting plot of well JZ9-1-A1

After being processed by the deconvolution algorithm in the paper, the corrected log-log pressure and pressure derivative curves are shown in Fig. 15a, b.

In Fig. 15, it can be found that the deconvolution method may favor us to detect more reservoir information, reservoir radial flow and boundary effects features are reflected.

The concave feature is displayed in the log-log pressure derivative curve, indicating the dual-porous media; the later pressure and pressure derivative curve shows an unwrapping property with the d_p/d_t amounting to 1, and then the production performance of adjacent wells has few influences on well JZ9-1-A1, and there may be a closed boundary around the wellbore nearby zone. Therefore, one model with well storage effects, dual-porous pseudo-steady flow and rectangular boundary is adopted for interpreting the test data of well JZ9-1-A1. The ultimate well test interpretation results are shown in Table 1.

As shown in Table 1, the deconvolution interpretation parameters of well JZ9-1-A1 are largely consistent with the results of production well test interpretation by the software Topaze. Fracture packing sand control develops a finite conductivity fracture with the half-length of 10.5 m in well JZ9-1-A1, whose formation is the dual-porous reservoir with a rectangular closed boundary. The result conforms to current geological recognition.

Table 1 Deconvolution well test interpretation fitting results of well JZ9-1-A1

Interpretation method	Model	Reservoir parameters									Boundary model
		Initial pressure	Wellbore storage coefficient	Fracture half length	Fracture conductivity	Effective permeability K/mD	Formation coefficient	Skin factor$s$	Radial composite	Detection radius R_{in}/m	
		P_i/MPa	C/ $m^3 \cdot MPa^{-1}$	m	$mD \cdot m$		Kh/mD·m		Double hole pseudosteady state		
Topaze production well test interpretation	Finite conductivity fracture + radial composite reservoir	26.5	0.8	10.5	820	15.92	318.4	−1.9	Inner radius Ri: 110 m; Mobility ratio M:7.6; Conductivity ratio D: 0.95	504	infinite
Deconvolution interpretation	Finite conductivity fracture + double-hole pseudo-steady state medium	26.5	0.75	10.5	890	15.82	316.4	−1.3	Energy storage ratio w: 0.24; Interporosity flow coefficient λ: 2.14×10^{-7}	590	Rectangular closure:L_1: 152; L_2: 304; L_3: 752; L_4: 914

6 Conclusions

1. Based on the varying law of bottom-hole flow pressure of low-permeability oil wells during the production pressure drop and shut-in pressure recovery, one well test interpretation model of convolution and deconvolution is established. The model favors to obtain the equivalent bottom-hole pressure response for the whole production period, and its solution of deconvolution pressure with good computational stability is given.
2. Through the application research of deconvolution method, it is found that the deconvolution well test interpretation can eliminate the wellbore storage effects' covering-up on radial flow characteristics of log-log pressure derivative. It may solve the tough problems in the well test interpretation such as unobvious boundary features, short shut-in period, slow pressure recovery due to the low-permeability reservoir;
3. The deconvolution well test interpretation can furtherly distinguish the late boundary effect features, such as composite boundary, closed boundary, the edge-bottom water distance and other complex boundary characteristics.
4. For some offshore production wells with short shut-in time, deconvolution well test interpretation results are closer to the real reservoir than the conventional well test method, which can make up for the deficiency of conventional well test interpretation technology.

Acknowledgements. The authors are grateful for funding from the Natural Science Foundation of China (Grant No. U1762212) and State Key Laboratory of Oil and Gas Enrichment Mechanisms (No: 2016ZX03027-004).

References

1. Wei, C., Quantang, F., Libin, Z.: Interpretation method for early time data of well test. Well Testing **18**(4), 17–19 (2009)
2. Hedong, S., Fangde, Z., Jia'en, L.: Analog deconvolution method of interpreting gas well test early data. Nat. Gas Ind. **22**(5), 75–77 (2002)
3. Liao, H., Yang, Z., Li, Y., et al.: Water production laws and controlling factors of gas reservoir in Xihu Sag. Lithologic Reservoirs **29**(6), 135–141 (2017)
4. Ren, C., Li, W., Li, S., et al.: Application and practice of deconvolution well testing interpretation technology deconvolution well testing interpretation technology in oil and gas field of the western China sea. Well Testing **26**(3), 22–24 (2017)
5. Yan, T., Liu, H.: Application of deconvolution well testing method in Eastsea oil & gas field. Well Testing **24**(2), 35–37 (2015)
6. Zhang, J., Yang, X., Zhang, H., et al.: Application of deconvolution well test interpretation in oHai complex oil and gas reservoir. Well Testing **24**(3), 21–23 (2015)
7. Jia, Y.L., Sun, G.F., Nie, R.S., et al.: Flow model and well test curves for quadruple-media reservoirs. Lithologic Reservoirs **28**(1), 123–127 (2016)
8. Zhu, C.Y., Cheng, S.Q., Tang, E.G., et al.: Well-test analyzing method with three-zone composite model for the polymer flooding. Petrol. Geol. Oilfield Develop. Daqing **35**(3), 106–110 (2016)

9. Liu, P., Tang, H., Lyu, D., et al.: A new method for single well controlled reserves by using pressure build up curve. Lithologic Reservoirs **22**(3), 106–109 (2010)
10. Liu, W., Liu, Y., Zhu, W., et al.: Improvement and application of ILK flow-rate deconvolution algorithm based on the second-order B-splines. Acta Petrol. Sinica **39**(3), 328–337 (2018)
11. Zhou, Z., Lu, X., Gao, X., et al.: Application of deconvolution well test technology in low permeability gas reservoir. Well Testing **26**(5), 36–38 (2017)
12. Yang, X., Zhang, J., Zhang, H., et al.: Guiding the high efficient development in complex oil and gas fields by using deconvolution well testing. Tech. J. Yangtze Univ. (Nat. Sci. Ed.), **12** (11), 67–70 (2015)
13. Li, Y., Li, B., Hu, Y., et al.: Application of deconvolution algorithm to early formation interpretation of gas wells. Acta Petrol. Sinica **31**(2), 298–301 (2010)
14. Yan, Z., Guo, K.L., Li, Y.P., et al.: Interference well test with numerical simulation design and program optimization at sea. Lithologic Reservoirs **27**(2), 98–102 (2015)
15. Wei, C., Chen, B., Liu, M., et al.: Interpretation of pressure transient well testing data of S gas well based on deconvolution technique. Well Testing **26**(3), 22–24 (2017)
16. Guo, L., Cai, B., Wang, X., et al.: An improved pressure derivative algorithm of the well test data. Well Testing **12**(1), 7–12 (2003)
17. Hu, X., Zheng, S., Gao, Y.: Deconvolution well test without the original formation pressure. Southwest Petrol. Uni. Nat. Sci. **32**(3), 89–92 (2010)
18. Gringarten, A.C.: From straight lines to deconvolution: the evolution of the state of the art in well test analysis. SPE 102079 (2008)
19. Zhang, X.: The application and algorithm of automatic type-curve matching in well testing. J. Southwest Petrol. Univ. (Sci. Technol. Ed.) **30**(4),135–137 (2008)
20. Umarov, S.R., Saidamatov, E.M.: A generalization of Duhamel's principle for differential equations of fractional order. Doklady Mathematics **75**(1), 12–16 (2007)
21. Groulx, Dominic: Analytical solution to the unsteady one-dimensional conduction problem with two time-varying boundary conditions: Duhamel's theorem and separation of variables. Heat Mass Trans. **46**(7), 150–157 (2010)
22. Onur, M., Cinar, M., Ilk, D., et al.: An investigation of recent deconvolution methods for well-test data analysis. SPE 102575 **13**(2), 226–247 (2008)
23. Levitan, M., Crawford, G.E., Andrew, H.: Practical considerations for pressure-rate deconvolution of well-test data. SPE 90680 **11**(1), 1093–1096 (2006)
24. Zeng, Y., Kang, X., Tang, E., et al.: Well test analysis method for polymer flooding after water flooding in triple-layer reservoir. Lithologic Reservoirs **30**(5), 2–7 (2018)

Unconventional Reservoirs Require Unconventional Thinking: Using Fracturing Model as an Example

Z. Peter Wang[1(✉)] and Yongping Zhang[2]

[1] IUT Goup, LLC, 170046, Austin, TX 78717, USA
zwang@iutgroup.com
[2] Production Research Institute of Daqing Oilfield Co., PetroChina, Daqing, China
zhangyongping@petrochina.com.cn

Abstract. The current widely adopted Discrete Fracture Network (DFN) model has its obvious flaws. This paper presents a conceptual model which may lead us having a fresh view on the mechanical and fluid flow mechanisms during and post fracturing. The proposed Fracturing Impacted Volume (FIV) model allows us not only to avoid the inputs data setting up blindingly with the DFN model but also to think the unconventional object in an unconventional way which we are not seeing that with the concept of DFN model. Instead, FIV model emphasizes both the fracturing fluid penetration and fracture initialization as well as propagation at equal weight rather than focusing on mechanically fracturing. Through a serial comprehensive analysis of all sorts of data including core testing and field observation and testing e.g. DFIT data published and unpublished, a conceptual model, namely FIV has been proposed in this paper. From this conceptual model point of view, the entire fracturing is a process of the reservoir system (microfracture and matrix) pressurization along with fracturing in unconventional reservoirs. This paper demonstrates a good correlation between reservoir permeability or leak off and reservoir effective pressure change through a serial core samples tests of tight sandstone and shale, and field observations which include the result of DFIT and other diagnostic tests. According to the proposed model, a different interpretation of shale fracturing fluid flow mechanism is presented in this paper. Finally, based on the concept of FIV model, a couple real cases have been studied and considerable positive results have been achieved with this model. The FIV model could provide alternative solutions to problems that the DFN models have run into. With the basic concept in this FIV model, fracturing simulation will become much effective, and production simulation results will be more consistent with field history data because the conceptual model provides a better angle for us to understanding what an unconventional reservoir may have been through and changed during the fracturing, post fracturing and to the production period down the road. It is truly fresh air for us in unconventional fracking and production simulation arena. It is the time for us going back to the fundamental and basics on this.

Keywords: Fracture impacted volume · Dynamic properties · Flow and fracturing mechanism · Pressure dependent permeability

J. Lin (ed.), *Proceedings of the International Petroleum and Petrochemical Technology Conference 2019*, pp. 45–62, 2020.
https://doi.org/10.1007/978-981-15-0860-8_4

1 Introduction

There are significant differences between of unconventional and conventional reservoirs as is known. This paper will focus those factors which distinguish it from conventional reservoir such as pressure dependent permeability/leak off and etc. and the conceptual model from fracturing design and simulation point of view. Some people view that fracture network are formed through the hydraulic fracture propagating and encountering natural fractures [1], which is considered as a very similar concept as SRV back to 20,006 or so. This is not uncommon view on this. Then a common seen conceptual model, the Discrete Fracture Network (DFN) or fracture network model has been used for simulation purpose. Even though the fracture network model was not well defined, many scholars tend to quantitatively describe the existing DFN model. Where as in unconventional reservoir, the concept of the limited in number and spatially distributed sizable natural fractures intersecting with the hydraulic fractures and creating either regular or irregular shaped fractures, DFN or fracture network is adopted by many. However, in practice, this model uses finite number of natural fractures which must be less than numerical grid numbers; and it uses the dual-porosity dual-permeability model for fluid flow mechanism, which has been implemented for conventional reservoirs for a few decades. The imperfect in using both the fracture network model and the concept is so obvious, and caution must be taken while applying this technique.

The unavoidable fact in using DFN model is that one has to face many unknown parameters in the model such as the size, total number and distribution, conductive and connectivity of natural fractures among others. Scholars tend to deliberately define those parameters randomly with very little or no data. Some use the rock outcrop to define the natural fracture geometry for reservoir; some others interpret from the microseismic event cloud, counting the number and location to interpret as reopened natural fractures. But neither of the methods can overcome the blindness or the randomness of the natural fracture distribution. Outcrop can be used to observe many geological properties of the rock, but not for fractures because the rocks had gone through totally different orogeny movement in geological history between the two, outcrop and reservoir, evidently. Microseismic was used as another tool to locate natural fractures. The large number of non-calibrated microseismic events would introduce numerous errors and quite low accuracy, which was used in interpreting the fracture network and discrete fracture as SRV [2, 3]. The less credibility model introduces its unavoidable and unconquerable problems to the fracture interpretations and its related computations. And some may take micro-fissures data from cores or image logs which are in very small/micro scale and drag them to numerical grid scale in their numerical model which is in macroscale.

Furthermore, conventional dual-porosity dual-permeability mode is also debatable for unconventional reservoir because the fluid flow mechanisms difference between the shale/tight sandstone reservoirs and the conventional reservoirs like in carbonate reservoirs. Erdal Oskan (2012) indicated that, while the matrix absolute permeability is lower than 10–6 mD, the contributions of natural fracture network is close to none [4], the dominate factor is the flow within the matrix. Ian Walton (2017) has summarized from 2000 wells in Barnett shale with its production history profile for a conclusion: Natural Fracture Add Little to Shale Gas Reservoir Productivity [5]. Similarly, massive matched production history from unconventional reservoir suggests that, the error in used permeability is easily in 2–4 order of magnitude [6]. In the application side of DFN, there are contradictory results between fracture network data and other parameters. Mayerhofer (2014) uses the SRV model to investigate inter-well relationship between a fracturing well and a producing well in Marcellus shale reservoir, and there is an extensive un-matching result; comparing to the better matching results with the simple fracture model. There are numerous examples of un-consistency of production data matching or sometime contrary with this fracture network model.

Hence, to overcome the above limitation from the current fracture network model and to avoid further problems, a novel conceptual model is presented in this paper. It is needed to overcome the mentioned flaws and be more practical and certainly will lead more robust and/or improved unconventional oil and gas fracturing and developing models. An important parameter of this new model discussed in this paper is the concept of permeability is a pressure dependent variable, or the Pressure-Dependent Permeability (PDP) [7]. Warpinski [8] uses summarized the data collected at northwestern Colorado low permeability gas sandstone shows similar relationship in the following figure.

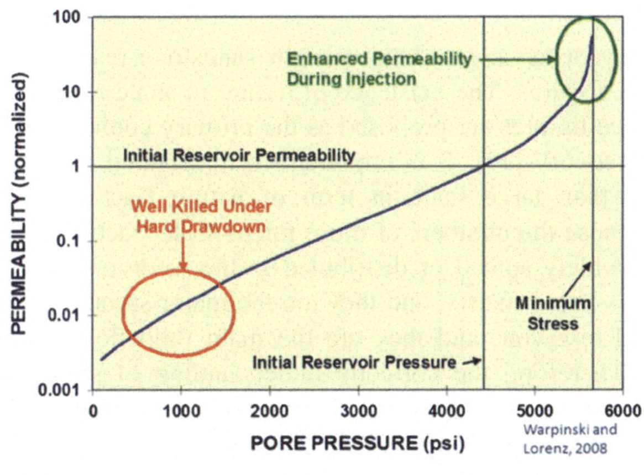

Fig. 1 Normalized permeability is a function of pore pressure [8]

Mittal [9] states that the permeability of the natural fracture increases during the hydraulic fracturing operation. When the pressure reaches the critical point, the permeability increases rapidly. Of course, similar phenomenon have been found from the lab testing and in the field observation as green circle shows in Fig. 1. Even though those investigations were made for tight sandstone reservoirs, it is believed to be true for as shale as well.

Many reservoir simulating models assume the permeability of the reservoir does not change, but it is no longer the case for unconventional modeling. Mohammad [10] proposed that there is an exponential relationship between permeability and reservoir pressure of the induced-fractured zone and hydraulic fractured-zone. The hydraulic fracture and the induced fractures have high geomechanics impacts as of other parameters, and it is even more sensitive to induced-fracture (i.e. the natural fracture is not been propped).

2 Fracturing Impacted Volume (FIV)

To overcome the imperfect and impractical fracture network model like DFN, a model to take two objects in a reservoir into account becomes necessary at certain degree so that the model can be applied much more effective and easier without losing much accuracy; and those two objects in a reservoir are: (1) the sizable fracture such hydraulic made ones or identifiable ones by some mean in the field in the scale like feet or meters, (2) countless microfractures, natural fissures, organic or inorganic pores, clay mineral dewatering microcracks/pore and the matrix—lump sum as the reservoir system with a localized or nonuniform permeability, Ks depend upon the data availability. This paper presents the conceptual model of FIV with the said features. More calculation and model details are to be presented in later articles with more data and investigation works come out.

2.1 FIV Model

From the laboratory observation, shale or tight sandstone reservoir usually fill with numerous naturally fissures. The existence of fissure in shale reservoir is important in shale reservoir, since fissures are proposed as the primary contributor to principle pore and permeability network [11]. It is important to understand that we are talking the micro-scale rather than large scale in term of nature fractures in unconventional reservoirs. It is because the numbers of those micro-scale fracture/fissures are so huge and they are also widely spread or distributed in the reservoir, actually they are the major part of the so-called matrix, and they are the major storage of the oil and gas in the unconventional reservoir, and they are the main fluid flow pathways as well at microscale level. Therefore, the correctly understanding of scale for nature fissure become so important in term of fracturing and reservoir simulation model. As D. Willis, an expert from Oil and Gas in Google Cloud pointed out during ATCE panel speeches in 2018, many of our problems we are facing in oil E&P are because of the misuse of scales. In this section we will present some lab testing and flied observation results for the tight sandstone and shale reservoirs, which demonstrated the correlation between the reservoir permeability and the pressure for shale reservoir system and even tight sandstone reservoir, and presents the view point on fracture mechanism from the proposed conceptual model, FIV.

Significance of natural microfractures in unconventional shale reservoirs is important because microfractures are commonly proposed as a principal pore and permeability network in the production of hydrocarbons from mudrocks (shale) according to Loucks and etc. [11] (Fig. 2). In microscale of observation, microfractures are primarily dominated intensive distributed, at least are much larger (a few orders of magnitude higher) in number and wider in distribution than those in the DFN model. These types of microfracture can not only be the gas/oil storage, but also be the fluid flow paths. For such a huge number of fissure and micro-pores, addressing unconventional reservoir stimulation with slickwater, the adjacent area near the main fracture is energized by the stimulation process, and the pressure within the naturally or induced fissure/fractures and micro-pores rapidly increase triggered by the pumping fluid. This spatial volume that is energized by the stimulation process, which is defined as Fracturing Impact Volume (FIV) and the corresponding model is defined in later part in this article.

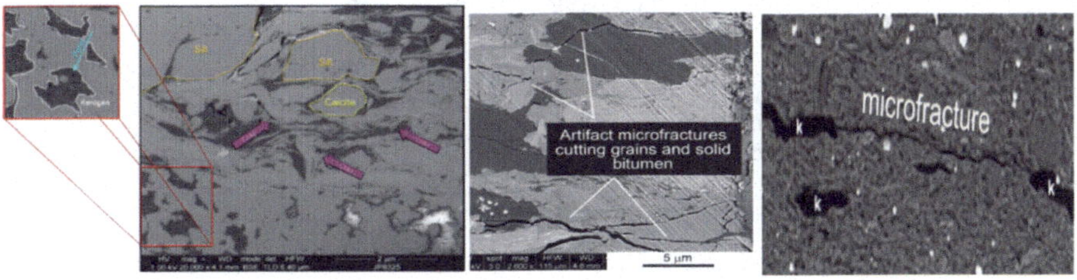

Fig. 2 Fissures in unconventional reservoir

The FIV model takes fully consideration of the pressure dependence reservoir system permeability and fluid loss. The micro discrete structural changes during the pore pressure charging process, which increases the connectivity of inter fissures, and between the fissure and the formation. The pressurized fracturing fluids (slickwater) enter the matrix which accommodate a lot micro fracture (fissure) and creates the FIV, and FIV is equivalent of increasing pore pressure and the initial production pressure difference. This is another mechanism of the unconventional reservoir stimulation.

During the stimulation, increase the micro-fracture and micro-pore pressure is easier than creating additional fracture, and the process happens earlier and lasts longer. As the fluid pressure within the pore gradually increase, the effective stress act on matrix frame work decrease, the connectivity in micro-fracture and micro-pore increases but the pressure is not exceeding the critical failure stress. In other word, there is initial fluid loss, but due to the limited contact area at early time, the fluid loss is small. With the initiation and propagation of fracture, the contact area increases. This increases the fluid loss to the matrix formation. During the stimulation process, because the fracturing fluid (slickwater) flows easily, it enters the formation matrix rapidly under greater pressure difference and break the pressure balance of the original formation; on other hand, the matrix-system which contains massive micro-fracture (fissure) was pressurized by stimulation, which increases the permeability and energize the formation in the FIV domain.

Figure 3 is the sketch for Fracturing Impacted Volume concept model, injecting with the better penetrability fluid and the hydraulic fracture is shown within the two solid red lines in very middle. The pressure difference increases as the pore pressure increase, fracturing fluid enters micro fractures to form the energized pressure zone (purple line in figure). This volume is the Fracture Impacted Volume. Micro-Fracturing Impacted Volume is related with micro-fracture density. The more of micro-fracture, the larger of FIV, and vice versa.

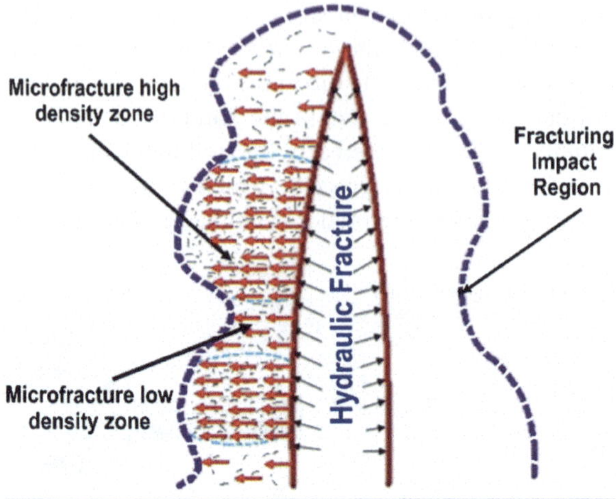

Fig. 3 Sketch for fracture impacted volume

Theoretically, SRV, is created by the fracture net pressure, which is the fluid pressure minuses the minimum horizontal stress/closure pressure. When the net pressure is equal to zero, the fracture closes.

$$P_{net} = P_{frac} - \sigma_{\min} \tag{1}$$

$$P_{net} > \sigma_{\min} \tag{2}$$

FIV is different than SRV. It is the pressure transfer with in the liquid system, the pressure difference between the fluid within the fracture and the pore pressure, ΔP. During the stimulation, the matrix system energized by fracture fluid. The system permeability increases before it reaches the critical failure pressure. In unconventional reservoir stimulation, before the rock breaks, the average pressure is Pt, which is pressure of the area of reservoir system fluid loss increase rapidly. The initial reservoir pressure is Pi, ΔP is the additional pressure increase, as in equation:

$$\Delta P = P_t - P_i \tag{3}$$

$$\sigma_{eff} = \sigma - P_{pore} \tag{4}$$

In the above equation, σ_{eff} is the effective stress acing on the matrix frame. The dispersion process with in the microfracture and micro-pore provides additional driving force for flow back. It is believed that this driving mechanism is the most outstanding and unique parameter in unconventional reservoir. After/during rock failure.

$$\Delta P_{flu} = P_{frac} - P_i \tag{5}$$

$$\Delta P_{flu} > P_{net} \tag{6}$$

where ΔP keeps increasing and reaches the maximum value, ΔP_{eff}, the fracture starts to initiate. The pressure in fracture, P_{frac}, is just a little higher than the pressure in reservoir pore pressure during penny fracture phase and then increase significantly while the fracture keeping growth. When the fracture propagates, it becomes the fracture extension pressure. It is worth noting that, during the whole process of pressure rising from original pore pressure to fracture initiation pressure, all liquids are a loss into the formation rather than creating fracture. The fracturing impacted volume FIV discussed in this paper is mainly referring to the system of reservoir rock prior fracture and during the fracturing process, the formation system of micro-fracture and micro-pore diffusion dominated pressurized on both sides of the fracked fracture. In other words, due to the influence of the changing the micro-pore and micro-fracture pressure during fracturing, part of fracturing increased production is mainly contributed due to the process of micro-fracture and the system pressurization, and the range of increasing production is the FIV-affected volume of fracturing which associate with the model of FIV.

2.2 Model Rationality

Because the penetration of fracturing fluid (most of the cases with slickwater) forms the pressurized zone around the main fracture in the reservoir system, which is the volume of FIV at that moment. Why the FIV model can be used to interpret what may happen during hydraulic fracturing than the Fracture Network/DFN will be mainly discussed from the following aspects.

First, from the whole system point of view, the proportion of fracturing fluid flowback in unconventional reservoirs is relatively low or quite low comparing with conventional ones, which is the basic fact. Based on this fact, where does the fracturing fluid go during a few hours of fracturing? There are two proposed different explanations: one is that fracturing fluids are all or most of them creating fractures to form fracture networks. That is, in addition to creating hydraulic fractures, it is to open natural fractures with relatively large scale (in feet or meters), so that macroscale model can be used. Another explanation is believed that big portion of fracturing fluids—slick water penetrates a considerable part of the reservoir system alone the hydraulic fractures. This should be a reasonable interpretation of why slickwater does not any other fluid, and why nanoscale permeability rock can produce unbelievable amount of oil and gas from the unconventional rock which were believed impossible. Is this all because the fracturing creates limited large-scale fracture intersecting with sizable nature fracture. If so, it will leave more questions than answers. If is not because the connectivity of microfractures and micropores with large cardinal number and wide distribution in the reservoir through the stimulation, it is non-explainable that millions barrels of oil in the microscale storage space can be produced. Based on this conceptual model, we can certainly have wide open mind to interpret what may have been happened during and after fracturing.

In addition to the above basic scientific and logical analysis, more rationality is to present as following paragraph. The rationality of the FIV model is discussed from the aspects of the field test data, the core experiment in the laboratory and the simulation results of the field DFIT diagnostic tests.

Warpinski and Lorenz field test shows result from many wells in the Piceance basin of low permeability sandstone gas in 2008 [8], with the increase of pore pressure, the permeability increases exponentially with increasing pressure; just open the existing micro-fractures, and did not create new cracks, just to improve the connectivity between micro-fracture and micro-pores, the system permeability in the reservoir increased significantly. These findings and technical ideas of this paper are common.

The Core Experiment Proves the Rationality. Figures 4 and 5 present the results of PDP (Pressure Dependent Permeability)/PDL (Pressure Dependent Leakoff core experiments. Figure 4 shows the relationship between effective fluid pressure difference (the difference between injection fluid pressure and original pore pressure) and fracture surface area water absorption under different confining pressures. The correlation between these two parameters is a simple reflection of permeability. Under different confining pressures, the water absorption per unit surface area is less affected by confining pressure when the confining pressure is above a certain level. However, with the increase of the effective fluid pressure difference, the water absorption capacity of the unit surface area increases gradually. When the effective fluid pressure difference reaches 5 MPa, the water absorption capacity of the unit surface area increases dramatically. In Fig. 5, the relationship between dimensionless fluid pressure difference and dimensionless permeability is obtained by using the derivative of red curve in the left diagram, and it is obvious that with the increase of fluid pressure difference, the connectivity between micro-fractures improves and the effective permeability increases.

As a result of liquid injection, the pore pressure increases and the effective pressure of the reservoir system (the overlying rock pressure or the difference between the

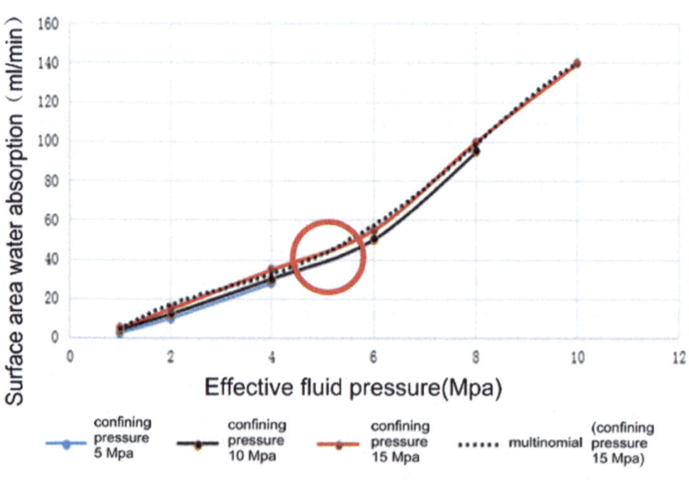

Fig. 4 Relationship between effective fluid pressure difference and surface water absorption. (Modified from [12])

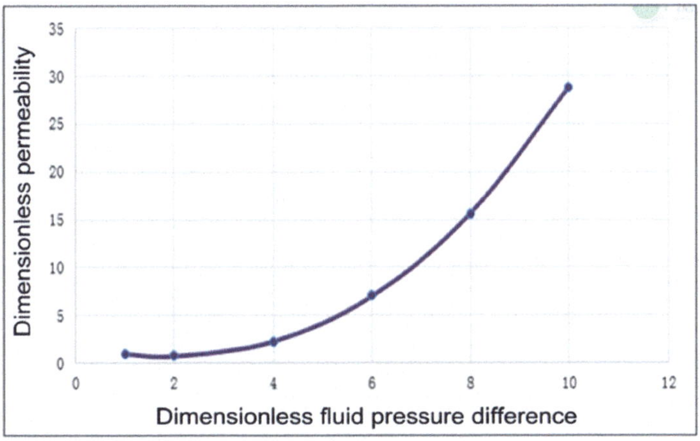

Fig. 5 Relationship between dimensionless fluid pressure difference and dimensionless permeability

confining pressure and the pore pressure in the rock) decreases, in reservoir system, the compressibility between grains decreases and the tension between particles increases, and the connectivity of tiny pores and fine fractures in rock becomes better, which leads to the increase of permeability, and the change of permeability will inevitably lead to the change of permeability of the whole system. According to many core experiments, it is found that there is an exponential relationship between core permeability and effective pressure difference:

$$K = K_i e^{d_f \Delta p} \tag{7}$$

Here K Is the reservoir system efficiency permeability, K_i is the initial reservoir system permeability, and ΔP is the effective fluid pressure difference ($P\text{-}P_{\text{initial}}$), and d_f is the exponential decay constant, which is defined by the particle size, sorting, and compaction of the rock. Need to note that the effective fluid pressure difference is not the effective stress, the effective fluid pressure difference is the difference between the injection fluid pressure and the original pore pressure, and the effective stress (the force loaded on the rock frame) is the difference between the stress and the pore pressure. The relationship between effective stress and system permeability is discussed below.

Figure 6 shows the core experimental data of several unconventional oil and gas fields from the Barnett and Utica sandstone gas fields, Eagle Ford and Marcellus shale

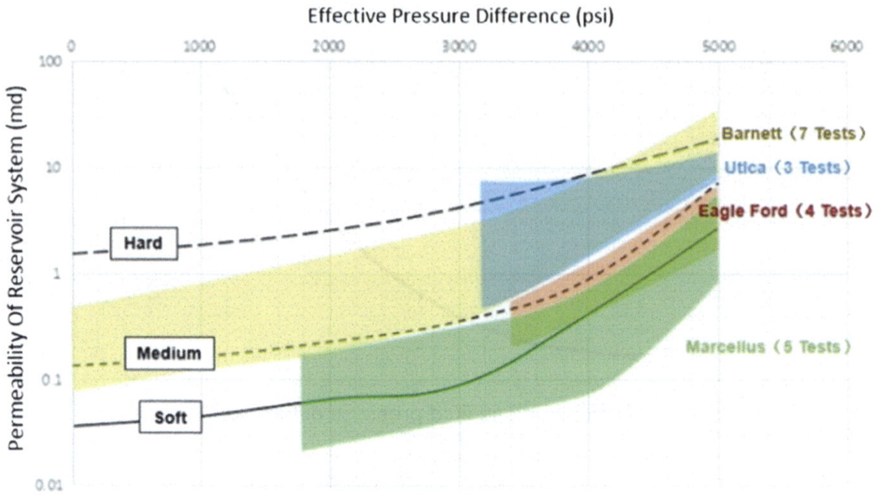

Fig. 6 Relationship between the permeability of reservoir system and effective pressure difference in unconventional oil and gas fields [13]

basins, to determine the relationship between the reservoir system permeability and the effective stress difference. There is an exponential relationship between them. With the increase in effective pressure difference, the effective stress decreases, and the system permeability increases. The reason is that when the effective pressure difference increases, the effective stress of the rock frame of the micro-fracture decreases, the connectivity of the micro-fracture is improved, and the effective permeability of the system is improved. With the decrease in effective stress, the system permeability of Utica sandstone gas reservoir in the blue area increases one order of magnitude, showing a decreasing exponential decreasing relationship. The results show that this type of sandstone reservoir has a good effect of micro-fracture and micro-pore pressure leading diffusion, and other shale reservoirs also have a similar relationship, which is a significant negative correlation between system permeability and effective stress, and a positive correlation between system permeability and effective fluid pressure difference.

In the diagram, the reservoir rock changes from soft to hard from bottom to top, and when the lithology is soft, the brittle index is lower, the density of micro-fracture is smaller, the less the total amount of fracturing fluid (slick water) enters the reservoir micro-fracture, the smaller the pressurization area controlled by micro-fracturing is. Therefore, the volume FIV affected by fracturing in green area is relatively small. For hard lithologic such as Barnett shale, brittleness index is higher, micro-fracture density is larger, and the pressurized area of microfracture density becomes larger, for example, the volume FIV of Barnett shale affected by fracturing in yellow area will also become larger. It should be noted that when shale and ultra-low permeability sandstone are separated from the high-pressure reservoir and fluid saturation environment, many micro-fractures will be completely closed and can no longer be energized or reopened.

Therefore, the relationship between the injection pressure or effective pressure difference measured in laboratory and the permeability of reservoir system is generally low. Normally, the permeability measured in laboratory is not more than two orders of magnitude with the increase of injection pressure difference, but most should be in one order of magnitude.

Field Diagnostic Test Proves the Rationality. The proposed Fracturing Impacted Volume model for micro-fracture and micro-pore pressurization can be verified by field test. Figure 7 shows a monitored pressure fitting curve in the DFIT diagnostic test of Vaca Muerta unconventional reservoir gas wells [14], the permeability of the DFIT diagnostic test is one order of magnitude greater than the core test. The black line represents the original monitored pressure. The blue line represents the pressure curve fitted using DFIT pore pressure and DFIT permeability. The red line represents the pressure curve fitted with low pore pressure and core permeability. The green line represents a pressure curve fitted using low pore pressure and DFIT permeability.

The blue line fits the best, while the green line fit the worst. Because the pore pressure of blue line is the higher DFIT pore pressure, the permeability is the per-

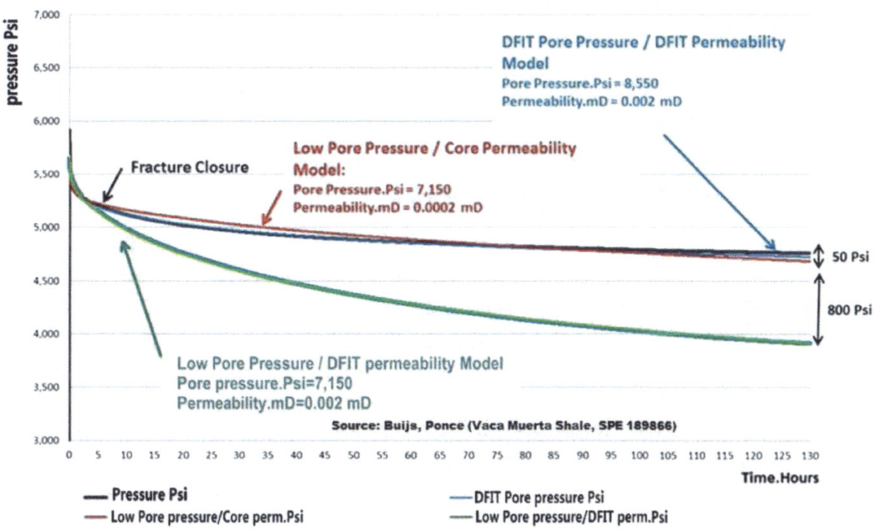

Fig. 7 Pressure fitting curve for monitoring a gas well [14]

meability is obtained by DFIT; As the pore pressure increases, the effective stress decreases, the connectivity between micro-fractures and micro-pores improves, and the effective permeability increases. The pore pressure and effective permeability increase simultaneously. Only by considering the both changes at the same time, we could become more reliable to the underground, and we can fit the monitoring pressure curve better.

The red curve in Fig. 7 seems to be not too bad in term of history matching of pressure. But in this paper, Buijs also compares the G-function curves in two cases

[14]. However, the interpretation of permeability measured by core and pore pressure measured by gas display, G function is completely wrong, while using both the permeability and pore pressure obtained by DFIT and the closed pressure explained by G function of that pore pressure are in very shape.

This case study shows that permeability obtain by other means is at least one magnitude lower than one obtained with DFIT, similar trend to the pore pressure. This shows the validity of the fracturing impact the permeability and pore pressure even with very small amount of water. The matrix is energized and activated, and the effective permeability increases due to increment of pore pressure. It is absolutely not ignorable if 75,000 barrels of water/fluid injected into reservoir within one stage of fracturing. What we have done to the wide distributed formation beyond creating some fractures. If that amount of water all used for creating fractures, our fracture space will not be just a few meters in spacing. **This is so obvious, and so basics**.

Because of the strong positive correlation between reservoir filtration and system permeability, in the initial stage of fracture closure, reservoir filtration shows pressure dependence, which also shows pressure dependence of system permeability. In the process of fracture main fracture opening and extending, fracture rock micro-fracture, micro-pore pressing, make the micro-fracture connectivity in the reservoir become better; At this time, no matter whether micro-cracks have micro-opening or slippage, there exists the phenomenon of increasing permeability and filtration loss, which increases the driving mechanism in subsequent production. Driving mechanism provides energy for subsequent fracturing.

2.3 The Role of FIV Model in Fracturing

Because of the filtration effect of fracturing fluid, a pressurized zone is formed around the fractures in the reservoir system, that is, fracturing affects the volume of FIV. The rationality and applicability of this interpretation is not only to avoid the trouble of distinguishing matrix from natural fracture in the calculation of fracture mesh model, but also to put forward a new way of thinking about fracturing model. The effect of micro fracture on the whole fracturing effect is considered more than the previous double porosity and double permeability model. For the productivity calculation model, the great effect of micro fractures (or capillary fractures) on the productivity after fracturing is fully considered. The change of conceptual model, many working ideas should also change with it.

Fracture mesh model is a common model for unconventional oil and gas fracturing, in which the volume of fracturing modification and the flow conductivity of fracture net are the key indexes to evaluate the effect of fracturing operation. In this model, the geometric parameters of the fracture mesh are obtained by simulating the expansion law of the fracture net and the flow of fracturing fluid and the movement of proppant in the fracture mesh, and the fracturing operation scheme is selected. Shale with natural fractures is the key point of fracture mesh fracturing. The model holds that the smaller the horizontal stress difference is, the easier it is to form fracture mesh, the larger the operation discharge is, the larger the total amount of fracturing fluid is, the larger the volume range of reservoir reconstruction is. The higher the seams' diversion capacity, the higher the productivity.

Obviously, under the influence of fracturing volume FIV model and the whole flow system combination model, the existence of fracture network is not the key and focus of the problem. Fracturing influence volume model is a new idea of unconventional fracturing. The goal of fracturing is different from fracture mesh model (whether to form fracturing standard for fracture net and so on) and to pursue optimal production. The model does not think that the larger the displacement, the larger the liquid scale, the better the fracturing effect is. Due to the slow process of micro-fracture and micro-pore pressure leading diffusion, the model usually uses small displacement and fluid flow to fracture, which does not require formation of fracture network, but increases pressure in micro-fractures and micro-pores. Thus, the microstructure of reservoir rock is improved, and its connectivity is improved (I.E. the matrix is punched and activated), the effective permeability of reservoir is increased, and the filtration is increased. In addition, the parameters optimization of fracturing design and fracturing operation for well spacing, fracture spacing and so on are also different from the fracture mesh model.

In a word, the influence volume model of fracturing has fundamentally changed the guiding ideology of unconventional fracturing and put forward different requirements for the optimum design and production mode of fracturing.

3 Application Example

To investigate the rationality of the above theory and method, a horizontal well fracturing in a gas field is applied and calculated. The horizontal section of the well is 500 m long, fracturing is divided into four sections, and the depth is about 2000 m. The specific physical parameters of the reservoir are shown in the Table 1.

According to the above test parameters, a model of 300, 300 and 100 m is established for long, wide and high respectively, wherein the interlayer is respectively

Table 1 Physical parameters of rock in a horizontal well

	Young modulus/GPa	Poisson ratio	Mean porosity/%	Mean permeability/mD
Interlayer	5.65	0.24	0.17	/
Air layer	15	0.21	0.17	0.001

25 m and the gas layer is selected to be 50 m, combined with the above fracturing effect volume FIV model, considering the dependence of permeability on effective pressure difference, the reservoir numerical simulation method is used to simulate the horizontal well fracturing.

When shale gas is produced, the production is defined in the inner side, the pressure is defined outside, and the pressure relief radius is the distance to which the pressure disturbance propagates at the current moment, the conditions for the solution are as follows:

$$\Psi|t = 0 = \Psi_i(p = p_i) \tag{6}$$

$$r\frac{\partial \Psi}{\partial r}\bigg| r = r_w = \frac{q_{sc}p_{sc}T}{\pi K_o h Z_{sc} T_{sc}} \tag{7}$$

$$\Psi|r = r_c = R(t) = \Psi_i(p = p_i) \tag{8}$$

Through the pressure distribution formula of unsteady seepage reservoir and MATLAB programming, the distribution curve of formation pressure in different time is obtained. As the production time increases, a pressure drop funnel is formed from the wellbore to the far end. According to the distribution of formation pressure at different production times and different distances from the wellbore (Fig. 8), combined with the change of reservoir permeability and pressure difference mentioned above, reservoir permeability distribution at different distances from wellbore and different production times can be obtained.

Figure 9 shows the permeability field of reservoir system in the early stage of fracturing production. Permeability changes in the near well zone after fracturing. According to the model of fracturing effect volume, the permeability of reservoir system increases because of the better connectivity between micro-fractures and micro-pores after fracturing. The closer it is to the well, the better the diffusion effect is, and the more obvious the permeability is. With the increase of distance, the diffusion effect of micro-fracture and micro-pore pressure leading to diffusion decreases gradually, and the increase of permeability becomes smaller.

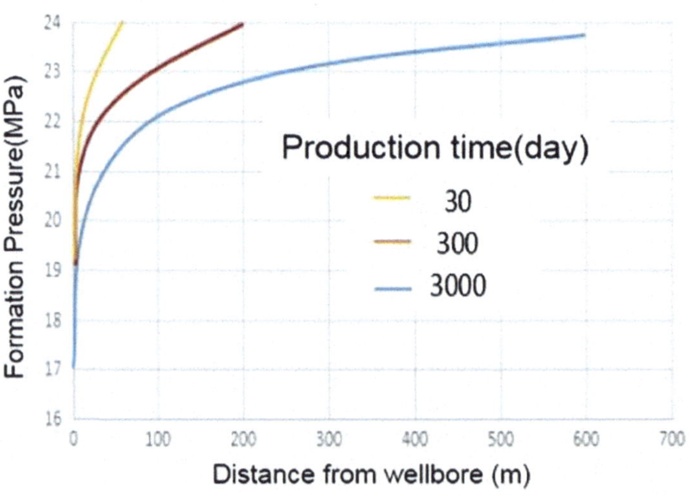

Fig. 8 Distribution of formation pressure at different distances from the well

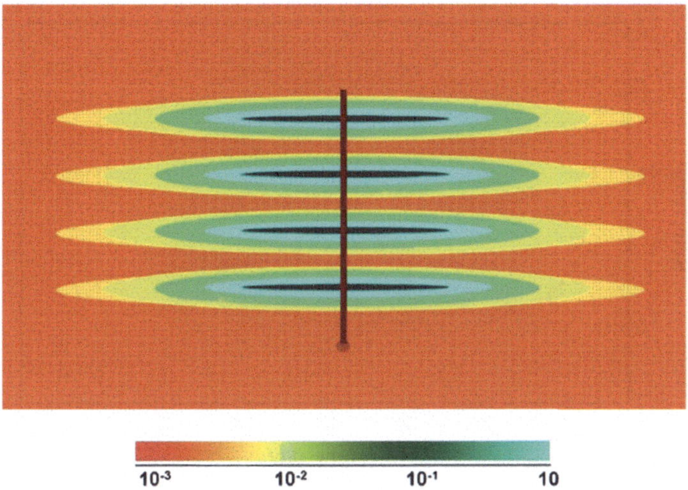

Fig. 9 Distribution of reservoir permeability at different distances from the well

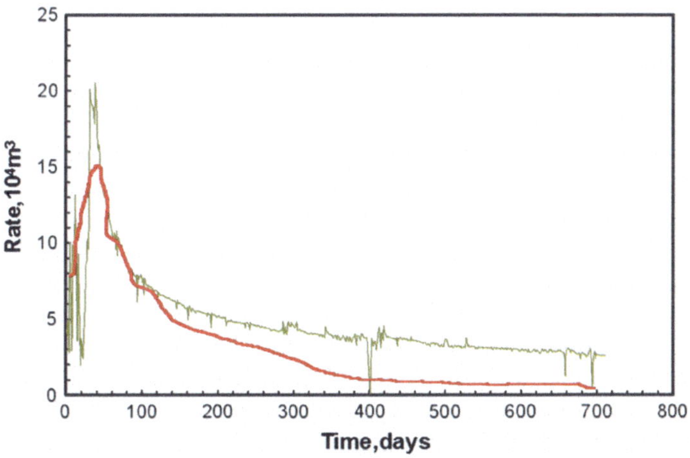

Fig. 10 Production fitting without considering the effect of fracturing on volume FIV effect

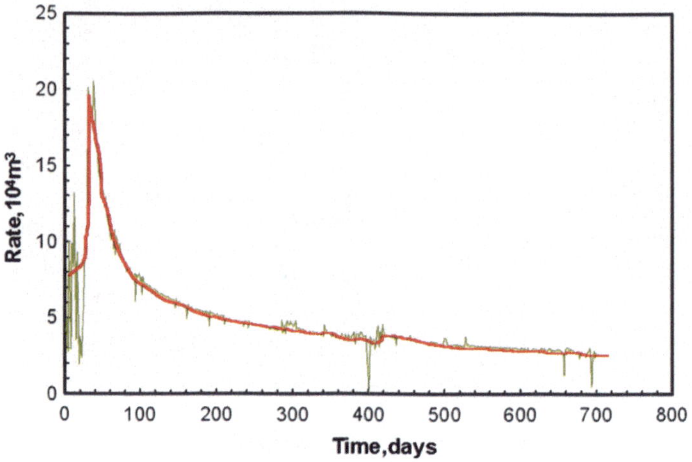

Fig. 11 Production fitting considering the effect of fracturing on volume FIV

Figure 10 shows the production matching of not considering Fracturing Impacted Volume effect, the curve trend and numerical has not been fitting. In early stage of opening the well for production, due to not considering Fracturing Impacted Volume effect, pore pressure and system permeability are overall low, lead to the single well production cannot be matched. Such as at the initial production stage, the average formation pressure of near wellbore area is 18 MPa, average permeability of near wellbore area is about 0.001 mD. When considering FIV effect, it must be considered fully the increasing condition of pore pressure and system permeability after the formation was stamped, the average formation pressure of near wellbore area is 20 MPa, average permeability of near wellbore area is about 0.01 mD. The parameter setting is more suitable for the real underground situation after fracturing. Figure 11 shows the production matching of considering Fracturing Impacted Volume effect (fixed permeability near wellbore area), the production curve matching is good.

As Cluff, etc. pointed out in their article of 2007, there is a minimum representative volume element in a fracturing simulation. The smallest representative volume unit is obviously larger than the minimum representative volume unit of the conventional reservoir, because the latter considers the existence of micro-cracks. The numerical modeling requires 2–4 magnitude orders of the system permeability of the gas permeability to match the flow rate and the final increasing. It is necessary to have a higher permeability pathway through the shale [6]. The example above is also fully showing that existing the phenomenon of matrix stamping activation, the connectivity of micropores is better and the system permeability increases, which was evidenced in the capacity simulation. The concept of Fracturing Impacted Volume model is consistent with the fitting results of actual production data, also indirectly shows that significance guiding of the model proposed in this paper of future fracturing and productivity evaluation.

4 Conclusion

Tremendous numbers of horizontal wells have been fractured and huge amount of production data are available with over decade production history of unconventional reservoirs. We have a lot of more information that we had a decade ago. It is no supervise for us to have some new understanding on the unconventional reservoir fracturing and production related simulation.

The drawbacks in using Discrete Fracture Network (DFN) or fracture network model are so obvious and unavoidable as discussed in this paper, an alternative mode is certainly necessary. We all know that we cannot get those parameters properly for the DFN model as discussed above in this paper, and trial and error method is not feasible neither due to the unlimited possibilities.

On the other hand, in the FIV model, regardless the size, number of the nature fractures and drilling/hydraulic induced fractures, all those non-quantitively characterized micro and macro fractures can be treated as the reservoir system properties, which may be measured through DIFT or other means including simulation method to estimate the alone with the main fracture created hydraulically. This is what has been called "the greatest truth for complicated things is simple, and most likely it is the best answer" as an old saying.

In the fracturing process, a large amount of slickwater was used as pad fluid to enhance the original permeability and the leak off increasing significantly, which increases the additional driving mechanism for subsequent production and provides additional potential energy and activation very localized pore/fracture space of the rock in the microscale of the reservoir. In this paper, the concept of FIV model and its flow system composite model is hinted. With this model and its related concepts, certainly the fracturing design and production mode can be improved significantly for unconventional reservoir development and EOR as well.

In terms of the mechanism of unconventional fracturing: FIV model and its concept are different from the concept and model of fracture network fracturing or DFN, the latter over emphases the importance of fracture networks or complex fractures and needs to quantitatively define them. That is an obvious and old pitfall as in conventional reservoir for decades—detailed static parameters characterization dilemmas because the so-called static parameters are not never static in many cases. It is believed that dynamic is always the prevailed factor for those parameters from the beginning to nowadays when the fracturing is implemented. In the FIV region, the reservoir permeability is enhanced and microcracks/micropores are activated through pressurization of fracturing, which stimulates oil and gas in microscale alone the main hydraulic fractures. That may be the most important mechanism of the fracturing all about for unconventional reservoirs.

In terms of fracturing simulation and production model: In the framework of in using FIV model, people don't need to puzzle their brains about how to set the amount of natural fracture, the position, the scale and their conductivity of nature fractures in the model, and it is not necessary and no way to distinguish which inch of the pay is the matrix and which microfracture/fissures are the natural ones or induced ones in the mesh model calculation. In this paper, FIV fracturing simulation and reservoir simulation model and its associated concepts are considered irreplaceable role.

Acknowledgements. Authors would like to appreciate Dr. Mike Smith for his constant discussions and suggestions during the paper drafting phase. And would like to appreciate engineers B. Wang from Stim & EOR Tech, LLC of IUT Group, W. Zhang and Y. Chen from Huaxia XiaoNeng PetroTec, Ltd. for their assisting on data and information gathering and some editing.

References

1. Kresse, O., Cohen, C., Weng, X., el al.: Numerical modeling of hydraulic fracturing in naturally fractured formations. US Rock Mech./Geomech. Symposium, San Francisco (45), 26–29 (2011)
2. Wang, Y., Wang, Z.P., el al.: Years of tight oil fracturing: what have we learned. Soc. Petrol. Eng. SPE181810 MS (2016)
3. Zhonghe, P.. Wang, Y.W., el al.: Years of tight oil fracturing: what have we learned. Soc. Petrol. Eng. SPE181810 PPT (2016)
4. Ozkan, E.: Releasing shale-gas potential with fractured horizontal wells. Soc. Petrol. Eng (2012)
5. Parshall, J. (ed.): Natural Fractures Add Little to Shale Gas Reservoir Productivity. SPE Distinguished Lecturer Ian Walton, JPT (2017)
6. Cluff, R., Shanley, K., Miller, M.: Three things we thought we understood about shale gas, but were afraid to ask. AAPG Ann. Meet (2007)
7. Ehlig-Economides, C., Vera, C.F.: Diagnosing Pressure-Dependent-Permeability in Long-Term Shale Gas Pressure and Production Transient Analysis
8. Warpinski, N.R., Lorenz, J.C.: Analysis of the multiwell experiment data and results: implications for the basin-centered gas model. GeoSci. World J. (1) (2008)
9. Mittal, R., Oruganti, Y., McBurney, C.: Re-fracturing simulations: pressure-dependent srv and shear dilation of natural fractures. In: Unconventional Resources Technology Conference (2015, July 20). https://doi.org/10.15530/urtec-2015-2154943
10. Eshkalak, M.O., Aybar, U., Sepehrnoori, K.: An Integrated Reservoir Model for Unconventional Resource, Coupling Pressure Dependent Phenomena
11. Loucks, Robert: Natural microfractures in unconventional shale-oil and shale-gas systems: real hypothetical, or wrongly defined. GCAGS Journal **5**, 64–72 (2016)
12. Ayal, M., Wanniarachchi, W., Gamage, R.P., el al.: Investigation of depth and injection pressure effects on breakdown pressure and fracture permeability of shale reservoirs: an experimental study. Appl. Sci (2017)
13. Britt, L.K., Smith, M.B., Klein, H.H.: Production benefits from complexity—effects of rock fabric, managed drawdown, and propped fracture conductivity. Soc. Petrol. Eng (2016)
14. Buijs, H., Ponce, J.: An engineered approach to hydraulically fracture the Vaca Muerta shale. Soc. Petrol. Eng. SPE189866 (2018)

Analysis of Cementing Quality in South Sudan Oilfield

Jindong Wang(✉)

Drilling Producing and Technical Services Company of Daqing Drilling and Exploration Engineering Company, PetroChina, Daqing 163000, Heilongjiang, China
383099747@qq.com

Abstract. Block 3/7 of South Sudan formed high permeability and low pressure layers after long-term exploitation. It affects the effective cementation of the first and the second interface after cementing operation. The reasons for poor cementing quality in high permeability and low pressure layers are mainly due to micro-leakage in drilling and cementing, high dehydration of cement slurry, formation fluid invasion in the process of waiting on cement (WOC), shrinkage of cement sheath, etc. This paper explains the adverse effect of high permeability and low pressure on the cementing quality by studying the formation mechanism of mud cake weak interface in high permeability and low pressure layers, the effect on weak interface of excessive dehydration of cement slurry, formation fluid seepage and corrosion. Continuously improving the cementing process and cement slurry system, developing and applying the bentonite low density cement slurry system, optimizing the practice of drilling engineering such as the drilling fluid system, the length of rat hole, centralizer placement design, wiper trip and ream in Block 3/7 of South Sudan ensure that the mud cake is thin and compact, and that the borehole is gauged. The requirements of South Sudan COMPANY are finally met for cementing quality.

Keywords: South Sudan · High permeability and low pressure (HPLP) · Weak interface · Cementing quality

J. Lin (ed.), *Proceedings of the International Petroleum and Petrochemical Technology Conference 2019*, pp. 63–69, 2020.
https://doi.org/10.1007/978-981-15-0860-8_5

1 Introduction

After many years of exploitation, the formation pressure in Block 3/7 of South Sudan cannot be compensated effectively, which has formed high permeability and low pressure layers adversely affecting the cementing quality. Through the continuous improvement of cementing technology, cement slurry system and drilling engineering, the cementing quality problem of high permeability and low pressure has been effectively solved.

2 Mechanism of Cementing Quality Influenced by High Permeability and Low Pressure Layers

The effect of high permeability and low pressure layer [1] on cementing quality is mainly reflected in three aspects. Firstly, mud cake weak interface is easily formed at high permeability and low pressure. Secondly, excessive dehydration of cement slurry forms weak interface. Thirdly, the seepage and corrosion of formation fluid strengthen the influence degree of weak interface.

The annular cemented transition zone with loose structure and increased porosity and permeability in cement sheath and casing cementation, and the mud cake transition zone between cement sheath and formation are collectively called weak interfaces. As shown in Fig. 1.

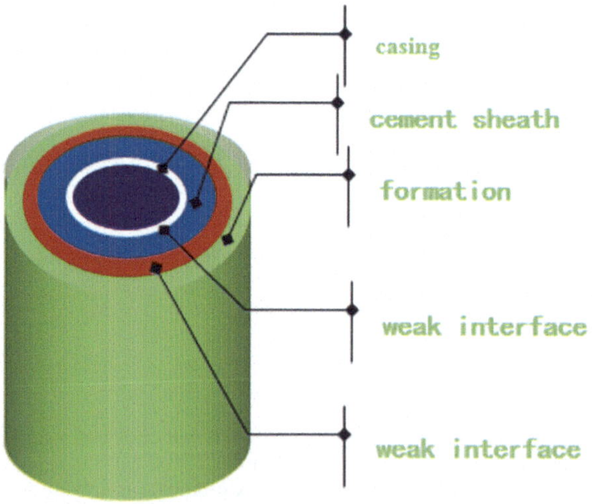

Fig. 1 Weak interface diagram

2.1 Formation of Mud Cake Weak Interface [2] of High Permeability and Low Pressure Layers

In the high permeability zone, when there is a difference of high pressure, the dehydration of drilling fluid increases and the solid particles move towards the borehole wall. Because of the high permeability, small particles gradually enter the stratum with filtrate, and large particles on the stratum build bridges and remain on the stratum surface. With the continuous effect of drilling pressure difference, the small particles build bridges on the large particles, and the smaller particles gather on the small particles, forming a multi-layer mud cake structure. Therefore, in the case of no leakage, with the increase of permeability, the more layers the mud cake forms, the thicker the mud cake gets.

Laboratory tests indicate that the cementing quality shows a trend of deterioration with the increase of mud cake thickness. When the mud cake thickness is greater than 4 mm, the Bond Index (BI) is less than 0.4, and the cementing quality is not qualified. As shown in Fig. 2.

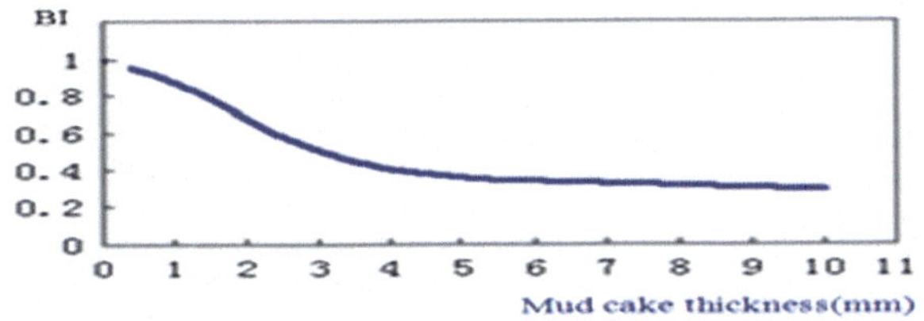

Fig. 2 Relationship between mud cake thickness and cementing quality (BI)

With the further increase of porosity and permeability, the radius of pore throat becomes larger. The particles, regardless of their sizes, will leak into the formation under the action of high pressure difference. There is no bridging on the surface of stratum and the effective mud cake cannot be formed. The stratum is filled with gradually lost drilling fluid, but the reverse permeability and carrying of the formation make the mud cake lose the plugging effect. Some of the large particles in the drilling fluid bridge each other to form thin mud cake in the stratum with high permeability, but the mud cake quality is very poor. This kind of thin mud cake results in poor cementing quality.

2.2 Excessive Dehydration of Cementing Slurry Forms Weak Interface

The mud cake quality becomes worse after the micro-leakage [3] of drilling fluid in high permeability and low pressure layer. Increasing or excessive dehydration of cement slurry results in inadequate hydration of cement, affects strength development of cement stone and deteriorates interfacial cementing quality.

2.3 Seepage and Corrosion of Formation Fluid Aggravate Weak Interface

The dissolution. Calcium hydroxide in cement stone is dissolved in formation, and calcium hydroxide after dissolution is removed by formation fluid. This continuous effect causes the cement particle cementation matrix to be destroyed. The porosity and permeability of cement stone increase unceasingly, resulting in serious weak interface problems. As shown in Fig. 3.

Fig. 3 Static curing of tap water

The chemical corrosion. Calcium hydroxide in cement stone is replaced with chlorine salt, carbonate and magnesium salt dissolved in formation fluid, thus forming solid phase which is not easily dissolved by water or has no cementation property. Calcium sulphoaluminate hydrate was formed by reaction with sulphate, which destroyed the interfacial cementation of cement stone. As shown in Fig. 4.

The flow rate of formation fluid determines the strength of the dissolution and chemical corrosion [4] mentioned above. Laboratory simulation tests show that cementing quality tends to become worse with the increase of the velocity of formation fluid. When the formation fluid velocity reaches 2000 ml/h, the acoustic amplitude is more than 30%, and the cementing quality is not qualified.

Fig. 4 Dynamic curing of formation water

3 Cementing Quality Analysis and Measures in Block 3/7 of South Sudan

Block 3/7 belongs to high yield and high permeability oil field. After years of exploitation, the formation pressure system of reservoir has changed greatly, and the trend of low pressure and under-pressure is obvious. Since the resumption of production in April 2017, the data of formation pressure, permeability and borehole diameter in several wells with unqualified cementing quality have confirmed the existence of high permeability low pressure layers and the influence of weak interface of mud cake on cementing quality.

3.1 Main Problems

Firstly, borehole shrinkage is obvious. Since the resumption of production, there has been a general situation of diameter reduction in the reservoir section of Block 3/7, especially in the Palouge area. According to the statistics of 21 vertical wells with 9-7/8″ bit in 2017, the average openhole diameter is 252.8 mm, and the average openhole diameter enlargement rate is 0.38%. Compared with 2010, the average openhole diameter of 18 vertical wells in this area was 269 mm, and the average openhole diameter enlargement rate was 6.9%. This is equivalent to adding 8 mm weak interface to the wellbore.

The lead slurry system and the length of sealing section need to be improved. The lead slurry of 1.50 g/cm^3 is Class G cement low density, which has large amount of water evolution and is unstable. If the sealing section is too long, the effect of U-Tube [5] will be increased. Under the action of negative pressure, the lead slurry has water evolution in casing, moves upward and mix with pre-tail slurry, which affects cementing quality of sealing section above the top of reservoir. After the resumption, the designed length of sealing section increase and the top of lead slurry returns into the previous casing. The cementing quality is not qualified. As shown in Table 1.

Table 1 U-Tube effect data for a well in Block 3/7

Time (min)	Liquid volume in (m³)	Pump Output (hp)	Pump Pressure (psi)	Surface Pressure (psi)	Equivalent Circulating Density		Free Fall Height (m)
					TD (g/cc)	Fracture Zone (g/cc)	
0.00	0.000	89.60	459.06	158.74	1.31	1.25	0.00
0.02	0.026	89.60	389.02	156.74	1.31	1.25	0.00
0.63	0.795	89.21	269.91	160.81	1.31	1.25	0.00
1.56	1.690	88.79	282.06	181.96	1.31	1.25	0.00
14.23	16.694	90.99	176.40	60.12	1.30	1.25	0.00
15.02	17.489	91.43	161.74	53.41	1.30	1.25	0.00
15.82	18.264	91.88	147.02	36.70	1.30	1.25	0.00
16.61	19.078	92.32	132.31	23.99	1.30	1.25	0.00
17.41	19.873	92.76	117.60	[illegible]	1.30	1.25	0.00
18.20	20.668	93.10	108.32	0.00	1.30	1.25	1.8
19.00	21.463	93.24	108.32	0.00	1.30	1.25	6.5
19.79	22.258	93.29	108.32	0.00	1.30	1.25	13.1
20.69	23.053	93.30	108.32	0.00	1.30	1.25	20.8
21.36	23.848	93.29	108.32	0.00	1.30	1.25	26.6
22.18	24.643	93.27	108.32	0.00	1.30	1.25	36.6
22.97	25.438	93.24	108.32	0.00	1.30	1.25	44.8
23.77	26.233	93.22	108.32	0.00	1.30	1.25	53.0
23.89	26.354	93.22	108.32	0.00	1.30	1.25	54.2
23.90	26.371	93.21	147.47	0.00	1.30	1.25	54.4
24.66	27.028	93.49	274.80	0.00	1.31	1.25	65.0
25.36	27.823	93.88	274.80	0.00	1.31	1.25	84.3
26.15	28.618	94.51	274.80	0.00	1.31	1.24	115.2
26.96	29.413	95.50	274.80	0.00	1.30	1.23	154.1
41.69	53.261	94.27	660.43	0.00	1.38	1.33	103.9
41.92	54.056	93.65	660.43	0.00	1.38	1.33	91.9
42.25	54.851	93.07	660.43	0.00	1.39	1.34	77.1
42.58	55.646	92.72	660.43	0.00	1.39	1.34	60.7
42.91	56.441	92.51	660.43	0.00	1.40	1.35	43.2
43.24	57.236	92.36	660.43	0.00	1.41	1.36	25.1
43.58	58.030	92.23	660.43	0.00	1.41	1.36	6.6
43.91	58.825	91.60	771.18	[illegible]	1.45	1.36	0.00
44.24	59.620	89.87	864.12	223.68	1.47	1.39	0.00
44.57	60.415	87.14	939.15	278.72	1.48	1.40	0.00
44.90	61.210	84.51	1,004.47	344.04	1.50	1.41	0.00
45.23	62.005	82.56	1,067.46	407.02	1.51	1.43	0.00
45.56	62.800	80.32	1,130.33	469.90	1.53	1.44	0.00
45.89	63.595	78.09	1,192.94	532.51	1.54	1.45	0.00
46.22	64.390	75.68	1,260.57	600.13	1.56	1.47	0.00

3.2 Measures to Improve Cementing Quality

In order to balance pressure, center, displace completely and pack off, the following measures are carried out.

Firstly, stick to wiper trip and ream after drilling, ensure mud cake on the borehole wall to be clean and reduce the hidden trouble of poor cementing quality caused by weak interface of mud cake.

Secondly, adjust the lead slurry length of sealing section to 200 m, use pre-hydrated bentonite cement slurry system [6] with the density of 1.50 g/cm^3. It usually absorbs 5.3 ml of water per gram of bentonite with a volume expansion of more than 15 times. The bentonite low density cement slurry system is that self-dispersion of bentonite occurs after water adsorption, suspended to support the settling cement particles to maintain the stability of the cement slurry system.

Thirdly, according to the characteristics of high permeability formation, improve the drilling fluid system to control water loss, viscosity, solid content and rheological properties, so that it can form thin and dense mud cake and meet the needs of good cementing quality.

Fourthly, use drilling fluid interface reinforcements. Adjust the drilling fluid performance before running casing, add the interface enhancer and make it permeate and diffuse into the mud cake, and improve the cementing quality of cement interface.

Fifthly, set centralizers strictly according to the design to ensure that the casing is centered and control strictly the length of rat hole. Rat hole should not exceed 2 m for wells with a vertical depth of about 1500 m.

Sixthly, develop and apply anti-seepage and anti-channeling cement slurry system to improve cementing quality in high permeability and low pressure formations. Develop and apply modified Class G cement to improve the erosion resistance, compactness and acoustic impedance of cement stone and improve cementation quality of the interface.

4 Conclusion

At the production site, the main ways to solve the cementing quality problem of the weak interface in the high permeability and low pressure layers are as follows.

4.1 Cement Slurry

On the premise of complete cement hydration, control water-cement ratio and reduce water volume. Reduce hydration heat of cementing slurry system. Improve the ability of cement slurry to prevent channeling. Improve interfacial compactness and reinforce interfacial acoustic impedance. Reduce the ratio of Ca/Si in cement to improve the corrosion resistance, the long-term strength and compactness of cement stone.

4.2 Drilling Engineering

Improve drilling fluid impermeability and other properties to form thin and dense mud cake. Increase the strength of interface. Stick to wiper trip and ream, set centralizers reasonably. Improve formation pressure system.

Acknowledgements. Jia Fushan, Deputy Director of Market Develop Centre of No. 1 Drilling Producing and Technical Service Company of Daqing Drilling and Exploration Engineering Company, PetroChina. Bao Xiangwen, Cementing Manager of Daqing Cementing Project of Block 3/7 of South Sudan.

References

1. Duan, D., Gong, C., Wang, S.: Cementing technology of low pressure and high permeability formation in Shanbei. Pet. Drill. Technol. **23**, 40–42 (1995)
2. Yi, Z., Liu, A.: Effect of wall shear stress on cementing quality in annular flow. Explor. Eng. **3**, 50–51 (2005)
3. Qi, G., Wang, Z.: Fundamentals of cementing technology (2nd edn). Pet. Ind. **7**, 78–80 (2016)
4. Vidick, B., Krummel, K.: Impact of formation type on cement bond logs. SPE/IADC **96022**, 31–32 (2005)
5. Liu, A., Deng, J.: Cementing technology of caracoles oilfield in Venezuela. Gas Ind. **3**, 53–54 (2005)
6. Liu, C., Huang, B., Xu, T.: Theory and application of oil and gas well cement injection. Pet. Ind. **9**, 130–131 (2001)

CO_2 Flooding and Geological Storage Potential Evaluation Method for Low Permeability Reservoirs in Dagang Oilfield

Hai-ying Cheng[1], Xin-wei Liao[2(✉)], Ming-jun Cai[1], Yang Zhang[1], and Rong-tao Li[2]

[1] Petrochina Dagang Oilfield, Tianjin, China
{chenghaiying, caimjun, zhangyang08}@petrochina.com.cn
[2] China University of Petroleum, Beijing, China
xinwei@cup.edu.cn, lirongtao2015@126.com

Abstract. Permanently sequestering carbon dioxide during carbon dioxide flooding is the most realistic way to meet carbon dioxide emission reduction obligations in developing countries. In order to effectively evaluate the potential of carbon dioxide flooding and geological storage of low permeability reservoirs in Dagang Oilfield, it is necessary to establish a reliable CO_2 flooding and geological storage potential evaluation method. Based on the characteristics of low permeability reservoirs in Dagang Oilfield, this paper establishes a CO_2 enhanced oil recovery (EOR) and geological storage potential evaluation method based on fractional flow theory. Through theoretical analysis and numerical simulation method, the carbon dioxide storage coefficient, carbon dioxide storage and CO_2 flooding efficiency of low permeability reservoirs in Dagang Oilfield under carbon dioxide miscible and immiscible flooding conditions were determined, and these values were analyzed and evaluated. The results show that the CO_2 flooding efficiency and storage coefficient of miscible flooding are significantly higher than that of immiscible flooding. In addition, as the amount of carbon dioxide injected increases, the carbon dioxide flooding oil increase and the amount of storage increase. Through the evaluation of 272 oil layers in the low-permeability oil area of Dagang Oilfield, the enhanced oil recovery and geological storage potential of CO_2 in this area were analyzed. It can be seen from the analysis and evaluation results that CO_2 miscible flooding is an effective displacement and storage method. The conclusions drawn can

J. Lin (ed.), *Proceedings of the International Petroleum and Petrochemical Technology Conference 2019*, pp. 70–79, 2020.
https://doi.org/10.1007/978-981-15-0860-8_6

provide technical reference for the implementation of CO_2 flooding and storage in low permeability reservoirs in Dagang Oilfield.

Keywords: CO_2 flooding · CO_2 geological storage · Potential evaluation method · Low permeability reservoir

1 Introduction

With the increasing global warming and climate deterioration, greenhouse gas emissions, especially CO_2 emissions, have attracted more and more attention of countries. Achieving the geological storage of CO_2 in reservoirs can not only greatly reduce the emission of CO_2 in the atmosphere, but also greatly improve the oil recovery. Research on CO_2 flooding and its storage has been carried out earlier, and a more complete theory of displacement and storage has been formed [1–12]. In 1997, Taber et al. summarized the screening criteria for CO_2 flooding reservoirs based on the successful case of CO_2 enhanced oil recovery [2]. In 2002, Shaw et al. screened the reservoir parameters suitable for CO_2-EOR on the basis of summarizing the previous research results, and graded the target reservoir by setting the optimal value and parameter weight [5, 6]. The potential of CO_2 flooding enhanced recovery was evaluated in the Alberta reservoir [8]. CO_2 sequestration is widely evaluated using CO_2 utilization coefficient in American and European countries, which is defined as total sequestration amount divided by cumulative oil production [9–12]. Shen Pingping et al. also proposed a similar approach in 2009 [13]. At present, China has a relatively late start in this field, especially for the potential of carbon dioxide storage in China's major oil areas, it is necessary to further establish a reasonable evaluation system. There are a large number of low permeability reservoirs in Dagang oilfield. After decades of exploitation, the effect of water injection development has been worse year by year. Comparatively speaking, CO_2 flooding has the advantages of low injection pressure and fast formation energy replenishment, and can obtain lower solvent/crude oil mobility ratio when miscible flooding is realized, which improves sweep efficiency and realizes geological storage of CO_2 while increasing production and tapping potential. Based on the establishment of CO_2 enhanced oil recovery and geological storage potential evaluation method, 272 oil layers in the low-permeability oil area of Dagang Oilfield are used as the evaluation objects, and the CO_2 enhanced oil recovery and geological storage potential in this area are analyzed.

2 Method

2.1 Numerical Simulation Calculation Models

The current CO_2 flooding enhanced recovery and geological storage numerical simulation calculation models mainly include four types: component model, black oil model, streamline model and split model. The advantage of the split model is that the three-dimensional problem is transformed into a one-dimensional flow problem, that the data volume needs to be reduced, and the operation speed is faster. The leakage characteristics of the split model are simplified. Considering that Dagang Oilfield CO_2 flooding enhanced oil recovery and geological storage potential evaluation workload is large, detailed data is difficult to obtain, we adopted the split model.

According to the principle of conservation of mass, the following conservation equations are established:

$$\frac{\partial C_{ij}}{\partial t_D} + \frac{\partial F_{ij}}{\partial X_D} = 0 \tag{1}$$

$$C_{ij} = C_{i0}S_0 + C_{i1}S_1 + C_{i2}S_2 \tag{2}$$

$$F_{ij} = C_{i0}f_0 + C_{i1}f_1 + C_{i2}f_2 \tag{3}$$

Here, C_{ij} is concentration of component i in j phase; S_j is j phase saturation; i is component of fluid; $i = 0$ is water component; $i = 1$ is crude oil component; $i = 2$ is injection gas component; t is time; X is length distance; f is split flow; D represents dimensionless.

Based on the traditional model, the calculation model considers the following main influencing factors in combination with the characteristics of Dagang oilfield reservoir: (1) viscous fingering and heterogeneity; (2) gas injection gravity differentiation; (3) introducing the minimum miscibility pressure calculation method to define the state of CO_2 flooding phase; (4) the difference of area sweep coefficient between miscible flooding and immiscible flooding; (5) the influence of interfacial tension effect on relative permeability change.

From this model, the CO_2 flooding enhanced oil recovery range can be determined, and the amount of CO_2 storage in the reservoir can be calculated from the following formula.

$$\begin{aligned} M_{CO_2t} = \rho_{CO_2r} \times [R_f \times A \times h \times \varphi \times (1 - S_{wi}) \\ - V_{iw} + V_{pw}] \end{aligned} \tag{4}$$

Here, M_{CO_2t} is CO_2 storage potential, t; ρ_{CO_2r} is CO_2 density in the formation, t/m^3; R_f is oil recovery; A is reservoir area, m^2; h is reservoir thickness, m; φ is reservoir porosity; S_{wi} is reservoir irreducible water saturation; V_{iw} is injected water volume, m^3; V_{pw} is producted water volume, m^3.

2.2 Model Reliability Analysis

Based on the CO_2 flooding enhanced recovery calculation model, a numerical simulation calculation program was developed. The program runs fast and the calculation results show that the reservoir engineering theory can be effectively reflected. Taking the typical reservoir of Dagang Oilfield as the research object, as shown in Fig. 1, the calculation results of the calculation software using this study are close to the calculation results of the commercial software Eclipse, and the difference between the two is within 10%, and the precision is high.

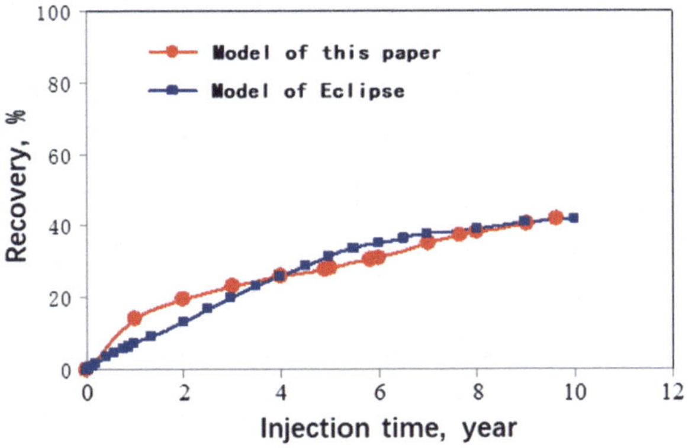

Fig. 1 The calculation results of this study and the commercial software Eclipse

2.3 Analysis of Influencing Factors

2.3.1 Influence of Reservoir Thickness

As shown in Fig. 2, as the thickness of the reservoir increases, the recovery of crude oil decreases. Mainly because the greater the thickness, the gravity differentiation causes the sweep volume coefficient to decrease.

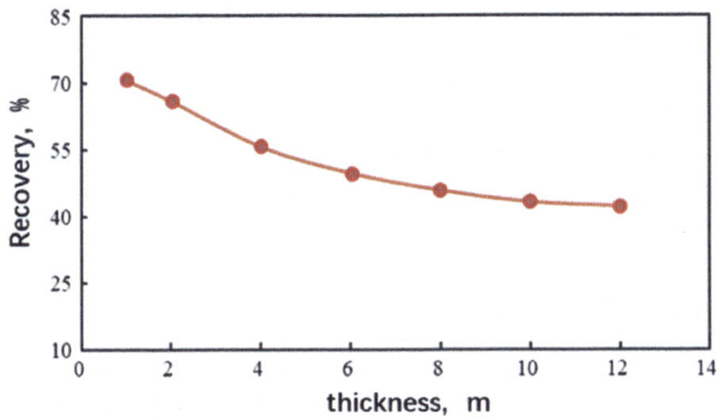

Fig. 2 Relationship between recovery factor and thickness

2.3.2 Influence of Reservoir Heterogeneity

As shown in Fig. 3, as the heterogeneous coefficient increases, the gas injection process leads to the intrusion of gas, and the volume coefficient of the sweep decreases, which in turn reduces the recovery factor.

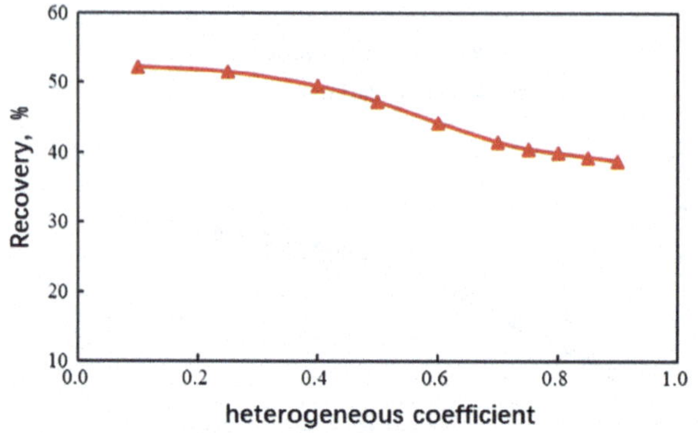

Fig. 3 Relationship between recovery factor and heterogeneous coefficient

2.3.3 Influence of Reservoir Crude Oil Viscosity

As shown in Fig. 4, as the viscosity of the crude oil decreases, the flow performance of the crude oil becomes better, the relative permeability of the oil phase increases, and the recovery factor increases exponentially.

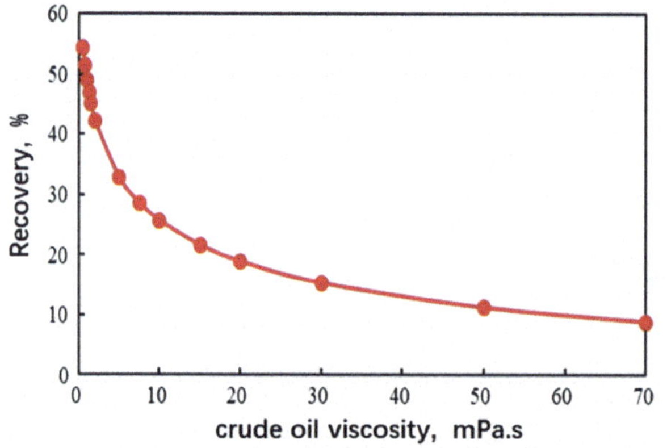

Fig. 4 Relationship between recovery factor and crude oil viscosity

2.3.4 Influence of Reservoir Oil Saturation

As shown in Fig. 5, as the oil saturation increases, the recovery factor has an upward trend with a large increase. It indicates that the better the oil bearing of the reservoir, the better the development of gas flooding.

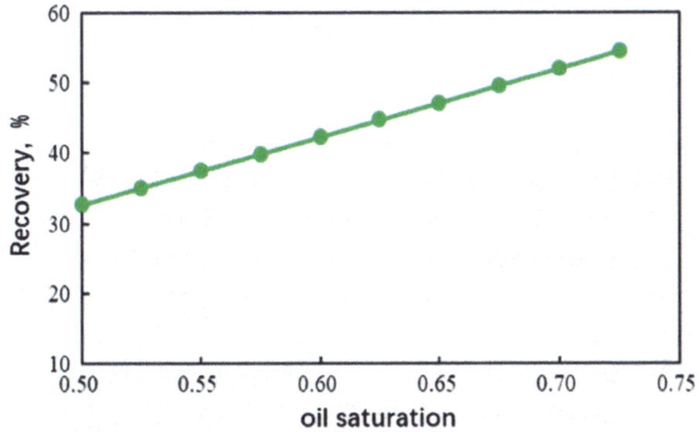

Fig. 5 Relationship between recovery factor and oil saturation

2.3.5 Influence of Reservoir Permeability

As shown in Fig. 6, as the reservoir permeability increases, the recovery rate increases, but when the permeability reaches a certain level, it is prone to gas enthalpy, resulting in a downward trend in oil recovery.

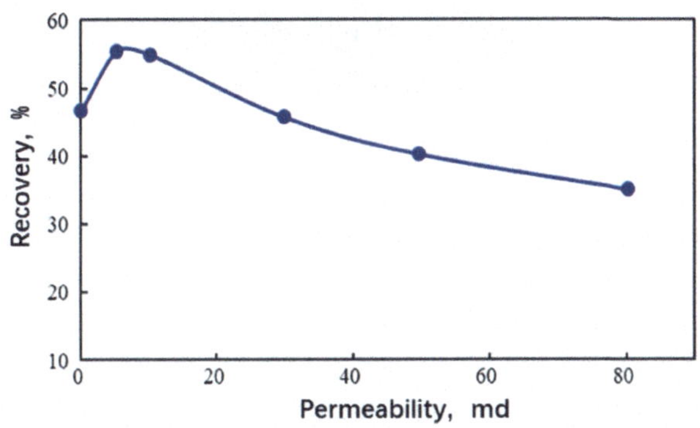

Fig. 6 Relationship between recovery factor and permeability

Based on the above analysis, the sensitivity rankings of factors affecting CO_2 flooding enhanced oil recovery are: crude oil viscosity > permeability > oil saturation > heterogeneity > reservoir thickness.

3 CO_2 EOR and Storage Potential Evaluation

3.1 Evaluation Process

We designed and developed the software, including four basic modules: basic data input, solution design, simulation calculation analysis, and report output. Based on the software, CO_2 flooding and geological storage potential evaluation can be performed through the process shown in Fig. 7.

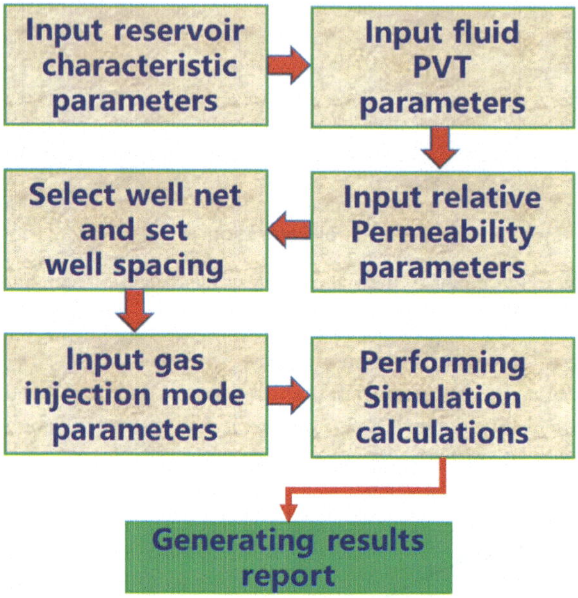

Fig. 7 Potential evaluation process

3.2 CO_2 EOR and Storage Potential Evaluation

3.2.1 CO_2 EOR Potential Evaluation

Through the evaluation of the potential of CO_2 flooding enhanced recovery in the low-permeability reservoirs developed in Dagang Oilfield, the results show that the geological reserves of the miscible flooding are great potential, and the number of miscible flooding blocks is 16. These low-permeability reservoirs can enhance oil recovery by 8.6% by CO_2 flooding, of which the average miscible flooding can be increased by 15.5%, and the average of immiscible flooding can be improved. 5.6%.

The evaluation results of the CO_2 flooding enhanced oil recovery potential of the undeveloped low-permeability reservoirs in Dagang Oilfield indicate that 144 blocks can achieve miscible flooding, accounting for the total amount. 65%. These low-permeability reservoirs can enhance oil recovery by an average of 9.9% by CO_2 flooding, of which the average recovery of miscible flooding is 12.3%, and the average of immiscible flooding increased oil recovery by 5.4%.

3.2.2 CO_2 Storage Potential Evaluation

Figures 8 and 9 show the results of CO_2 geological storage potential of 272 oil layers in the low-permeability oilfield of Dagang Oilfield. Figure 8 shows the CO_2 geological storage potential of the miscible flooding area, and Fig. 9 shows the CO_2 geological storage potential of the immiscible flooding area.

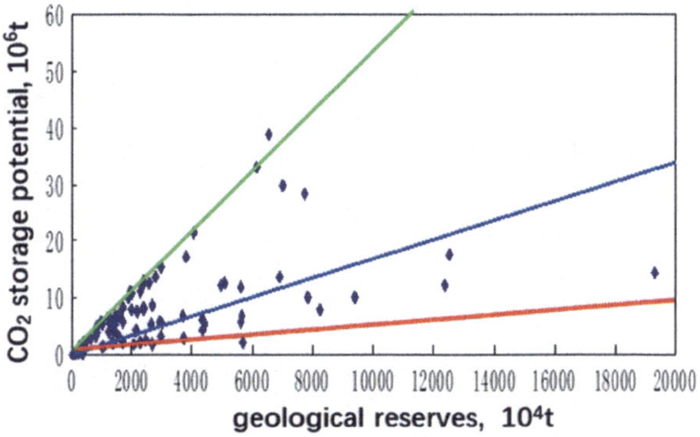

Fig. 8 Storage potential of miscible flooding area

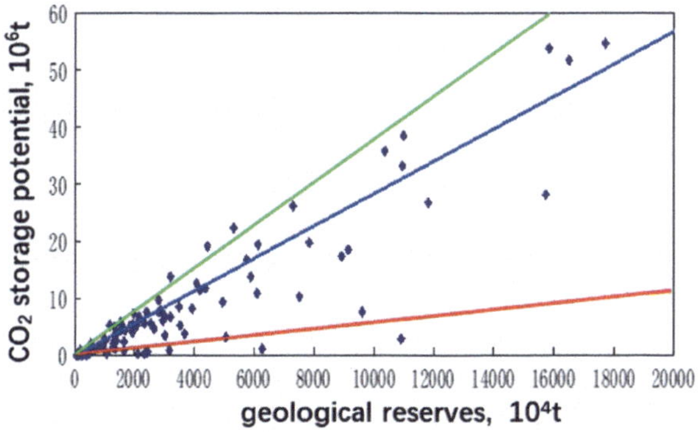

Fig. 9 Storage potential of immiscible flooding area

Shen Pingping et al. [9] predict storage potentials as below:

$$M_{CO_2t} = R_{CO_2} \cdot N \tag{5}$$

In which, M_{CO_2t} is CO_2 storage potentials, t; R_{CO_2} is sequestration storage coefficient, dimensionless; N is geological reserves, t.

According to the formula (5), the sequestration storage coefficients of Figs. 8 and 9 can be obtained. Here, we divide the sequestration storage coefficients into three levels of the highest value, the median value and the lowest value, as shown in Table 1.

Table 1 Sequestration storage coefficients

Values	Miscible area	Immiscible area
The highest value	0.50	0.41
The median value	0.17	0.28
The lowest value	0.06	0.06

4 Conclusion

(1) This paper establishes a method for evaluating the enhanced recovery and storage potential of CO_2 flooding in actual reservoirs of Dagang Oilfield. This method can effectively evaluate CO_2 flooding and geological sequestration potential of low permeability reservoirs in similar oilfields.
(2) By evaluating the CO_2 flooding potential of low-permeability reservoirs in Dagang Oilfield, the results show that Dagang Oilfield has a huge potential for CO_2 flooding to enhance oil recovery.
(3) By evaluating the geological storage potential of CO_2 flooding in low-permeability reservoirs in Dagang Oilfield, the results show that Dagang Oilfield has a considerable amount of CO_2 geological storage, and the highest sequestration storage coefficients can reach 0.5.

Acknowledgements. At the completion of the article, I would like to thank Professor Liao Xinwei of China University of Petroleum (Beijing). We thank the supports from Chinese National Major Science and Technology Projects (2017ZX05009004-005) and Chinese National Major Science and Technology Projects (2017ZX05030002-005). We also thank the help of experts from Dagang Oilfield for their correlation study.

References

1. Robl, F.W., Emanuel, A.S., Van Meter, O.E.: The 1984 Natl. petroleum council estimate of potential EOR for miscible processes. J. Pet. Technol. **20**(8), 875–882 (1986)
2. Taber, J.J., Martin, F.D., Seright, R.S.: EOR screening criteria revisited—part 1: introduction, to screening criteria and enhanced recovery field projects. SPE Reserv. Eng. **12**(3), 189–198 (1997)
3. Stevens, S.H., Kuuskraa, V.A., Taber, J.J.: Sequestration of CO_2 in depleted oil and gas fields: barriers to overcome in implementation of CO_2 capture and storage (disused oil and gas fields). IEA, USA (1999)
4. Winter, E.M., Dergman, P.D.: Availability of depleted oi1 and gas reservoirs for disposal of carbon dioxide in the United States. Energy Convers. Manag. **34**(6), 1177–1187 (2001)
5. Bradshaw, J., Bachu, S.: Screening, evaluation, and ranking of oil reservoirs suitable for CO_2-flood EOR and carbon dioxide sequestration. J. Can. Pet. Technol. **41**(9), 51–61 (2002)
6. Bachu, S., Stewart, S.: Geological sequestration of anthropogenic carbon dioxide in the Western Canada sedimentary basin: suitability analysis. J. Can. Pet. Technol. **41**(2), 32–40 (2002)
7. Kovseek, A.R.: Screening: criteria for CO_2 storage in oil reservoirs. Pet. Sci. Technol. **20**(7–8), 841–866 (2003)
8. Bachu, S., Shaw, J.: Evaluation of the CO_2 sequestration capacity in Alberta's oil and gas reservoirs at depletion and the effect of underlying aquifers. J. Can. Pet. Technol. **42**(9), 31–41 (2003)
9. Bachu, S., Shaw, J., Robert, M.: Estimation of oil recovery and CO_2 storage capacity in CO_2 EOR incorporating the effect of underlying aquifers. SPE, 89340 (2004)

10. Bradshaw, J., Bachu, S., Bonijoly, D., et al.: A taskforce for review and development of standards with regards to storage capacity measurement. http://www.cslforum.org/documents/TaskforceStorageCapacityEstimationVersion2.pdf (2005)
11. Bradshaw, J., Bachu, S., Bonijoly, D., et al.: Estimation of CO_2 storage capacity in geological media. http://www.cslforum.org/documents/PhaseIIReportStorageCapacityMeasurementTaskForce.pdf (2007)
12. Bradshaw, J., Bachu, S., Bonijoly, D., et al.: Comparison between methodologies recommended for estimation of CO_2 storage capacity in geological media. http://www.cslforum.org/documents/PhaseIIIReportStorageCapacityEstimationTaskForce0408.pdf (2008)
13. Shen, P., Liao, X., Liu, Q.: Methodology for estimation of CO_2 storage capacity in reservoirs. Pet. Explor. Dev. **36**(2), 216–220 (2009)

Research on Fracture Initiation Mechanisms of Hydraulic Fracturing Guided by Multi-radial Boreholes

Yu-xin Chen, Zhan-qing Qu(✉), Tian-kui Guo, Yu Bai, and Ji-wei Wang

College of Petroleum Engineering, China University of Petroleum, Huadong 266580, China
chen_yuxin@163.com, quzhq@upc.edu.cn

Abstract. At present, the technology of hydraulic fracturing guided by radial boreholes has gradually been applied to the development of low permeability reservoirs, but the related fracture initiation mechanism is insufficient. In order to clarify the fracture initiation mechanism under the guidance of radial boreholes, it is necessary to study the stress distribution around wellbore with the existence of radial boreholes. The effects of in situ geo-stress, wellbore internal pressure and fracturing fluid percolation effect on geo-stress field distribution are taken into account, a mechanical model of multi-radial boreholes with any number, vertical depth and azimuth angle (the angle between the radial borehole and the horizontal maximum geo-stress) is established. By using Matlab to carry out sample analysis of the model, the influence of various parameters on the maximum principal stress have been studied, and the fracture initiation position has been predicted by combining with tensile fracture criterion. Studies have shown that the maximum principal stress increases with the increase of the horizonal geo-stress difference and the diameter of the radial borehole. When the horizonal geo-stress difference is small, the maximum principal stress increases first and then decreases with the increase of the azimuth angle; when the horizonal geo-stress difference is large, the maximum principal stress decreases with the increase of the azimuth angle. When multi-radial boreholes existing, the maximum principal stress increases with the increase in the number of radial boreholes and the decrease in the vertical spacing between radial boreholes. The maximum principal stress generally decreases as the phase angle (angle between radial boreholes) increases, but anomalous phenomena occur when the horizonal geo-stress

J. Lin (ed.), *Proceedings of the International Petroleum and Petrochemical Technology Conference 2019*, pp. 80–100, 2020.
https://doi.org/10.1007/978-981-15-0860-8_7

difference is small and the radial borehole diameter is large. This study provides theoretical support for guiding fracture by multi-radial boreholes. In order to make the radial boreholes better guide the fracture, the optimal design should be optimized by optimizing the azimuth angle of the borehole, increasing the diameter of the borehole, reducing the borehole spacing, increasing the number of radial boreholes, and reducing the phase angle between boreholes.

Keywords: Hydraulic fracturing · Multi-radial boreholes · Fracture initiation · Maximum principal stress

1 Introduction

The combination of radial wells and hydraulic fracturing is an innovative technology that is being applied to the development of various oil fields and has achieved good results. Compared with conventional hydraulic fracturing, this technology has the following advantages: 1. The radial well reduces the fracture pressure of the formation. 2. Within a certain range, the radial well can guide the initiation and extension of fractures, facilitating the establishment of communication with the target area. 3. Radial wells increase the contact area between the well and the reservoir, which is beneficial to obtain better fracturing effects.

At present, the main problem of radial wells hydraulic fracturing technology is that the mechanical mechanism of fracture initiation and propagation under the guidance of radial wells is not clear, so many scholars have done some theoretical research on it. Fallahzadeh and Shadizadeh (2013) built the calculation model of fracture initiation pressure for deviated casing perforation and studied the stress distribution of arbitrary dip direction and azimuth around the casing wellbore. Guo et al. (2017) studied on initiation mechanisms guided by vertical two radial boreholes. However, the research on initiation mechanisms of hydraulic fracture guided by multi-radial boreholes with different azimuth is virtually blank now.

Therefore, this paper studies the mechanism of fracture initiation under the guidance of radial wells, and establishes the mechanical model of the stress field in the near-wellbore zone under the existence of radial boreholes with arbitrary number, vertical depth and azimuth angle. Besides, the influence of many factors on the maximum principal stress is analysed, which provides a reliable scientific basis for the effective implementation of radial well hydraulic fracturing technology.

2 Stress Distribution Model Under the Condition of Radial Boreholes

2.1 Thought of Model Establishment

Radial boreholes are considered as small open wellbores that intersect the main wellbore vertically. Each wellbore affects the in situ geo-stress field independently. In this paper, wellbore internal pressure and fracturing fluid percolation effect on geo-stress field distribution of each well are considered. Firstly, the stress expressions in the respective wellbore coordinate system are obtained, and then transformed into the main wellbore coordinate system through several coordinate changes, and finally the stress expression at any position around the wellbore is obtained by superimposing with the in situ geo-stress field. In the specific model establishment steps, the stress distribution model of the near-wellbore zone under condition of single radial well is established firstly, and then extended to the multi-radial wells condition, and finally mechanical model of multi-radial boreholes with any number, vertical depth and azimuth angle is established.

2.2 Derivation of Stress Distribution Model Under the Condition of Single Radial Borehole

To facilitate the derivation, the following assumptions are made: (1) The formation is intact, and the wellbores do not intersect with natural cracks, caves and fracture zones; (2) Rock is a homogeneous isotropic linear elastomer; (3) physicochemical actions between the fracturing fluid and the formation rock do not change the mechanical properties of the rock; (4) Ignore the influence of the radial borehole on the in situ geo-stress component.

In the derivation, the tensile stress is determined to be positive and the compressive stress is negative. The in situ stress can be expressed by three mutually perpendicular stresses, which are the maximum principal stress σH, the minimum principal stress σh and the overburden stress σv. The three axes of the rectangular coordinate system (x, y, z) are in the direction of three principal stresses respectively. The z1 axis is along the direction of the radial borehole axis. The established coordinate systems are shown in Fig. 1. β is the azimuth angle of the radial borehole 1, which is the angle between the direction of the borehole axis and the direction of the maximum horizontal principal

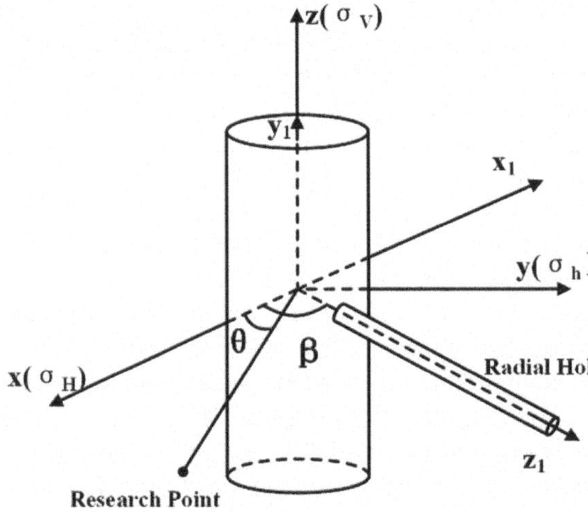

Fig. 1 Coordinate systems of single radial borehole

stress. θ is angle between research point and the maximum horizontal principal stress. Rectangular coordinate systems of main wellbore and radial borehole are (x, y, z), (x1, y1, z1), and the corresponding cylindrical coordinate systems are (r, θ, z), (r1, θ1, z1).

2.2.1 Stress Caused by in Situ Geo-Stress

The in situ stress can be expressed by six in situ stress components, as shown in the Fig. 2. Since the main wellbore is in the vertical direction, the stress components of the in situ stress in the coordinate system (x, y, z) are as:

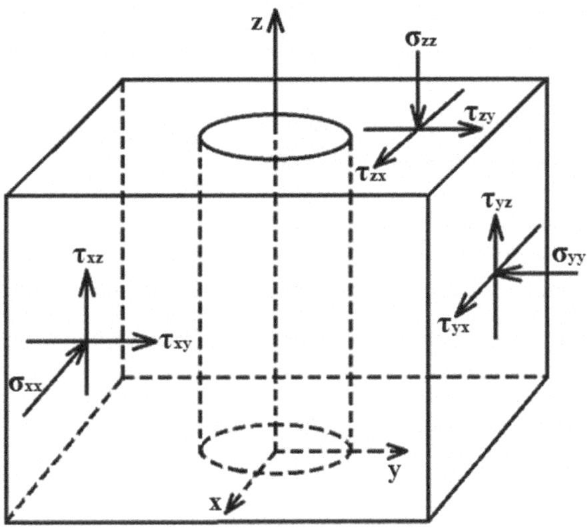

Fig. 2 Schematic diagram of in situ stress components

$$\begin{aligned} \sigma_{xx} &= \sigma_H \\ \sigma_{yy} &= \sigma_h \\ \sigma_{zz} &= \sigma_v. \end{aligned} \tag{1}$$

By substituting the above in situ stress components into the Haimson and Fairhurst formula, the stress components caused by in situ stress in the main wellbore cylindrical coordinate (r, θ, z) can be obtained as follows:

$$
\begin{aligned}
\sigma_r^s &= \frac{1}{2}(\sigma_H + \sigma_h)\left(1 - \frac{R^2}{r^2}\right) + \frac{1}{2}(\sigma_H - \sigma_h)\left(1 + \frac{3R^4}{r^4} - \frac{4R^2}{r^2}\right)\cos 2\theta \\
\sigma_\theta^s &= \frac{1}{2}(\sigma_H + \sigma_h)\left(1 + \frac{R^2}{r^2}\right) - \frac{1}{2}(\sigma_H - \sigma_h)\left(1 + \frac{3R^4}{r^4}\right)\cos 2\theta \\
\sigma_z^s &= \sigma_v - 2v\cos 2\theta(\sigma_H - \sigma_h)\frac{R^2}{r^2} \\
\tau_{r\theta}^s &= \frac{1}{2}(\sigma_H - \sigma_h)\left(1 - \frac{3R^4}{r^4} + \frac{2R^2}{r^2}\right)\sin 2\theta \\
\tau_{\theta z}^s &= 0 \\
\tau_{rz}^s &= 0
\end{aligned} \tag{2}
$$

Of which: R and r are the radiuses of vertical main wellbore and the distance between research point and the axis of main wellbore.

2.2.2 Stress Caused by Internal Pressure

In the process of hydraulic fracturing, because the fluid in the wellbore has a certain pressure, it will produce a stress component along the wellbore direction.

(1) Stress caused by internal pressure of main wellbore

The stress generated by individual wellbore pressure can be given according to the elastic solution formula of Lame.

$$
\begin{aligned}
\sigma_r &= -\frac{R^2}{r^2}P_w \\
\sigma_\theta &= \frac{R^2}{r^2}P_w \\
\sigma_z &= c\frac{R^2}{r^2}P_w
\end{aligned} \tag{3}
$$

Of which: c is correction factor and its value is between 0.9 and 1, Pw is internal pressure, r is radius of borehole.

In the main wellbore cylindrical coordinate system (r, θ, z), the stress produced by the internal pressure of the main wellbore is the form of the above formula. Substituting relevant parameters, the results are named σri0, σθi0, σzi0.

(2) Stress caused by internal pressure of radial wellbore 1

According to the formula (3), the expression of stress components produced by internal pressure of radial well 1 can be obtained in cylindrical coordinate system (r1,θ1,z1). Then, the stress components are converted to the rectangular coordinate system of radial well 1 (x1, y1, z1). The specific conversion formula is as follows:

$$
\begin{aligned}
\sigma_r &= \sigma_x \cos^2\theta + \sigma_y \sin^2\theta + 2\tau_{xy}\sin\theta\cos\theta \\
\sigma_\theta &= \sigma_x \sin^2\theta + \sigma_y \cos^2\theta - 2\tau_{xy}\sin\theta\cos\theta \\
\sigma_z &= \sigma_z \\
\tau_{r\theta} &= (\sigma_y - \sigma_x)\sin\theta\cos\theta + \tau_{xy}(\cos^2\theta - \sin^2\theta) \\
\tau_{\theta z} &= -\tau_{zx}\sin\theta + \tau_{yz}\cos\theta \\
\tau_{zr} &= \tau_{zx}\cos\theta + \tau_{yz}\sin\theta.
\end{aligned} \tag{4}
$$

The stress expression in the rectangular coordinate system of radial well 1 (x1, y1, z1) is obtained and then transformed into the main well rectangular coordinate system (x, y, z). The transformation formula is as follows:

$$
\begin{aligned}
\sigma'_x &= \sin^2\beta\sigma_x + \cos^2\beta\sigma_z - 2\sin\beta\cos\beta\tau_{zx} \\
\sigma'_y &= \cos^2\beta\sigma_x + \sin^2\beta\sigma_z + 2\sin\beta\cos\beta\tau_{zx} \\
\sigma'_z &= \sigma_y \\
\tau'_{xy} &= -\sin\beta\cos\beta\sigma_x + (\cos^2\beta - \sin^2\beta)\tau_{zx} + \sin\beta\cos\beta\sigma_z \\
\tau'_{yz} &= \cos\beta\tau_{xy} + \sin\beta\tau_{yz} \\
\tau'_{zx} &= -\sin\beta\tau_{xy} + \cos\beta\tau_{yz}.
\end{aligned} \tag{5}
$$

Finally, the stress components are converted into the cylindrical coordinate system of the main wellbore (r, θ, z).

$$
\begin{aligned}
\sigma_r^{i1} &= \frac{-R_1^2 P_w(\sin\beta\cos\theta - \cos\beta\sin\theta)^2(x_1^2 - y_1^2)}{(x_1^2 + y_1^2)^2} \\
&\quad + \frac{cR_1^2 P_w(\cos\beta\cos\theta + \sin\beta\sin\theta)^2(x_1^2 + y_1^2)}{(x_1^2 + y_1^2)^2} \\
\sigma_\theta^{i1} &= \frac{-R_1^2 P_w(\sin\beta\sin\theta + \cos\beta\cos\theta)^2(x_1^2 - y_1^2)}{(x_1^2 + y_1^2)^2} \\
&\quad + \frac{cR_1^2 P_w(\cos\beta\sin\theta + \sin\beta\cos\theta)^2(x_1^2 + y_1^2)}{(x_1^2 + y_1^2)^2} \\
\sigma_z^{i1} &= \frac{R_1^2 P_w(x_1^2 - y_1^2)}{(x_1^2 + y_1^2)^2} \\
\tau_{r\theta}^{i1} &= \frac{R_1^2 P_w\left(\sin\beta\cos\beta(\cos^2\theta - \sin^2\theta) - (\cos^2\beta - \sin^2\beta)\sin\theta\cos\theta\right)(x_1^2 - y_1^2)}{(x_1^2 + y_1^2)^2} \\
&\quad + \frac{cR_1^2 P_w\left(\sin\beta\cos\beta(\cos^2\theta - \sin^2\theta) - (\cos^2\beta - \sin^2\beta)\sin\theta\cos\theta\right)(x_1^2 + y_1^2)}{(x_1^2 + y_1^2)^2} \\
\tau_{\theta z}^{i1} &= \frac{-2R_1^2 P_w(\sin\beta\sin\theta + \cos\beta\cos\theta)x_1 y_1}{(x_1^2 + y_1^2)^2} \\
\tau_{zr}^{i1} &= \frac{2R_1^2 P_w(\sin\beta\cos\theta - \cos\beta\sin\theta)x_1 y_1}{(x_1^2 + y_1^2)^2}
\end{aligned} \tag{6}
$$

Of which:

$$x_1 = \frac{r\sin\theta}{\cos\beta} - (r\cos\theta + r\sin\theta\tan\beta)\sin\beta$$

$$y_1 = z$$

z is the z coordinate of the research point in (x, y, z) system.

2.2.3 Stress Caused by Fracturing Fluid Percolation

(1) Stress caused by fracturing fluid percolation in main wellbore

In the process of hydraulic fracturing, the fluid pressure in the wellbore Pw is generally higher than the pore-fluid pressure. Therefore, the fracturing fluid will be radially filtered from the wellbore to the formation under the pressure difference and increase the stress in the rock surrounding the wellbore. When single wellbore exists, the expression of stress component around the wellbore can be provided in its cylindrical coordinate system by formula:

$$\begin{aligned}\sigma_r &= A + \frac{B}{r^2}\\ \sigma_\theta &= A - \frac{B}{r^2}\\ \sigma_z &= C\end{aligned} \tag{7}$$

Of which:

$$A = \delta(P_w - P_p)\left(\phi - \frac{\alpha(1-2v)}{2(1-v)}\right)$$

$$B = \delta R^2(P_w - P_p)\frac{\alpha(1-2v)}{2(1-v)}$$

$$C = \delta(P_w - P_p)\left(\phi - \frac{\alpha(1-2v)}{1-v}\right).$$

In formula, Pp is initial pore pressure, α is biot porous elastic coefficient, ν poisson's ratio of the rock, ϕ rock porosity, r is borehole radius, and δ penetrability coefficient.

Substituting relevant parameters of main wellbore, the stress components generated by the main wellbore fracturing fluid percolation can be calculated directly. The results are expressions in main wellbore cylindrical coordinate system (r, θ, z) and named with σrp0, σθp0, σzp0.

(2) Stress caused by fracturing fluid percolation in radial wellbore 1

From Eq. (7), the expression of stress components generated by fracturing fluid percolation can be obtained in radial borehole 1 cylindrical coordinate system (r1, θ1, z1). Combining with the transformation relationship formulas (4 and 5) and solving the equations, the stress components expression can be transformed into forms in the rectangular coordinate system (x1, y1, z1) and the rectangular coordinate system (x, y, z) successively. Then the expression is converted into the cylindrical coordinate system of the main borehole (r, θ, z), and the final expression of the stress components caused by the fracturing fluid percolation in radial well 1 can be obtained.

$$\begin{aligned}
\sigma_r^{p1} &= (\sin\beta\cos\theta - \cos\beta\sin\theta)^2 \\
&\quad\times \frac{A\left(x_1^4 - y_1^4\right) + B_1\left(x_1^2 - y_1^2\right) + 2Ay_1^2\left(x_1^2 + y_1^2\right)}{\left(x_1^2 + y_1^2\right)^2} \\
&\quad + C(\cos\beta\cos\theta + \sin\beta\sin\theta)^2 \\
\sigma_\theta^{p1} &= (\sin\beta\sin\theta + \cos\beta\cos\theta)^2 \\
&\quad\times \frac{A\left(x_1^4 - y_1^4\right) + B_1\left(x_1^2 - y_1^2\right) + 2Ay_1^2\left(x_1^2 + y_1^2\right)}{\left(x_1^2 + y_1^2\right)^2} \\
&\quad + C(\cos\beta\sin\theta - \sin\beta\cos\theta)^2 \\
\sigma_z^{p1} &= \frac{A\left(x_1^2 + y_1^2\right)^2 - B_1\left(x_1^2 - y_1^2\right)}{\left(x_1^2 + y_1^2\right)^2} \\
\tau_{r\theta}^{p1} &= \left[(\cos^2\beta - \sin^2\beta)\sin\theta\cos\theta - \sin\beta\cos\beta(\cos^2\theta - \sin^2\theta)\right] \\
&\quad\times \frac{A\left(x_1^4 - y_1^4\right) + B_1\left(x_1^2 - y_1^2\right) + 2Ay_1^2\left(x_1^2 + y_1^2\right)}{\left(x_1^2 + y_1^2\right)^2} \\
&\quad + C\left[(\sin^2\beta - \cos^2\beta)\sin\theta\cos\theta + \sin\beta\cos\beta(\cos^2\theta - \sin^2\theta)\right] \\
\tau_{\theta z}^{p1} &= 2B_1(\sin\beta\sin\theta + \cos\beta\cos\theta)\frac{x_1 y_1}{\left(x_1^2 + y_1^2\right)^2} \\
\tau_{zr}^{p1} &= 2B_1(-\sin\beta\cos\theta + \cos\beta\sin\theta)\frac{x_1 y_1}{\left(x_1^2 + y_1^2\right)^2}
\end{aligned} \tag{8}$$

2.2.4 The Total Stress Components Around the Wellbore Under the Condition of Single Radial Borehole

Based on the influence of the above three factors, the total stress component expression can be obtained by superimposing the stress components in each part [formulas (3, 6, 7, 8)] and shown in formula (9).

$$\begin{aligned}
\sigma_r &= \sigma_r^s + \sigma_r^{i0} + \sigma_r^{p0} + \sigma_r^{i1} + \sigma_r^{p1} \\
\sigma_\theta &= \sigma_\theta^s + \sigma_\theta^{i0} + \sigma_\theta^{p0} + \sigma_\theta^{i1} + \sigma_\theta^{p1} \\
\sigma_z &= \sigma_z^s + \sigma_z^{i0} + \sigma_z^{p0} + \sigma_z^{i1} + \sigma_z^{p1} \\
\tau_{r\theta} &= \tau_{r\theta}^s + \tau_{r\theta}^{i0} + \tau_{r\theta}^{p0} + \tau_{r\theta}^{i1} + \tau_{r\theta}^{p1} \\
\tau_{\theta z} &= \tau_{\theta z}^s + \tau_{\theta z}^{i0} + \tau_{\theta z}^{p0} + \tau_{\theta z}^{i1} + \tau_{\theta z}^{p1} \\
\tau_{rz} &= \tau_{rz}^s + \tau_{zr}^{i0} + \tau_{zr}^{p0} + \tau_{zr}^{i1} + \tau_{zr}^{p1}
\end{aligned} \tag{9}$$

2.3 Derivation of Stress Distribution Model Under the Condition of Multi-Radial Boreholes

The stress distribution model around the wellbore under the condition of single radial borehole has been obtained by the previous part derivation. Now it is extended to the condition of multi-radial boreholes. In the process of deriving the stress distribution model under the condition of multi-radial boreholes, the same assumptions which taken in the single-radial borehole model are adopted. To obtain the mechanical model of multi-radial boreholes with any number, vertical depth and azimuth angle, the model under the condition of two radial is studied firstly.

2.3.1 Derivation of Stress Distribution Model Under the Condition of Two Radial Boreholes

A new radial borehole 2 is added. The phase angle and the vertical spacing between the new radial borehole 2 and the radial hole 1 are φ and L respectively. The establishment of the coordinate system is shown in Fig. 3. The rectangular coordinate system (x2, y2, z2) is obtained by translating the rectangular coordinate system (x3, y3, z3) the length L downward. The rectangular coordinate systems (x2, y2, z2) and (x, y, z) have the same origin. Because the interaction between boreholes is neglected, the total stress field can be obtained by only calculating the stresses generated by internal pressure and fracturing fluid percolation of the new radial borehole 2 and then superimposing them into the stress field under the condition of single radial borehole.

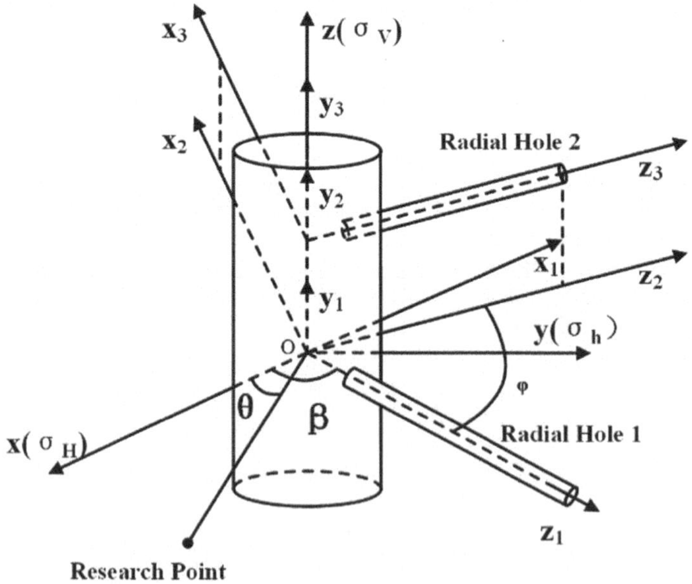

Fig. 3 Coordinate systems of two radial boreholes

(1) Stress caused by internal pressure of radial wellbore 2

First, according to the formula (3), the stress expression caused by the internal pressure of the radial borehole 2 in its cylindrical coordinate system (r2, θ2, z2) can be obtained. secondly, using Eq. (4), it is transformed into the radial borehole 2 rectangular coordinate system (x3, y3, z3). Then, subtracting L from all the y terms, the stress components expression in (x2, y2, z2) are obtained. Since the origins of (x2, y2, z2) and (x, y, z) coincide with each other, the stress components can be transformed from (x2, y2, z2) into (x, y, z). Because the phase angle between the radial borehole 2 and the radial borehole 1 is φ, it can be known that the angle between the axial direction of the radial borehole 2 and the direction of maximum principal stress is β + φ, and the transformation formula can be given by:

$$\begin{aligned}
\sigma'_x &= \sin^2(\beta+\varphi)\sigma_x + \cos^2(\beta+\varphi)\sigma_z - 2\sin(\beta+\varphi)\cos(\beta+\varphi)\tau_{zx} \\
\sigma'_y &= \cos^2(\beta+\varphi)\sigma_x + \sin^2(\beta+\varphi)\sigma_z + 2\sin(\beta+\varphi)\cos(\beta+\varphi)\tau_{zx} \\
\sigma'_z &= \sigma_y \\
\tau'_{xy} &= -\sin(\beta+\varphi)\cos(\beta+\varphi)\sigma_x + \left(\cos^2(\beta+\varphi) - \sin^2(\beta+\varphi)\right)\tau_{zx} \\
&\quad + \sin(\beta+\varphi)\cos(\beta+\varphi)\sigma_z \\
\tau'_{yz} &= \cos(\beta+\varphi)\tau_{xy} + \sin(\beta+\varphi)\tau_{yz} \\
\tau'_{zx} &= -\sin(\beta+\varphi)\tau_{xy} + \cos(\beta+\varphi)\tau_{yz}.
\end{aligned} \tag{10}$$

Finally, the stress components are transformed from (x, y, z) to (r, θ, z) by formula (4), and the final expression of stress components caused by internal pressure of radial wellbore 2 is obtained.

$$\begin{aligned}
\sigma_r^{i2} &= \frac{-R_2^2 P_w(\sin(\beta+\varphi)\cos\theta - \cos(\beta+\varphi)\sin\theta)^2\left(x_2^2 - (y_2-L)^2\right)}{(x_2^2 + (y_2-L)^2)^2} \\
&\quad + \frac{cR_2^2 P_w(\cos(\beta+\varphi)\cos\theta + \sin(\beta+\varphi)\sin\theta)^2\left(x_2^2 + (y_2-L)^2\right)}{(x_2^2 + (y_2-L)^2)^2} \\
\sigma_\theta^{i2} &= \frac{-R_2^2 P_w(\sin(\beta+\varphi)\sin\theta + \cos(\beta+\varphi)\cos\theta)^2(x_2^2 - (y_2-L)^2)}{(x_2^2 + (y_2-L)^2)^2} \\
&\quad + \frac{cR_2^2 P_w(\cos(\beta+\varphi)\sin\theta + \sin(\beta+\varphi)\cos\theta)^2(x_2^2 + (y_2-L)^2)}{(x_2^2 + (y_2-L)^2)^2} \\
\sigma_z^{i2} &= \frac{R_2^2 P_w(x_2^2 - (y_2-L)^2)}{(x_2^2 + (y_2-L)^2)^2} \\
\tau_{r\theta}^{i2} &= \frac{R_2^2 P_w\left(\sin(\beta+\varphi)\cos(\beta+\varphi)\left(\cos^2\theta - \sin^2\theta\right) - \left(\cos^2(\beta+\varphi) - \sin^2(\beta+\varphi)\right)\sin\theta\cos\theta\right)(x_2^2 - (y_2-L)^2)}{(x_2^2 + (y_2-L)^2)^2} \\
&\quad + \frac{cR_2^2 P_w\left(\sin(\beta+\varphi)\cos(\beta+\varphi)\left(\cos^2\theta - \sin^2\theta\right) - \left(\cos^2(\beta+\varphi) - \sin^2(\beta+\varphi)\right)\sin\theta\cos\theta\right)(x_2^2 + (y_2-L)^2)}{(x_2^2 + (y_2-L)^2)^2} \\
\tau_{\theta z}^{i2} &= \frac{-2R_2^2 P_w(\sin(\beta+\varphi)\sin\theta + \cos(\beta+\varphi)\cos\theta)x_2(y_2-L)}{(x_2^2 + (y_2-L)^2)^2} \\
\tau_{zr}^{i2} &= \frac{2R_2^2 P_w(\sin(\beta+\varphi)\cos\theta - \cos(\beta+\varphi)\sin\theta)x_2(y_2-L)}{(x_2^2 + (y_2-L)^2)^2}
\end{aligned} \tag{11}$$

(2) Stress caused by fracturing fluid percolation of radial wellbore 2

The processing method of this part is the same as that of the internal pressure of radial wellbore 2. Firstly, the expression of stress components caused by fracturing fluid percolation of radial wellbore 2 is obtained in (r2, θ2, z2) by using formula (7). Using the same series of stress transformations as in the radial borehole 2 internal pressure analysis, the final expression of the stress components caused by the fracturing fluid percolation of the radial borehole 2 can be obtained in main borehole cylindrical coordinate system (r, θ, z).

$$\begin{aligned}
\sigma_r^{p2} &= (\sin(\beta+\varphi)\cos\theta - \cos(\beta+\varphi)\sin\theta)^2 \\
&\quad\times \frac{A\left(x_2^4-(y_2-L)^4\right)+B_2\left(x_2^2-(y_2-L)^2\right)+2A(y_2-L)^2\left(x_2^2+(y_2-L)^2\right)}{\left(x_2^2+(y_2-L)^2\right)^2} \\
&\quad + C(\cos(\beta+\varphi)\cos\theta + \sin(\beta+\varphi)\sin\theta)^2 \\
\sigma_\theta^{p2} &= (\sin(\beta+\varphi)\sin\theta + \cos(\beta+\varphi)\cos\theta)^2 \\
&\quad\times \frac{A\left(x_2^4-(y_2-L)^4\right)+B_2\left(x_2^2-(y_2-L)^2\right)+2A(y_2-L)^2\left(x_2^2+(y_2-L)^2\right)}{\left(x_2^2+(y_2-L)^2\right)^2} \\
&\quad + C(\cos(\beta+\varphi)\sin\theta - \sin(\beta+\varphi)\cos\theta)^2 \\
\sigma_z^{p2} &= \frac{A\left(x_2^2+(y_2-L)^2\right)^2-B_2\left(x_2^2-(y_2-L)^2\right)}{\left(x_2^2+(y_2-L)^2\right)^2} \\
\tau_{r\theta}^{p2} &= \left[\left(\cos^2(\beta+\varphi)-\sin^2(\beta+\varphi)\right)\sin\theta\cos\theta - \sin(\beta+\varphi)\cos(\beta+\varphi)\left(\cos^2\theta-\sin^2\theta\right)\right] \\
&\quad\times \frac{A\left(x_2^4-(y_2-L)^4\right)+B_2\left(x_2^2-(y_2-L)^2\right)+2A(y_2-L)^2\left(x_2^2+(y_2-L)^2\right)}{\left(x_2^2+(y_2-L)^2\right)^2} \\
&\quad + C\left[\left(\sin^2(\beta+\varphi)-\cos^2(\beta+\varphi)\right)\sin\theta\cos\theta + \sin(\beta+\varphi)\cos(\beta+\varphi)\left(\cos^2\theta-\sin^2\theta\right)\right] \\
\tau_{\theta z}^{p2} &= 2B_2(\sin(\beta+\varphi)\sin\theta + \cos(\beta+\varphi)\cos\theta)\frac{x_2(y_2-L)}{\left(x_2^2+(y_2-L)^2\right)^2} \\
\tau_{zr}^{p2} &= 2B_2(-\sin(\beta+\varphi)\cos\theta + \cos(\beta+\varphi)\sin\theta)\frac{x_2(y_2-L)}{\left(x_2^2+(y_2-L)^2\right)^2}
\end{aligned} \tag{12}$$

(3) The total stress component around the wellbore under the condition of two radial boreholes

By superimposing the stresses generated by internal pressure and fracturing fluid percolation of the new radial borehole 2 into the stress field under the condition of single radial borehole, the total Stress Components around the Wellbore under the condition of two radial boreholes can be obtained.

$$\begin{aligned}
\sigma_r &= \sigma_r^s + \sigma_r^{i0} + \sigma_r^{p0} + \sigma_r^{i1} + \sigma_r^{p1} + \sigma_r^{i2} + \sigma_r^{p2} \\
\sigma_\theta &= \sigma_\theta^s + \sigma_\theta^{i0} + \sigma_\theta^{p0} + \sigma_\theta^{i1} + \sigma_\theta^{p1} + \sigma_\theta^{i2} + \sigma_\theta^{p2} \\
\sigma_z &= \sigma_z^s + \sigma_z^{i0} + \sigma_z^{p0} + \sigma_z^{i1} + \sigma_z^{p1} + \sigma_z^{i2} + \sigma_z^{p2} \\
\tau_{r\theta} &= \tau_{r\theta}^s + \tau_{r\theta}^{i0} + \tau_{r\theta}^{p0} + \tau_{r\theta}^{i1} + \tau_{r\theta}^{p1} + \tau_{r\theta}^{i2} + \tau_{r\theta}^{p2} \\
\tau_{\theta z} &= \tau_{\theta z}^s + \tau_{\theta z}^{i0} + \tau_{\theta z}^{p0} + \tau_{\theta z}^{i1} + \tau_{\theta z}^{p1} + \tau_{\theta z}^{i2} + \tau_{\theta z}^{p2} \\
\tau_{rz} &= \tau_{rz}^s + \tau_{zr}^{i0} + \tau_{zr}^{p0} + \tau_{zr}^{i1} + \tau_{zr}^{p1} + \tau_{zr}^{i2} + \tau_{zr}^{p2}
\end{aligned} \tag{13}$$

2.3.2 Derivation of Stress Distribution Model Under the Condition of Multi-radial Boreholes

On the basis of the previous article, the model is now extended to the condition of n + 1 radial boreholes. Radial well 1 is used as the reference radial boreholes. The phase angle and vertical spacing between each radial borehole and radial borehole 1 are φ1, φ2, …, φn and L1, L2, …, Ln respectively (Fig. 4).

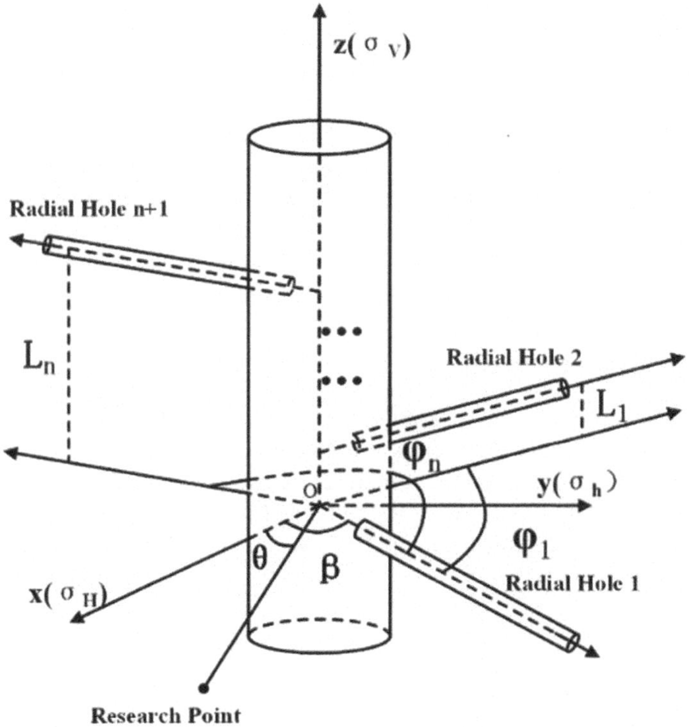

Fig. 4 Schematic diagram of multi-radial boreholes

Since ignoring the interaction between boreholes, the additional stress field produced by each radial borehole can be obtained by substituting the parameters of each borehole into the formulas (11 and 12). Finally, the stress distribution model under the condition of multi-radial boreholes can be obtained by superimposing each additional stress field with the stress field model of single radial well.

$$\begin{aligned}
\sigma_r &= \sigma_r^s + \sigma_r^{i0} + \sigma_r^{p0} + \sigma_r^{i1} + \sigma_r^{p1} + \sigma_r^{i2} + \sigma_r^{p2} + \ldots + \sigma_r^{in} + \sigma_r^{pn} \\
\sigma_\theta &= \sigma_\theta^s + \sigma_\theta^{i0} + \sigma_\theta^{p0} + \sigma_\theta^{i1} + \sigma_\theta^{p1} + \sigma_\theta^{i2} + \sigma_\theta^{p2} + \ldots + \sigma_\theta^{in} + \sigma_\theta^{pn} \\
\sigma_z &= \sigma_z^s + \sigma_z^{i0} + \sigma_z^{p0} + \sigma_z^{i1} + \sigma_z^{p1} + \sigma_z^{i2} + \sigma_z^{p2} + \ldots + \sigma_z^{in} + \sigma_z^{pn} \\
\tau_{r\theta} &= \tau_{r\theta}^s + \tau_{r\theta}^{i0} + \tau_{r\theta}^{p0} + \tau_{r\theta}^{i1} + \tau_{r\theta}^{p1} + \tau_{r\theta}^{i2} + \tau_{r\theta}^{p2} + \ldots + \tau_{r\theta}^{in} + \tau_{r\theta}^{pn} \\
\tau_{\theta z} &= \tau_{\theta z}^s + \tau_{\theta z}^{i0} + \tau_{\theta z}^{p0} + \tau_{\theta z}^{i1} + \tau_{\theta z}^{p1} + \tau_{\theta z}^{i2} + \tau_{\theta z}^{p2} + \ldots + \tau_{\theta z}^{in} + \tau_{\theta z}^{pn} \\
\tau_{rz} &= \tau_{rz}^s + \tau_{zr}^{i0} + \tau_{zr}^{p0} + \tau_{zr}^{i1} + \tau_{zr}^{p1} + \tau_{zr}^{i2} + \tau_{zr}^{p2} + \ldots + \tau_{zr}^{in} + \tau_{zr}^{pn}
\end{aligned} \tag{14}$$

3 Influence of Radial Borehole Parameters on Fracture Initiation

According to the stress distribution model and rock fracture criterion, the maximum principal stress curves under different conditions are plotted by using matlab, and then the influences of various parameters on fracture initiation are analysed.

In the specific calculation and analysis, the selected basic parameters are: σv = −42 Mpa, σH = −35 Mpa, σh = −26 Mpa, Pp = 15 Mpa, Pw = 32 Mpa, c = 0.9, α = 0.6, δ = 1, ϕ = 0.1, ν = 0.25, R = 75 mm, R1 = R2 = 15 mm, r = 200 mm, z = 75 mm. For each particular example, there are some changes in some parameters, which are explained in detail in the corresponding part.

3.1 Rock Fracture Criterion

When the tensile stress (i.e. the maximum principal stress) of the rock around borehole exceeds its tensile strength, the rock will produce initial fracture. This criterion is the maximum principal stress criterion, which can be used as the criterion for judging the initiation of rock fracture. According to the related theory of elasticity mechanics, the calculation formula of the three principal stresses can be obtained and is indicated as formula(15):

$$\begin{aligned}
\sigma_1 &= \sigma_r \\
\sigma_2 &= \frac{1}{2}\left[(\sigma_\theta + \sigma_z) + \sqrt{(\sigma_\theta - \sigma_z)^2 + 4\tau_{\theta z}^2}\right] \\
\sigma_3 &= \frac{1}{2}\left[(\sigma_\theta + \sigma_z) - \sqrt{(\sigma_\theta - \sigma_z)^2 + 4\tau_{\theta z}^2}\right].
\end{aligned} \tag{15}$$

Since the model stipulates that the tensile stress is positive and the compressive stress is negative, σ2 is the maximum tensile stress around the borehole. The concrete expression is as follows:

$$\sigma_{\max} = \frac{1}{2}\left[(\sigma_\theta + \sigma_z) + \sqrt{(\sigma_\theta - \sigma_z)^2 + 4\tau_{\theta z}^2}.\right] \tag{16}$$

3.2 Influence of Horizonal Geo-Stress Difference on Fracture Initiation

In order to study the effect of horizontal geo-stress difference on fracture initiation, the maximum principal stresses under the conditions of serial geo-stress differences are calculated. This factor analysis uses model of the single borehole condition. The radial well azimuth beta is 30° and z is 30 mm. 3, 6, 9, 12 and 15 Mpa are selected for stress difference. The other parameters are same with the selected basic parameters. The result is shown in the Fig. 5.

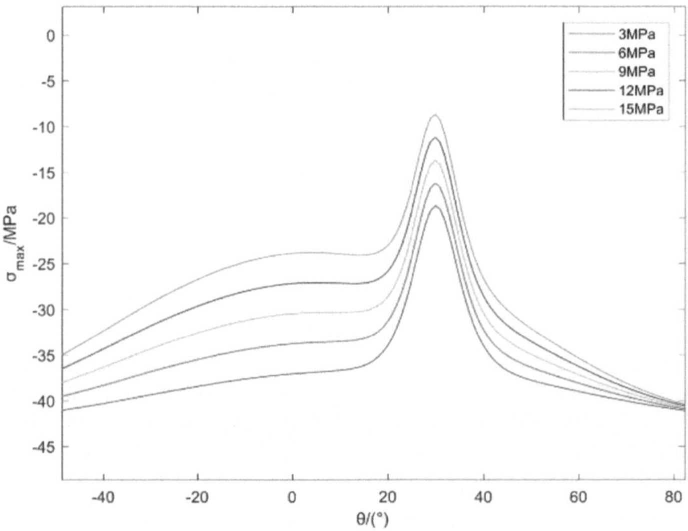

Fig. 5 σ max change law curve of different geo-stress difference with variable θ values

It can be seen from the figure that the maximum principal stress changes with the change of θ, and the maximum value of maximum principal stress appears at the radial well orientation where θ = β = 30°. Therefore, fracture is most likely to start at the location of radial borehole. As the stress difference increases, the maximum value of maximum principal stress increases. It is indicated that as the stress difference increases, the tensile force required for tensile failure decreases. The greater the geo-stress difference is, the smaller fracture initiation pressure is needed, and the more easily the rock breaks.

3.3 Influence of the Radial Borehole Diameter on Fracture Initiation

By changing the borehole diameter of each radial well and calculating the maximum principal stress under each borehole diameter, the effect of radial borehole diameter on fracture initiation is studied. The factor analysis is carried out by using single radial borehole model. The radial well azimuth beta is 30° and z is 30 mm. The radial borehole diameters in each case are set to 10, 20, 30, 40 mm respectively. The other parameters are same with the selected basic parameters (Fig. 6).

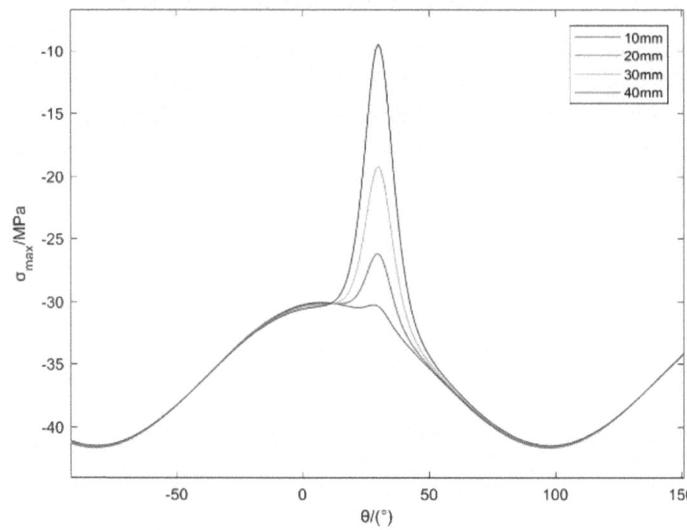

Fig. 6 σ max change law curve of different radial borehole diameters with variable θ values

As the diameter of the radial borehole increases, the maximum value of the maximum principal stress increases. At the same time, the increase area of maximum principal stress expands obviously, so it can be concluded that the action scope of stress field produced by radial boreholes expands with the increase of radial well diameter. When conditions permit, the diameter of the radial well can be appropriately increased to reduce the fracture initiation pressure and enhance the ability of the radial borehole to guide the fracture.

3.4 Influence of the Radial Borehole Azimuth on Fracture Initiation

Single radial borehole model is adopted to analyze the azimuth factor. In order to study the effect of radial borehole azimuth on fracture initiation, different radial borehole azimuths β are selected to calculate the maximum principal stress distribution around the main wellbore.

Firstly, the case of large horizontal geo-stress difference is calculated. In this case, all the parameters adopt the basic parameters but z changes to 50 mm, and the horizontal geo-stress difference is 9 MPa. The taken radial borehole azimuth is 0°, 10°, 20°, 30°, 40°, 50°, 60° respectively. The results of each azimuth are shown in Fig. 7.

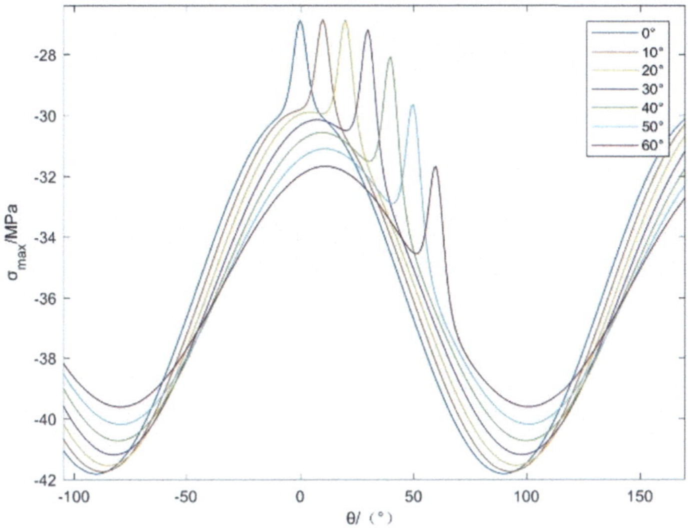

Fig. 7 σ max change law curve of different radial borehole azimuths with variable θ values (case 1)

It can be seen from the figure that the maximum value of maximum principal stress decreases with the increase of radial well azimuth when the horizontal geo-stress difference is large.

Then, the case of small horizontal stress difference is calculated. In this case, the horizontal stress difference is changed to 3 MPa, while the other parameters remain unchanged. The azimuth of different radial wells is calculated. The results are as follows (Fig. 8).

It can be seen from the figure that the maximum principal stress increases first and then decreases with the increase of the azimuth angle when the horizontal geo-stress difference is small.

3.5 Influence of the Radial Boreholes Number on Fracture Initiation

The multi-radial boreholes model was used to analyze the number factor. Each time a new radial borehole is added, the new borehole is located 20 mm below the previous bottom radial borehole and has the same azimuth with previous boreholes. All parameters adopt the basic parameters. Researcher calculates the maximum principal stress distribution around main wellbore in the cases of 1, 2, 3, and 4 radial boreholes. The results are as follows (Fig. 9).

It can be concluded that as the number of radial boreholes increases, the maximum value of maximum principal stress around main wellbore increases obviously. Therefore, increasing radial boreholes should be considered as a measure to reduce fracture initiation pressure.

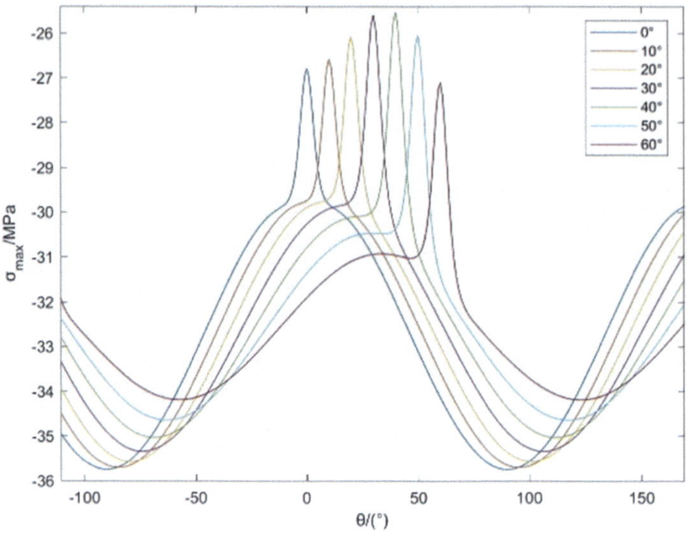

Fig. 8 σ max change law curve of different radial borehole azimuths with variable θ values (case 2)

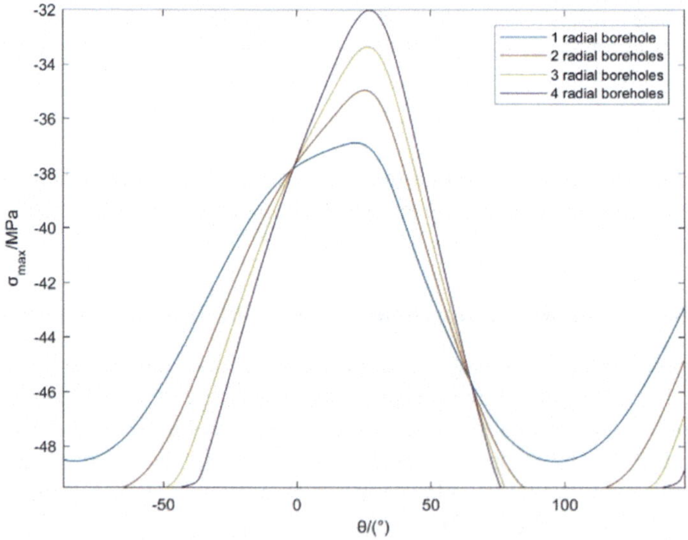

Fig. 9 σ max change law curve of different radial borehole numbers with variable θ values

3.6 Influence of the Radial Boreholes Vertical Spacing on Fracture Initiation

In this part, the model of two radial boreholes is applied to research the factor. In different cases, the vertical spacing L between two radial boreholes are 100, 150, 200, 250, 300, 350, 400 mm respectively and the corresponding z is half of L. The radial boreholes azimuth are both 30°. The rest parameters take the basic parameters. The results of the maximum principal stress distribution around the well are calculated as follows (Fig. 10).

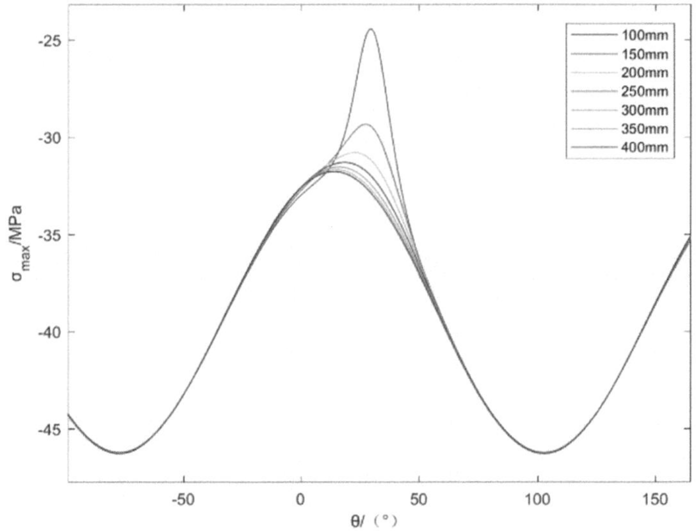

Fig. 10 σ max change law curve of different radial boreholes spacings with variable θ values

It can be seen that with the decrease of vertical spacing between radial wells, the maximum value of maximum principal stress will increase significantly. Therefore, in the production practice, the radial boreholes spacing can be decreased properly to reduce the fracture initiation pressure.

3.7 Influence of the Phase Angle Between Boreholes on Fracture Initiation

Two radial boreholes model is adopted to analyze the phase angle factor. In order to study the effect of phase angle on fracture initiation, the maximum principal stresses under different phase angle conditions are calculated. In each case, the phase angles between the two radial boreholes are 0°, 10°, 20°, 30°, 40°, 50°, 60° respectively and all azimuth angles are 10°. First, we use the basic parameters set before to calculate the cases of different phase angles. The results are as follows (Fig. 11).

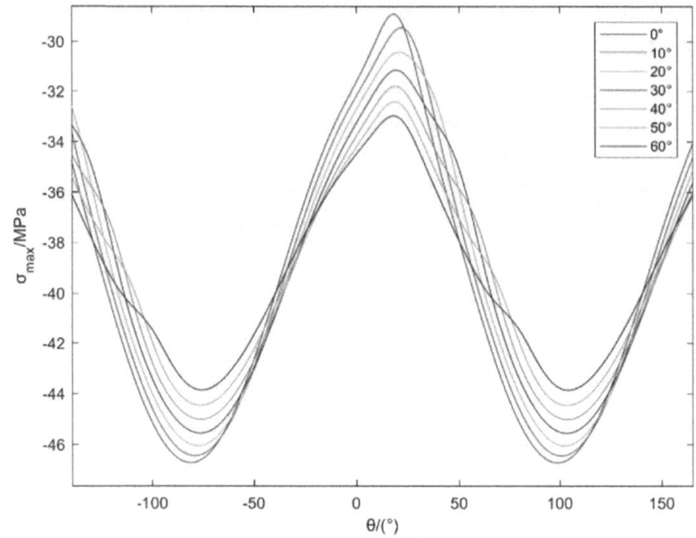

Fig. 11 σ max change law curve of different radial boreholes phase angles with variable θ values (case 1)

It can be concluded from the figure that the maximum value of maximum principal stress decreases gradually with the increase of the phase angle between two radial boreholes, which is not conducive to fracture initiation. But this rule applies only to general situations.

Then, the horizontal stress difference is reduced to 3 MPa (σH = −29 Mpa, σh = −26 Mpa), and the radiuses of two radial boreholes are increased to 30 mm. The other parameters remain unchanged. The results calculated in different phase angles are shown below (Fig. 12).

In the figure, the maximum value of maximum principal stress in the case of 10° is greater than that in case of 0°. It can be seen that the maximum value of maximum principal stress does not necessarily decrease with the increase of the phase angle when the horizontal geo-stress difference is small and the diameter of the radial borehole is large. In condition of horizonal geo-stress difference and radial borehole diameter are large, anomalous phenomena which does not conforms to the law occur.

4 Conclusion

The effects of in geo-stress, wellbore internal pressure and fracturing fluid percolation effect on formation stress field distribution are taken into account, a mechanical model of multi-radial boreholes with any number, vertical depth and azimuth angle is established. Combining the model with rock tensile fracture criterion, we calculate the maximum principal stress value of the formation around the wellbore in different conditions and study the influence of many factors. The results are as follows.

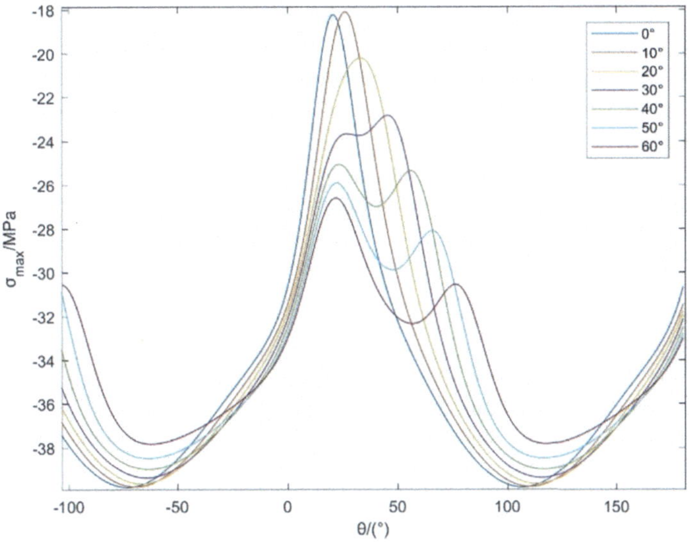

Fig. 12 σ max change law curve of different radial boreholes phase angles with variable θ values (case 2)

1. The maximum value of maximum principal stress around the main wellbore appears at the position of radial boreholes, and fracture is likely to start at the location of radial boreholes. This shows that using radial boreholes to guide fracture initiation has a theoretical basis.
2. The horizontal geo-stress difference affects the difficulty of fracture initiation. As the stress difference increases, the tensile force required for tensile failure decreases. The greater the geo-stress difference is, the smaller fracture initiation pressure is needed, and the more easily the rock breaks.
3. As the radial borehole diameter increases, the maximum value of the maximum principal stress increases. Therefore, in a tight oil layer where fracture initiates hardly, it can be considered to facilitate fracture initiation by increasing the diameter of the radial borehole.
4. When the horizontal geo-stress difference is large, the maximum value of maximum principal stress decreases as the azimuth of the radial borehole increases. However, when the horizontal geo-stress difference is small, the maximum value of maximum principal stress will increase first and then decrease with the increase of the azimuth angle.
5. With the increase of radial borehole number, the maximum value of maximum principal stress around the main borehole increases obviously. In production operations, increasing the number of radial boreholes properly should be considered as a way to make fractures initiate more easily.
6. The maximum value of maximum principal stress around the main borehole increases with the decrease of vertical spacing between radial boreholes. In fracturing operation, the vertical spacing can be appropriately reduced to facilitate

fracture initiation, but the stability of boreholes should be taken into account simultaneously.

7. Generally, the maximum value of maximum principal stress decreases with the increase of phase angle between boreholes, and the fracture is more difficult to initiate. However, when the diameters of radial boreholes are large enough and the geo-stress difference is small, abnormal phenomena will occur, which does not conform to the rule in general conditions.

References

Fallahzadeh, S.H., Shadizadeh, S.R.: A new model for analyzing hydraulic fracture initiation in perforation tunnels. Energy Sour. **35**(1), 9–21 (2013)

Guo, T., Liu, B., Qu, Z., et al.: Study on initiation mechanisms of hydraulic fracture guided by vertical multi-radial boreholes. Rock Mech. Rock Eng. **50**(7), 1767–1785 (2017)

Research and Application of Clean Operation Technology for Rod and Tubing

Hua Huang(✉), Tao Wang, Jin-xia Wang, Xiao-ping Yuan, Peng-fei Lei, Guang-qing Gu, Lu-lu Ping, and Wen-jing Zhang

Engineering Technology Institute of Huabei Oil Field, Hebei, China
cyy_huangh@petrochina.com.cn

Abstract. Oil carried out from well hole by rod or tubing is unavoidable to pollute the well site in pulling operation of production or water injection wells. When cleaning oil and hard paraffin on rod or tubing, the anti-pollution technology including washing, mechanic oil wiper cause incomplete cleaning and formation damage for wells with low pressure, water sensitivity reservoir and severe paraffin trouble. On account of efficient pollution prevention and control, a whole process of clean operation technology for source control was developed for characters of wellbore and wellhead. An online closed style heat cleaning device and technology for lifting rod and tubing was developed. According to operation characteristics of lifting rod, a down hole short range circulating path is designed, which realizes jet cleaning and steam cycle control by closed device and double-ended pressing. The characters of nozzle is optimized to realize thorough coverage and instant cleaning of oil and paraffin without influence on rod job efficiency. Aiming at instantly cleaning tubing during pulling operation, integrated tool of wellbore stream flushing and oil wiper, tool of internal wall rotary scraping and process of negative pressure liquid collecting are designed. Utilizing high-speed steam flowing through annular space close to both sides of tubing, internal and external walls are cleaned rapidly. The flow velocity of steam is 8 times of direct steam flowing in the tubing. The liquid is recovered in a closed manner by wellhead negative pressure liquid collector. The process of washing internal and external of tubing can be carried out at the same time in pulling work. Field application indicates that online closed style heat cleaning technology had characteristics of convenient use, reliable performance and well

J. Lin (ed.), *Proceedings of the International Petroleum and Petrochemical Technology Conference 2019*, pp. 101–109, 2020.
https://doi.org/10.1007/978-981-15-0860-8_8

site adaptation. Clean operation and environment protection intention is realized that oil and water is prevented from falling down in pulling operation. The technology possesses a good prospect of application in oilfield green mine construction.

Keywords: Down hole operation · Rod · Tubing · Closed style · Clean · Green operation

1 Introduction

With the implementation of the new environmental protection law, more stringent requirements for pollution prevention and control are put forward in oilfield production. As one of the important links in oilfield production and maintenance, it is necessary and urgent to eliminate the environmental pollution caused by operation. Oil and water adhering to the outer surface of sucker rod and the inner and outer wall of tubing are the main pollutants during pulling work.

In recent years, the main anti-pollution measures adopted in domestic oilfields are hot washing, rod and tube scraping, wellhead fluid collection and so on. Although hot washing technology can remove crude oil and hard paraffin adhering to sucker rod and tubing surface, it can cause reservoir damage to formation with leakage and water sensitivity. The integral sucker rod scraper commonly used in oilfields is not suitable for the sucker rod with paraffin scraper and centralizer. The multipartite sucker rod scraper has poor scraping effect on hard paraffin and cannot meet the cleaning requirements of pulling rod. The sucker rod and tubing often need to be washed by steam on site. The falled oil and water during the cleaning process will cause damage to environment. If the sucker rod and tubing are pulled back to the tubing plant for centralized cleaning, much transportation costs will be increased. The existing wellhead collector can only recover most of the sewage in the tubing, and there is still sewage splashing around the wellhead during operation [1–6].

Aiming at environmental protection in the current operation process, a series of on-line closed heat cleaning technology and matching tools for sucker rod and tubing are developed to solve the problems of centralized recovery of waste oil and sewage, cleaning of sucker rod and tubing in the operation process and realize clean production in pulling operation.

2 Research Strategy

For wells with low pressure and water sensitivity formation that cannot adapt hot washing technology, a new type of closed cleaning device is developed.

(1) By introducing thermal field, the problem of blind spots and incomplete cleaning of mechanical oil scraping could be solved, and the steam which could be well controlled is selected as heat transfer medium.
(2) In order to improve the cleaning efficiency and ensure the cleaning effect, thermal cleaning is carried out in the wellbore space at the lower part of the wellhead, which is carried out simultaneously with pulling and uncoupling operations.

(3) Turn passive wellhead liquid collecting into pressure controlled collecting.

The cleaning and recovery problem of oil and water brought by rod and tubing is solved on site.

3 Online Closed Style Heat Cleaning Technology for Rod

The online closed style heat device for sucker rod consists of a sucker rod flusher, an tubing-casing connector and a negative pressure device. By down hole short range circulation, the steam flow control and booster flushing are realized by using the wellhead closed style cleaning and pressure control device.

3.1 Down Hole Short Range Circulation and Online Heat Cleaning Device

In order to solve the problem of space occupied by sucker rod cleaning and liquid scattering, the wellbore space is utilized to realize integrally thermal cleaning for rod body, and the device volume is minimized. A tubing-casing connector matched with the length of sucker rod is designed. Before pulling the rod, the lower part of the tubing hanger is connected with tubing-casing connector and put into down hole again. The negative pressure device is connected with one side of the casing gate to establish a controllable short range steam circulation channel.

The thermal flusher device is connected to tubing lift sub and consists of a steam anti-leakage mechanism, a control valve, a tubing tee joint and an internal booster flushing nozzles. The nozzles with inclination of 30° are distributed in two rows of annular chamber, and the diameter is adjustable. In order to ensure safety control, special steam control valves are used to meet the requirements of high temperature and high pressure resistance, and sucker rod blowout preventer sub is provided to meet the requirements of well control (Table 1).

Table 1 Main technical parameters for online closed cleaning of sucker rod

Parameter	Value
Steam temperature, °C	130–150
Steam pressure, MPa	0.7–1.0
Depth of connector, m	<20
OD(connector), in	$2^7/_8$ or $3^1/_2$
OD(tubing), in	$2^7/_8$ or $3^1/_2$

3.2 Technological Principle

The heat source sub of the control valve is connected with the steam hose of the boiler truck, so that the steam can flow rapidly along the annular space between rod and tubing, continuously flushing the oil attached to body of sucker rod. Oil enters the annulus through connector and goes up. By the negative pressure, the oil water mixture is forced to be pumped out and discharged into the recovery tank through the casing gate. During pulling operation, the steam flushes the rod body at a high speed. During screwing off and laying operation, the steam continues to clean the lower sucker rod.

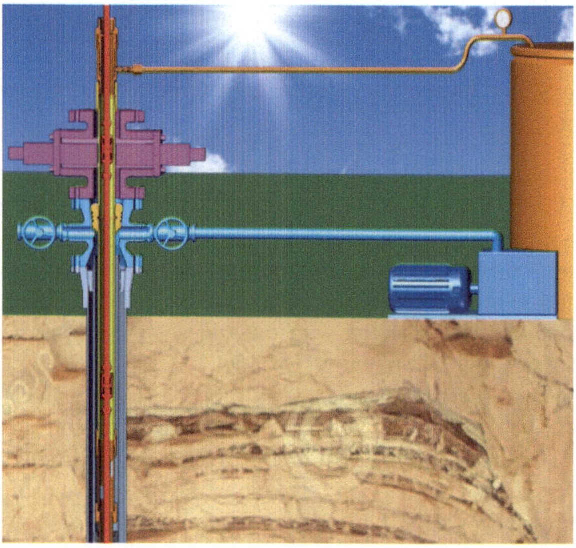

Fig. 1 Online closed style heat cleaning device for sucker rod

The device turns the horizontal open cleaning on pipe bridge into vertical closed style integrally cleaning without blind spots. There is no restriction of sucker rod specifications or paraffin scrapers. The operation doesn't affect time-effect of pulling rods. There is no leakage of oil and sewage (Fig. 1).

4 Online Closed Style Heat Cleaning Technology for Tubing

The online closed style heat cleaning device for tubing is composed of steam pipeline, cleaning cap, inner wall steam scraping tool, wellhead closed liquid collecting device, etc. Closed style cleaning for both sides of tubing wall and pulling operation are completed simultaneously.

4.1 Closed Cleaning and Collecting Device for Tubing

The inner wall steam scraping tool is composed of tubing coupling locking mechanism, centralizer, rotary nozzle and scraper. The tubing coupling locking mechanism has elastic expansion slips which locates the lower tubing coupling and stops tool to ensure continuous operation. Steam is utilized as downward driving force of tools. Compared with direct steam flowing into the tubing, the steam flow velocity is 8 times at the same flow rate. All steam thoroughly flows through the annular gap and is effectively utilized.

Table 2 Technical parameter of inner wall steam scraping tool

Tubing OD	Tool OD, mm	Tool length, mm	Expansion slips control mode
73.02 mm	58	230	Press tool downward to retract and the jaw resets automatically
88.9 mm	72	230	

In addition, the flow passage, rotating nozzle, scraper and synchronous rotation mechanism are designed. The rotating nozzle uses self-lubricating material to reduce friction, and forces steam to directly flush the inner wall of the tubing. The mechanical scraping and steam flushing are utilized together to improve the cleaning effect.

The closed liquid collecting device at the wellhead is pneumatic-driven, which can move along the slide way to adjust the position of buckling and releasing. The liquid collecting mechanism is split barrel-shaped, and the liquid collecting arm is controlled by the cylinder with strong controllability (Table 2).

4.2 Working Principle

Before pulling operation, the inner wall steam scraping tool is put into the pipe, and the upper end of the tubing is coupled with a cleaning cap. When lifted out of wellhead, the tubing is screwed off and lifted up 50–100 mm. Then the liquid collecting device is fastened. The inner wall scraping steam tool is driven down by steam which flows through the cleaning cap. The oil, paraffin and dirt on the inner wall are flushed and scraped. The liquid collecting device is connected with the vacuum pump, which provides the suction to promote steam flow. All the dirt is recovered in a closed manner.

The process turns horizontal open cleaning on pipeline bridge into vertical closed cleaning, which greatly improves cleaning timeliness of inner wall, effectively saves steam consumption and fuel consumption, and solves the problem of on-site cleaning of tubing (Fig. 2).

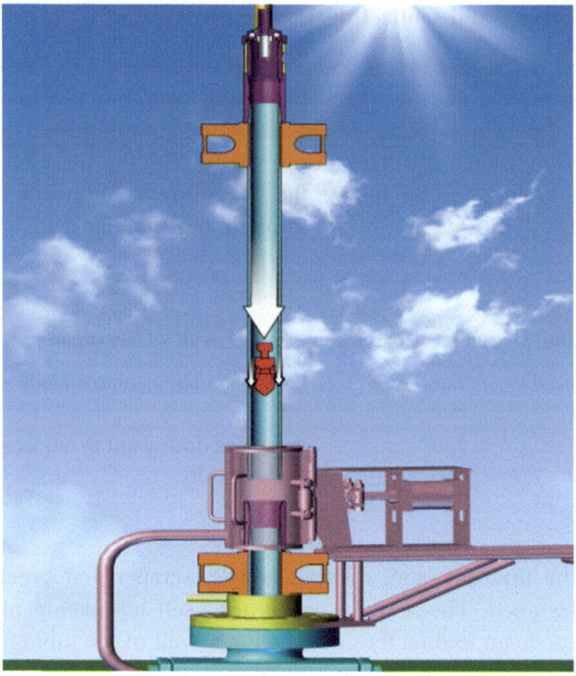

Fig. 2 Online closed thermal cleaning device for tubing

5 Skid-Mounted Cleaning Device

By the integration of vacuum device, liquid collecting device and boiler, the power and heating source are contained. It improves adaptability to the field, realizes quick installation, and reduces equipment cost. The popularization and application of the technology could be more convenient (Fig. 3 and Table 3).

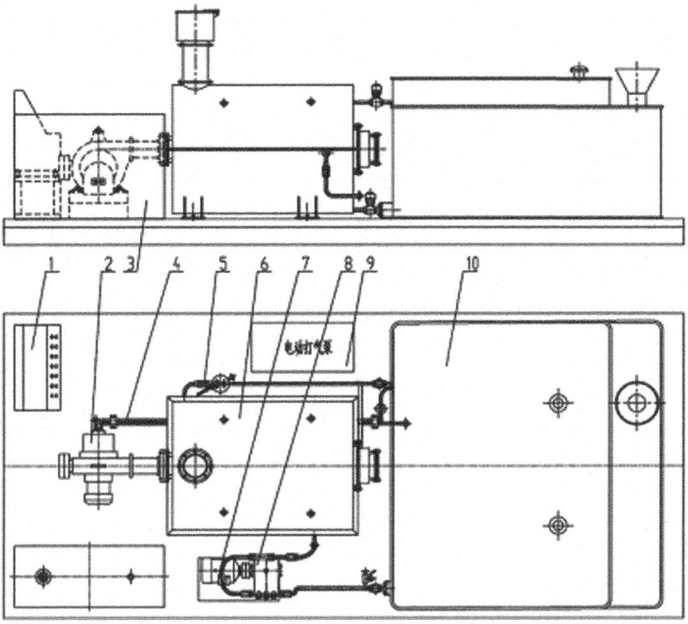

Fig. 3 Sketch of skid-mounted structure. (1, Control Cabinet 2, Burner 3, Generator 4, Diesel Pipeline 5, Boiler Exhaust Pipeline 6, Vehicle Special Boiler 7, Motor Drive 8, Ceramic Piston Pump 9, Electric Air Pump 10, Oil and Water Cabinet)

Table 3 Main technical parameters of skid-mounted cleaning device

Parameter	Value
Max working temperature, °C	200
Max pressure of boiler, MPa	6
Max flow rate, m^3/h	1.26
Working medium	Clear water
Working voltage, V	380

6 Field Application

Taking well T47-6 as an example, this well was an oil production well with KY21/65 wellhead, D73.02 mm NU tubing. Depth of plunger was 2132 m. There was a tubing drain in the string. The sucker rod assembly was D22 mm scraper rod and D19 mm sucker rod. The scraper rod had a centralizer joint. When mechanical oil scraper was used in the same block, oil was left at the coupling and centralizer, and the effect of scraping was poor. Paraffin deposition was serious in tubing from wellhead to depth of 1500 m, and thickness of black hard paraffin was 3–5 mm.

Table 4 Application chart

Technology	Specification	Pulling time/h	Water consumption/m^3	Fuel consumption/L
Online closed style heat cleaning for rod	D19 mm/D22 mm rods (with scraper and centralizer)	4.2	4.5	150
Online closed style heat cleaning for tubing	D73.02 mm tubing	3.8	8	260

The sucker rod and tubing had been cleaned by applying on-line closed style heat cleaning device. When sucker rod was pulled, the steam pressure was 0.8–1 MPa, the steam temperature was 130–140 °C. The pulling speed was about 60 rods per hour, and the total time was 4.2 h. After cleaning, there was no oil left on body of rod, and no sewage was took out of wellhead. The pulling speed of the sucker rod was comparable to the conventional process, and there was no obstruction when the scraper or centralizer was passing through.

During tubing pulling operation, the steam pressure was 1.5–2 MPa, the steam temperature was 120–140 °C. The pulling speed was about 60 pipes per hour, and the total time was 3.8 h. There was no crude oil or paraffin in the inner and outer walls of the tubing, and all the cleaning liquid was collected without leakage. The pulling speed of the tubing was comparable to the conventional process.

The technology is being popularized and applied on a large scale (Table 4).

7 Conclusion

In view of the well with serious water sensitivity, low pressure or paraffin deposition in reservoir, an online closed style heat cleaning device for sucker rod and tubing is developed, and a set of cleaning operation technology has been formed.

(1) The online closed style heat cleaning technology for sucker rods is developed. The sucker rods are cleaned comprehensively by short range down hole circulation and steam flow direction control. It is not restricted by rod body specifications, centralizers or paraffin scrapers, and does not affect the working time.

(2) The developed tubing inner wall cleaner with automatic positioning function greatly improves the efficiency of closed style heat cleaning of tubing and realizes efficient utilization of steam. With the pneumatic control of slide way type negative pressure liquid collector on wellhead, the oil and water are recovered while inner and outer wall of tubing is cleaning. The pulling operation is carried out at the same time.

(3) Field application shows that the technology is convenient to operate, reliable in performance, adaptive to the field. The goal of cleaning and environmental protection operation has been achieved while no oil or sewage falls out in operation. The technology has broad prospects for application in the construction of green mines in oilfields.

References

1. Bingguo, Du, Ruqiang, Li, Dangming, Zheng: Development and application of a closed style anti-pollution device of splashing tubing lift. Oil Field Equip. **38**(10), 75–77 (2009)
2. Hongfu, Ma.: Development and application of new environmental protection workover supporting device. Well Test. **20**(3), 64–66 (2011)
3. Xiaofang, Yang: Development and application of hydrodynamic scraper. China Petrol. Mach. **41**(1), 94–95 (2013)
4. Guotian, L.I.: Research and application of green workover operation supporting devices. Environ. Prot. Oil Gas Fields **23**(2), 34–35 (2013)
5. Weisheng, H., Yuzheng, T., Baochen, F., Jianye, W.: Research and application of pollution prevention and control measures in the down hole operation of Qi-40 steam-driven block. Environ. Prot. Oil Gas Fields **19**(Supplement), 61–63, 69 (2009)
6. Jincang, Liu: Development of rod absolute oil tool in workover operations of pumping wells. Oil Drill. Prod. Technol. **31**(Supplement 2), 140–141 (2009)

Research on the Integrated Technology for Acidizing and Water Injection

Lulu Ping(✉), Tao Wang, Pengfei Lei, Chunsheng Zhang, Ran Li, and Yao Wang

Engineering Technology Institute of Huabei Oil Field, Hebei, China
cyy_pll@petrochina.com.cn

Abstract. The acidizing and water injection of the injection well require two pipe strings to complete. During the operation, it is easy to cause secondary pollution of the reservoir. In order to avoid the problems above and shorten the duty cycle. A new integrated process for acidizing operation and water injection of the injection well has been developed. In order to achieve water injection directly after the acidizing, weak current controlled acid sprayer and integrated packer which can achieve acidizing and water injection are developed. Also, we have experimented breakdown test with the tools and optimized the design parameters. This process has been successfully used in multiple wells in the Gao 20 block and the Yan 63 block of Yanling Oilfield. Through laboratory tests and field applications, the integrated technology of acidizing operation and water injection breaks through the problems that conventional packers are not suitable for high-pressure acidizing construction and the water distributor can't meet the requirements of high pressure acidizing. it can avoid operations such as pressure releasing, wellhead replacement and the up and down of the pipe string when acidizing, efficiently complete the purpose of acidizing operation and water injection. The new integrated process for acidizing operation and water injection can radically shorten the cycle time, reduce the pollution of the reservoir during the operation and guarantee the effect of acidizing. It is of great significance to reduce the cost and increase efficiency of old oilfields.

Keywords: Injection well · New integrated process for acidizing operation and water injection · Sprayer · Packer · Reduce the cost and increase efficiency

J. Lin (ed.), *Proceedings of the International Petroleum and Petrochemical Technology Conference 2019*, pp. 110–117, 2020.
https://doi.org/10.1007/978-981-15-0860-8_9

1 Introduction

The Yan 63 block of Yanling Oilfield is a multilayer and low permeability sandstone reservoir, taking the waterflood development as the main way. The reservoir property of it is poor. Because of scaling near the well, the water injection pressure is high during the process of water injection development, result in low liquid yields and low production.

Layering acidizing is the main way of lowering injection pressure and increasing water injection volume. Because of the tool's pressure-bearing level and the rate of flow of the pipe string of the conventional layering water injection can't ensure the acidizing treatment, it's required that tripping in the acidizing pipe string, releasing the pressure after modification and changing the pipe string of water injection or the wellhead equipment should to be implemented. Moreover, the pressure relief and the up and down of the pipe string's lead the backflow of impurities in the stratums to block the narrow passage, which is easy to cause secondary pollution of the reservoir.

In order to reduce the damage to reservoir and ensure the acidizing effect to the greatest extent, a combined scheme of acidizing and water injection is proposed. Through repeated design improvements and indoor experiment, research has formed a sour note for integration tools and successfully implemented construction of acidizing and water injection to five wells. It marks that the technology can be applied to the scene and meets the demand of lowering injection pressure and increasing water injection, at the same time, further substantially shorten the operation cycle and protect the reservoir.

2 Difficulty Analysis

(1) How to combine separate layer acidizing and water injection tools closely to form an integrated process string.
(2) The pressure of acidizing construction is high, and the pressure index of the conventional packer is not suitable for acidizing construction with high pressure [1].
(3) In the process of acidizing and water injection, the jacking force is generated, which is easy to cause packer unsealing [2].
(4) After the acidizing construction, the repeated backwashing process will easily lead to the difficulty in resetting the backwashing valve and the lax sealing.

3 Integrated Technology of Acidizing and Water Injection for Water Injection Well

3.1 Structure and Working Principle of Integrated Pipe String

The integrated string of acidizing and water injection mainly includes hydraulic anchor, acidizing and water injection packer, acid sprayer, water distributor, centralizer, ball

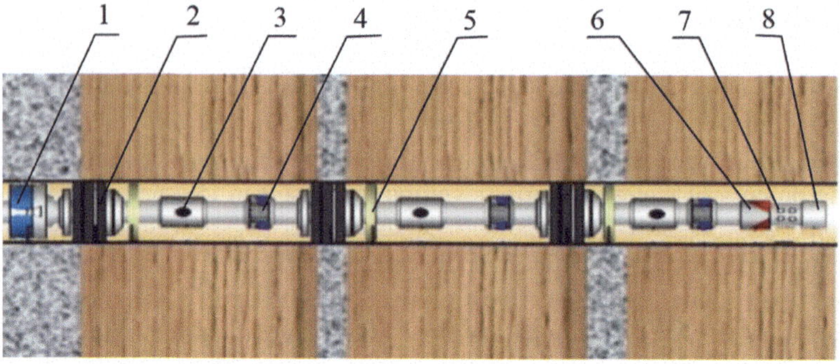

Fig. 1 Schematic diagram of the integrated pipe string structure. (1, hydraulic anchor 2, acidizing and water injection packer 3, acid sprayer 4, water distributor 5, centralizer 6, ball seat 7, sieve tube 8, bull plug)

seat, sieve tube and pipe plug (Fig. 1). When the pipe string enters, the acid sprayer and water distributor are closed [3].

The process principle:

(1) Pitch and set the multistage packer.
(2) In the process of transforming a certain layer, open the corresponding acid sprayer through the downhole measuring and adjusting instrument (or through ground hydraulic control technology), inject acid solution through the wellhead protector and carry out acidizing construction.
(3) After acidizing, the lower tester (or liquid control mode) will close the acid injection port of this layer, open the sliding sleeve of the lower acid injection device, and then acidize. The next layer is the same as this.
(4) Test and adjust all the water distributors with the instrument and inject water.

3.2 Technical Indexes and Characteristics of the Integrated Pipe String

Technical Index: Setting pressure −15 MPa; Working pressure −50 MPa; 114 mm diameter; internal diameter −50 mm; Working temperature −120 °C; the unsealing pressure is 20 MPa.

Technical features: Combined with conventional separate injection technology closely and its measuring technology is mature, which using mechanical anchor to balance jacking force of the string and using the top packer and lower packer to layer step by step. Using the measuring instrument of concentric water distributor as the switch tool of acid sprayer and water distributor, the technical problem of combining separate injection and separate acidizing pipe string is solved, and the purpose of acidizing is realized.

Applicable Formation: Temperature <150 °C; perforated layer; large difference of physical property; geology requirement for precision injection.

Applicable Well Conditions: In the case of static wellhead, carry out acidizing operation, then directly water injection construction.

4 Key Supporting Tools' Structure and Principle

4.1 Acid Sprayer with Micro-electric Control Switch

The acid sprayer of micro-electric control switch is mainly composed of upper joint, mandrel barrel, outer barrel, adjusting sleeve, transition joint, coordination body of adjusting sleeve, single-flow valve assembly, fixing sleeve and lower joint (Fig. 2).

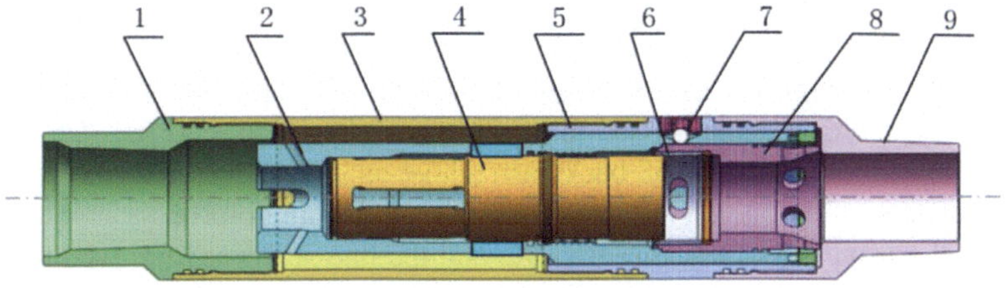

Fig. 2 Structure of acid sprayer with micro-electric control switch. (1, upper joint 2, mandrel barrel 3, outer barrel 4, adjusting sleeve 5, transition joint 6, coordination body of adjusting sleeve 7, single-flow valve assembly 8, fixing sleeve 9, lower joint)

Working principle: Using the electric control switch to control the acid sprayer, its control parts produce the linear motion along the axis. The positive and negative power supply's voltage in the ground control the forward and backward movement of the acid sprayer's piston so as to control the opening and closing of the acid sprayer's liquid flow channel. We can monitor the stroke of the rotating mechanism and determine the opening and closing state of the sliding sleeve through the current condition of micro-motor (loading/idling) in the ground.

4.2 Integrated Hydraulic Tubing Anchor of Acidizing and Water Injection

Integrated hydraulic tubing anchor of acidizing and water injection is mainly composed of upper joint, upper center pipe, shell joint, shell, lock block, upper cone, slip seat, fixing pin of the slip, slip body and lower center pipe (Fig. 3).

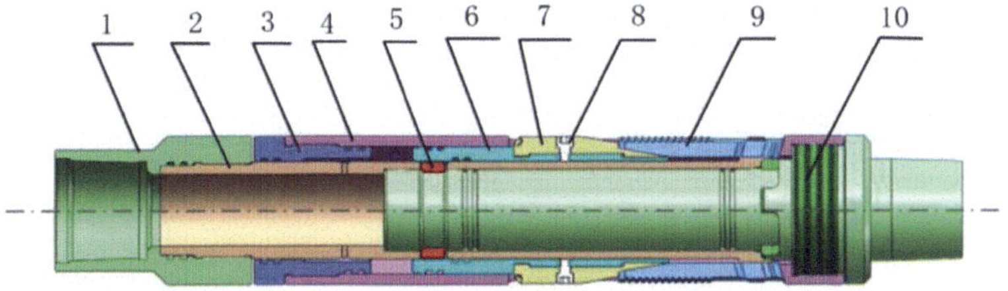

Fig. 3 Structure of the integrated hydraulic tubing anchor. (1, upper joint 2, upper center pipe 3, shell joint 4, shell 5, lock block 6, upper cone 7, slip seat 8, fixing pin of the slip 9, slip body 10, lower center pipe)

Working principle: Under the hydraulic pressure, the slip seat pushes the slip body to extend and anchor on the casing pipe. The tubing anchor is connected to the top packer to balance the string's jacking force generated during acidizing or water injection, and to solve the problem that the packer is easy to unseal due to the moving up of the string.

4.3 The Integrated Tool of Acid Injector and Packer Used to Realize Multiple Stage Setting and Releasing

This integrated tool of acid injector and packer mainly consists of acid injection mechanism, pressure guide part, unsealing mechanism of the packer, seal assembly, well cleaning part, setting mechanism of the packer (as shown in Fig. 4). Aiming at the problem of reset and sealing of backwash valve in the process of repeated well's backwash after stratified acidizing construction, annular micro-process embossing was designed at the front end of backwash valve, and the line seal was improved to be metal soft seal. The rear end is designed with sink and vulcanized rubber ring, and the optimized interference is 22%. Increase the diameter of the return spring by 15% and increase the spring stiffness. The packer is set through the oil pipe to make the hydraulic cylinder mechanism generate axial pressure to compress the rubber cylinder. Because the conventional single hydraulic cylinder generates axial force of 6–8t, a two-stage hydraulic cylinder mechanism is designed. Under the setting pressure, the axial force can reach more than 14t (as shown in Table 1), making the seat seal more reliably [4–6].

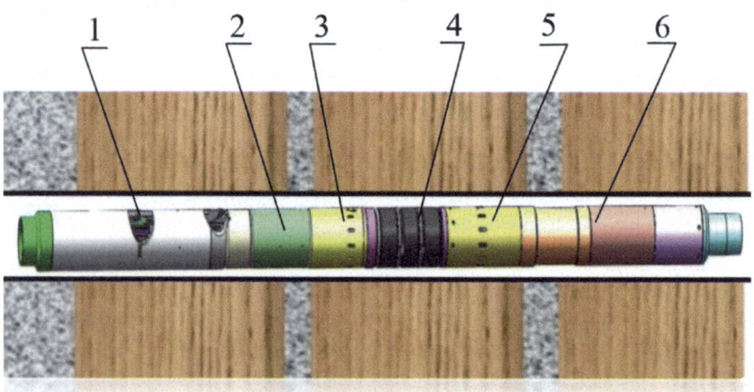

Fig. 4 The integrated tool of acid injector and packer. (1, acid injection mechanism 2, unsealing mechanism of the packer 3, pressure guide part 4, seal assembly 5, well cleaning part 6, setting mechanism of the packer)

Table 1 The corresponding axial load under the setting pressure

Axial setting pressure/MPa	Corresponding axial load/kN
4	32
5	40
6	50
8	66
10	82
12	100
15	124
17	142

5 Laboratory Test

5.1 Aim of Experiment

In the process of acidizing, the packer is subjected to the force of the squeezed acid, which may lead to loss of seal or abnormal unsealing. The setting test, pressure bearing test and unsealing test of the packer are carried out in the oil immersion laboratory to test whether the setting reliability, unsealing reliability and pressure bearing performance of the packer can meet the requirements of high-pressure construction.

5.2 Experimental Results

After be put into the test well, the packer is pressed and set, and the setting pressure of the device is 15 MPa, which meets the design requirements. The annulus was compressed to 50 MPa. The pressure was stabilized for 30 min, and the pressure drop was less than 0.3 MPa (Fig. 5).

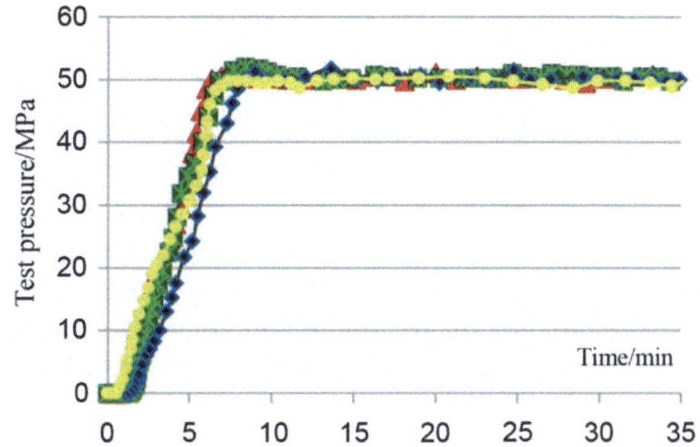

Fig. 5 Experimental curve

6 Field Application

In November 2017, field tests and applications were carried out in Gao 20 block and Yan 63 block of Yanling Oilfield. Gao 20 block and Yan 63 block of Yanling Oilfield belong to the sandstone reservoir developed by water injection of low permeability. The well depth is about 3000 m. At present, there are more than 90 injection Wells for stratified water injection with injection pressure of 20–25 MPa. It takes 2–6 months for the well to be closed before the operation, and the pressure inside the well cannot be released thoroughly by the pressure relief. The integrated process of acidizing and water injection was applied to meet the demand of geological reconstruction. Taking Yan63-52x well as an example, four-stage and four-stage separate injection was carried out. The injection thickness was 49 m, the tubing pressure was 12 MPa and the daily injection volume was 20 m^3. The 2# and 3# layers did not absorb water at all, which requiring stratified acidizing. The process string was four-stage and two-stage separate injection integrated string. The highest oil pressure was 30 MPa. After acidizing, water injection was directly restored. The testing and adjustment time was 3 days, the acidizing time was 2 days and the water injection time was 12 days. Five wells were tested in the field, and the success rate was 100%. Up to now, compared with before, the average daily increase of injection is 143%, reaching the longest term of injection allocation of 247 days, with a cumulative increase of 17,000 m^3 (as is shown in Table 2).

Table 2 Application chart

Well's number	Well pressure before construction/MPa	Daily water injection before construction/m^3	Well pressure after construction/MPa	Daily water injection after construction/m^3
Yan63-3	25.5	10	9	20
Yan63-155	29	20	4	35
Yan60-31X	31	15	9	60
Yan63-52X	14	20	3	50
Yan60-27	26	35	8.8	40

7 Conclusion

(1) The acid sprayer with micro-electric control switch and hydraulic tubing anchor solved the problems of the former acid sprayer being unable to close after opening, restricting the pass of the testing and controlling meter and the packer being easy to be unsealed due to the upward movement of the pipe string.

(2) A new integrated tool of acid sprayer and packer has been developed, which can bear a pressure of 50 Mpa and solve the problem that water injection packer is not suitable for high-pressure acidizing construction.
(3) the formation of acidizing operation and water injection technology realized that a set of pipe string system achieve the dual functions of stratified acidizing and stratified water injection, which can effectively reduce the damage to the stratum, significantly shorten the operating time. Its economic benefit is remarkable and the environmental protection degree is high which does not have the residual acid backflow.
(4) The integrated acidizing and water injection process has been successfully applied in the field, which not only meets the requirements of depressurizing and increasing injection, but also further achieves the effect of greatly shortening the operating cycle and protecting the reservoir.

References

1. Hongbo, P., Helin, P., Yaqiang, G., Songtao, Y., Xiaowei, W., Rongjie, C.: Hydraulic fracturing evaluation methods and analysis of objectives. Inner Mongolia Petrochem. Ind. **14** (2009)
2. Ying, L., Wencai, L., Shaojiang, C., Feng, L.: Development and application of Y241 high pressure squeeze packer. China Petrol. Mach. **09** (2009)
3. Xiangyang, R., Guohui, P., Yuanqiang, L.: Application of tube column technology of new lamination acidizing in Jiangsu oilfield. Fault-Block Oil Gas Field **04** (2008)
4. Zhang, R., Yan, G., Deng, L.: The first application of ball separate-acidization technology in Tazhong 4 Oilfield. J. Yangtze Univ. (Nat. Sci. Edn.) Sci. Eng. V. **01** (2009)
5. Chunde, L., Xinjun, X., Min, R.: Application of ball separate-acidization technology in limestone strata of Dagang oilfield. Oil Drill. Prod. Technol. **S1** (2002)
6. Haibo, Z., Yongxiang, T., Zongbo, C., Entong, J., Haimin, G.: Study on configuration and selection of acidizing string for separated layer. Well Test. **05** (2010)

Influence of Formation Temperature on Minimum Miscibility Pressure of CO_2-Crude Oil System

Jiaqi Wu, Xinwei Liao(✉), and Peng Dong

State Key Laboratory of Petroleum Resources and Prospecting of China, University of Petroleum (Beijing), Beijing, China
2518321568@qq.com, xinwei@cup.edu.cn, 18993726513@163.com

Abstract. The minimum miscibility pressure (MMP) is the key parameter to determine whether the crude oil is miscible with CO_2 in the reservoir. It is found that the formation temperature has a significant influence on the MMP, so it is very important to establish the CO_2 MMP prediction correlation with the formation temperature as the main parameter. This correlation is used to explore the influence law of formation temperature on the MMP deeply. Based on the slim tube experiment, 10 representative crude oil samples were selected to conduct experiment to determine the MMP of each sample at different test temperatures. The experimental results of each group were analyzed and the influence of formation temperature on the MMP was summarized, Based on that, a pure CO_2 MMP correlation based on Multiple-Linear-Regression modeling (MLR) technique mainly considering the influence of formation temperature is established. The prediction error of correlation is calculated to be 1.71%, which shows appreciable accuracy. Further, separately taking the results of the slim tube experiment and the MMP correlation as the starting point, discuss and analyze the effect of formation temperature on the MMP of CO_2-crude oil system. The analysis results show that the formation temperature has a significant positive correlation with the MMP of the CO_2-crude oil system within a certain temperature range. It is also found that the results of the experimental analysis of the slim tube experiment are basically consistent with the analysis of the results of the proposed CO_2 MMP correlation. Finally, the study of the

J. Lin (ed.), *Proceedings of the International Petroleum and Petrochemical Technology Conference 2019*, pp. 118–132, 2020.
https://doi.org/10.1007/978-981-15-0860-8_10

influence of formation temperature on the MMP of CO_2-crude oil system can greatly improve the accuracy and reliability of the MMP prediction of CO_2-crude oil system, and provides an important basis for the screening of CO_2 flooding blocks in reservoirs.

Keywords: CO_2 miscible displacement · MMP · Slim tube experiment · Prediction correlation · Formation temperature

1 Introduction

Miscible gas injection has developed into another imperative enhanced oil recovery (EOR) approach after polymer flooding, water flooding, and steam flooding. The minimum miscibility pressure (MMP) between the injected gas and the crude oil in the reservoir is an important parameter for the technology of gas injection and recovery of crude oil. The MMP refers to a minimum pressure at which the injected gas and the reservoir crude oil reach the mixed phase after multiple contact or one contact [1]. CO_2 is widely used due to its low cost, abundant reserves, and stability, which can reduce engineering implementation costs. The MMP of CO_2-crude oil can be used as the preliminary screening basis for the CO_2 flooding implementation block of the reservoir. Therefore, determining the MMP of CO_2 and known reservoir crude oil is an important part of CO_2 flooding technology.

There are some main methods for determining the MMP: traditional experimental method, the empirical formula method, and the numerical simulation method. The multiple mixing cell method [2] proposed later. The experimental method to determine the MMP is the most traditional method. The experimental methods include the thin tube experiment method, the bubble rising method, the vapor density method, and the interfacial tension method [3]. The rapid capillary test method is proposed later [4], and the slim tube experiment is an experimental method commonly used indoors and the accuracy of the measurement result is high. Different experimental methods have their own characteristics, and the experimental principles, operating procedures and methods of different experimental methods are also different. The shortcoming of the experimental method is that the experimental time cost is relatively high, however, the measured MMP results are relatively accurate. Considering the high time cost of the experiment, many scholars summarized and analyzed the experimental results of the MMP of different crude oil samples, and they obtained the MMP prediction correlation using mathematical fitting method. The typical prediction correlations for the MMP are Alston correlation, Yell-Met correlation, and GA correlation [5], and the typical prediction correlations to determine the MMP is relatively simple, but the limitation is that each type of correlation has certain applicable conditions. The scope of application is naturally narrow.

To establish the MMP correlation of the CO_2-crude system, the analysis of the influencing factors of the MMP is a very important task. The MMP of CO_2-crude oil system is affected by many factors, including crude oil composition, injection gas composition, formation temperature, and so on. Alston [6] pointed out the effect of reservoir temperature and C_{5+} composition in crude oil on the MMP. Yelling and Metcalfe [7] found through experiments that the composition of crude oil has no significant effect on the MMP, and the MMP has a positive and positive linear correlation with the formation temperature within a certain range. Shokir [8] proposed that the volatile light hydrocarbon component (C_2–C_4) in crude oil has a positive correlation effect on the MMP, and pointed out that the formation temperature has a significant influence on the MMP of the CO_2-crude system, but it has not been described in depth. Cronquist [9] considered that MMP is related to the formation temperature T, the relative molecular mass M_{C5+} of pentane and heavier fractions.

The author aims to study the influence of formation temperature on the MMP of CO_2-crude system, using the MMP correlation gained by multiple linear regression modeling technique [10] and the slim tube experiment to illustrate and analyze the influence of formation temperature.

2 Methods

2.1 Mechanisms of CO_2 Miscible Displacement

First of all, the author briefly introduces the mechanism of miscible displacement. The miscible displacement can be divided into one contact miscible displacement and multiple contact miscible displacement. The one-contact miscible process means that at a certain reservoir pressure and temperature. The injected gas is in contact with the crude oil and can be completely mixed. In this process, the oil-gas mixture is always maintained as a single phase. The one-contact miscible process does not produce a transition zone of the oil-gas displacement phase. The cost of using a single-contact miscible drive is usually high. The multiple-contact miscible displacement refers to the continuous contact between the gas and the formation crude oil at the beginning of the injection and the extraction between the oil and gas components. After the gas is in contact with the crude oil for a long time, most of the light and medium components in the crude oil are extracted by the gas. In this process, there is an obvious transition phase or transition zone between the oil and gas phases. When the reservoir pressure is low, this miscible phase transition band will become larger in comparison. The lowest miscible pressure of CO_2 and crude oil forming a miscible phase is higher. When the reservoir pressure is not very high, CO_2 and the formation crude oil have to be contacted several times to reach the miscible state.

The CO_2 miscible displacement can only happen after the CO_2 is miscible with the crude oil. When the average pressure of the reservoir reaches the MMP of the CO_2-crude system, the CO_2 miscible displacement can be formed. The formation temperature is generally high, and CO_2 must be in contact with the formation crude oil for multiple phases to achieve a miscible state. In this process, crude oil and injected CO_2 will be extracted from each other. The proportion of the light hydrocarbon component (C_2–C_4) in crude oil will continue to decrease. Besides, the gas-liquid interfacial tension between CO_2 and crude oil drops to zero during oil-gas miscible phase. The following Fig. 1 [11] shows the process of CO_2 and reservoir crude oil being contacted multiple times to form a miscible phase under formation temperature and formation pressure. Generally speaking, the mechanism of CO_2 miscible displacement to improve oil recovery mainly includes: reducing crude oil density, expanding crude oil, reducing oil and gas interfacial tension, evaporating and extracting crude oil components [12].

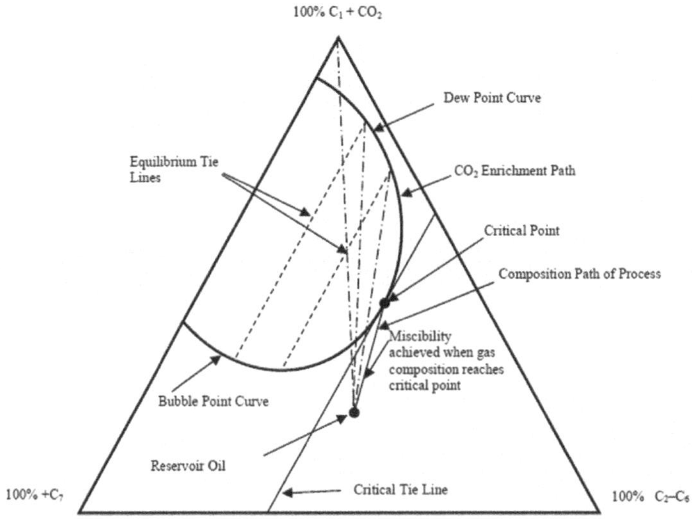

Fig. 1 Multiple-contact miscibility process

3 Experimental

The authors use the classical slim tube test method to determine the MMP of CO_2 and representative crude oil in 10 different regions. The 10 groups of crude oil were sampled and analyzed before the start of the experiment, and the composition and physical properties of the crude oil were also analyzed. At the same time, the crude oil components needed to be split and merged. After multiple fitting calculations, the crude oil components were divided into 7 sets of pseudo components: H_2S, CO_2, C_1 & N_2, C_2, C_{3+}, C_{7+} and C_{24+}.

3.1 Experiment Process

The experiment was carried out in a slim tube model filled with porous media. The basic parameters of the slim tube model are shown in Table 1. The injected gas is CO_2 with a purity of 99.95%. The experimental temperature and pressure were set according to the formation temperature and formation pressure of the crude oil sample, and the experimental pressure was set above the crude oil bubble point pressure. The experimental flow is shown in Fig. 2. Before performing the experiment, the whole experimental equipment should be vacuumed, and the vacuum should be maintained for a period of time, then the saturation process of the crude oil is carried out. The oil

Table 1 Slim tube model basic data

Parameters	Values
Maximum temperature	423.15 K
Highest pressure	55 MPa
Length	20 m
Inside diameter	3.86 mm
Outside diameter	6.35 mm
Gas permeability	5.43 μm^2

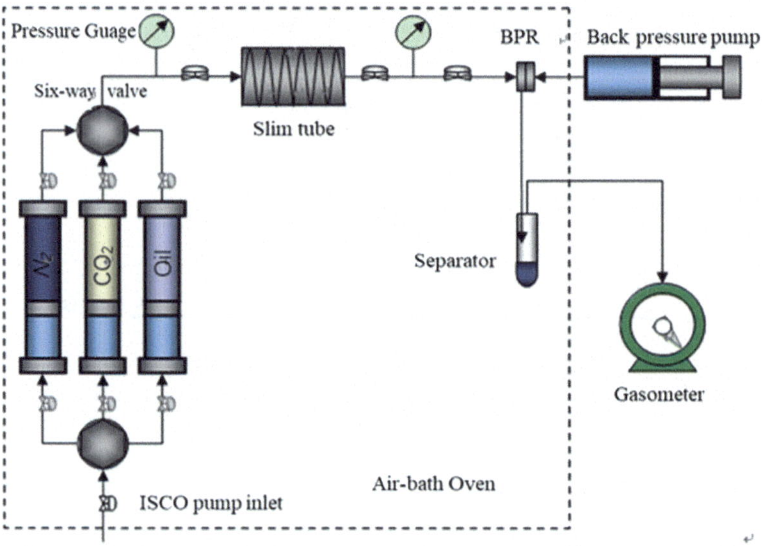

Fig. 2 Slim tube experiment procedure

saturation process is generally about 5–6 h, and the specific time is determined according to the viscosity of the crude oil sample.

The gas injection is carried out at the corresponding experimental pressure and experimental temperature. When the cumulative pump exceeds 1.20 times the pore volume (1.2PV) or no more oil is produced, the displacement is stopped. By setting 6 sets of experimental pressures to obtain the relationship between the oil displacement efficiency and the displacement pressure, there is a significant inflection point on the curve, and the pressure indicated by the abscissa corresponding to the point is the lowest miscible pressure, as shown in Fig. 3.

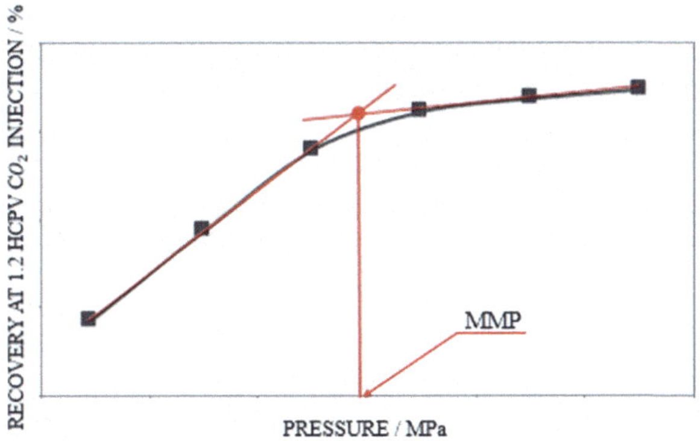

Fig. 3 Experimental image to determine MMP

Table 2 Results from slim tube experiment

Samples	T/k	MMP/MPa
Sample 1	345.71	18.0
Sample 2	356.10	22.3
Sample 3	378.78	25.3
Sample 4	358.54	20.2
Sample 5	327.59	17.3
Sample 6	365.99	25.1
Sample 7	327.59	17.3
Sample 8	360.40	27.8
Sample 9	307.74	8.7
Sample 10	363.10	32.1

3.2 Experimental Result

Table 2 shows the results of the slim tube experiment.

The MMP test results of the 10 groups of crude oil samples in Table 2 will be used as the error detection data of the MMP correlation of the CO_2-crude oil system, and the accuracy of the MMP correlation is verified by error analysis. At the same time, 10 sets of experimental data will also be used for in-depth study of the influence of formation temperature on the MMP of CO_2-crude oil system.

4 Prediction Correlation of CO_2 MMP

4.1 MMP Correlation of Pure CO_2 Miscible Flooding

This section aims to establish a MMP correlation for a pure CO_2 miscible displacement. The MMP correlation is obtained using multiple linear regression modeling technique, and the established correlation is composed of crude oil composition and formation temperature. Before establishing the correlation, the relevant parameters of crude oil samples under 100 pure CO_2 miscible displacement conditions were collected from actual engineering and related literature [13, 14]. At the same time, after multiple fitting calculations, it is determined that the crude oil components of each group are divided into H_2S, CO_2, C_1 & N_2, C_2, C_{3+}, C_{7+} and C_{24+}, and 7 pseudo-components are used as correlation parameters of crude oil components. The crude oil composition parameter

Table 3 Correlation coefficient between MMP and correlation parameters

	H_2S	CO_2	C_1 & N_2	C_2	C_{3+}	C_{7+}	C_{24+}	T	MMP
H_2S	1	−0.204	−0.187	−0.219	−0.021	0.124	−0.022	0.077	−0.005
CO_2	−0.204	1	0.259	0.378	−0.132	−0.453	−0.438	0.019	0.245
C_1 & N_2	−0.187	0.259	1	0.436	−0.247	−0.631	−0.338	−0.088	0.530
C_2	−0.219	0.378	0.436	1	−0.154	−0.542	−0.369	−0.081	0.144
C_{3+}	−0.021	−0.132	−0.247	−0.154	1	0.015	−0.328	−0.095	−0.378
C_{7+}	0.124	−0.453	−0.631	−0.542	0.015	1	−0.003	0.041	−0.455
C_{24+}	−0.022	−0.438	−0.338	−0.369	−0.328	−0.003	1	0.101	0.063
T	0.077	0.019	−0.088	−0.081	−0.095	0.041	0.101	1	0.573
MMP	−0.005	0.245	0.530	0.144	−0.378	−0.455	0.063	**0.573**	1

and the formation temperature T constitute parameters of the MMP correlation. Correlation analysis was carried out on the collected 100 groups of MMP data, and the correlation coefficients between the seven pseudo-components, the formation temperature T and the MMP of the CO_2-crude system was obtained. The correlation coefficients is shown in Table 3.

It can be seen from Table 3 that the correlation coefficient between the formation temperature and the MMP of pure CO_2 and crude oil is 0.573, which is the highest value among the parameters, indicating the effect of formation temperature on the MMP is stronger than other influencing factors. Based on the correlation coefficients between each correlation parameter and the MMP, a multiple linear regression modeling method is used to obtain a MMP correlation between pure CO_2 and crude oil. The correlation includes H_2S, CO_2, C_1 & N_2, C_2, C_{3+}, C_{7+} and C_{24+} and formation temperature T. The MMP correlation between pure CO_2 and crude oil is shown in following formula (1).

$$
\begin{aligned}
\mathrm{MMP} = & -139.142 - 1.370(\mathrm{H_2S}) + 1.352(\mathrm{CO_2}) \\
& + 1.726(\mathrm{C_2}\,\&\,\mathrm{N_2}) + 0.948(\mathrm{C_2}) - 0.998(\mathrm{C_{3+}}) \\
& - 1.045(\mathrm{C_{7+}}) + 1.308(\mathrm{C_{24+}}) + 0.123(\mathrm{T})
\end{aligned}
\tag{1}
$$

4.2 MMP Correlation Verification of Pure CO_2 Miscible Flooding

As mentioned above, the multiple linear regression modeling method was used to process the relevant parameter data of 100 groups of crude oil samples to obtain the MMP between pure CO_2 and crude oil. To verify the reliability of the MMP correlation, the formula (1) was used to calculate the MMP of the 10 sets of crude oil samples used in the previous slim tube experiment, and compared with the MMP of each group of crude oil measured by the slim tube experiment, the comparison results are shown in

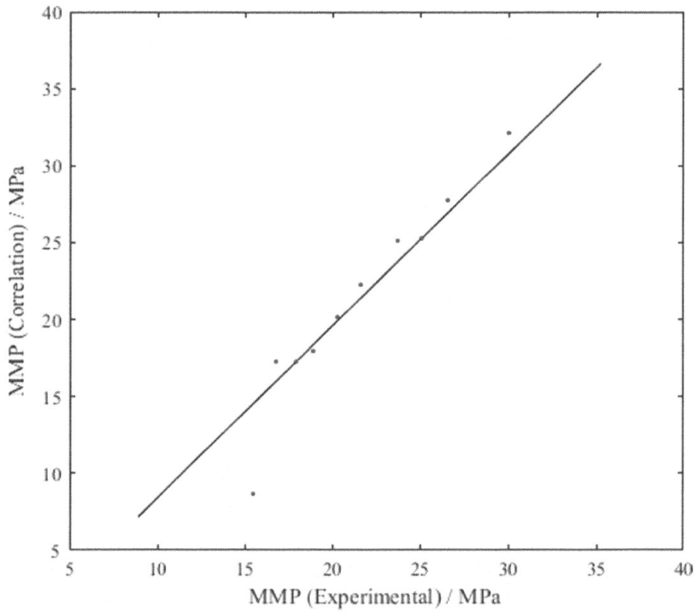

Fig. 4 Comparison of experimental data and correlation data for MMP

following Fig. 4. It can be seen from the Fig. 4 that the overall error between the correlation calculation result of the MMP of the 10 groups of crude oil samples and the results of the slim tube test is very low. After calculation, the relative error between the correlation calculation result and the experimental data is 0.93%. It shows that the MMP correlation has good accuracy. The results obtained by the MMP correlation can be used as the MMP prediction value of pure CO_2 and crude oil within a certain range, and the prediction result has certain reliability. The established MMP correlation will also be used for in-depth study of the effect of formation temperature on the MMP of the CO_2-crude system.

At the same time, in order to further verify the reliability of the MMP correlation, according to the crude oil composition of the crude oil sample and the formation temperature T, the MMP value of each group of crude oil samples is calculated by the established correlation, and the calculation results are compared with the MMP data

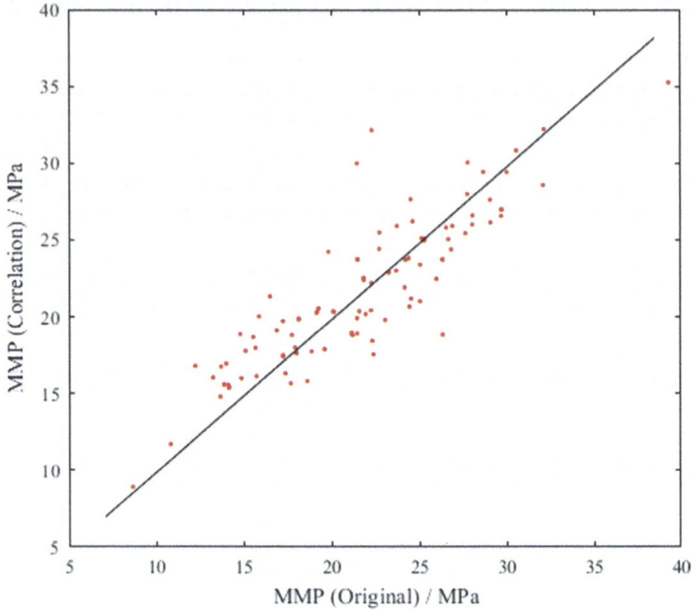

Fig. 5 Comparison between experimental data and correlation data for MMP

collected by the literature [13, 14]. By comparing the results to determine the accuracy of the correlation regression, the comparison results are shown in Fig. 5. It can be seen from the Fig. 5 that the calculated values of the MMP correlations of the 100 groups of crude oil samples are in good agreement. After calculation, the relative error between the correlation calculation result and the collected raw data values is 1.71%, and the error range is within the allowable range. Therefore, it can be determined that the MMP correlation obtained by multiple linear regression modeling method has certain reliability.

5 Results and Discussion

5.1 Overall Analysis of the Effects of Formation Temperature

There are many factors affecting the MMP of CO_2-crude oil system. The main factors are reservoir crude oil composition, formation temperature and composition of injected gas. This part aims to study the influence of formation temperature on the MMP of pure CO_2-crude oil system.

Figure 6 shows the relationship between the formation temperature of the collected 100 sets of crude oil samples and the corresponding CO_2 flooding MMP data. It can be seen that when the formation temperature is in the range of 300–420 K, due to the

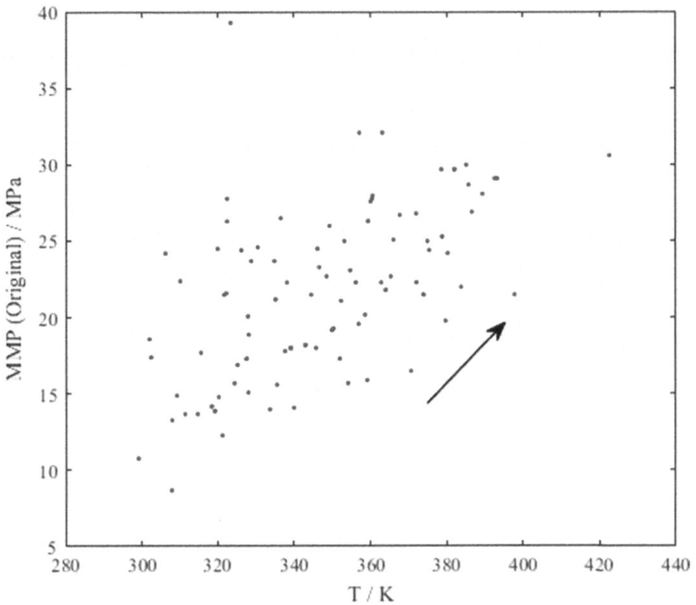

Fig. 6 MMP and temperature of raw data

difference in crude oil composition, with the increase of formation temperature, the trend of the MMP of CO_2-crude system is not the same.

However, the MMP of CO_2-crude system has a significant overall trend with respect to formation temperature.

5.2 Experimental Analysis of the Effects of Formation Temperature

Taking into account the time cost of the experiment, this section will simultaneously

Table 4 Experimental results of MMP at different experimental temperatures

	310 K	330 K	350 K	370 K	390 K
Sample 2, MMP (MPa)	14.7	15.2	19.9	23.8	26.4
Sample 4 MMP (MPa)	12.6	16.2	19.1	22.2	27.5
Sample 6 MMP (MPa)	17.7	20.6	23.3	25.5	22.6

conduct a slim tube experiment of CO_2 flooding in three sets of crude oil samples. The purity of CO_2 used in the experiment is 99.95%. Five different test temperatures were set to determine the MMP of the three groups of crude oil samples. And the variation of the MMP of the three groups of crude oil samples at different test temperatures was analyzed. Based on this, the influence of formation temperature on the MMP with CO_2-crude system was discussed.

The crude oil used in the slim tube experiment uses the crude oil sample 2, sample

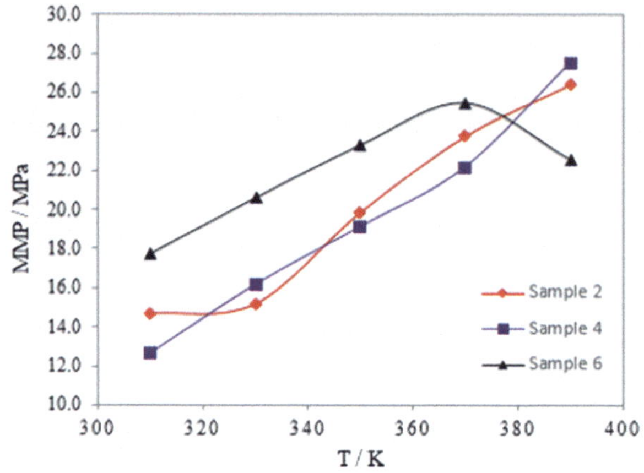

Fig. 7 Experimental results of MMP

4, and sample 6 mentioned in the third part. The crude oil sample information is shown in Table 1. Under the same conditions, the slim tube experiments of each sample set 5 different experimental temperatures to simulate different formation temperatures. The experimental temperatures set from low to high are: 310, 330, 350, 370, and 390 K. For details of the experimental setup and experimental procedure, we can see the Pilot Experiment section above. The results of the slim tube experiments of the three groups of crude oil samples in the five sets of experimental temperatures are shown in Table 4.

According to the results of the slim tube experiment of the three reservoir crude oil samples in Table 4, the variation curves of the MMP and the experimental temperature of the three sets of crude oil samples can be made in the same coordinate system, as shown in Fig. 7. It can be seen that in the range of the formation temperature of 310–390 K, the MMP of CO_2 and reservoir crude oil samples 2 and 4 increases with the formation temperature. The MMP of sample 6 increases with the increase of formation temperature in the range of formation temperature from 310 to 370 K and decreases after the formation temperature is higher than 310 K.

Under other conditions, as the formation temperature increases, the density and viscosity of carbon dioxide will become smaller, and the ability of carbon dioxide to extract light hydrocarbon components from crude oil will decrease. Carbon dioxide and reservoir crude oil need to be at a higher displacement pressure to reach the miscible state. Therefore, under normal conditions, the formation temperature increases, and the MMP of CO_2 and reservoir crude oil also increases.

As can be seen in Fig. 7, the MMP of CO_2 and sample 2 varies from 0.160 MPa/K in the formation temperature range of 310–390 K, the MMP of CO_2 and sample 4 varies from 0.179 MPa/K in the formation temperature range of 310–390 K, the MMP of CO_2 and sample 6 varies from 0.10 MPa/K in the formation temperature range of 310–370 K, after the formation temperature exceeds 370 K, the MMP decreases as the formation temperature increases, the MMP at a formation temperature of 390 K is close to the MMP at a formation temperature of 350 K. It can be seen that the MMP of the CO_2-crude system does not always increase as the formation temperature increases. In a certain temperature range, the MMP of the CO_2-crude system is positively correlated with the formation temperature. When this temperature range is exceeded, the MMP begins to decrease with the increase of formation temperature. At this time, the

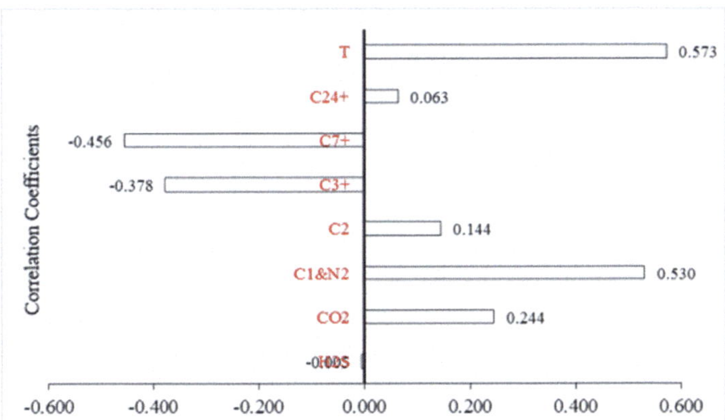

Fig. 8 Correlation coefficients between correlation parameters and MMP

influence of formation temperature on the MMP of CO_2-crude system has not played a leading role and there is no correlation or a small positive correlation between the formation temperature and the MMP.

5.3 Correlation Analysis of the Effects of Formation Temperature

In addition to using the above experimental methods to discuss the effect of formation temperature on the MMP of the CO_2-crude system, at the same time, the author also analyzed the correlation between the formation temperature of the 100 sets of crude oil samples collected and the MMP of the CO_2-crude system, and compared the analysis results with the results of the slim tube experiments. According to the correlation coefficients between the correlation parameters and the MMP obtained from the data of 100 sets of crude oil samples in Table 2, a correlation coefficient map is made, as shown in Fig. 8.

As can be seen from Fig. 8, the correlation coefficients between the five parameters (CO_2, C_1 & N_2, C_2, C_{24+} composition, formation temperature T) and the MMP are positive numbers, which represent the positive correlation between the two. The correlation coefficient between the formation temperature and the MMP is 0.573. This positive correlation coefficient indicates that the formation temperature has a positive correlation with the MMP, which is also consistent with the relationship coefficient of the formation temperature T in the correlation formula (1). Although the formation temperature T is the highest compared to the MMP, the correlation coefficient between the formation temperature and the MMP is not very high by value, and it can be seen that the correlation coefficient between the C_1 & N_2 component and the MMP in the crude oil of the reservoir also reaches 0.530. It can be concluded that the formation temperature is only one of the main influencing factors of the MMP, or that the formation temperature has a dominant influence on the MMP of the CO_2-crude system within a certain temperature range. There is a positive correlation between the two, this is also in line with the experimental results of the MMP and formation temperature of the three reservoir crude oil samples given in Fig. 6.

5.4 Comparison of Analysis Results of Experimental and MMP Correlation

The results of the thin tube experiments of the three sets of reservoir crude oil samples at different experimental temperatures and the MMP prediction correlation of the CO2-crude system returned from the 100 samples of the collected crude oil samples were used to analyze the influence of formation temperature on the MMP. The results of the two analyses are compared and the comparison results show that the results of the slim tube experiment and the MMP correlation are basically consistent.

The results of the slim tube experiment showed that the MMP of the crude oil sample 2 and sample 4 increased with the increase of the formation temperature, the MMP of the sample 6 increased with the formation temperature in the range of the formation temperature of 310–370 K, and then decreased. The experimental results show that the MMP of the CO_2-crude system increases only with the increase of the formation temperature within a certain formation temperature range. At this time, the influence of formation temperature on the MMP plays a leading role. Outside this range, the influence of formation temperature on the MMP of some reservoir crude oils will be lower than other influencing factors, so that the effect of formation temperature on the MMP is not obvious.

The correlation analysis of the MMP prediction of CO_2-crude oil system shows that the formation temperature T has a positive correlation with the MMP of the CO_2-crude system. The correlation coefficient between the two is 0.573, however, the correlation coefficient between the C_1 & N_2 component and the MMP in the crude oil of the reservoir also reached 0.530, indicating that the effect of formation temperature on the MMP has not yet reached a completely dominant position. The formation temperature has a positive correlation with the minimum miscible pressure within a certain range, when it exceeds this range, the effect of some crude oil components in the crude oil of the reservoir will be stronger than the effect of the formation temperature. This change cannot be reflected in the MMP correlation, which has been reflected in the results of the slim tube experiment described above.

6 Conclusions

Based on the theory of reservoir engineering, reservoir physics and enhanced oil recovery technology, the author conducted a series of slim tube experiments at different test temperatures for 10 groups of crude oil samples. Based on the analysis and processing results of the relevant parameters of 100 sets of crude oil samples collected, the MMP correlation of pure CO_2-crude oil system was established by multiple linear regression method (MLR). The effects of formation temperature on the MMP of CO_2-crude oil system were discussed and analyzed from the results of the slim tube experiment and the MMP correlation. The main conclusions are as follows:

1. The MMP correlation prediction error of pure CO_2-crude oil system established by multiple linear regression method (MLR) is 1.71%, and the correlation has good reliability and can be used to predict the MMP of pure CO_2 flooding in different reservoir temperatures at different formation temperatures.
2. The influence of formation temperature on the MMP of carbon dioxide-crude oil system is analyzed from the results of the slim tube experiment and the MMP correlation. The experimental results are basically consistent with the results of correlation analysis. The formation temperature only plays a dominant role in the MMP of the CO_2-crude oil system within a certain temperature range, and has a significant positive correlation with the MMP. Above this temperature range, the influence of formation temperature does not play a major role. At this time, there is no correlation or a small positive correlation between the formation temperature and the MMP.

Acknowledgements. This work was supported by National Science and Technology Major Project of China (2017ZX05030002-005) and National Science and Technology Major Project of China (2017ZX05009004-005). The authors sincerely thank the colleagues at State Key Laboratory of Petroleum Resources for their help.

References

1. Johnson, J.P, Pollin, J.S. (eds.) Measurement and correlation of CO_2 miscibility pressures. In SPE/DOE Enhanced Oil Recovery Symposium. Pub Place: Society of Petroleum Engineers; Year Published (1981)
2. Ahmadi, K., Johns, R.T.: Multiple-mixing-cell method for MMP calculations. SPE J. **16**(04), 733–742 (2011)
3. Al-Kadem, M.S, Al-Mashhad, A.S, Al-Dabbous, M.S, Sultan, A.S. (eds.) Integrating Peng Robinson EOS with association term for better minimum miscibility pressure estimation. In: SPE Kingdom of Saudi Arabia Annual Technical Symposium and Exhibition. Pub Place: Society of Petroleum Engineers; Year Published (2018)
4. Yan, W., Michelsen, M.L, Stenby, E.H. (eds.) Calculation of minimum miscibility pressure using fast slimtube simulation. In: SPE Improved Oil Recovery Symposium. Pub Place: Society of Petroleum Engineers; Year Published (2012)
5. Glaso, O.: Generalized minimum miscibility pressure correlation (includes associated papers 15845 and 16287). Soc. Petrol. Eng. J. **25**(06), 927–934 (1985)

6. Alston, R.B., Kokolis, G.P., James, C.F.: CO_2 minimum miscibility pressure: a correlation for impure CO_2 streams and live oil systems. Soc. Petrol. Eng. J. **25**(02), 268–274 (1985)
7. Yellig, W.F., Metcalfe, R.S.: Determination and prediction of CO_2 minimum miscibility pressures (includes associated paper 8876). J. Petrol. Technol. **32**(01), 160–168 (1980)
8. Shokir, E.M.E.: CO_2–oil minimum miscibility pressure model for impure and pure CO_2 streams. J. Petrol. Sci. Eng. **58**(1–2), 173–185 (2007)
9. Cronquist, C. (ed.) Effects of permeability variation and production rate on recovery from partial water drive gas reservoirs. In: Fall Meeting of the Society of Petroleum Engineers of AIME. Pub Place: Society of Petroleum Engineers; Year Published (1973)
10. Alomair, O., Iqbal, M. (eds.) CO_2 minimum miscible pressure (MMP) estimation using multiple linear regression (MLR) technique. In: SPE Saudi Arabia Section Technical Symposium and Exhibition. Pub Place: Society of Petroleum Engineers; Year Published (2014)
11. Menouar, H. (ed.) Discussion on carbon dioxide minimum miscibility pressure estimation: an experimental investigation. In: SPE Western Regional & AAPG Pacific Section Meeting 2013 Joint Technical Conference. Pub Place: Society of Petroleum Engineers; Year Published (2013)
12. Chen, P., Kalam, M.Z., Al Kindi, S.A., Shtepani, E., Thakur, G. (eds.) Maximize EOR potential by optimizing the miscible CO_2 injection parameters in carbonate reservoirs. In: Offshore Technology Conference. Pub Place: Offshore Technology Conference; Year Published (2018)
13. Harmon, R.A., Grigg, R.B.: Vapor-density measurement for estimating minimum miscibility pressure (includes associated papers 19118 and 19500). SPE Reservoir Eng. **3**(04), 1215–1220
14. Alcocer, C.F, Menzie, D.E. (eds.) Enhanced oil recovery by nitrogen injection: the effect of increased temperature and the amount of gas in solution. In: SPE Enhanced Oil Recovery Symposium. Pub Place: Society of Petroleum Engineers; Year Published (1984)

Research on Wellbore Pressure Distribution Model for Injection Wells During Foam Flooding

Ke Sun[1], Guoqing Han[1](✉), Zhiyong Zhu[1], Haiying Cheng[2], Mingjun Cai[2], Yang Zhang[2], Yishuang Wang[1], and Shuhao Wu[1]

[1] China University of Petroleum, Beijing, China
{skcbtb,hanguoqing}@163.com, {920555527,1589084280, 790749033}@qq.com
[2] Petrochina Dagang Oilfield, Tianjin, China
{chenghaiying,caimjun,zhangyang08}@petrochina.com.cn

Abstract. Foam flooding technology is an effective measure of EOR in oilfield development. In order to properly design the injection project, it is necessary to study the pressure distribution state of foam along the wellbore in injection wells. Based on the existing flow law research of foam fluids in the wellbore, this paper introduced temperature field calculation of heat transfer between the wellbore and foam in the theoretical study. Considering both rheological properties and PVT characteristics of foam fluids, the wellbore pressure distribution model was established. The iterative calculation method was used to calculate the whole wellbore pressure distribution of foam injection wells. Based on the new model, the effects of parameters such as foam flow rate of liquid and gas, injection pressure, injection temperature and tubing diameter on the wellbore pressure distribution of foam were analyzed. Meanwhile, the wellbore temperature and pressure distribution profiles of one case well were calculated, which meet the requirements of engineering application. In this paper, the whole wellbore pressure distribution model which introduced foam characteristics and temperature field calculation of heat transfer was established, and relevant influencing factors were analyzed. It has the important guiding significance for establishment of a reasonable production system for foam flooding oilfield development.

J. Lin (ed.), *Proceedings of the International Petroleum and Petrochemical Technology Conference 2019*, pp. 133–151, 2020.
https://doi.org/10.1007/978-981-15-0860-8_11

Keywords: Foam · Injection wells · Temperature field calculation · Wellbore pressure distribution model

1 Introduction

Foam is a multiphase disperse system in which gas is dispersed in liquid. In a foam fluid, the liquid is continuous phase and the gas is discontinuous phase. Foam belongs to a typical kind of non-Newtonian fluids, which has the advantages of low friction, low density, low filter loss, strong backflow capacity, strong ability to carry solid particles, and little damage to the reservoir. It is widely used in various engineering technologies such as drilling,completion, stimulation improvement, workover and EOR in the development of the oil and gas fields, which has achieved good application results [1, 2]. Foam flooding is a displacement method in which different kinds of gases are mixed with foam agents as displacement mediums [3]. As fluidity-control fluid in the displacement process, foam can improve the fluidity ratio, prevent water channeling or fingering during water flooding, expand the flooding area of water and gas, and enhance oil and gas recovery. Therefore, foam flooding technology has a great development prospect in oilfield development.

The application of foam fluids in the petroleum industry began in the 1950s. Scholars have done many experimental and theoretical studies on the rheological properties, stability, and hydraulic calculation of foam. Since 1960, Mitchell [4], Blauer [5], Princen [6], Kraynik [7] and others have successively studied the change law of foam structure, viscosity and shear stress with the change of foam quality. In addition, scholars such as Krug [8], Lord [9] and Sanghani and Ikoku [10] also made important contributions to the hydraulic calculation methods of foam fluids. Among them, Lord's method has been widely used in petroleum engineering, but it requires the determination of high-precision average friction coefficient.

The calculation of the wellbore pressure distribution of foam is the basis for optimal design of injection project during foam flooding, and its importance is obvious. The author investigated the research progress on calculation methods of the foam-filled wellbore pressure distribution and found that all research objects are stable foam with foam quality concentrated between 55 and 95%. Through analysis and summary, it is found that the current research has following three limitations: a. Most temperature field calculation models are too simple. Only the geothermal gradient is introduced into the calculation model, and the heat transfer law between the wellbore and foam is not considered. Since gas characteristics are extremely sensitive to temperature changes, there is a large calculation error existing. b. Due to the unclear rheological properties of foam fluids and the insufficient consideration of its PVT characteristics, various physical parameters during the calculation process are simplified, which ultimately leads to a large deviation between the calculation results and the actual field data. c. The calculation processes of most models are extremely complicated, which are not satisfied the requirements of engineering application and difficult to promote in the field. In the study of this paper, the author fully considered rheological properties and PVT characteristics of foam fluids, introduced the heat transfer law calculation between the wellbore and foam, perfected temperature field calculation, and established a new model of wellbore pressure distribution. The iterative calculation method was used to calculate the pressure distribution of the whole injection wellbore during foam flooding.

2 Wellbore Pressure Distribution Model

It is assumed that the research object is stable foam of one-dimensional single-phase flow ($0.55 \leq \Gamma \leq 0.95$), the flow process ignores the slip-off motion between gas and liquid, and the physical parameters of foam are the same on any flow cross section.

According to the momentum conservation principle, the following general wellbore pressure gradient equation can be established:

$$\left(\frac{dP}{dH}\right)_{total} = \left(\frac{dP}{dH}\right)_{gravity} + \left(\frac{dP}{dH}\right)_{friction} + \left(\frac{dP}{dH}\right)_{acceleration} \tag{1}$$

where $\left(\frac{dP}{dH}\right)_{total}$, $\left(\frac{dP}{dH}\right)_{gravity}$, $\left(\frac{dP}{dH}\right)_{friction}$ and $\left(\frac{dP}{dH}\right)_{acceleration}$ are total pressure gradient (MPa/m), gravity pressure gradient (MPa/m), friction pressure gradient (MPa/m) and acceleration pressure gradient (MPa/m) respectively.

For one-dimensional single-phase stable foam flow, the effect of the acceleration term can be ignored, there is:

$$\left(\frac{dP}{dH}\right)_{total} = \left(\frac{dP}{dH}\right)_{gravity} + \left(\frac{dP}{dH}\right)_{friction}. \tag{2}$$

If the direction of injection (from top to bottom) is positive, the expression is as follows:

$$\frac{dP}{dH} = \rho_F g \sin\theta - \lambda \frac{\rho_F}{D} \frac{v_F^2}{2} \tag{3}$$

where P is pressure (MPa), H is wellbore depth (m), ρ_F is foam density (kg/m^3), g is gravitational acceleration (m/s^2), θ is angle between the wellbore and the horizontal plane (°), λ is friction coefficient, D is tubing inner diameter (m), v_F is foam flow velocity (m/s).

Equation (3) is the mathematical model for calculating the wellbore pressure distribution of foam. Obviously, the key is to get the value of ρ_F, λ and v_F as precisely as possible. Since the compressibility of gas phase in foam is extremely strong, the three parameters can be affected by the changes of pressure and temperature along the wellbore to varying degrees. In other words, the parameters of foam are constantly changing, so how to select and evaluate the values of these three parameters becomes the key to accurately calculate the pressure distribution of the whole wellbore.

3 Calculation of Key Foam Parameters

3.1 Foam Density

For the convenience of calculation, it is assumed that the liquid phase in foam is incompressible, only the gas phase can be compressed, and the mass ratio of gas and liquid in foam is constant. According to the real gas state equation:

$$pV_g = \frac{m_g}{M_g} ZRT. \tag{4}$$

The following expressions can be given:

$$\rho_g = \rho_{gst}\left(\frac{Z_{st}}{Z}\right)\left(\frac{p}{p_{st}}\right)\left(\frac{T_{st}}{T}\right) \tag{5}$$

$$V_g = V_{gst}\left(\frac{Z}{Z_{st}}\right)\left(\frac{p_{st}}{p}\right)\left(\frac{T}{T_{st}}\right) \tag{6}$$

where p and p_{st} are gas pressure (MPa) and gas pressure under standard conditions (0.101 MPa), V_g and V_{gst} are gas volume (m^3) and gas volume under standard conditions (m^3), m_g is gas quality (kg), M_g is molar quality of gas (kg/kmol), Z is compressibility factor of gas, R is gas constant (8.31×10^{-3} MPa $\cdot$ m^3 $\cdot$ kmol^{-1} $\cdot$ K^{-1}), T and T_{st} are gas temperature (K) and gas temperature under standard conditions (273.15 K), ρ_g and ρ_{gst} are gas density (kg/m^3) and gas density under standard conditions (kg/m^3).

The general expression for foam density is as follows:

$$\rho_F = \frac{m_l + m_g}{V_l + V_g} = \frac{\rho_l \cdot V_l + \rho_g \cdot V_g}{V_l + V_g} \tag{7}$$

where m_l, V_l and ρ_l are foam liquid phase quality (kg), foam liquid phase volume (m^3) and foam liquid phase density (kg/m^3) respectively.

Substitute Eqs. (5) and (6) into Eq. (7) to obtain the expression of foam density:

$$\rho_F = \frac{\rho_l \cdot V_l + \rho_{gst} \cdot V_{gst}}{V_l + V_{gst}\left(\frac{Z}{Z_{st}}\right)\left(\frac{p_{st}}{p}\right)\left(\frac{T}{T_{st}}\right)}. \tag{8}$$

Set the gas-liquid ratio of foam under standard conditions: $R_{GLst} = \frac{V_{gst}}{V_l}$, then Eq. (8) can be changed into following expression:

$$\rho_F = \frac{\rho_l + \rho_{gst} \cdot R_{GLst}}{1 + R_{GLst}\left(\frac{Z}{Z_{st}}\right)\left(\frac{p_{st}}{p}\right)\left(\frac{T}{T_{st}}\right)}. \tag{9}$$

Foam quality is defined as the gas volume content in the foam under certain pressure and temperature conditions, that is:

$$\Gamma = \frac{V_g}{V_g + V_l} = \frac{1}{1 + \frac{1}{R_{GL}}}. \tag{10}$$

Set Γ_{st} as foam quality under the standard conditions, then:

$$\Gamma_{st} = \frac{V_{gst}}{V_{gst} + V_l} = \frac{1}{1 + \frac{1}{R_{GLst}}}. \tag{11}$$

When Eq. (6) and Eq. (11) are substituted into Eq. (10), there is:

$$\Gamma = \frac{\Gamma_{st}}{\Gamma_{st} + (1 - \Gamma_{st})\left(\frac{Z_{st}}{Z}\right)\left(\frac{p}{p_{st}}\right)\left(\frac{T_{st}}{T}\right)}. \tag{12}$$

Substitute Eq. (11) into Eq. (9), and the following expression of foam density can be obtained:

$$\rho_F = \frac{\rho_l(1 - \Gamma_{st}) + \rho_{gst} \cdot \Gamma_{st}}{1 - \Gamma_{st} + \Gamma_{st}\left(\frac{Z}{Z_{st}}\right)\left(\frac{p_{st}}{p}\right)\left(\frac{T}{T_{st}}\right)}. \tag{13}$$

Gas compressibility factor Z is a function of pressure and temperature, and the remaining parameters are known quantities. Therefore, given a set of pressure and temperature data, the foam density at this state can be calculated according to the above equation.

3.2 Friction Coefficient

Foam is a typical kind of non-Newtonian fluids, and the rheological model must be considered in hydraulic calculation. It is found that the most commonly used rheological models are Bingham model, Power-law model and Herschel-Buckley model. Most scholars regard foam as a compressible non-Newtonian power law fluid. The author also considers that the rheological model of foam is more consistent with power law model.

In 1981, Sanghani and Ikoku simulated and tested the rheological properties of foam in the wellbore and annulus. According to the regression analysis of experimental data, the relationship between foam quality Γ and generalized fluid consistency coefficient K' and fluidity index n was obtained: (Table 1).

Table 1 Relationship between Γ, K' and n

Γ	$K'(\text{Pa} \cdot \text{s}^n)$	n
$96\% < \Gamma \leq 98\%$	4.529	0.326
$92\% < \Gamma \leq 96\%$	5.880	0.290
$75\% < \Gamma \leq 92\%$	$34.330\Gamma - 20.732$	$0.7734 - 0.643\Gamma$
$55\% < \Gamma \leq 75\%$	$2.538 + 1.302\Gamma$	0.295

The relationship between generalized fluid consistency coefficient K' and consistency coefficient K is:

$$K' = K\left(\frac{3n+1}{4n}\right)^n. \tag{14}$$

For power-law fluids, generalized Reynolds number Re'_{PL} is introduced to determine the flow state of foam in the wellbore:

$$Re'_{PL} = \frac{8^{1-n}\rho_F D^n v_F^{2-n}}{K\left(\frac{3n+1}{4n}\right)^n}. \tag{15}$$

For power-law fluid flowing in a circular tube, the theory of Hanks [11] is used to determine the generalized critical Reynolds number Re'_{PLc}:

$$Re'_{PLc} = 6464\frac{n(n+2)^{\frac{n+2}{n+1}}}{(3n+1)^2}. \tag{16}$$

If $Re'_{PL} \leq Re'_{PLc}$, the flow state is laminar, or if $Re'_{PL} > Re'_{PLc}$, the flow state is turbulent.

If the foam flow is laminar, friction coefficient λ can be calculated according to the following formula:

$$\lambda = \frac{64}{Re'_{PL}}. \tag{17}$$

If the foam flow state is turbulent, the semi-empirical formula of Dodge and Metzner [12] can be used to calculate friction coefficient λ:

$$\frac{1}{\sqrt{f}} = \frac{4.0}{(n')^{0.75}} \lg\left[Re'_{PL} \cdot f^{\left(1-\frac{n'}{2}\right)}\right] - \frac{0.4}{(n')^{1.2}}. \tag{18}$$

Here, f is Fanning friction coefficient ($\lambda = 4f$), and n' is flow characteristic index, for power-law fluids, fluidity index $n = n'$.

The above expression is an implicit equation. For simplifying, Dodge and Meltzer concluded the Blasius-type empirical formula of the turbulent friction coefficient of power-law fluids from the relationship graph based on the experimental data:

$$f = \frac{a}{\left(\mathrm{Re}'_{PL}\right)^{b}}. \tag{19}$$

In the expression, a and b both are functions of the fluidity index n, The relationship between them is shown in the following (Table 2):

Table 2 Coefficient values in Blasius formula

n	a	b
0.2	0.2584	0.349
0.3	0.2470	0.325
0.4	0.2848	0.307
0.6	0.2960	0.281
0.8	0.3044	0.263
1.0	0.3116	0.250
1.4	0.3212	0.231
2.0	0.3304	0.213

In addition, the simplified modified Blasius formula can be also used to calculate:

$$f = \frac{a}{4\left(\mathrm{Re}'_{PL}\right)^{b}}. \tag{20}$$

Here, the value of a and b are calculated according to the following expressions:

$$a = 0.314n^{0.105} \tag{21}$$

$$b = 0.2495n^{-0.217}. \tag{22}$$

3.3 Foam Flow Velocity

During steady injection process of foam, the gas phase is compressed to change the flow rate of foam due to changes in temperature and pressure, which in turn causes changes in the pressure distribution along the wellbore. Thus, a reasonable calculation of the distribution of foam flow rate along the wellbore is a prerequisite for calculating pressure distribution accurately.

The flow velocity of stabilized foam in the wellbore is:

$$v_F = \frac{Q_g + Q_l}{\frac{\pi}{4}D^2} \tag{23}$$

where Q_g and Q_l are volume flow rate of gas phase and liquid phase in foam respectively (m^3/s).

For the gas phase portion in foam, the following expression can be obtained from Eq. (6):

$$Q_g = Q_{gst}\left(\frac{Z}{Z_{st}}\right)\left(\frac{p_{st}}{p}\right)\left(\frac{T}{T_{st}}\right) \tag{24}$$

where Q_{gst} is volume flow rate of gas phase in foam under standard conditions (m^3/s).

Substitute Eq. (24) into Eq. (23), then:

$$v_F = \frac{4\left[Q_{gst}\left(\frac{Z}{Z_{st}}\right)\left(\frac{p_{st}}{p}\right)\left(\frac{T}{T_{st}}\right) + Q_l\right]}{\pi D^2}. \tag{25}$$

According to Eq. (25), the foam flow velocity under any temperature and pressure conditions along the wellbore can be obtained.

4 Temperature Field Model

In fact, the heat transfer law of wellbore temperature field during foam injection process is complicated. The gas phase in foam is extremely sensitive to temperature changes. It is obvious that only the change of formation temperature cannot accurately reflect the heat transfer law in wellbore. Therefore, calculation of the wellbore temperature field should not only conform to actual heat transfer law, but also the influence of some secondary factors should be neglected to simplify calculation process. For the convenience of research, the author makes the following assumptions: a. The gas and liquid phases are in a thermodynamic equilibrium state. b. The heat transfer in the wellbore is steady-state, while the heat transfer in the surrounding strata is unsteady. c. The heat transfer between wellbore and formation is radial, without considering the heat transfer along the shaft direction, and ignoring the frictional heat generated during foam injection. d. The change of thermodynamic parameters along well-depth direction with the temperature is not considered.

In the injected wellbore, heat will be transferred in the following links due to temperature difference: Injected foam—tubing wall—casing annular space—casing wall—cement ring—formation, as shown below (Fig. 1).

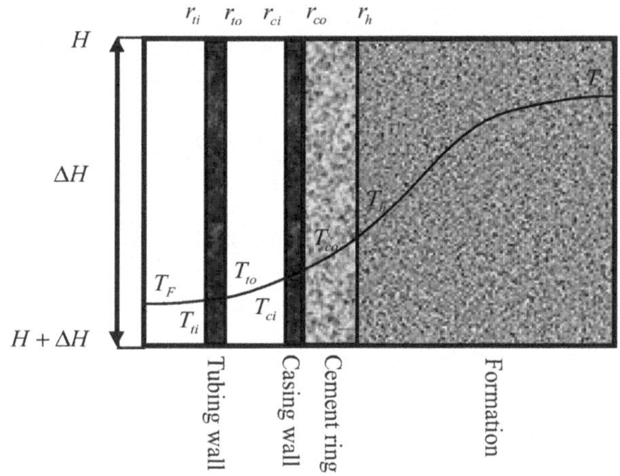

Fig. 1 Structure and temperature conduction diagram of injection well

In the process of injecting, heat flow Q_w exchanged radially between tubing and the wellbore near-well formation (to the interface between cement ring and formation), that is, thermal variation of the wellbore. The radial heat transfer process of any micro-element segment dH satisfies the steady-state heat transfer equation, then:

$$dQ_w = 2\pi r_{to} U_t (T_h - T_F) \cdot dH \tag{26}$$

where r_{to}, U_t, T_F and T_h are outer radius of tubing (m), total heat transfer coefficient based on the outer surface of tubing ($\text{kJ} \cdot \text{m}^{-2} \cdot \text{h}^{-1} \cdot {}^{\text{o}}\text{C}^{-1}$), foam temperature (°C)and temperature at the interface between cement ring and formation (°C).

In fact, total heat transfer coefficient U_t is a function of well depth and time. It is usually assumed that the thermodynamic parameters of the wellbore do not change with temperature, and the parameters under average temperature are used for approximate calculation.

The heat change of foam Q_F in this micro-segment through the heat transfer process is:

$$dQ_F = W_{inj} C_p \cdot dT_F \tag{27}$$

where W_{inj} and C_p are mass flow rate of injection foam (kg/h) and constant pressure specific volume of injection foam ($\text{kJ} \cdot \text{kg}^{-1} \cdot {}^{\text{o}}\text{C}^{-1}$).

The differential equation of formation radial unsteady heat transfer [13] is:

$$\frac{\partial^2 T_e}{\partial r^2} + \frac{1}{r}\frac{\partial T_e}{\partial r} = \frac{1}{\alpha}\frac{\partial T_e}{\partial t} \tag{28}$$

where T_e, α and t are formation temperature (°C), thermal diffusion coefficient of formation (m^2/s) and injection time (h).

In order to simplify the process of calculation, the semi-analytical method of Ramey [14] was adopted. By introducing non-dimensional transient heat transfer function $f(t_D)$, radial heat flow of interface between cement ring and formation Q_f is:

$$dQ_f = \frac{2\pi\lambda_e(T_e - T_h)\cdot dH}{f(t_D)}. \tag{29}$$

In the expression above, λ_e is thermal conductivity of formation $(\mathrm{kJ\cdot m^{-1}\cdot h^{-1}\cdot {}^{o}C^{-1}})$.

$f(t_D)$ can be calculated according to Hasan [15] formula:

$$t_D = \frac{\alpha t}{r_h^2} \tag{30}$$

$$f(t_D) = \begin{cases} 1.1281\sqrt{t_D}(1 - 0.3\sqrt{t_D}), (t_D \le 1.5) \\ (0.4063 + 0.5\ln t_D)\left(1 + \frac{0.6}{t_D}\right), (t_D > 1.5) \end{cases} \tag{31}$$

where t_D and r_h are dimensionless time and well-hole radius (m).

When injection time is long enough, according to the study by Sagar [16], $f(t_D)$ can be directly calculated by the following formula:

$$f(t_D) = f_{t_D} = -10.70866 r_h + 3.53 \tag{32}$$

According to the principle of energy conservation, there is:

$$Q_F = Q_w = Q_f. \tag{33}$$

Set ground temperature as T_G and ground temperature gradient as G_T, then:

$$T_e = T_G + G_T H \sin\theta. \tag{34}$$

In combination (26), (27), (29), (33) and (34), the following expressions can be obtained:

$$\frac{dT_F}{dH} = K(T_G + G_T H \sin\theta - T_F) \tag{35}$$

$$K = \frac{2\pi r_{to} U_t \lambda_e}{W_{inj} C_p[\lambda_e + r_{to} U_t f(t_D)]}. \tag{36}$$

According to the above expressions, it is clear to find that temperature gradient is mainly affected by three factors: fluid temperature, ground temperature and well depth. The coefficient K takes effects of fluid properties, wellbore structures, formation properties and injection parameters into account.

5 Case Study

According to the new model, corresponding calculation program was compiled. The calculation model of foam temperature was solved by classical fourth-order Runge-Kutta method, and the wellbore pressure distribution of foam was calculated by iterative method.

The basic parameters of one injection well W in a nitrogen foam flooding block was shown in the table below (Table 3).

Table 3 Basic parameters of injection well W

Project	Unit	Value
Well-type	/	Vertical well
Depth	m	2850
Well-hole diameter	m	0.24
Tubing inner diameter	m	0.0620
Casing outer diameter	m	0.1397
Surface temperature	°C	20
Geothermal gradient	°C/m	0.03
Injection temperature	°C	20
Liquid density of foam	kg/m^3	1100
Liquid flow rate of foam	m^3/h	8.34
Gas flow rate of foam	Nm^3/h	420
Injection pressure	MPa	2.5

5.1 Calculation of Foam Temperature Along the Wellbore

Assuming injection time was long enough, Sagar's correlation was used to calculate non-dimensional transient heat transfer function. Substitute into parameters related the wellbore and heat transfer process, and temperature distribution of foam along the wellbore was obtained as shown in Fig. 2.

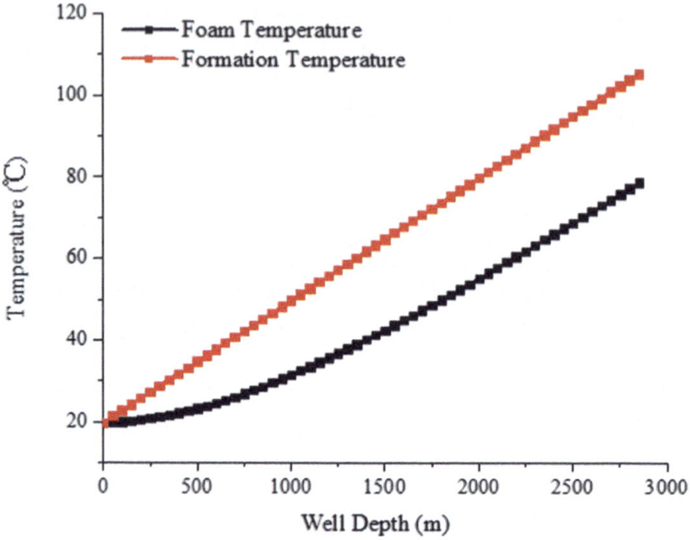

Fig. 2 Temperature distribution of foam along the wellbore

5.2 Calculation of Pressure Distribution Along the Wellbore

The pressure distribution of foam along the wellbore was calculated, as shown in Fig. 3.

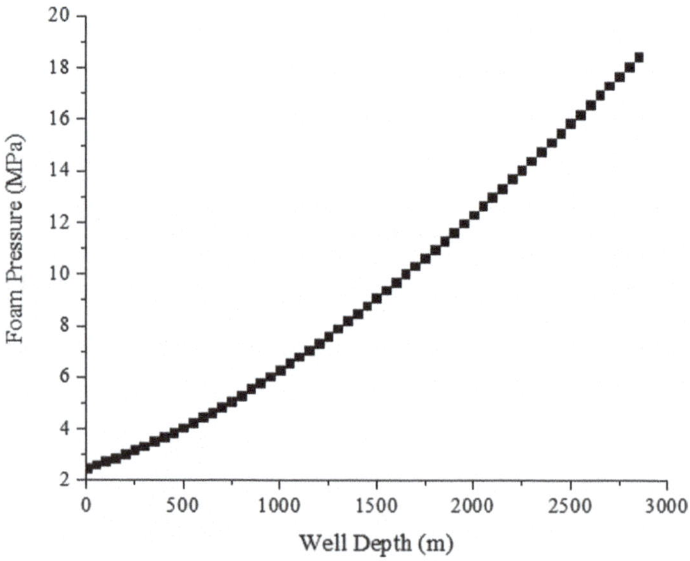

Fig. 3 Pressure distribution of foam along the wellbore

5.3 Comparison of Calculation Results

In order to verify the accuracy of the calculation results, commercial software WELLFLO was used to calculate under the same injection conditions. The following figure shows the pressure distribution curves obtained by these two models respectively (Fig. 4).

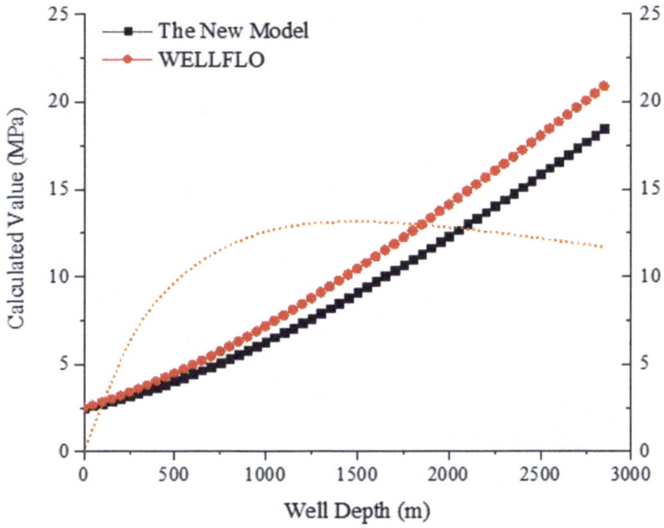

Fig. 4 Pressure distribution and relative error calculated by two models

As can be seen from the above figure, the results calculated by new model is closed to the results from WELLFLO, and the average relative error is 11.1%, which can meet the requirements of engineering calculation.

Because a simple linear geothermal gradient is used to calculate the temperature distribution of foam in WELLFLO, which makes the calculated value of foam temperature is larger than the actual situation, the results of wellbore pressure distribution of foam is also larger. Apparently, the analysis is consistent with the calculation results obtained in this paper.

6 Sensitivity Study

In fact, it is difficult to accurately calculate the wellbore pressure distribution of foam injection wells because the calculated parameters and conditions are numerous and the mechanism of flow changes is extremely complicated. Factors changed in the external environment such as formation temperature, formation pressure, well structure and physical properties of foam itself can affect the calculation of pressure distribution to varying degrees. Based on the established model and the calculated parameters of example application, four sensitivity factors including injection pressure, injection temperature, liquid flow rate & gas flow rate and tubing diameter were selected to analysis in this paper.

6.1 Injection Pressure

In this paper, six groups of different wellhead injection pressures of 1.5, 2.0, 2.5, 3.0, 3.5 and 4.0 MPa were selected to obtain the calculation results of wellbore pressure distribution of foam as shown in Fig. 5.

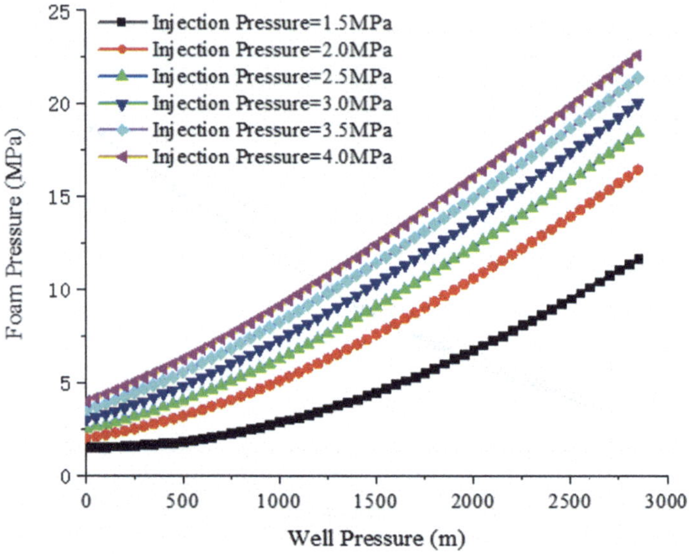

Fig. 5 Wellbore pressure distribution of foam under different injection pressures

As can be seen from the above figure, the total pressure difference of foam in the wellbore increases with the increase of injection pressure. When injection pressure is 1.5 MPa, the bottom hole pressure is 11.70 MPa, and the total pressure difference is 9.20 MPa. When injection pressure is increased to 4.0 MPa, the bottom hole pressure is 22.64 MPa and the total differential pressure is 20.14 MPa. Thus, the injection pressure increased by 2.5 MPa and the total differential pressure increased by 10.94 MPa. The reason for this phenomenon is that with the increasing of wellhead pressure, the gas volume in foam is compressed more and more, which leads to the increasing of total pressure difference. It can be inferred that in order to achieve a better injection effect, injection pressure should be increased as much as possible under the same conditions.

6.2 Injection Temperature

In this paper, five different foam injection temperatures of −10, 0, 10, 20 and 30 °C were selected. The wellbore pressure distribution of foam was calculated under the same conditions, and the calculation results as shown in the following figure were obtained (Fig. 6).

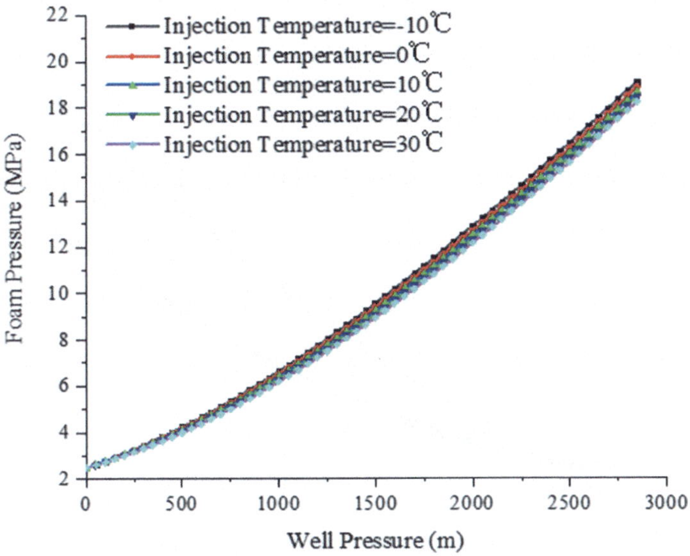

Fig. 6 Wellbore pressure distribution of foam under different injection temperatures

As can be seen from the above figure, the bottom hole pressure decreases with the increase of foam injection temperature. When the injection temperature is −10 °C, the bottom hole pressure is 19.10 MPa. When the injection temperature is 30 °C, the bottom hole pressure is 18.26 MPa. Compared with injection pressure, the change of injection temperature has less effect on the wellbore pressure distribution.

6.3 Liquid Flow Rate & Gas Flow Rate

Under other conditions unchanged, the flow rate of liquid in foam was only changed. Six groups of different liquid flow rates of 5.34 m^3/h, 6.34 m^3/h, 7.34 m^3/h, 8.34 m^3/h, 9.34 m^3/h and 10.34 m^3/h were selected to calculate the wellbore pressure distribution of foam respectively, and the calculation results as shown in the following figure were obtained (Fig. 7).

If only the flow rate of gas in foam was changed, six groups of different gas flow rates of 400 Nm^3/h, 410 Nm^3/h, 420 Nm^3/h, 430 Nm^3/h, 440 Nm^3/h and 450 Nm^3/h are selected to calculate the wellbore pressure distribution of foam respectively, and the calculation results as shown in the following figure were obtained (Fig. 8).

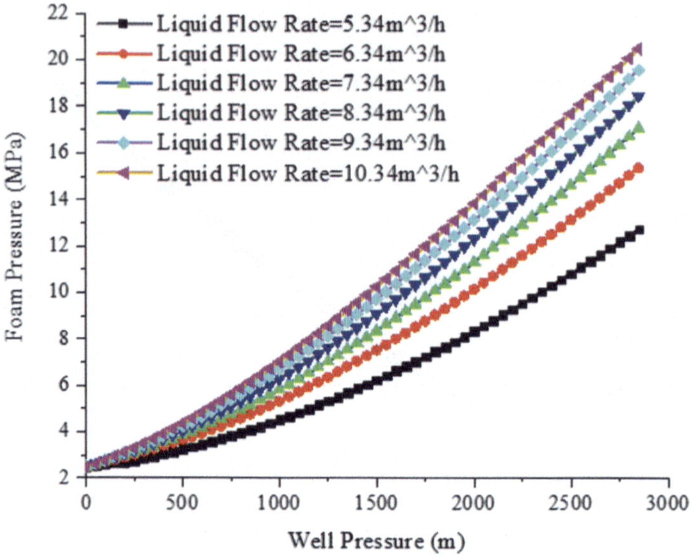

Fig. 7 Wellbore pressure distribution of foam under different liquid flow rates

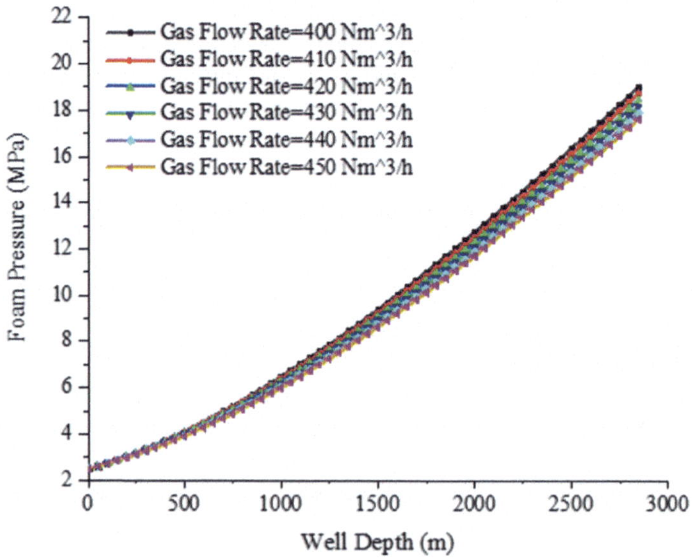

Fig. 8 Wellbore pressure distribution of foam under different gas flow rates

It can be seen from the above two figures that the bottom hole pressure increases with the increase of liquid flow rate and decreases with the increase of gas flow rate. When liquid flow rate increased from 5.34 to 10.34 m^3/h, the bottom hole pressure increased from 12.74 to 20.52 MPa. As gas flow rate increased from 400 to 450 Nm^3/h, the bottom pressure decreased from 19.02 to 17.64 MPa. Apparently, the change of liquid phase flow is more sensitive to the effect of foam pressure distribution in wellbore.

6.4 Tubing Diameter

Under the condition that the injection parameters are unchanged, six groups of different tubing inner diameters of 0.046, 0.054, 0.062, 0.070, 0.078 and 0.086 m were selected to calculate the wellbore pressure distribution of foam, as shown in Fig. 9.

It can be seen from the figure above that the bottom hole pressure increases with the increase of tubing diameter. The inner diameter of tubing increased from 0.046 to 0.086 m, and the bottom pressure increased from 14.72 to 20.39 MPa. In order to achieve a better injection effect, the tubing with a larger diameter should be selected in the early completion process.

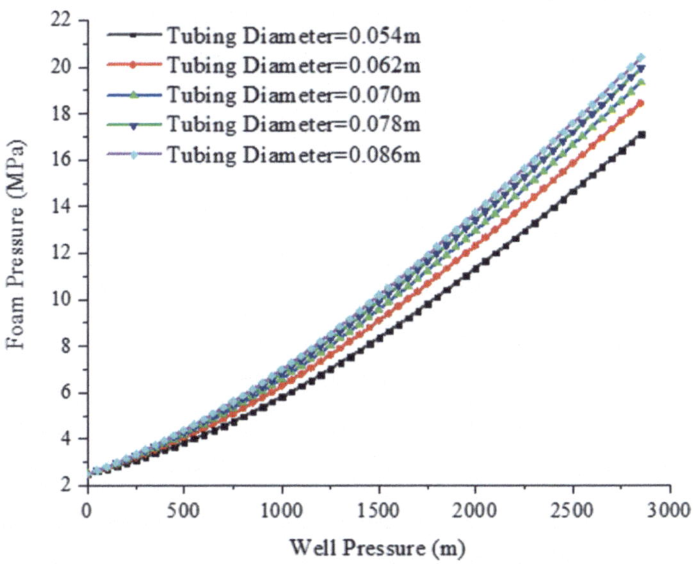

Fig. 9 Wellbore pressure distribution of foam under different tubing diameters

7 Conclusion

1. In this paper, the heat transfer calculation between the wellbore and foam was introduced in the study,and the temperature field calculation model was perfected. The rheological properties and PVT characteristics of foam fluids were considered,

and the calculation model of wellbore pressure distribution was established. Using iterative method, calculation of the whole wellbore pressure distribution during foam injection was realized.

2. Based on this model, the corresponding calculation program was compiled, and the wellbore temperature and pressure distribution of the case well were calculated. Compared with the calculation results of WELLFLO, the average relative error was 11.1%. The reason of error was analyzed and the new model meets the requirement of engineering calculation.
3. This paper analyzed sensitivity parameters and found that the total foam differential pressure of the wellbore increases with the increase of injection pressure. In addition, the bottom pressure decreases with the increase of injection temperature, increases with the increase of liquid flow rate, decreases with the increase of gas flow rate, and increases with the increase of tubing diameter.

Acknowledgements. This work was supported by the National Natural Science Foundation of China (Grant No. 51574256) and National Science & Technology Major Project (Grant No. 2017ZX05009-003).

References

1. Li, Z.: Application of Foam Fluid in Oil and Gas Exploitation, 1st edn, p. 1. Petroleum Industry Press (2010)
2. Ma, Wenying, Wang, Zhonghua, Cao, Pinlu: Application of foam fluid in oil field. Oilfield Chem. **27**(2), 221–226 (2010)
3. Wang, Qiwei: Development status and prospect of foam flooding. Oil Drill. Prod. Technol **35**(2), 94–97 (2013)
4. Mitchell, B.J.: Viscosity of Foam. University of Oklahoma, Norman (1960)
5. Blauer, R.E., Mitchell, B.J., Kohlhaas, C.A.: Determination of laminar, turbulent and transitional foam flow losses in pipes. In: SPE 4885 Presented at the 44th Annual California Regional Meeting of the Society of Petroleum Engineers of AIME, pp. 4–5. San Francisco, California, USA, April 1974
6. Princen, H.M.: Rheology of foams and highly concentrated emulsion. I. Elastic properties and yield stress of a cylindrical model system. J. Colloid Interface Sci. **91**(1), 160–175 (1983)
7. Kraynik, A.M.: Foam flows. Ann. Rev. Fluid Mech. **20**(20), 325–357 (1988)
8. Kurg, J.A., Mitchell, B.J.: Pressure needed for foam drilling. Oil Gas J. **70**(6), 61–64 (1972)
9. Lord, D.L.: Analysis of dynamic and static foam behavior. J. Petrol. Technol. **33**(1), 39–45 (1981)
10. Sanghani, V., Ikoku, C.U.: Rheology of foam and its implication in drilling and cleanout operations. J. Energy Resour. Technol. **105**(3), 362–371 (1983)
11. Hanks, R.W.: A theory of laminar flow stability. A. I. Ch. E. J. **15**(1), 25–37 (1969)
12. Dodge, D.W., Metzner, A.B.: Turbulent flow of non-newtonian systems. A. I. Ch. E. J. **5**(2), 189–204 (1959)

13. Chen, Lin, Sun, Lei, Peng, Caizhen: Stimulation of temperature field distribution in CO_2 injection wellbore. Fault Block Oil Gas Field **16**(6), 82–84 (2009)
14. Ramey Jr., H.J.: Wellbore heat transmission. J. Petrol. Technol. **14**(4), 427–435 (1962)
15. Hasan, A.R., Kabir, C.S.: Aspects of wellbore heat transfer during two-phase flow. SPE Prod. Facil. **9**(3), 211–216 (1994)
16. Sagar, R., Doty, D.R.: Predicting temperature profiles in a flowing well. SPE Prod. Eng. **6** (4), 441–448 (1991)

Study on Corrosion Prediction of CO_2 Flooding Well

Jie Zhang[1], Shu-hao Wu[2], Ming-jun Cai[1], Jie Yang[2], Min Liu[1], Zhi-yong Zhu[2], Ke Sun[2], and Guo-qing Han[2(✉)]

[1] Petrochina Dagang Oilfield, Tianjin, China
{dg_zhangjie, caimjun, dgyt_liumin}@petrochina.com.cn
[2] China University of Petroleum, Beijing, China
{790749033, 1397065707, 920555527}@qq.com, {skcbtb, hanguoqing}@163.com

Abstract. CO_2 flooding is one of the effective measures to enhance oil recovery. In a CO_2 flooding well, CO_2 will cause severe acid corrosion to the tube, which will lead to safety accidents such as tube failure. Therefore, study on the pipeline corrosion caused by CO_2 and pipe optimization is particularly important for safe production. This paper provided a study on the mechanism of CO_2 corrosion, and a well-known commercial software was implemented to Simulate and predict corrosion status. First, the variables, such as the temperature distribution, pressure and flow rate are simulated and analyzed. On this basis, the Waard model is used to predict the influence of these parameters on the corrosion rate and find the corrosion risk point. By comparing with the measured data, the method in this paper can accurately simulate the CO_2 corrosion rate of wellbore. Studies have shown that CO_2 corrosion is more likely to occur in the middle of the tube. The corrosion rate first increases with the increasing of temperature, but begins to decrease after the temperature reaches a certain value. The corrosion rate is positive proportional to the partial pressure and injection speed of CO_2, and is inversely related to the pipe diameter. The methods presented in this paper provide better guidance on production practices.

Keywords: Vertical wellbore · CO_2 corrosion · Sensitivity analysis · Corrosion risk point · Corrosion rate

---

J. Lin (ed.), *Proceedings of the International Petroleum and Petrochemical Technology Conference 2019*, pp. 152–165, 2020.
https://doi.org/10.1007/978-981-15-0860-8_12

1 Introduction

CO_2 flooding has been widely used in oil fields as a tertiary oil recovery technology, However, its accompanying problem of CO_2 corrosion has become increasingly prominent, which has become an important factor restricting the efficient production of oil fields. There are numerous fatal accidents caused by CO_2 corrosion occurring in the whole world. For example, the oil wells are abandoned due to corrosion in South Texas in the United States, Ekofisk oil field [1] in Norway, etc., which caused enormous economic losses. For example, although the AISI 410 stainless steel material is used in the CO_2 injection process of the Sacroc oil field [2] in the United States, the wellhead equipment is still severely corroded. The CO_2 content of Houston's North Personville oil and gas field [3] is as high as 2.5 (mol)%. The N80 casing is corroded and perforated in less than one year, and the corrosion rate is 5.6 mm/a. Although anti-corrosion measures which included internal coating anti-corrosion layer and corrosion inhibitor and regular pigging were adopted in the CAROLINE sour gas gathering pipeline [4] of Shell Canada Co., Ltd, The leakage accident still happened in 5-32S pipeline after six month in early 1994 in consequence of the wet Hydrogen sulfide and CO_2 contained in natural gas. The subsequent measurements confirmed that the corrosion rate of the acid gas to the steel pipe was as high as 30 mm/a. It can be seen that CO_2 corrosion has become a problem that cannot be ignored. Scholars at home and abroad have conducted a lot of research on CO_2 corrosion.

Zhao et al. [5] utilized the OLGA software that is widely used in the world to simulate and predict the corrosion condition. By comparing the results given by the actual test and software prediction, the software was proved that it can accurately simulate the predicted Carbon dioxide corrosion rate, and can give the prediction of temperature, pressure, flow rate, liquid holdup, flow pattern changes and the trend of pipeline corrosion rate along the pipeline. However, the downhole string is different from the gathering pipeline in temperature, temperature gradient, pressure and so on. Thus, this method cannot predict the corrosion of the downhole string.

Based on the NORSOK corrosion model [6], Ma et al. [7] considered the corrosion under high temperature conditions, and established a corrosion prediction model Which is more suitable for actual oilfield. The sensitivity analysis of Carbon dioxide corrosion was carried out for ambient temperature, CO_2 partial pressure, pH value and flow rate. Ossai [8] established a wellhead corrosion prediction model based on MATLAB, and studied the effects of temperature, CO_2 partial pressure, flow rate and pH on the wellhead corrosion rate. By using field data,multivariate linear regression analysis of these parameters shows that the results are in good agreement with the results of the De Waard model. Based on the mass transfer coefficient and the similarity between heat and mass transfer mechanism, Adsani et al. [9] considered the bubble flow, slug flow and intermittent flow, and established a prediction model of CO_2 corrosion rate in vertical multiphase flow. The model finally gives the relationship between the corrosion rate and the flow rate. However, they are not able to give a determination of the corrosion risk point, meanwhile, the influence of injection rate of CO_2 on corrosion rate cannot be studied by using the model, so it is difficult to give specific guidance to the site.

In this paper, the vertical wellbore is studied, and the corrosion mechanism is analyzed. The relationship between corrosion rate and temperature, CO_2 partial pressure, injection speed and pipe diameter are obtained. The corrosion risk point is found, which can be used for wellbore corrosion prediction and Anticorrosion.

2 Carbon Dioxide Corrosion Mechanism

Dry Carbon dioxide cannot corrode, but only when dissolved in water to form carbonic acid, which will lead to wellbore corrosion. At the same pH, it will corrode the wellbore more seriously than hydrochloric acid [10]. The basic chemical reaction formula is as follows:

$$CO_2 + H_2O \rightarrow H_2CO_3 \tag{1}$$

$$H_2CO_3 + Fe \rightarrow FeCO_3 + H_2 \tag{2}$$

Anode reaction

$$Fe \rightarrow Fe^{2+} + 2e^- \tag{3}$$

Cathodic reaction

$$H_2CO_3 \rightarrow H^+ + HCO_3{}^- \tag{4}$$

$$H^+ + 2e^- \rightarrow H_2 \uparrow \tag{5}$$

CO_2 is dissolved in water to form carbonic acid, and then carbonic acid is ionized in water to form H^+, which forms an acidic environment in the wellbore, thereby causing corrosion to the wellbore.

As can be seen from the reaction formula that H_2CO_3 reacts with iron under anaerobic conditions to form precipitate $FeCO_3$, and the corrosion product $FeCO_3$ adsorbed on the metal surface in the form of a film. When the $FeCO_3$ film is densely distributed on the metal surface, it will protect the wellbore and will reduce the corrosion of the wellbore to some extent. While the membrane is not densely distributed, it not only fails to provide effective protection, but also exacerbates local corrosion of the wellbore.

3 Formation Mechanism of Iron Carbonate Film

The formation of the iron carbonate film layer mainly depends on the deposition rate. The faster the deposition rate, the more corrosion products are formed, and the greater the thickness and density of the resulting film. Correspondingly, there are roughly two ways to destroy the iron carbonate film layer. One is that the high-speed fluid washes away the corrosion products from the surface of the substrate under the action of the flow rate; the other way is that the membrane is dissolved by the action of acid. Both will result in the product membrane being no longer intact, as well as being vacant, then it is no longer able to protect the wellbore, and even leads to further corrosion and rate of corrosion medium and steel.

Pots [11] hypothesized that the final state of formation of the iron carbonate film depends on the equilibrium between the rate at which the film is destroyed and the deposition rate of the iron carbonate film. The specific form of expression is as follows:

$$ST = \frac{R_{FeCO_3(s)}}{CR} \tag{6}$$

where $R_{FeCO_3(s)}$ is the deposition rate of iron carbonate, CR is the rate of corrosion.

When the deposition rate of the iron carbonate film layer is greater than or equal to the rate at which the iron carbonate film layer is destroyed, the formed film layer is sufficiently dense, even if it is very thin, it still can protect the wellbore well. When the deposition rate of the iron carbonate film layer is smaller than the rate at which the iron carbonate film layer is destroyed, the formed film layer is thick. However, since it is not dense enough, the formed pores cannot stop the further invasion of carbonic acid, so that the wellbore cannot be protected. As shown in Table 1.

Table 1 The relationship of ST value and protection efficiency

Value	Meaning
ST > 1	Protected
ST = 1	Protected
ST < 1	No protection

Sun and Nesic [12] supplemented this study by conducting experiments that found that in addition to these two cases, there was a middle ground. The research results are shown in Table 2.

Table 2 The relationship of ST value and corrosion type

Value	Meaning
ST $\leq$ 0.3	No protective film No local corrosion Severe uniform corrosion
0.3 < ST < 3	Only part of the protective film Local corrosion Uniform corrosion
ST $\geq$ 3	A dense protective film No localized corrosion No slight uniform corrosion

In addition, by combining mass balance equation, electrochemical reaction model and the dissolution equation of each iron, the thickness, porosity and protection effect of the formed iron carbonate film layer can be predicted directly [13].

The mass balance equation of iron carbonate affected by the damage of membrane is expressed as follows:

$$\frac{\partial \varepsilon}{\partial \mathrm{t}} = -\frac{M_{\mathrm{FeCO_3}(s)}}{\rho_{\mathrm{FeCO_3}(s)}} R_{\mathrm{FeCO_3}(s)} - CR\frac{\partial \varepsilon}{\partial x} \quad (7)$$

The electrochemical reaction model is as follows

$$i = \pm i_0 \cdot 10\frac{\pm E - E_{\mathrm{rev}}}{b} \cdot \prod_{s=1}^{ns} (1 - \theta_s) \quad (8)$$

i is the current density, i_0 is the exchange current density, Am^{-2}, E is the metal surface potential. Erev is a reversible potential, V. "±" means that if it is an anodic reaction, the sign on the right side of the equation is positive, and if it is a cathodic reaction, the sign on the right side of the equation is negative. b is the Tafel slope, s may be adsorbed corrosion inhibitor or corrosion product protective film. θ_s is the area fraction of the surface of the steel where the electrochemical reaction does not occur because the substance s covers the surface.

Uniform chemical reaction rate:

$$R_j = k_f \prod_{r=1}^{n_r} C_r - k_b \prod_{p=1}^{n_p} c_p \quad (9)$$

where k_f, k_b is the reaction rate constant of a chemical reaction forward and backward. C_r, C_p is the concentration of reactants and products.

4 De Waard Model

(1) Early De Waard [14] model

The De Waard model is a semi-empirical model, and it was first proposed by De Waard in 1975, and its mathematical expression is

$$\log V = 7.96 - \frac{2320}{t + 273} 0.00555t + 0.67 \log PCO_2 \tag{10}$$

where V is the corrosion rate, PCO_2 is the CO_2 partial pressure, MPa, t is the temperature, °C.

It can be seen from the formula that the corrosion rate only changes with temperature and CO_2 partial pressure, and the influencing factors are too single, which cannot reflect the real corrosion condition Accurately, and the error is too large compared with the Field data.

(2) Modified De Waard [15] model

De Waard proposed a revised prediction model in 1991. Based on the previous model, combined with the experiment and the actual situation on site, a series of correction factors were added to consider the effect of pH value, iron ion concentration, corrosion product film and other factors on corrosion rate.

The modified De Waard model is given by:

$$\log V_{corr} = 5.8 - \frac{1710}{t + 273} + 0.67 \log PCO_2 \tag{11}$$

where, t is the temperature, °C, V_{corr} is the corrosion rate before correction, mm/y.

$$V = V_{corr} \times F_a \times F_b \times \cdots \times F_i(1) \tag{12}$$

where V is the corrected corrosion rate, mm/y. $F_a \ldots F_i$ is the influence factor.

(3) De Waard [16] 95 model

The De Waard 95 model further considers the influence of the medium transport process and the liquid flow velocity on the corrosion rate. The mathematical expression of the corrosion rate prediction model is given by:

$$\frac{1}{V_{corr}} = \frac{1}{V_r} + \frac{1}{V_m} \tag{13}$$

$$\log V_r = 4.93 - \frac{1119}{t + 273} + 0.58 \log PCO_2 - 0.34(PH_{act} - PH_{CO_2}) \tag{14}$$

$$V_m = 2.45 \frac{U^{0.8}}{U^{0.2}} PCO_2 \tag{15}$$

$$PH_{CO_2} = 3.82 + 0.00384t - 0.5 \log PCO_2 \quad (16)$$

where V_{corr} is the corrosion rate, mm/a; V_r is the reaction rate, mm/a; V_m is the mass transfer rate, mm/a; t is the medium temperature, °C; P_{CO_2} is the partial pressure of CO_2, MPa; PH_{act} is the actual PH value; PH_{CO_2} is the pH of the CO_2 saturated solvent; U is the liquid phase flow velocity of the medium, m/s; d is the diameter of the pipe, m.

5 Analysis and Evaluation of Corrosion Conditions of CO_2 Flooding Injection Wells

The injection well model based on the actual situation is shown in Fig. 1.

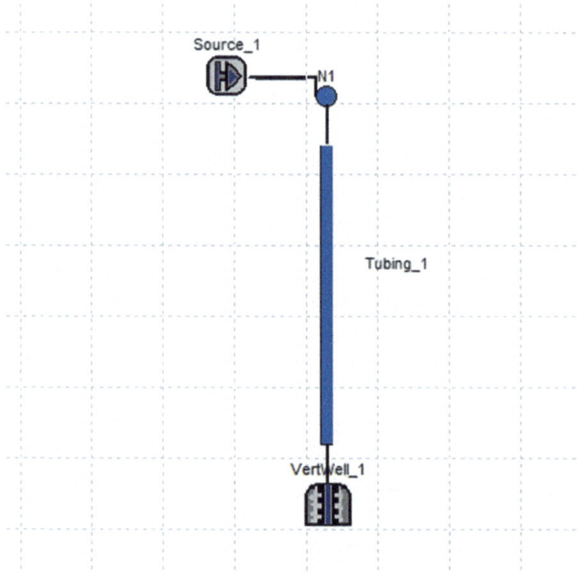

Fig. 1 Vertical wellbore model diagram

The average formation pressure is 39.03 MPa, the pressure coefficient is 1.0947, the average formation temperature is 123 °C, and the geothermal gradient is 3.39 °C/100 m, which indicates that it is a normal temperature and pressure system. The pure liquid CO_2 injected, and the injection mass flow rate is 30 kg/h. The minimum injection is 12 MPa to reach the minimum miscible pressure of 38 MPa. If the formation pressure is higher than the minimum miscible pressure, the mixed phase is formed. with the disappearance of interfacial tension, thus it is easier to displace [17]. The corrosion model uses a classic, widely used semi-empirical prediction model, namely De Waard 95 model. The specific data is shown in Table 3.

Table 3 Basic data

Parameter	Value
Wellhead temperature	20 °C
Wellhead pressure	12 MPa
Well depth	3830 m
Formation temperature	123 °C
Formation pressure	39.03 MPa
Mass flow	20 kg/h
Moisture content	0.02%
Pipe diameter	76.2 mm

6 Model Verification

Comparing the simulation results of the temperature and pressure distribution along the wellbore based on the software with the actual operation and the actual results of the field, the software can simulate the CO_2 corrosion of the vertical wellbore (Figs. 2 and 3).

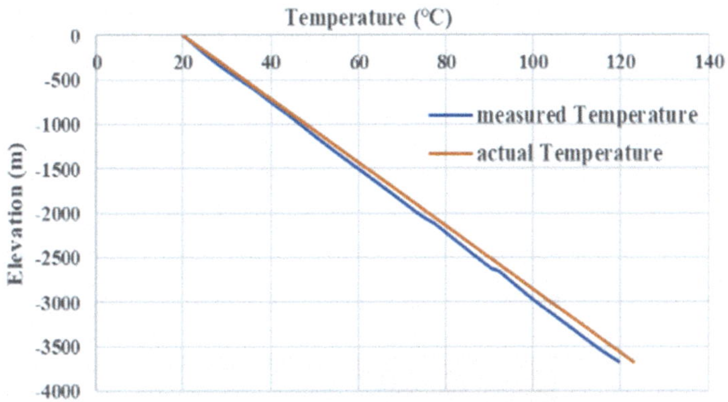

Fig. 2 Calculated temperature distribution versus the actual temperature distribution

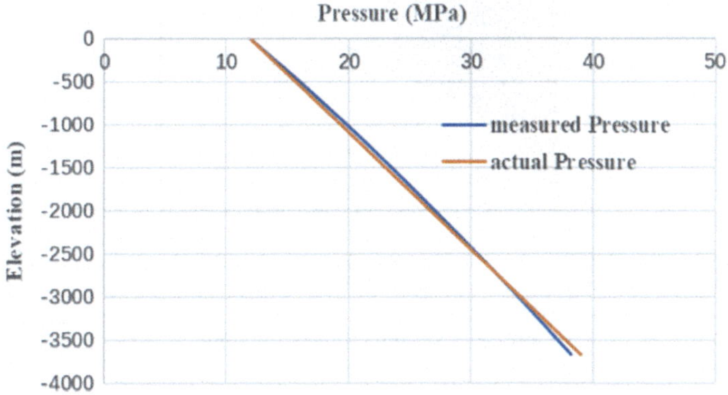

Fig. 3 Calculated pressure distribution versus actual pressure distribution

7 Analysis of Corrosion Law of CO_2 Flooding Well

1. Corrosion risk point analysis

As can be seen from Fig. 4, the highest corrosion rate is neither located at the bottom of the tubing nor at the wellhead. It is in the middle of the tubing at a distance from the wellhead. This position corresponding to the highest corrosion rate is called the corrosion risk point. At the point of corrosion risk, the corresponding temperature is 40 °C and the pressure is 18 MPa.

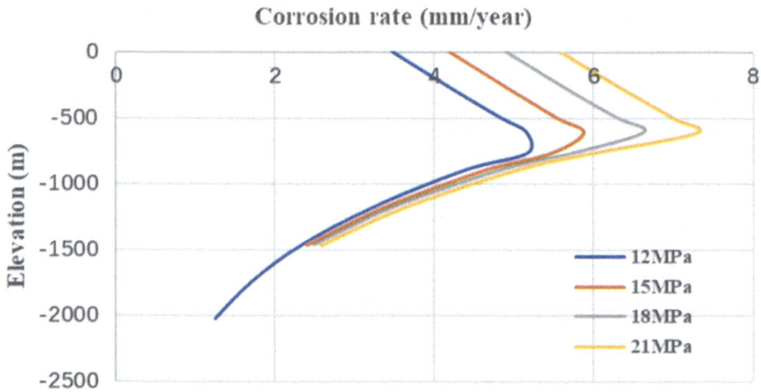

Fig. 4 Relationship between corrosion rate and wellbore depth

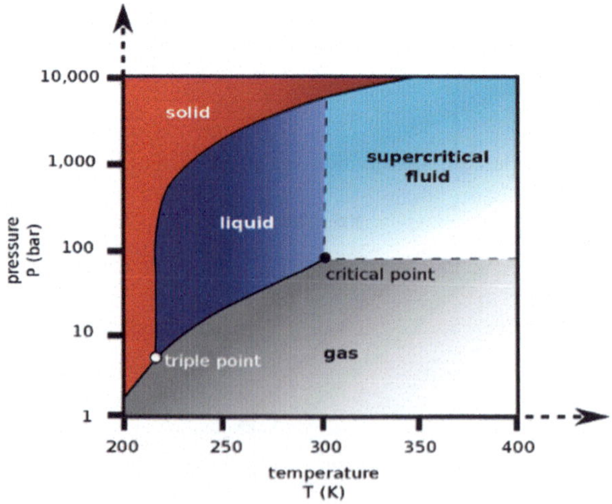

Fig. 5 CO_2 three-phase diagram

It can be seen from Fig. 5 that carbon dioxide will change to a supercritical state when the temperature is higher than the critical temperature (31.26 °C) and the pressure is higher than the critical pressure (72.9 atm). Currently, the gas-liquid two-phase interface disappears and the density is close to that of the liquid. The viscosity is close to that of gas, and the diffusion coefficient is 100 times that of liquid, which has considerable solubility. It no longer has corrosive properties and even can achieve corrosion inhibition [18], which reduces the corrosion rate [18].

2. Effect of temperature on corrosion rate

It can be seen from Fig. 6 that in a certain temperature range, the corrosion rate of the wellbore increases with the increase of temperature gradually, reaching a peak at 40 °C, and then the corrosion rate decreases remarkably with further increase of temperature. Therefore, the effect of temperature on CO_2 corrosion can be described as follows: the corrosion rate increases first and then decreases gradually.

The analysis shows that in a certain temperature range, the reaction corrosion rate and mass transfer corrosion rate are increased with the increase of temperature, and the corrosion rate is also increased. However, when the temperature is higher than a certain value, the deposition rate is higher than the corrosion rate, and the deposition rate is dominant. The formed product film is complete and very dense, which has a good hindrance to the further corrosion invasion. Since it has a good protective effect on wellbore, the corrosion rate decreases gradually.

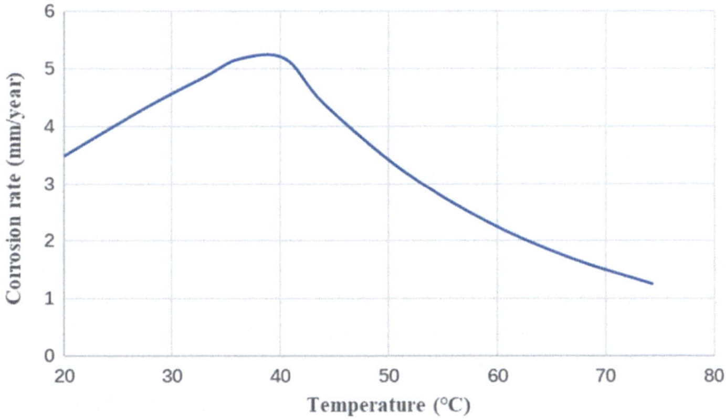

Fig. 6 The relationship of corrosion rate and temperature

3. Effect of CO_2 partial pressure on corrosion rate

The sensitivity analysis of CO_2 partial pressure is carried out, and the partial pressure range is 10–25 MPa. Figure 7 shows that with the increase of CO_2 partial pressure, the corrosion rate increases. This is because the higher the partial pressure, the higher the ratio of CO_2 dissolved in aqueous solution, and the lower the PH value of aqueous solution, thus the corrosion is aggravated. In the low temperature section (i.e., near the well head), the CO_2 partial pressure has a great influence on the corrosion rate, but in the high temperature section (i.e., near the bottom hole), the CO_2 partial pressure has little effect on the corrosion rate (Fig. 8).

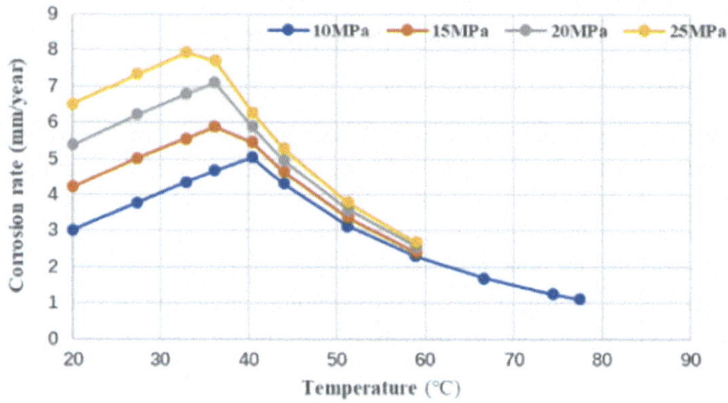

Fig. 7 The relationship between corrosion rate and CO_2 partial pressure

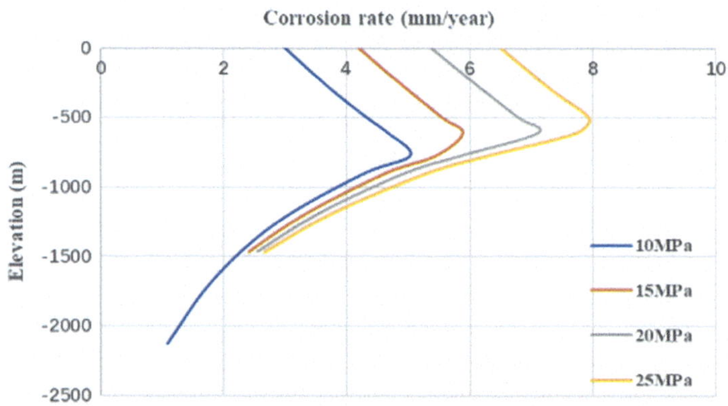

Fig. 8 The relationship between CO_2 partial pressure and corrosion rate at different well depths

4. Injection speed

Sensitivity analysis of injection speed was performed, and the interval of injection speed is 20–50 kg/h. It can be seen from Fig. 9 that the corrosion rate increases as the injection speed increases.

The analysis shows that the effect of the injection rate on the corrosion rate is essentially the effect of the flow rate on the corrosion rate, but studying the injection rate is more conducive to the guidance of the field.

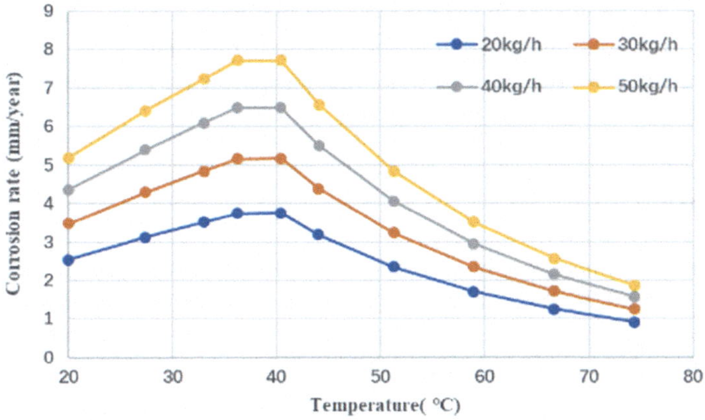

Fig. 9 The relationship of corrosion rate and injection speed

The higher the injection speed, the higher the flow rate in the wellbore. The high-speed fluid scouring will hinder the formation of the corrosion product film, and sometimes even scrape off the corrosion product film, making it incomplete and vacant. As a result, the substrate is completely exposed to acid, which leads to the increase of corrosion rate.

5. Pipe diameter

Sensitivity analysis of the pipe diameter was conducted, and the pipe diameter range is 40–100 mm. As can be seen from Fig. 10, the smaller the pipe diameter, the more serious the corrosion.

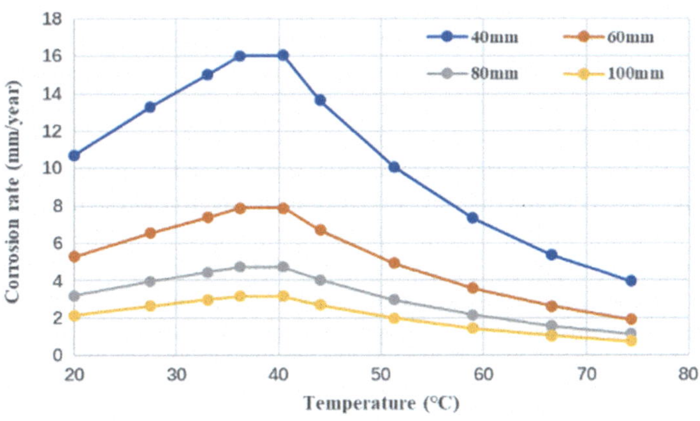

Fig. 10 Relationship of corrosion rate and pipe diameter

Analysis of the De Waard 95 model indicates that the corrosion rate is composed of reactive corrosion rate and mass transfer corrosion rate. The pipe diameter mainly affects the mass transfer rate of corrosion. The smaller the pipe diameter, the larger the mass transfer rate, which leads to the greater corrosion rate.

8 Conclusion

As CO_2 corrosion is increasingly valued by oilfield workers and researchers, this paper establishes a vertical wellbore model with mature commercial software. Based on the De Waard model, the effects of different parameters on CO_2 corrosion rate are studied. The research results are summarized as follows.

1. The corrosion rate is mainly affected by temperature, CO_2 partial pressure, flow rate, and pipe diameter. The corrosion rate first increases with the increase of temperature, but when the temperature is higher than a certain value, the corrosion rate decreases. And the corrosion rate is proportional to the CO_2 partial pressure and injection speed, and inversely proportional to the tube diameter.
2. The effect of temperature on the corrosion rate is mainly achieved by affecting the balance between the corrosion rate of the iron carbonate film layer and the deposition rate. The effect of CO_2 partial pressure on the corrosion rate is mainly achieved by affecting the solubility of CO_2 in water, i.e., the pH value, and the influence on corrosion rate at high temperature stage and low temperature stage is inconsistent. The effect of the injection rate on the corrosion rate is mainly achieved by the scouring of the protective film with high-speed fluid. The diameter of the pipe mainly affects the mass transfer rate of corrosion.
3. The distribution of corrosion rate along the wellbore is not gradually not gradually increasing or decreasing, but there is an inflection point in the middle of the wellbore, which has the highest corrosion rate and is called the corrosion risk point. not gradually increasing or decreasing, but there is an inflection point in the middle of the wellbore, which has the highest corrosion rate and is called the corrosion risk point. The prediction method of corrosion risk point is given in this paper. In addition, the reason that the corrosion risk point is located in the middle of the tubing is analyzed. Because this point is the critical point of temperature and pressure, CO_2 becomes supercritical carbon dioxide and its properties change.
4. In this paper, the mechanism of CO_2 corrosion and the formation mechanism of iron carbonate film are analyzed. It is found that the damage of CO_2 corrosion to the wellbore depends largely on whether a dense and complete vacancy-free protective corrosion product film is formed. If it is not formed, the corrosion will intensify. Otherwise, the corrosion rate is suppressed, and the wellbore can be effectively protected from corrosion.

Acknowledgements. This work was supported by the National Natural Science Foundation of China (Grant No. 51574256) and National Science and Technology Major Project (Grant No. 2016ZX05072).

References

1. Houghton, C.J., Westermark, R.V.: North Sea Downhole Corrosion: Identifying the Problem; Implementing the Solutions. Society of Petroleum Engineers (1983). https://doi.org/10.2118/11669-pa
2. Newton, W.H.: SACROC CO/sub 2/project—corrosion problems and solutions. United States (1984)
3. Gunaltun, Y., Kopliku, A.: Field Data Collection, Evaluation and Use for Corrosivity Prediction and Validation of Models—Part I: Collection of Reliable Field Data for Validation of Prediction Models. NACE International, Houston (2006)
4. Bich, N.N., Goerz, K.: Caroline Pipeline Failure: Findings on Corrosion Mechanisms in Wet Sour Gas Systems Containing Significant Co. NACE International, Houston (1996)
5. Zhao, X., Wang, Z., Dong, J.: CO_2 corrosion simulation based on OLGA. In: 2013 China International Pipeline Conference and China Academic Conference on Pipeline and Tank Corrosion and Protection (2013)
6. Olsen, S.: CO_2 Corrosion Prediction by Use of the NORSOK M-506 Model—Guidelines and Limitations. NACE International, Houston (2003)
7. Ma, D., Wang, B., Wu, S.: Prediction of CO_2 corrosion in oil and gas production process. J. Southwest Pet. Univ. (J. Nat. Sci.) **32**(3), 000137–000140 (2010)
8. Ossai, C.I.: Predictive modelling of wellhead corrosion due to operating conditions: a field data approach. ISRN Corros. **2012**(Article ID 237025), 8 p (2012)
9. Adsani, E., Shirazi, S.A., Shadley, J., Rybicki, E.: CO_2 Corrosion Rate Prediction in Vertical Multiphase Flow. Society of Petroleum Engineers (2005). https://doi.org/10.2118/95185-ms
10. Song, F., Kirk, D.W., Graydon, J.W., Cormack, D.E.: Prediction for CO_2 Corrosion of Active Steel under a Precipitate. NACE International, Houston (2004)
11. Pots, B.F.M., Hendriksen, E.L.J.A.: CO_2 Corrosion Under Scaling Conditions—The Special Case of Top of Line Corrosion in Wet Gas Pipelines. NACE International, Houston (2000)
12. Sun, Y., George, K., Nesic, S.: The Effect of Cl^- and Acetic Acid on Localized CO_2 Corrosion in Wet Gas Flow. NACE International, Houston (2003)
13. Nesic, S., Xiao, Y., Pots, B.F.M.: A Quasi 2-D Localized Corrosion Model. NACE International, Houston (2004)
14. De Waard, C., Milliams, D.E.: Carbonic acid corrosion of steel. Corrosion **31**(5), 177–181 (1975)
15. De Waard, C., Lotz, U., Milliams, D.E., et al.: Predictive model for CO_2 corrosion engineering in wet natural gas pipelines. Corrosion **47**(12), 976–985 (2012)
16. De Waard, C., Lotz, U., Dugstad, A.: Influence of liquid flow velocity on CO_2 corrosion: a semi-empirical model (1995)
17. Elsharkawy, A.M., Poettmann, F.H., Christiansen, R.L.: Measuring Minimum Miscibility Pressure: Slim-Tube or Rising-Bubble Method? Society of Petroleum Engineers (1992). https://doi.org/10.2118/24114-ms
18. Turgoose, S., John, G., Flynn, M., Kadir, A.A., Economopoulos, G., Dicken, G.: Corrosion Inhibition in Supercritical Carbon Dioxide Systems Containing Water. NACE International, Houston (2014)

Safety Evaluation of Oil Drilling Rig System by the Extension Theory and Analytic Hierarchy Process

Xiao-jiao Zhao[1,2](✉), Zhan Qu[2,3], Hui-bo Zhao[4], and Heng Fan[1]

[1] School of Electronic Engineering, Xi'an Shiyou University, Xi'an 710065, China
{zhaoxsyu, 723515507}@qq.com

[2] The Key Laboratory of Well Stability and Fluid and Rock Mechanics in Oil and Gas Reservoir of Shanxi Province, Xi'an 710065, China
zhqu@xsyu.edu.cn

[3] School of Petroleum Engineering, Xi'an Shiyou University, Xi'an 710065, China

[4] Xi'an Center of Geological Survey, Xi'an 710054, China
315695573@qq.com

Abstract. Oil drilling rig is an important equipment for oil and gas drilling operations. There is a great risk in the process of its use. If the risk factors are not well managed, it will cause great loss. In order to prevent accidents, the safety evaluation index and evaluation model of the oil drilling rig system are established by the extension theory. We qualified and normalized each index, determined the classical domain and the joint domain, then calculated the correlation coefficient of the evaluation index by the association function of the extension set. The weight of all the system indexes is determined by the improved analytic hierarchy process. By the characteristic value of the level variable, the grade of the failure probability of the oil drilling rig system is judged. The feasibility of the model is verified by experiments, and the safety evaluation of the oil drilling rig system is carried out. It provides a new and effective way to study the safety evaluation of the oil drilling rig system.

Keywords: Extension theory · Analytic hierarchy process · Oil drilling rig system · Safety evaluation · Matter element model

---

J. Lin (ed.), *Proceedings of the International Petroleum and Petrochemical Technology Conference 2019*, pp. 166–180, 2020.
https://doi.org/10.1007/978-981-15-0860-8_13

1 Introduction

Oil drilling rig is a large and comprehensive set of industrial equipment in the oil industry. Once it fails, it will affect the production, cause significant economic losses, major accidents may occur such as derrick collapse, blowout, fire and so on (Wang and Chen 2011). Therefore, accurate evaluation of oil drilling rig system is very important for safety production and production efficiency improvement.

Scholars have used different methods to evaluate the drilling system or drilling operation. The methods mainly include the fuzzy evaluation method (Wang and Chen 2011; Li et al. 2008; Duan 2017), rough set theory and neural network (Jian et al. 2017), analytic hierarchy process (Zhao et al. 2013), Bayesian method (Xu 2015), and so on. There are still some shortcomings in these methods. Fuzzy evaluation method is complicated in calculation and subjective in determining index weight vectors. Rough set theory is not suitable for dealing with large capacity dynamic data, and its fault tolerance and data generalization ability are low. Bayesian method requires lots of data and complex computation, some of which are subjective probability. When the learning samples are insufficient, the calculation accuracy of neural network method cannot be guaranteed. As a comprehensive evaluation method, extension theory can explain the characteristics of objective things by establishing the matter element model of each evaluation index, and then solve the incompatible problems from both qualitative and quantitative perspectives. However, the extension theory has not been used in the evaluation of oil drilling rig yet. In this paper, the applicability is studied. The matter-element model of each evaluation index is established by the extension theory, the safety of the system is analysed from qualitative and quantitative aspects, and the weight coefficient of the evaluation index is determined using the improved AHP. The weight coefficient changes dynamically with the magnitude of evaluation index combining the above two methods, then we can accurately determine the grade of the failure possibility of the oil drilling rig. It provides a new way to evaluate the oil drilling system.

2 Model

2.1 Extension Theory

Extenics is a new subject founded by Professor Cai in 1983. It is an evaluation method combining the theory of matter element and the theory of extension set (Cai 1994). In recent years. It has been widely used in coal mines, tunnels, highways and buildings etc. (Cai 1994; Liang et al. 2010, Luan et al. 2014). The essence of extension theory is to evaluate the two aspects of the research object from feasibility and optimization, to make qualitative analysis by the extenics of matter elements, to make quantitative calculation by the association function of the extension set theory, and to make a judgement in conformity with the actual object by combining qualitative and quantitative analysis.

Matter-element is the logical cell of extenics, represented by ordered three tuple $R = (N, C, V)$. N, C and V are the three elements of matter R, where N represents things, C represents characteristics, and V represents the value of N about characteristic C. The concrete steps of the extension theory are as follows.

2.2 Determination of Index Weight by Improved AHP Method

AHP is a system analysis method combining qualitative analysis with quantitative calculation. It is suitable for determining the index weight of evaluation problems of multi-level and multi structure. It can make complex problems and qualitative problems to be quantified. The steps are shown as follows (Cai 1994; Yang 1999; Li et al. 2012; Saaty 1980, 1990).

Step 1 Determination of index weight

$$E = \begin{bmatrix} e11 & e12 & \cdots & e1n \\ e21 & e22 & \cdots & e2n \\ \vdots & \vdots & \vdots & \vdots \\ en1 & en2 & \cdots & enn \end{bmatrix} \tag{1}$$

E is a square matrix of order $n \times n$; the element e_{ij} is the judgement value of i sub target S_i to the relative importance of j sub target S_j. The assignment of e_{ij} is based on the AHP scaling principle.

Step 2 Determination of relative weight

The judgement matrix can be drawn through two comparisons, and the relative weight of the next level element for the upper level element can be obtained by using the eigenvalue method. The eigenvalue method is to solve the eigenvalue of A of judgement matrix.

$$Aw = \lambda_{\max} w \tag{2}$$

$\lambda_{\max}$ is the largest eigenvalue of the matrix A, and w is the corresponding eigenvector. w is normalized to be used as a weight vector.

Step 3 Determination of the synthetic weight of each index

What we get above is the weight vector of a set of elements on its first layer. What we need to get is the relative weight of each element to the total target, especially the relative weight of each evaluation index in the lowest level to the total target, which is the synthetic weight.

Assuming that the ranking weight vector of n_{k-1} elements on the $k - 1$ level relative to the total target has been calculated:

$$w(k-1) = (w1^{(k-1)}, w2^{(k-1)}, \ldots, wn^{(k-1)})^T \tag{3}$$

On the k level, the weight vectors of the n_k elements for the j elements on the $k - 1$ level are:

$$pj^{(k)} = (p1j^{(k)}, p2j^{(k)}, \ldots, pnkj^{(k)})^T \tag{4}$$

The weight of the element that is not dominated by the j element is 0.

$$p^{(k)} = (p1^{(k)}, p2^{(k)}, \ldots, p_{nk-1}^{(k)}) \tag{5}$$

This is the matrix of $n_k \times n_{k-1}$, which represents the ordering of the elements on the k layer relative to the elements on the k − 1 layer. Then the elements of the aggregate synthesis of the elements on the k level are:

$$wi^{(k)} = P^{(k)} P^{(k-1)}, \ldots, w^{(2)} \tag{6}$$

$w^{(2)}$ is the sort vector of the second layer to the total target.

2.3 Safety Evaluation Model for Drilling Rig System

We propose an objective safety evaluation model of petroleum drilling rig system based on extension theory and AHP theory, as shown below.

(1) *Construct the safety evaluation index system of drilling rig system.*
(2) *Determine matter element safety evaluation and weights using improved AHP.*

In this step, all indicators in the evaluation system are calculated, and the qualitative indicators are transformed into quantitative indicators. Delphi index method is used to quantify each index and determine the evaluation matter element. The weight coefficients of all indexes are calculated by improved analytic hierarchy process.

(3) *The determination of the classical domain of the oil drilling rig system*

The classical domain can be expressed as:

$$R_j = (N_j, C_i, V_{ji}) = \begin{bmatrix} N_j & C_1 & V_{j1} \\ & C_2 & V_{j2} \\ & \vdots & \vdots \\ & C_n & V_{jn} \end{bmatrix} = \begin{bmatrix} N_j & C_1 & <a_{j1}, b_{j2}> \\ & C_2 & <a_{j2}, b_{j2}> \\ & \vdots & \vdots \\ & C_n & <a_{jn}, b_{jn}> \end{bmatrix} \tag{7}$$

R_j is a j matter elements; N_j is a divided j failure level; C_i is a i evaluating indicator; V_{ji} is value range of the failure level N_j corresponding to the evaluation index C_i, the classical domain $V_{ji} = <a_{ji}, b_{ji}>$, $i = 1, 2, \ldots, n$, $j = 1, 2, \ldots, m$.

(4) *The determination of the area of the oil drilling rig system*

The section domain R_p is set up by the evaluation index C_i in the whole evaluation system.

$$R_p = (N_p, C_i, V_{pi}) = \begin{bmatrix} N_p & C_1 & V_{p1} \\ & C_2 & V_{p2} \\ & \vdots & \vdots \\ & C_n & V_{pn} \end{bmatrix} = \begin{bmatrix} N_p & C_1 & <a_{p1}, b_{p2}> \\ & C_2 & <a_{p2}, b_{p2}> \\ & \vdots & \vdots \\ & C_n & <a_{pn}, b_{pn}> \end{bmatrix} \tag{8}$$

N_p is the whole failure grade of the drilling rig system; V_{pi} is the value range taken by the level N_j according to C_i, and $<a_{ji}, b_{ji}> \subset <a_{pi}, b_{pi}>$.

(5) *Determination of matter element to be evaluated for oil drilling rig system*

For the things to be evaluated T_x, the data information obtained from detection and collection is represented by matter element R_x.

$$R_x = (T_x, C_i, V_i) = \begin{bmatrix} T_x & C_1 & V_1 \\ & C_2 & V_2 \\ & \vdots & \vdots \\ & C_n & V_n \end{bmatrix} \tag{9}$$

T_x is the things to be evaluated. V_i is all the data of the T_x about the index C_i.

(6) *Correlation degree of each evaluation index on failure grades of petroleum rig system*

The single index correlation degree of each evaluation index for each failure grade is

$$K_j(V_i) = \begin{cases} \frac{-\rho(V_i, V_{ji})}{\left|V_{ji}\right|} & (V_i \in V_{ji}) \\ \frac{\rho(V_i, V_{ji})}{\rho(V_i, V_{pi}) - \rho(V_i, V_{ji})} & (V_i \notin V_{ji}) \end{cases} \tag{10}$$

$$\rho(V_i, V_{ji}) = \left| V_i - \frac{a_{ji} + b_{ji}}{2} \right| - \frac{b_{ji} - a_{ji}}{2} \tag{11}$$

$$\rho(V_i, V_{pi}) = \left| V_i - \frac{a_{pi} + b_{pi}}{2} \right| - \frac{b_{pi} - a_{pi}}{2} \tag{12}$$

V_{ji} is the value range that the failure level N_j corresponding to its evaluation index C_i.

(7) *The determination of the comprehensive correlation degree for the failure level of the matter element to be evaluated*

The comprehensive correlation degree of the matter element R_x to be evaluated on the failure level N_j is:

$$K_j(T_x) = \sum_{i=1}^{n} W_i K_j(V_i) \tag{13}$$

$K_j(T_x)$ is the comprehensive correlation degree of the matter element to be evaluated R_x to each failure level. W_i is the weight value of each evaluation index C_i.

(8) *Grade evaluation of failure probability of oil drilling rig system*

According to the principle of maximum correlation, if $K_{jo}(T_x) = \max(K_j(T_x))$, The failure level of the subject matter element R_x belongs to class j_o.

$$K_j^*(T_x) = \frac{K_j(T_x) - \min K_j(T_x)}{\max K_j(T_x) - \min K_j(T_x)} \quad (14)$$

$$j^* = \sum_{j=1}^{m} jK_j^*(T_x) / \sum_{j=1}^{m} K_j^*(T_x) \quad (15)$$

j^* is the level variable eigenvalue of the thing T_x.

3 Safety Evaluation Index System of Oil Drilling Rig System

The factors affecting the drilling accident are complex and diverse. This paper takes oil drilling rig as an example to analyse and evaluate. Referring to the statistical analysis results of historical accidents at home and abroad, and after the expert systematic demonstration and analysis, the risk of 4 aspects, such as derrick and base, operating mechanism, control mechanism, human and safety factors, and many specific evaluation indexes are determined. However, the contribution rate of these indexes to drilling rig accidents is different. According to the acquisition and quantitative difficulty of these indexes, 14 evaluation indexes such as structural derrick, base, electrical equipment and diesel engine are selected after consultation with relevant experts, and the safety evaluation index system of drilling rig is established, as shown in Table 1.

Table 1 Safety evaluation index system of drilling rig system

Drilling rig system A	Derrick and pedestal B_1	Derrick C_1
		Pedestal C_2
	Operating mechanism B_2	Electrical equipment C_3
		Diesel engine C_4
		Transmission C_5
		Drilling pump C_6
		Centrifugal pump C_7
		Hoisting and rotating device C_8
		Rig tool C_9
	Control mechanism B_3	Throttle device C_{10}
		Blowout device C_{11}
	Human and safety factors B_4	Security management C_{12}
		Quality of personnel C_{13}
		Safety facilities C_{14}

4 Case Study

Taking the oil drilling rig system in Changqing Oilfield as an example, this paper uses the proposed evaluation model to carry out safety evaluation to verify the feasibility of the model.

Step 1 Determination of evaluation criteria

At present, there is still no fixed standard for the safety evaluation of oil drilling rig at home and abroad. Some experts have done some research on the safety evaluation standards of oil drilling rig in China, but the evaluation standards in different periods and regions are not the same. On the basis of the previous research results, consulting many relevant scholars and experts, and referring to the standards of SY/T 5527-2001 and GB/T 23507.3-2009, this paper divides the evaluation index system in the safety evaluation index system of oil drilling rig into Grade I (low), Grade II (lower), Grade III (medium), Grade IV (higher), Grade V (high), see Table 2.

Table 2 Mapping table between evaluation indexes and failure probability levels of drilling rig system

Second level index	Third level index	I	II	III	IV	V
B_1	C_1	[20, 25]	(15, 20]	(8, 15]	[1, 8]	[0, 1)
	C_2	[0, 12)	[12, 18)	[18, 22)	[22, 28)	[28, 34]
B_2	C_3	(4.2, 5]	(3, 4.2]	(2.4, 3]	(1, 2.4]	(0, 1]
	C_4	[2, 2.4]	[1.6, 2)	[1.25, 1.6)	[1, 1.25)	[0, 1)
	C_5	[10, 12]	[8, 10)	[4, 8)	[2, 4)	[0, 2)
	C_6	[2.2, 3]	[1.5, 2.2)	[0.8, 1.5)	[0.5, 0.8)	[0, 0.5)
	C_7	[2, 2.5]	[1.6, 2)	[1.4, 1.6)	[0.8, 1.4)	[0, 0.8)
	C_8	[20, 25]	(15, 20]	(8, 15]	[1, 8]	[0, 1)
	C_9	[12, 14]	[8, 12)	[4, 8)	[2, 4)	[0, 2)
B_3	C_{10}	[16, 20]	(13, 16]	(8, 13]	[3, 8]	[0, 3)
	C_{11}	[15, 18]	[9, 15)	[4, 9)	[2, 4)	[0, 2)
B_4	C_{12}	[20, 25]	(15, 20]	(8, 15]	[1, 8]	[0, 1)
	C_{13}	[35, 40]	[30, 35)	[20, 30)	[10, 20)	[0, 10)
	C_{14}	[20, 25]	(15, 20]	(8, 15]	[1, 8]	[0, 1)

Step 2 Normalization of index

In order to eliminate the difference of the dimension of each index and facilitate calculation, the evaluation index is normalized according to formula (16).

$$\overline{V}_{mk} = \begin{cases} \frac{V_{mk} - V_{mk}^{\min}}{V_{mk}^{\max} - V_{mk}^{\min}} & \text{(For the targets, the greater the requirements are, the safer they are.)} \\ \frac{V_{mk}^{\max} - V_{mk}}{V_{mk}^{\max} - V_{mk}^{\min}} & \text{(For the targets, the smaller the requirements are, the safer they are.)} \end{cases} \quad (16)$$

V_{mk} is the evaluation standard value of the k level for the m index. $\overline{V}_{mk}$ is the standard value after the non-dimensional evaluation of the index. $V_{mk}^{\min}$ and $V_{mk}^{\max}$ are the minimum and maximum values of the evaluation criteria.

Step 3 Determination of the classical domain and the nodal domain

According to the formula (7), we can normalize the data in Table 1 to determine the classical domain R_j and node domain R_p of the evaluation index, as shown in Table 3.

Table 3 Classic domain and joint domain of failure probability evaluation indexes of oil drilling rig system

Second level index	Third level index	Classical domain R_j					Node domain R_p
		N_1	N_2	N_3	N_4	N_5	
B_1	C_1	(0.8–1)	(0.6–0.8)	(0.32–0.6)	(0.04–0.32)	(0–0.04)	0–1
	C_2	(0.65–1)	(0.47–0.65)	(0.35–0.47)	(0.18–0.35)	(0–0.18)	0–1
B_2	C_3	(0.84–1)	(0.6–0.84)	(0.48–0.6)	(0.2–0.48)	(0–0.2)	0–1
	C_4	(0.83–1)	(0.67–0.83)	(0.52–0.67)	(0.42–0.52)	(0–0.42)	0–1
	C_5	(0.83–1)	(0.67–0.83)	(0.33–0.67)	(0.17–0.33)	(0–0.17)	0–1
	C_6	(0.73–1)	(0.5–0.73)	(0.27–0.5)	(0.17–0.27)	(0–0.17)	0–1
	C_7	(0.8–1)	(0.64–0.8)	(0.56–0.64)	(0.32–0.56)	(0–0.32)	0–1
	C_8	(0.8–1)	(0.6–0.8)	(0.4–0.6)	(0.2–0.4)	(0–0.2)	0–1
	C_9	(0.86–1)	(0.57–0.86)	(0.29–0.57)	(0.14–0.29)	(0–0.14)	0–1
B_3	C_{10}	(0.8–1)	(0.65–0.8)	(0.4–0.6)	(0.15–0.4)	(0–0.15)	0–1
	C_{11}	(0.83–1)	(0.5–0.83)	(0.22–0.5)	(0.11–0.22)	(0–0.11)	0–1
B_4	C_{12}	(0.8–1)	(0.6–0.8)	(0.32–0.6)	(0.04–0.32)	(0–0.04)	0–1
	C_{13}	(0.88–1)	(0.75–0.88)	(0.5–0.75)	(0.25–0.5)	(0–0.25)	0–1
	C_{14}	(0.8–1)	(0.6–0.8)	(0.32–0.6)	(0.04–0.32)	(0–0.04)	0–1

$N_1(j = 1)$ indicates that the failure level is *I*, $N_2(j = 2)$ indicates that the failure level is *II*, $N_3(j = 3)$ indicates that the failure level is *III*, $N_4(j = 4)$ indicates that the failure level is *IV*; $N_5(j = 5)$ indicates that the failure level is *V*.

Step 4 Determination of matter-element to be evaluated

According to the field data, the index parameters of the four oil rigs are shown in Table 4.

Table 4 Indexes values of 4 oil drilling rig systems

Matter-element	C_1	C_2	C_3	C_4	C_5	C_6	C_7	C_8	C_9	C_{10}	C_{11}	C_{12}	C_{13}	C_{14}
T_1	16	14	3.8	1.8	9.2	1.9	1.2	12	7.2	9.6	12	18	32	17
T_2	22	8.4	4.6	2.2	9.6	2.8	1.8	16	9.4	18	17	22	36	18
T_3	5	24	1.8	1.2	3	0.7	1.2	7	3.6	2.2	1.6	0.8	7	0.4
T_4	9	19	2.8	1.5	5.2	1.2	1.58	11	6.2	9.8	1.2	0.4	4	0.6

Normalization of the data in Table 4 is used to get the evaluation element (R_1, R_2, R_3, R_4) of 4 oil drilling rigs, see Table 5.

Table 5 Four matter-element to be evaluated

Matter-element	C_1	C_2	C_3	C_4	C_5	C_6	C_7	C_8	C_9	C_{10}	C_{11}	C_{12}	C_{13}	C_{14}
R_1	0.64	0.59	0.76	0.75	0.77	0.63	0.48	0.48	0.51	0.48	0.67	0.22	0.8	0.68
R_2	0.88	0.75	0.92	0.92	0.8	0.93	0.72	0.64	0.67	0.9	0.94	0.16	0.9	0.72
R_3	0.2	0.29	0.36	0.5	0.25	0.23	0.48	0.28	0.26	0.11	0.46	0.03	0.2	0.02
R_4	0.36	0.44	0.56	0.63	0.43	0.4	0.63	0.44	0.44	0.49	0.75	0.02	0.1	0.02

Step 5 Calculation of the correlation degree of single index

According to the formulas (4)–(6), we calculate the correlation degree of each evaluation index for each matter-element to be evaluated. The following tables are the Correlation function values of indicators with failure levels of matter-element. See Tables 6, 7, 8, 9.

Table 6 Correlation function values of indicators with failure probability levels of matter-element T_1

Grade	I	II	III	IV	V
C_1	−0.308	0.200	−0.100	−0.471	−0.625
C_2	−0.128	0.333	−0.226	−0.369	−0.500
C_3	−0.250	0.333	−0.400	−0.538	−0.700
C_4	−0.242	0.500	−0.242	−0.479	−0.569
C_5	−0.207	0.375	−0.303	−0.657	−0.723
C_6	−0.213	0.435	−0.260	−0.493	−0.554
C_7	−0.400	−0.250	−0.143	0.333	−0.250
C_8	−0.400	−0.200	0.400	−0.143	−0.368
C_9	−0.417	−0.109	0.214	−0.310	−0.430
C_{10}	−0.400	−0.262	0.320	−0.143	−0.407
C_{11}	−0.327	0.485	−0.340	−0.577	−0.629
C_{12}	−0.725	−0.633	−0.313	0.357	−0.450
C_{13}	−0.286	0.385	−0.200	−0.600	−0.733
C_{14}	−0.273	0.400	−0.200	−0.529	−0.667

Table 7 Correlation function values of indicators with failure probability levels of matter-element T_2

Grade	I	II	III	IV	V
C_1	0.400	−0.400	−0.700	−0.824	−0.875
C_2	0.286	−0.286	−0.528	−0.615	−0.695
C_3	0.500	−0.500	−0.800	−0.846	−0.900
C_4	0.471	−0.529	−0.758	−0.833	−0.862
C_5	−0.130	0.187	−0.394	−0.701	−0.759
C_6	0.259	−0.741	−0.860	−0.904	−0.916
C_7	−0.222	0.500	−0.222	−0.364	−0.588
C_8	−0.308	0.200	−0.100	−0.400	−0.550
C_9	−0.365	0.345	−0.233	−0.535	−0.616
C_{10}	0.500	−0.500	−0.714	−0.833	−0.882
C_{11}	0.353	−0.647	−0.880	−0.923	−0.933
C_{12}	−0.800	−0.733	−0.500	0.429	−0.429
C_{13}	0.167	−0.167	−0.600	−0.800	−0.867
C_{14}	−0.222	0.400	−0.300	−0.588	−0.708

Table 8 Correlation function values of indicators with failure probability levels of matter-element T_3

Grade	I	II	III	IV	V
C_1	−0.750	−0.667	−0.375	0.429	−0.444
C_2	−0.554	−0.383	−0.171	0.353	−0.275
C_3	−0.571	−0.400	−0.250	0.429	−0.308
C_4	−0.398	−0.254	−0.038	0.200	−0.138
C_5	−0.699	−0.627	−0.242	0.500	−0.242
C_6	−0.685	−0.540	−0.148	0.400	−0.207
C_7	−0.400	−0.250	−0.143	0.333	−0.250
C_8	−0.650	−0.533	−0.300	0.400	−0.222
C_9	−0.698	−0.544	−0.103	0.200	−0.316
C_{10}	−0.863	−0.831	−0.725	−0.267	0.267
C_{11}	−0.446	−0.080	0.143	−0.343	−0.432
C_{12}	−0.963	−0.950	−0.906	−0.250	0.250
C_{13}	−0.773	−0.733	−0.600	−0.200	0.200
C_{14}	−0.975	−0.967	−0.937	−0.500	0.500

Table 9 Correlation function values of indicators with failure probability levels of matter-element T_4

Grade	I	II	III	IV	V
C_1	−0.550	−0.400	0.143	−0.100	−0.471
C_2	−0.323	−0.064	0.250	−0.170	−0.371
C_3	−0.389	−0.083	0.333	−0.154	−0.450
C_4	−0.351	−0.098	0.267	−0.229	−0.362
C_5	−0.482	−0.358	0.294	−0.189	−0.377
C_6	−0.452	−0.200	0.435	−0.245	−0.365
C_7	−0.315	−0.026	0.125	−0.159	−0.365
C_8	−0.450	−0.267	0.200	−0.083	−0.456
C_9	−0.488	−0.228	0.464	−0.254	−0.405
C_{10}	−0.388	−0.246	0.360	−0.155	−0.410
C_{11}	−0.242	0.242	−0.500	−0.679	−0.719
C_{12}	−0.975	−0.967	−0.937	−0.500	0.500
C_{13}	−0.886	−0.867	−0.800	−0.600	0.400
C_{14}	−0.975	−0.967	−0.937	−0.500	0.500

Step 6 Calculation of index weights by improved AHP

The contribution rates of the derrick and pedestal, the operating mechanism, the control mechanism, the human and the safety factors to the accidents of the oil drilling rig are determined through the reference literature, the related documents at home and abroad, the analysis of the engineering data and the comprehensive expert inquiry. The 9 scale representation method of the judgement matrix is expressed as follows (Table 10).

Table 10 Drilling system judgment matrix

A	B1	B2	B3	B4
B1	1	1/7	1/5	1/6
B2	7	1	3	2
B3	5	1/3	1	1/2
B4	6	1/2	2	1

The weight of the second level index can be W = (0.05, 0.48, 0.18, 0.29), and it is concluded that the influence of operational factors in the second level index is relatively large (Table 11).

Table 11 Judgement matrix of derrick and base

B1	C1	C2
C1	1	1
C2	1	1

The weight of the derrick and pedestal is W_{u1} = (0.50, 0.50), and the influence of the derrick and the base of B_1 is the same (Table 12).

Table 12 Running mechanism judgment matrix

B2	C3	C4	C5	C6	C7	C8	C9
C3	1	1	1	1/3	1/3	1/2	3
C4	1	1	1	1/3	1/3	1/2	3
C5	1	1	1	1/3	1/3	1/2	3
C6	3	3	3	1	1	2	9
C7	3	3	3	1	1	2	9
C8	2	2	2	1/2	1/2	1	6
C9	1/3	1/3	1/3	1/9	1/9	1/6	1

The index weight of the operating mechanism is W_{u2} = 0.087, 0.087, 0.087, 0.274, 0.274, 0.162, 0.029), and the influence of the drilling pump and the centrifugal pump in the operating mechanism is relatively large (Table 13).

Table 13 Control mechanism judgment matrix

B3	C10	C11
C10	1	1/3
C11	3	1

The index weight of the available control mechanism is W_{u3} = (0.25, 0.75). It is concluded that the impact of blowout preventer is relatively large in the control organization (Table 14).

Table 14 Human and safety factor judgment matrix

B4	C12	C13	C14
C12	1	1	3
C13	1	1	3
C14	1/3	1/3	1

The weight of human and safety factors is W_{u4} = (0.43, 0.43, 0.14). It is concluded that the impact of safety management and personnel literacy in human and safety factors is relatively large.

The weight of each index of the three level indexes C_1–C_{14} is w = (0.025, 0.025, 0.0418, 0.0418, 0.0418, 0.1315, 0.1315, 0.0778, 0.0139, 0.0450, 0.1350, 0.1247, 0.1247). It is concluded that the influence of electrical equipment, diesel engine, transmission device and throttle device in the three level indexes is relatively large.

Step 7 Calculation of comprehensive association degree

According to formula (7), the comprehensive correlation degree of T_1–T_4 for each failure level is calculated as follows:

$$K(T_1) = (-3.574, 2.992, -0.793, -3.619, -6.606)$$
$$K(T_2) = (1.887, -1.870, -6.589, -7.739, -9.580)$$
$$K(T_3) = (-8.423, -6.758, -3.798, 2.684, -0.618)$$
$$K(T_4) = (-6.266, -3.528, 0.696, -3.018, -2.339)$$

Step 8 Determination of failure grade

According to the maximum correlation degree principle, the failure grade of the drilling rig T_1 is grade II. The failure grades of T_2, T_3 and T_4 are I, IV and III respectively.

In order to determine the failure degree of the 4 drilling rigs more accurately, the grade variable eigenvalue j^* of 4 drill rig failure probability classes is calculated according to the formulas (8) and (9). The results of the grade variable eigenvalue and failure probability grade of the T_1–T_4 drilling rig are shown in Table 15.

Table 15 Failure rating evaluation of drilling rig system

Drilling rig	Extension evaluation		
	j_0	j^*	Evaluation grade
T_1	2	2.409	II tends to III
T_2	1	1.800	I tends to II
T_3	4	3.994	III tends to IV
T_4	3	3.496	III tends to IV

It can be seen from the table that the extension evaluation not only can judge the grade of the probability of failure of the drill, but also can reflect the degree of the failure of the drilling rig tending to another. It provides a new and feasible method for the evaluation of the probability of failure of the drill.

5 Conclusion

(1) The safety evaluation index system of oil drilling rig system is established, including four indexes and 14 two level indexes.
(2) The index weight is determined by the Delphi theory and the improved analytic hierarchy process, considering the fuzziness of human judgment. The weight of different categories of evaluation parameters is obtained, which is more in line with the construction site.

(3) The safety evaluation model of the oil drilling rig system is established, combing the extension theory and the improved AHP method. Considering the multi factor control function, the matter element model of each evaluation index is established. It can be used to calculate the comprehensive correlation degree of the failure grade for the oil drilling rig. The calculation of the grade variable characteristic value of the failure grade can more accurately reflect the degree of the failure probability tending to the other class. This method is rigorous in theory, simple in calculation and reasonable in evaluation results, which is proved to be a new and effective way for the safety evaluation of oil drilling rig.

(4) The practicability of the proposed safety evaluation model is verified by an example. Through the evaluation, the factors which have great influence on the failure of the oil drilling rig system are got, such as the throttle device in the control mechanism; the electrical equipment, the diesel engine and the transmission device in the operating mechanism. The conclusion can provide theoretical basis for policy making in engineering practice.

Acknowledgements. The present work is supported by National Natural Science Foundation of China (Grant No. 51674200, 51704233), and Research project of Shaanxi Provincial Department of Education (18JK0610).

References

Cai, W.: Matter Element Model and Its Application. Science and Technology Press, Beijing (1994)

Duan, Y.-H.: Risk evaluation method of drilling operation in gas well containing H_2S based on multifactor fuzzy identification and quantitative calculation. J. Saf. Sci. Technol. **13**(11), 149–154 (2017)

Jian, L.-I., Ke, L.-I., Wang, B.: An evaluation model of drilling safety based on combined rough set and neural network. J. Southwest Petrol. Univ. **39**(5), 120–128 (2017)

Li, Q., Yu, L.-L., Liu, Z.-K., et al.: Integrated drilling risk evaluation method and model establishment. Nat. Gas. Ind. **28**(5), 120–122 (2008)

Liang, G.-L., Xu, W.-Y., Tan, X.-L.: Application of extension theory based on entropy weight to rock quality evaluation. Rock Soil Mech. **31**(2), 535–540 (2010)

Luan, T.-T., Xie, Z.-H., Wu, Z.-Z., et al.: Extension evaluation and warning for waste dump landslide of open-pit mine. J. Cent. South Univ. Sci. Technol. **45**(4), 1274–1279 (2014)

Li, F.-W., Du, X.-L., Zhang, M.-J., et al.: Application of improved AHP in risk identification during open-cut construction of a subway station. J. Beijing Univ. Technol. **38**(2), 167–172 (2012)

Saaty, T.L.: The Analytic Hierarchy Process: Planning, Priority Setting, Resource Allocation. McGraw-Hill, NY, USA (1980)

Saaty, T.L.: Multi-criteria Decision Making: The Analytic Hierarchy Process: Planning, Priority Setting Resource Allocation. USA (1990)

Wang, C.-J., Chen, G.: Applications of risk factor fuzzy evaluation method (RFEM) to risk evaluation of drilling rig. J. Saf. Sci. Technol. **7**(9), 149–152 (2011)

Xu, Z.: Research and Implementation of Risk Evaluation Model for Drilling Operation Based on Bayesian Network. Southwest Petroleum University, China (2015)

Yang, Y.: The application of improved AHP method in the safety management evaluation for mine. Syst. Eng. Theory Pract. **6**(6), 121–125 (1999)

Zhao, Y.-S., Bai, J.-P., Liu, J.-M., et al.: Comprehensive safety analysis and evaluation on unconventional-gas drilling operation based on AHP. J. Saf. Sci. Technol. **9**(7), 139–143 (2013)

Characteristic Analysis of Miscible ZONE of Slim Tube Experiment of CO_2 Flooding

Jie Zhang[1], Xin-wei Liao[2(✉)], Ming-Jun Cai[1], Min Liu[1], and Rong-tao Li[2]

[1] Petrochina Dagang Oilfield, Tianjin, China
{dg_zhangjie, caimjun, dgyt_liumin}@petrochina.com.cn
[2] China University of Petroleum, Beijing, China
xinwei@cup.edu.cn, lirongtao2015@126.com

Abstract. The minimum miscible pressure of CO_2 and crude oil in CO_2 flooding process is a key indicator of whether it can reach the miscible state. The characteristics of oil-gas miscible zone provide an important basis for the study of CO_2 flooding minimum miscibility pressure. Taking the underground oil sample of an oil field in China as an example, the minimum miscibility pressure test of CO_2 and crude oil was carried out based on the slim tube experiment. The numerical simulation technique was used to fit the PVT data of crude oil, and the critical parameters of the pseudo component were obtained. Then, through the simulation slim tube experiment method, the multi-contact miscibility of CO_2 and crude oil was calculated. The characteristics of oil-gas interfacial tension and miscible zone are analyzed. The interfacial tension and miscible zone during the mixing process of CO_2 and crude oil are quantitatively characterized, and their variation trend with the volume and pressure of CO_2 injection is also described. It provides a reference for accurate prediction of the minimum miscible pressure of CO_2.

Keywords: CO_2 miscible flooding · MMP · Slim tube experiment · Miscible zone · Interfacial tension

J. Lin (ed.), *Proceedings of the International Petroleum and Petrochemical Technology Conference 2019*, pp. 181–189, 2020.
https://doi.org/10.1007/978-981-15-0860-8_14

1 Introduction

The research results at domestic and abroad show that CO_2 flooding enhanced oil recovery technology can make up for the shortages of water flooding and improve oil recovery more effectively, and CO_2 miscible flooding can greatly improve oil recovery. The experience shows that CO_2 flooding can increase oil recovery by 5–15% compared with water flooding, in which miscible flooding can increase by 8–15% and immiscible flooding can increase by 5–9% [1]. There are two main methods for determining the minimum miscibility pressure (MMP) [2–4], namely experimental methods and calculation methods. In the experimental methods, the slim tube experiment method is one of the most reliable methods. Among the calculation methods, there are mainly four types of methods, the correlation method [5–7], the component simulation method, the mixed cell method, and the line analysis method [8–10]. The component simulation method can determine the minimum miscibility pressure, and the phase behavior and miscibility zone distribution characteristics of crude oil components in the slim tube can also be obtained. In this paper, the minimum miscibility pressure of CO_2 and crude oil in an oil field in China is obtained by slim tube experiment. The slim tube model is established by numerical simulation method to further fit the minimum miscibility pressure (MMP) of the reservoir. The characteristics of oil-gas mixed phase zone and interfacial tension distribution are studied, which provides a reference for the prediction of CO_2 flooding minimum miscibility pressure.

2 Reservoir Fluid Properties and Slim Tube Experiment

2.1 Reservoir Fluid Properties

With PVT test for CO_2 flooding of the oil sample of A well in an oil field in China, properties of reservoir fluid are listed in Table 1.

Table 1 Oil sample PVT test results

Reservoir fluid type	Black oil
Saturation pressure	9.25 MPa
GOR	65.1 m^3/m^3
Formation oil volume factor	1.2523 m^3/m^3
Formation oil density	0.7295 g/cm^3
Formation oil viscosity	1.35 mPa s
Clearance oil density	0.8311 g/cm^3
Well fluid components	
$C_1 + N_2$	42.95 mol%
$CO_2 + C_2 \sim C_{10}$	22.71 mol%
C_{11+}	34.34 mol%

Establishing a reliable database of fluid phase parameters plays a key role in the numerical simulation of CO_2 flooding reservoirs. In order to improve the calculation efficiency, the actual components are usually combined and re-processed, where we combine into 10 pseudo-components, namely N_2, CO_2, C_1, C_2, C_3, C_4, C_5, C_6, $C_{7\text{-}27}$, C_{28+}. Relevant parameters for numerical simulation, furthermore, could be conducted using PR EOS for matching experimental data of the oil sample used. Figure 1 presents

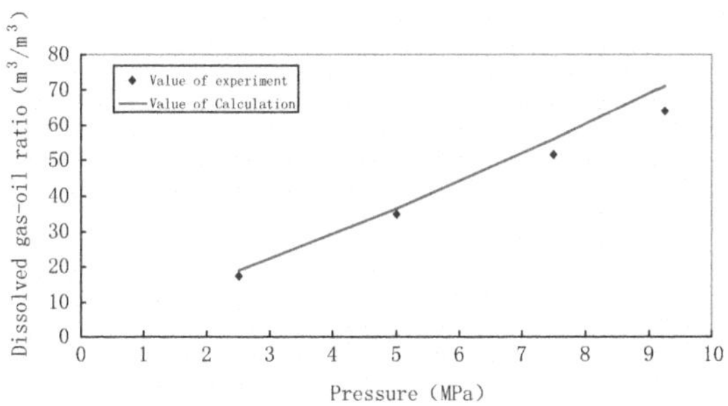

(a) Dissolved gas-oil ratio fitting results

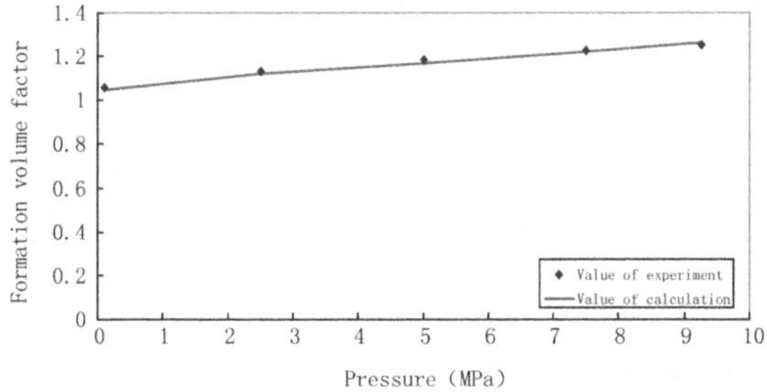

(b) Formation volume factor fitting results

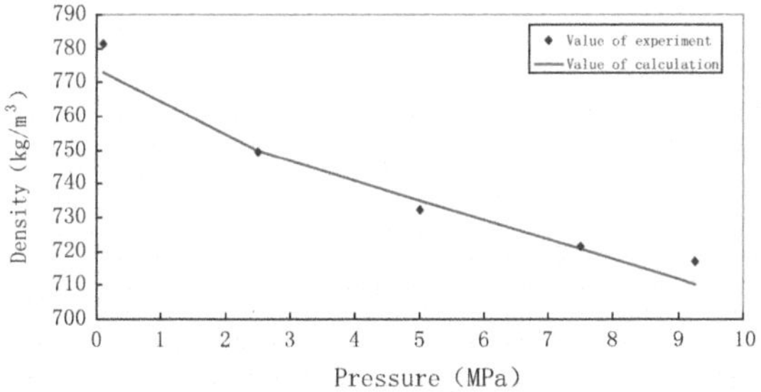

(c) Density fitting results

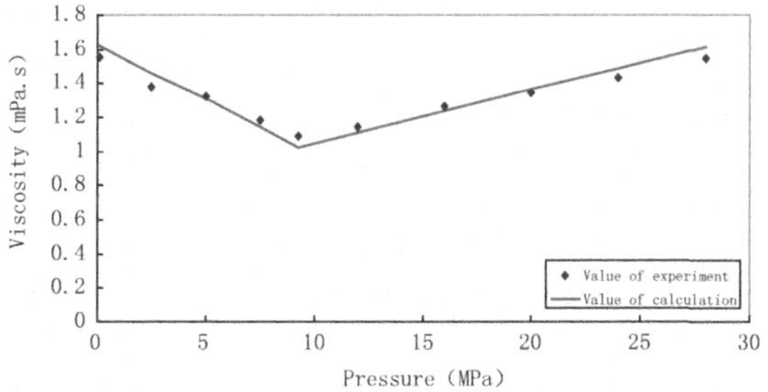

(d) Viscosity fitting results

Fig. 1 Matching results with experimental values and calculated values of fluid characteristics the matching curves between experimental values and calculated values. With the satisfactory fitting results, the matching curves could be reliable to determine reservoir fluid characteristics. The parameters of equation of state which can represent the real reservoir fluid characteristics are listed in Table 2.

Table 2 Pseudo-component state parameters of reservoir fluid

Component	Molar fraction/%	Molecular mass/g/mol	Critical pressure/MPa	Critical temperature/K	Omega A	Omega B	Z factor
N_2	0.858	28.01	33.944	126.2	0.089	0.151	0.29
CO_2	0.031	44.01	69.596	299.1	0.421	0.071	0.26
C_1	42.09	16.04	26.197	150.7	0.210	0.014	0.20
C_2	3.741	30.07	27.476	239.7	1.209	0.152	0.20
C_3	1.696	44.10	24.138	286.9	0.647	0.109	0.20
C_4	1.450	58.12	37.188	409.9	0.409	0.122	0.28
C_5	1.501	72.15	38.195	439.5	0.686	0.089	0.32
C_6	2.6	84	34.521	467.3	0.797	0.068	0.31
C_{7-27}	32.7	177.8	21.447	673.1	0.722	0.114	0.26
C_{28+}	13.333	564.2	4.796	975.4	1.363	0.146	0.13

2.2 Slim Tube Experiment

The authors use the classical slim tube test method to determine the MMP of CO_2.

2.2.1 Experiment Process

The experiment was carried out in a slim tube model filled with porous media. The RUSKA 2328-86 slim tube model is mainly composed of a slim tube with an outer diameter of 6.35 mm, an inner diameter of 3.86 mm and a length of 20.0 m. The maximum temperature of the slim tube is 150 °C, the highest pressure is 50 MPa, the gas permeability is 5.43 μm^2, and the porosity is 20%. The experimental steps include: (1) saturated oil in slim tube; (2) CO_2 displacement by constant pressure injection; (3) gradually increasing pressure repetition displacement process until inflection point appears on the recovery-displacement pressure curve. The experimental flow is shown in Fig. 2.

2.2.2 Experimental Result

Gas injection was carried out at different experimental pressures at the experimental temperature of 92 °C. When the cumulative pump exceeds 1.20 times the pore volume (1.2 PV) or no more oil is produced, the displacement is stopped. By setting 6 sets of experimental pressures to obtain the relationship between the oil displacement efficiency and the displacement pressure, there is a significant inflection point on the curve, and the pressure indicated by the abscissa corresponding to the point is the minimum miscibility pressure (MMP), that is 18.1 MPa, as shown in Fig. 3.

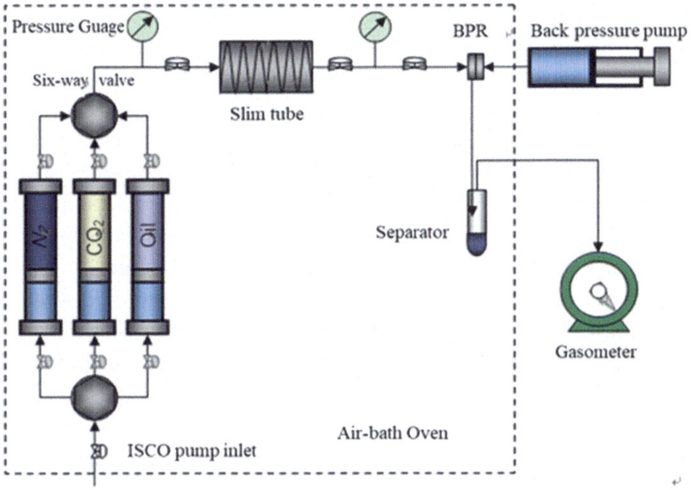

Fig. 2 Slim tube experiment procedure

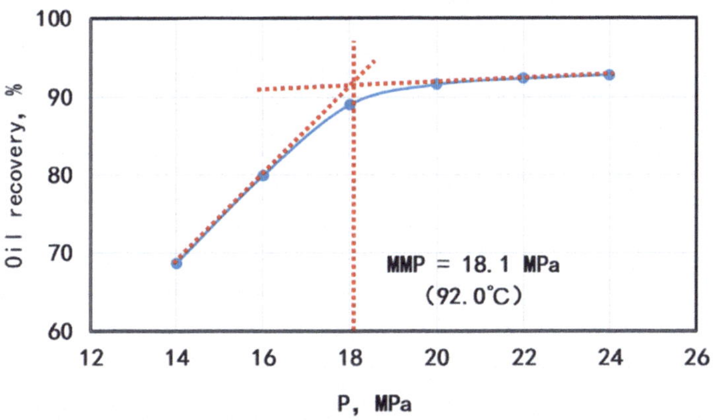

Fig. 3 MMP determined by experiment

3 Numerical Simulation of Slim Tube Model

3.1 MMP Determined by Numerical Simulation

The numerical simulation method is an important method to study the effect of miscible flooding and design miscible flooding. The component model commonly used in miscible flooding is simulated. One-dimensional component model is used in the simulation of the slim tube model. In the simulation, the length of the slim tube is divided into equal parts, and the cross-sectional area is square. The grid is divided only in the X direction. A injection well with constant injection volume is set in the first grid block, and a production well with constant pressure is set in the last grid block. Under reservoir conditions, the slim pipe is saturated with crude oil, and the composition of injection solvent used in simulation is the same as that of gas used in the slim tube experiment.

The cross section size of the slim tube model is 3.86 mm * 3.86 mm, and the other model parameters are the same as the actual slim tube model. The above PVT simulation results are used for formation crude oil. The whole displacement is carried out at a constant temperature of 92 °C ("formation temperature"). The slim tube model is 100% saturated oil, and then CO_2 is injected into each experimental pressure point for displacement. After 1.2 PV CO_2 is injected, the simulation is stopped. The oil recovery at different pressure points is obtained, and the relationship curve between oil recovery and pressure is drawn (Fig. 4). According to the miscibility criterion, the MMP was 18.6 MPa, and the error was 2.76% compared with the slim tube experiment.

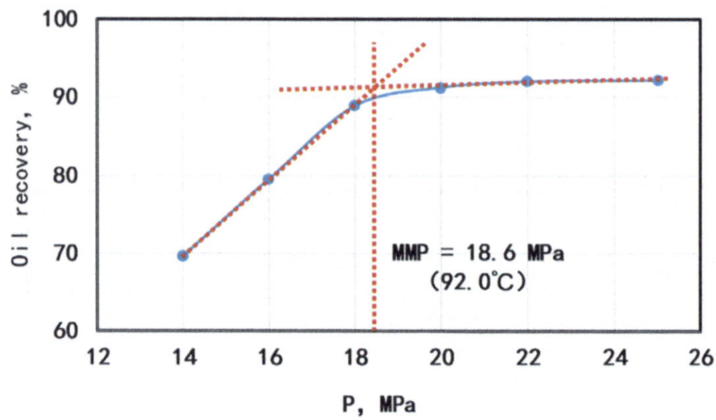

Fig. 4 MMP determined by numerical simulation

3.2 Characteristic Analysis of Miscible Zone

The numerical simulation of slim tube can not only determine the minimum miscible pressure by the relationship curve between recovery and pressure, but also observe the change process of interfacial tension and miscible zone in CO_2 flooding slim tube simulation, and further analyze the characteristics of CO_2 miscible zone by the change trend of interfacial tension in miscible zone. In the numerical simulation, the interfacial tension of oil and gas is calculated by the molar density, molar fraction and isotonic volume of oil and gas. The calculated relationship is as in Eq. 1.

$$\sigma = \left[\sum_{i=1}^{N_C} [V]_i \left(b_L^m x_i - b_V^m y_i \right) \right]^4 \tag{1}$$

Here, σ is interfacial tension, b_L^{m} is molar density of oil, b_V^{m} is molar density of gas, x is molar fraction of oil, y is molar fraction of gas, V is Isotonic volume. N_c Number of components.

When the simulated pressure is greater than the minimum miscible pressure of the reservoir fluid, CO_2 and the crude oil are in multiple contact miscible phases, and the residual oil saturation in the region affected by CO_2 is extremely low. As shown in Fig. 5, an oil-gas mixed phase zone can be observed, and the interfacial tension at the leading edge of the mixed phase zone is the maximum, which gradually decreases from the leading edge of the mixed phase zone, indicating that the interfacial tension between the CO_2 and the crude oil is high, which does not occur. One contact with the mixed phase. CO_2 extracts and mixes the light components in the crude oil, and the mixture passes to the trailing edge of the mixed phase belt. The mixture is continuously in contact with CO_2 to transfer mass. The interfacial tension of the oil and gas is gradually reduced, and finally the mixed phase is reached. drive. And the miscible zone moves forward with the displacement, the bandwidth gradually increases, and disappears at the outlet end of the slim tube model.

As shown in Fig. 5, during the formation of the mass transfer channel (before the mixed phase reaches the exit end of the slim tube model), the mixed phase bandwidth gradually increases with the increase of the number of CO_2 injected PV, and finally reaches a highest point. After the mass transfer channel is formed, the miscible band disappears. From the variation trend of the mixed phase bandwidth with pressure in Fig. 6, the mixed phase bandwidth gradually decreases with the increase of pressure, indicating that the higher the pressure, the faster the speed of the mixed phase of CO_2 and crude oil and the shorter the time.

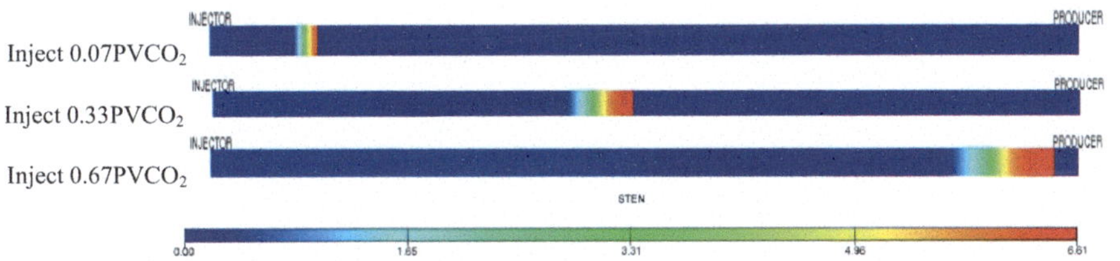

Fig. 5 Change phase of miscible zone at 20 MPa

It can be seen from Fig. 7 that the interfacial tension at the leading edge of the miscible zone increases with the increase of the CO_2 injection volume, and finally reaches a gentle state. The maximum value is reached before the miscible band disappears. The larger the simulated pressure, the more obvious the trend is, but the mixed-phase leading edge interface The tension value decreases with increasing pressure, indicating that the higher the pressure, the more likely the CO_2 and the crude oil to contact the mixed phase.

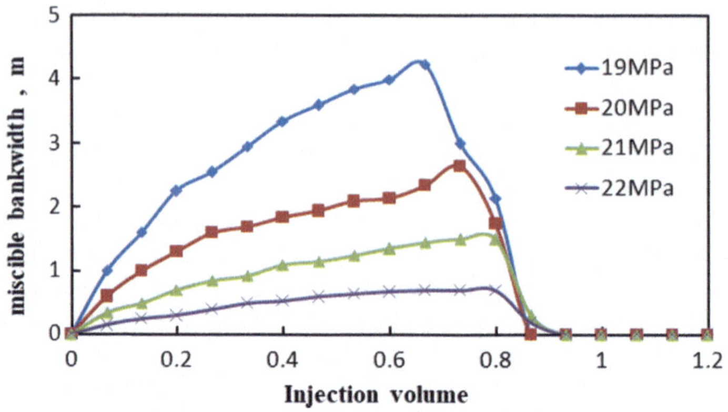

Fig. 6 Miscible bandwidth and injection volume curve

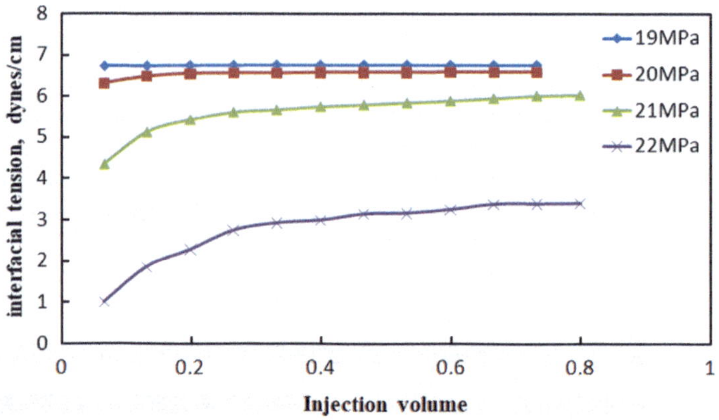

Fig. 7 Interfacial tension and injection volume curve

4 Conclusions

(1) In this paper, a numerical simulation method for component model simulation of slim tube experiments is established. The minimum miscibility pressure of CO_2 flooding crude oil can be determined by this method. The minimum miscibility pressures of the CO_2 flooding crude oil in the target reservoir were 19.1 MPa and 18.6 MPa, respectively, through the slim tube experimental method and the numerical simulation method of the slim tube. The relative error between the two was 2.76% with high precision.

(2) During the CO_2 miscible displacement process, the miscible zone moves forward, the bandwidth gradually increases and then disappears at the outlet end.

(3) Before the mass transfer channel is formed, the interfacial tension at the leading edge of the miscible zone and the miscible zone increases with the increase of the injected CO_2 volume, and decreases with the increase of the simulated pressure.

Acknowledgements. At the completion of the article, we would like to thank Professor Liao Xinwei of China University of Petroleum (Beijing). We thank the supports from Chinese National Major Science and Technology Projects (2017ZX05009004-005) and Chinese National Major Science and Technology Projects (2017ZX05030002-005). We also thank the help of experts from Dagang Oilfield for their correlation study.

References

1. Shen, P., Liao, X., Liu, Q.: Methodology for estimation of CO_2 storage capacity in reservoirs. Petrol. Explor. Dev. **36**(2), 216–220 (2009)
2. Liao, X., Shen, P.: Research on CO_2 sequestration and recycling as resource. In: 20-World Petroleum Congress (2011)
3. Guo, P., Yang, X.F.: Minimum Miscibility Pressure Study of Gas Injection in Reservoirs. Petroleum Industry Press, Beijing (2005) (in Chinese)
4. Brock, W.R., Bryan, L.A.: Summary Results of CO_2 EOR Field Tests, 1972-1987. SPE 18977 (1989)
5. Stalkup, F.I.: Miscible Displacement. Henry L. Doherty Memorial Fund of AIME, Society of Petroleum Engineers of AIME, New York (1983)
6. Johns, R.T., Orr Jr., F.M.: Miscible gas displacement of multicomponent oils. SPE J. **1**(01), 39–50 (1996)
7. Yellig, W.F., Metcalfe, R.S.: Determination and prediction of CO_2 minimum miscibility pressures. J. Petrol. Technol. **32**(01), 160–168 (1980)
8. Yun, W., Douglas, G.P.: Analytical Calculation of Minimum Miscibility Pressure: Comprehensive Testing and Its Application in a Quantitative Analysis of the Effect of Numerical Dispersion for Different Miscibility Development Mechanism. SPE 59378 (2000)
9. Kristian, J.: Global approach for calculation of minimum miscibility pressure. Fluid Phase Equilib. **153**, 251–263 (1998)
10. Esmaeilzadeh, F., Roshanfekr, M.: Calculation of minimum miscibility pressure for gas condensate reservoirs. Fluid Phase Equilibria. **249**, 75–81 (2006)

Characteristics Analysis and Slug Flow Model in Undulated Wellbore

Guo-qing Han[1,2], Shu-zhe Shi[1,2(✉)], Xiao-dong Wu[1,2], Cheng-cheng You[1,2], Xue-qi Cen[1], Zu-guo Zhang[1], Zhi-yong Zhu[1,2], Ke Sun[1,2], and Yi-chen Tao[1,2]

[1] State Energy Center for Shale Oil Research and Development, Snopec Petroleum Exploration and Production Research Institute, 31 Xueyuan Road, Haidian, Beijing 100083, China
{hanguoqing, cenxueqi2005, 15120058521, skcbtb, tyc_cupb}@163.com, {303070516, 42638521}@qq.com, wuxd308@263.net, {303070516

[2] China University of Petroleum-Beijing, 18 Fuxue Road, 102249 Changping, Beijing, China

Abstract. Wellbore trajectory of horizontal Wells often has undulations, and formation of gas-liquid slug with increase of gas flow rate will lead to instability of production. Therefore, by studying formation mechanism and motion characteristics of slug flow in undulating wellbore, we optimize lifting scheme. Based on formation mechanism of hydrodynamic slug flow, this paper studies flow characteristics of gas-liquid two-phase flow in undulating wellbore and establishes prediction model of gas-liquid two-phase pressure drop in undulating wellbore and slug tracking model. The prediction results of two-phase pressure drop model and slug tracking model established in this paper are basically consistent with the results of numerical simulation. At the same time, the model can also explain the transient characteristics of slug flow. By comparing the instantaneous pressure changes calculated by the model with results of experimental data, it is found that overall trend of pressure fluctuation curve is alike, and the model is highly accurate. Innovation point, technical contribution and significance: establishment of a continuous undulation pipeline slug track model, which is of great practical significance for optimizing well trajectory and improving stability of pipeline and equipment operation.

This paper was prepared for presentation at the 2019 International Petroleum and Petrochemical Technology Conference in Beijing, China, 27–29, March, 2019.

J. Lin (ed.), *Proceedings of the International Petroleum and Petrochemical Technology Conference 2019*, pp. 190–203, 2020.
https://doi.org/10.1007/978-981-15-0860-8_15

Keywords: Hilly terrain wellbore · Undulating slug flow · Pressure drop prediction · Slug tracking model

1 Introduction

The slug flow is a very common flow pattern produced during the gas-liquid two-phase flow process. It is also a flow pattern with the great impact on the stable operation of the pipeline and has the most complicated flowing law. At present, the wellbore trajectory of horizontal wells often has undulations, so that the gas and liquid in the wellbore periodically generate unsteady flow and gas-liquid slug formed with the increase of the gas flow rate will increase the instability of production. At the same time, in the wells with lifting measures, the slug will also greatly affect the lifting of oil and gas, such as the reduction of pump efficiency, which seriously affects the production of oil wells. Therefore, research on the flow in the wellbore should be carried out.

In 1986, Rothe and Crowley [1] constructed an undulating line with a transparent tube with an inner diameter of 38.1 mm, and studied the generation, dissipation and flowing law of slug flow.

In 1990, Scott and Kouba [2] proposed a slug flow model that can be used to predict the change of slug length in the undulating pipeline. However, the model does not consider the accumulation of liquid at the bottom of the low-elbow place and the change in liquid holdup and liquid film thickness during slug transport.

In 1994, Zheng et al. [3, 4] studied the influence of undulating terrain on the characteristics of gas-liquid two-phase flow, and proposed a simple slug tracking model. It can be used to describe the flow characteristics of all simplified slugs in a single undulating pipeline. In 1995, Henau and Raithby [5] completed slug flow experiment in continuous undulating pipelines and model verification, and proposed a transient two-phase flow model, which considered the friction that was ignored by the predecessors. In 2000, Zhang et al. [6, 7] proposed a steady-state model describing the growth and attenuation of slugs in undulating pipelines, which can predict the length of slugs and the frequency of slug generation at any position in the undulating pipe.

In 2005, Al-Safran [8, 9] and others conducted detailed experimental research on the motion law of gas-liquid two-phase slug flow in the undulating wellbore. By continuously adjusting the flow rate of gas and liquid, they divided the flow state appearing in the undulating wellbore into five different state. According to the experimental results, Al-safran has drawn the flow characteristics table of the liquid slug in the bottom elbow pipe with two angles and different gas-liquid superficial velocities, and the diagram of gas-liquid two-phase flow pattern in the descending pipe section.

Subsequently, Al-Safran et al. [10, 11] found through experimental and theoretical studies that in the undulating pipeline, the length of the ascending slug would fluctuate greatly due to the formation of the new slug at the bottom of the low-elbow, and the length of the slug after the undulating terrain might become 3 times of the original length. Al-Safran also built a probability/mechanism model.

Therefore, the effects of well trajectory, completion and wellhead throttling on the gas-liquid flow stability in a wellbore were studied through the pipeline experiments of different angles and the flow stability control technology lay a foundation for optimizing wellbore lifting scheme. The pressure drop model of gas-liquid two-phase flow in undulating wellbore is always the top priority in the calculation of the multiphase flow. It is of great practical significance for optimizing well trajectory design and improving the operation stability of pipeline and downstream oil-gas separation device.

2 Undulation Wellbore Unit Pressure Drop Model

For a continuous undulating wellbore, it has a similar flow pattern and flow characteristics as a unit of undulating pipeline with descending section-undulating section-ascending section-undulating section. Therefore, assuming that the flow rate of the gas-liquid fluid is continuous, the calculation of the pressure drop of the continuous undulating wellbore can be simplified to calculate the pressure drop of each unit undulating pipeline.

A typical undulation unit pipeline diagram is as follows (Fig. 1).

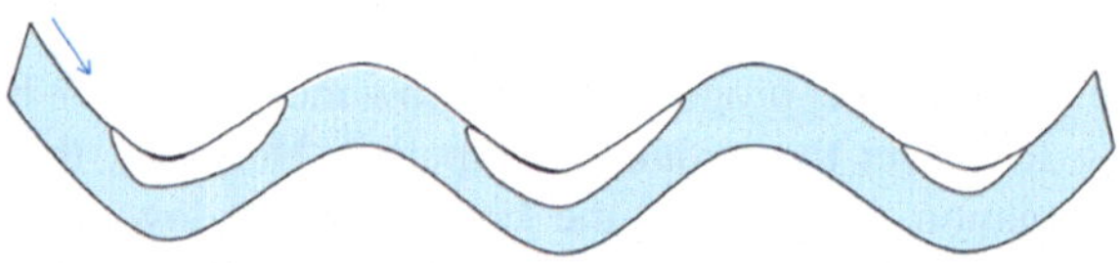

Fig. 1 Undulating pipeline diagram

2.1 Calculation of Gas-Liquid Two-Phase Pressure Drop in Descending Pipe Section

The pressure drop caused by the gas-liquid two-phase flow in the inclined pipe is calculated using the Beggs-Brill [12] method, which is now used to calculate the pressure drop in the descending section of the undulating wellbore. According to the Beggs-Brill pressure drop expression of inclined pipe, the pressure drop is obtained as

$$-\frac{dp}{dL}=\frac{\left[\rho_l H_l+\rho_g(1-H_l)\right]gsin\alpha+2\frac{\lambda M_m u_m}{\pi D^3}}{1-\frac{\left[\rho_l H_l+\rho_g(1-H_l)\right]u_m u_{sg}}{p}}. \tag{1}$$

Liquid holdup:

$$H_l(\alpha)=H_l(0)\psi \quad \psi=1+C\left[sin(1.8\alpha)-\frac{1}{3}sin^3(1.8\alpha)\right].$$

According to the horizontal liquid holdup $H_l(0)$ and coefficient C under stratified flow conditions that

$$H_l(0) = \frac{0.98E_l^{0.4846}}{N_{Fr}^{0.0868}} \quad C = (1 - E_l)In\left(\frac{4.7N_{ul}^{0.1244}}{E_l^{0.3692}N_{Fr}^{0.5056}}\right).$$

In addition, it should be noted that the angle α in the above formula takes a negative value.

2.2 Calculation of Gas-Liquid Two-Phase Pressure Drop in Ascending Pipe Section

The pressure drop caused by the gas-liquid two phase flow in the ascending section of the undulating wellbore is still calculated by the Beggs-Brill method. According to the Beggs-Brill method, the pressure drop of the ascending pipe is:

$$-\frac{dp}{dL} = \frac{\left[\rho_l H_l + \rho_g(1 - H_l)\right] g sin\theta + 2\frac{\lambda M_m u_m}{\pi D^3}}{1 - \frac{\left[\rho_l H_l + \rho_g(1 - H_l)\right] u_m u_{sg}}{p}}. \quad (2)$$

Liquid holdup:

$$H_l(\theta) = H_l(0)\psi \quad \psi = 1 + C\left[sin(1.8\theta) - \frac{1}{3}sin^3(1.8\theta)\right].$$

Then determine the horizontal liquid holdup $H_l(0)$ and coefficient C under the slug flow condition that

$$H_l(0) = \frac{0.845E_l^{0.5351}}{N_{Fr}^{0.0173}} \quad C = (1 - E_l)In\left(\frac{2.96E_l^{0.305}N_{Fr}^{0.0978}}{N_{ul}^{0.4473}}\right)$$

2.3 Calculation of Gas-Liquid Two-Phase Pressure Drop in the Undulating Pipe Section

When dealing with the pressure drop at the junction between the ascending and the descending of the undulating wellbore, many scholars have simplified it into an inclined pipe flow and then calculated the pressure drop. However, the gas-liquid two-phase flow through the elbow pipe will generate additional local resistance losses in addition to the frictional resistance. Therefore, the calculation of the pressure drop of the undulating pipe section must be considered separately.

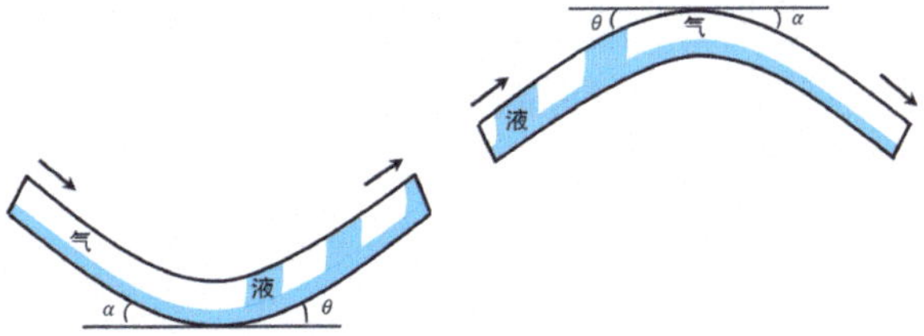

Fig. 2 Flow diagram of gas-liquid two-phase flow in undulation elbow pipe

The diagram of the gas-liquid flow in the bottom/top elbow section is shown in Fig. 2.

Chisholm [13] proposed a semi-theoretical and semi-empirical model that can be used to calculate the pressure drop of gas-liquid two-phase flow through an elbow pipe. The expression of the model is as follows:

$$\begin{aligned}\Delta p_{bm} &= \Delta p_{bl}\left\{1+\left(\frac{\rho_l}{\rho_g}-1\right)\left[\frac{2}{\xi_{bl}}x(1-x)\Delta\left(\frac{1}{s}\right)+x\right]\right\}\\ &= \xi_{bl}\frac{M_m^2}{2A^2\rho_l}\left\{1+\left(\frac{\rho_l}{\rho_g}-1\right)\left[\frac{2}{\xi_{bl}}x(1-x)\Delta\left(\frac{1}{s}\right)+x\right]\right\}\end{aligned} \tag{3}$$

where,

Δp_{bm} is local resistance pressure drop of gas-liquid mixture through curved tube, Pa;
Δp_{bl} is local resistance pressure drop across a curved tube when the two-phase fluid is all liquid, Pa;
ξ_{bl} is local resistance coefficient of the whole liquid phase passing through curved tube, dimensionless;
M_m is mass flow rate of the gas-liquid mixture, kg/s;
A is Cross-sectional area of the pipe, m^2;
ρ_l, ρ_g are density of liquid and gas phases respectively, kg/m^3;
x is mass gas holdup, decimal; $\Delta(1/s)$ is slip ratio increment, dimensionless.

In order to calculate the local pressure drop of the undulating pipe at different angles, the semi-empirical model of D. Chisholm is used to calculate the pressure drop generated by the gas-liquid two-phase flow through the 90° elbow pipe section. Then the local pressure drop of the elbow pipe at any angle is obtained by the resistance coefficient method. According to the Eq. (3), the local pressure drop generated when the gas-liquid flow through the 90° bend is

$$\begin{aligned}\Delta p_{bm(90^\circ)} &= \xi_{bl(90^\circ)}\frac{M_m^2}{2A^2\rho_l}\left\{1+\left(\frac{\rho_l}{\rho_g}-1\right)\left[\frac{2}{\xi_{bl(90^\circ)}}x(1-x)\Delta\left(\frac{1}{s}\right)+x\right]\right\}\\ \xi_{bl(90^\circ)} &= \lambda\frac{L_e}{D}\quad \frac{1}{\sqrt{\lambda}} = -2lg\left(\frac{\varepsilon}{3.7D}+\frac{2.51}{Re_l\sqrt{\lambda}}\right)\quad Re_l = \frac{\rho_l Q_m D}{A\mu_l}\end{aligned} \tag{4}$$

where,

λ is resistance coefficient along the path of single-phase liquid flow, dimensionless;
L_e is the equivalent length of curved pipe, m; ε is equivalent roughness of curved pipe, m;
R_{el} is full liquid phase Reynolds number, dimensionless; Q_m is volume flow of gas-liquid mixture, m^3/s.

For a 90° elbow pipe, the equivalent length L_e can be obtained by looking into the relationship chart between the relative equivalent length L_e/D and the relative radius of curvature R/D.

The equivalent length L_e can also be obtained approximately by the following relationship:

$$\frac{L_e}{D} = 0.004126\left(\frac{R}{D}\right)^6 - 0.141574\left(\frac{R}{D}\right)^5 + 1.913753\left(\frac{R}{D}\right)^4 - 12.92243\left(\frac{R}{D}\right)^3 + 45.750481\left(\frac{R}{D}\right)^2 - 79.088258\frac{R}{D} + 63.376796 \quad (5)$$

where,

R is curvature radius of elbow pipe, m;

D is pipe inner diameter, m.

In addition, for the 90° elbow pipe, D. Chisholm proposed an empirical formula for calculating the slip ratio increment $\Delta(1/s)$ based on the experimental data of the air-water two-phase mixture flowing through the elbow pipe. The relationship is expressed as follows:

$$\Delta\left(\frac{1}{s}\right) = \frac{1.1}{2 + \frac{R}{D}} \quad (6)$$

After all the parameters are known or calculated by the relevant equations, the local pressure drop generated by the gas-liquid two-phase flow passing through the 90° elbow pipe can be calculated by the Eq. (4). For non-90° elbow pipe, when the gas-liquid flow rate, the pipe diameter, the length of the elbow pipe, and the curvature are the same, and the ratio of the local resistance coefficient of the elbow at any angle β to the 90° elbow is obtained.

The equation can be approximately expressed as

$$\frac{\xi_\beta}{\xi_{90}} = 0.0186\left(\frac{\pi\beta}{180}\right)^3 - 0.159\left(\frac{\pi\beta}{180}\right)^2 + 0.685 \bullet \frac{\pi\beta}{180} + 0.2513 \quad (7)$$

and

$$\frac{\Delta p_{bm(\beta)}}{\Delta p_{bm(90^\circ)}} = \frac{\xi_\beta}{\xi_{90}}. \quad (8)$$

By combining the above equation with the Eq. (4), the local pressure drop generated by the gas-liquid two phase flowing through the bottom/top undulating pipe can be obtained, and the pressure drop is expressed as

$$\begin{aligned}\Delta p_{bm(\beta)} &= \frac{\xi_\beta}{\xi_{90}} \cdot \Delta p_{bm(90^\circ)} \\ &= \frac{\xi_\beta}{\xi_{90}} \cdot \lambda \frac{L_e M_m^2}{2DA^2 \rho_l} \left\{ 1 + \left(\frac{\rho_l}{\rho_g} - 1 \right) \left[\frac{2}{\xi_{bl(90^\circ)}} x(1-x)\Delta\left(\frac{1}{s}\right) + x \right] \right\}.\end{aligned} \tag{9}$$

In the equation, for the bottom/top undulating pipe, when the descending angle α and the ascending angle θ are both known, Here $\beta = \alpha + \theta$.

Using the method provided above, the pressure drop of the descending section, the ascending section and the elbow section of the wellbore can be calculated separately, then the pressure drop of the entire unit undulating pipeline can be expressed as:

$$\Delta p_{undulating} = \Delta p_{descending} + \Delta p_{ascending} + 2\Delta p_{elbow}. \tag{10}$$

When the pressure drop of each unit of the undulating pipeline is determined, the pressure drop through the entire undulating wellbore can be easily obtained.

2.4 Comparison and Verification of Pressure Drop Model

In order to verify the accuracy of the continuous undulation wellbore pressure drop prediction model, four groups of different parameters of two-phase flow and undulating pipeline are set here, and the model calculation results are compared with the results calculated by PiPesim which could evaluate the prediction of the model.

The schematic diagram of the continuous undulating wellbore model established by PiPesim is as follows (Fig. 3).

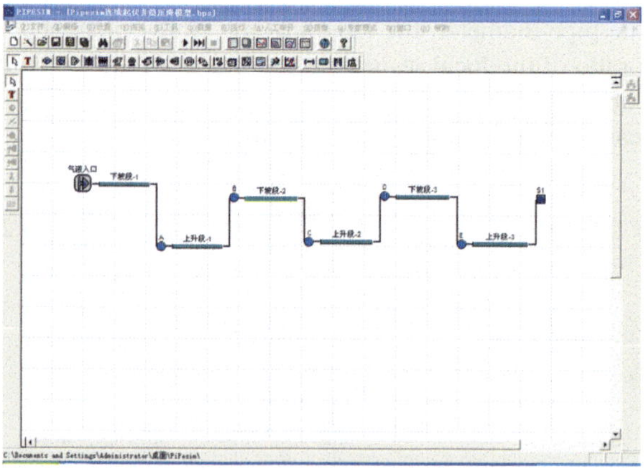

Fig. 3 Schematic diagram of continuous undulating pipeline model

For the above different continuous undulating pipelines, the pressure drop passing through the whole pipeline is calculated by using the PiPesim and the model proposed in this paper respectively. The calculation results are as follows (Table 1).

Table 1 PiPesim calculation results and model prediction results data table

	Diameter D(m)	Curvature R(m)	Inlet pressure(KPa)
1	0.06	0.6	6000
2	0.06	0.6	6000
3	0.06	0.6	6000
4	0.06	0.6	6000
Pressure(KPa)	Model result	PiPesim result	Model result
1	2410.49	3611.9	3589.51
2	3088.13	2922.69	2911.87
3	4404.54	1582.64	1595.46
4	4370.9	1621.96	1629.1

According to the data in the above table, the predicted values (outlet pressure, pressure drop) of the model is similar to those calculated by PiPesim, and the accuracy of the model is very high. In addition, it has been found that although the local pressure drop generated by a single elbow pipe accounts for a low proportion of the total pressure drop of the entire undulating pipeline, when the undulating rate of the pipeline that is the number of undulating units per mile of the pipeline is large, the total local pressure drop loss should not be ignored.

3 Slug Tracking Model of Undulating Wellbore

3.1 Slug Tracking Model Establishment

In order to obtain the flow characteristics of the slug flow better, and maintain the safety of oil and gas production, further study should be carried out on the flow parameters (slug rate, slug length, Taylor bubble length, pressure, liquid holdup, etc.) of the topographic slug flow.

Gomes [14] believes that only when the pipe angle is in the range of 86° to 90°, there would be a symmetrical liquid film around the Taylor bubble, and when the angle is below 86°, the Taylor bubble will move along the upper wall. The flow diagram of the topographic slug flow in the ascending section is as follows.

Figure 4 depicts the basic structure of the slug unit. A complete slug unit comprises liquid slug region and liquid film region, the liquid film region comprises a long bubble at the top and a liquid film at the bottom, which can usually be treated as a stratified flow. In liquid slug region, beside a large amount of liquid, some dispersed bubble is distributed at the front of the liquid slug, which is generally regarded as a bubble flow.

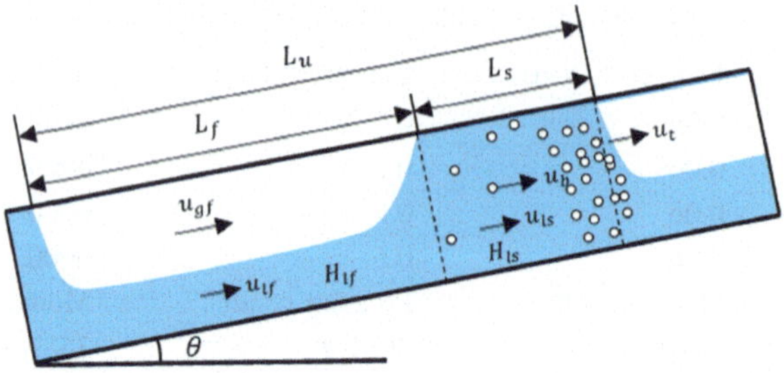

Fig. 4 Structure diagram of slug flow in the ascending section

(1) Liquid holdup:

The liquid holdup H_{ls} in the liquid slug region is calculated using the improved method proposed by Kaya, that

$$H_{ls} = 1 - \frac{u_{sg}}{u_b} \tag{11}$$

$$u_b = 1.208u_m + 1.41\left[\frac{(\rho_l - \rho_g)g\sigma}{\rho_l^2}\right]^{0.25}\sqrt{sin\theta} \tag{12}$$

where,

u_b is speed of dispersed bubbles in the liquid slug, m/s;
u_{sg} is superficial flow rate of gas phase, m/s;
u_m is gas-liquid mixture flow rate in the liquid plug, m/s.

The H_{lf} of the liquid film region can be obtained by solving the momentum equation of the stratified flow. Assuming that the thickness of the liquid film is always constant, the momentum equations of the liquid phase and the gas phase in the liquid film region are shown respectively:

$$-A_f\left(\frac{dp}{dx}\right) + \tau_i S_i - \tau_f S_f - A_f \rho_l g sin\theta = 0 \tag{13}$$

$$-A_g\left(\frac{dp}{dx}\right) - \tau_i S_i - \tau_g S_g - A_g \rho_g g sin\theta = 0. \tag{14}$$

Combine the above two formulas and eliminate the pressure term

$$\tau_f \frac{S_f}{A_f} - \tau_g \frac{S_g}{A_g} - \tau_i\left(\frac{S_i}{A_f} + \frac{S_i}{A_g}\right) + (\rho_l - \rho_g) g sin\theta = 0. \tag{15}$$

The calculation of the relevant parameters in the above formula is as follows
Shear stress

$$\tau_f = f_f \frac{\rho_l u_{lf}^2}{2}, \ \tau_g = f_g \frac{\rho_g u_{gf}^2}{2}, \ \tau_g = f_i \frac{\rho_g \left(u_{gf} - u_{lf}\right)^2}{2}$$

Reynolds number: $Re_f = \frac{\rho_l u_{lf} d_f}{\mu_l}$, $Re_g = \frac{\rho_g u_{gf} d_g}{\mu_g}$

Hydraulic diameter: $d_f = \frac{4A_f}{S_f}$, $d_g = \frac{4A_g}{\left(S_g + S_i\right)}$

The Fanning coefficient between the liquid film and the pipe wall, and between the gas and pipe wall is calculated according to the following formula.

When the laminar flow: ($Re \leq 2000$): $f = 16/Re$;

When the turbulent flow: ($Re > 2000$):

$$\frac{1}{\sqrt{f}} = 3.48 - 4lg\left(\frac{2\varepsilon}{D} + \frac{9.35}{Re\sqrt{f}}\right)$$

The Fanning coefficient of the gas-liquid film interface is constant, that $f_i = 0.0142$.

The cross-section area of the liquid film and the gas phase are shown respectively $A_f = \frac{1}{8}D^2(\alpha - sin\alpha)$, $A_g = \frac{1}{4}\pi D^2 - A_f$

Wet perimeter: $S_f = \frac{1}{2}D\alpha$, $S_g = \pi D - \frac{1}{2}D\alpha$, $S_i = Dsin\frac{\alpha}{2}$

Where the radian α is: $\alpha = 2arccos\left(1 - 2\frac{h_l}{D}\right)$.

Substituting the above relation into the implicit equation Eq. (15), the liquid film thickness h_l or the dimensionless parameter h_l/D of the liquid film can be obtained by the trial method. Then, according to the geometric relationship, the liquid holdup H_{lf} of the liquid film region is obtained:

$$H_{lf} = \frac{\alpha - sin\alpha}{2\pi} \quad (16)$$

(2) Speed parameter

When the holdup of liquid slug region and the liquid film region are known, the liquid phase velocity u_{ls} in the liquid slug can be obtained by the continuity equation in the cross section of the liquid slug region

$$u_m = u_{ls}H_{ls} + u_b(1 - H_{ls}) \quad (17)$$

U_b is the speed of the dispersed bubble in the liquid slug, which can be calculated by Eq. (12). Similarly, the continuity equation on the cross section in the liquid film region is:

$$u_m = u_{lf}H_{lf} + u_{gf}\left(1 - H_{lf}\right). \quad (18)$$

Due to the deceleration effect of gravity, the speed of liquid film in inclined pipe flow is generally lower than that of liquid slug. Therefore, for the slug unit in a stable flow, the volume of liquid entrainment in slug should be equal to the volume of liquid which is relative sliding in liquid film, and it can be concluded that:

$$(u_t - u_{ls})H_{ls} = \left(u_t - u_{lf}\right)H_{lf}. \tag{19}$$

By combining Eqs. (18) and (19), the speed of u_{lf} and u_{gf} of liquid and gas phase in the liquid film region can be obtained. Where u_t is the translational velocity of the liquid slug or the velocity of the Taylor bubble, which can be calculated using the relation proposed by Bendiksen:

$$u_t = 1.2u_m + (0.35sin\theta + 0.54cos\theta)\sqrt{gD}$$

(3) Length parameter

The length of the liquid slug L_s is calculated directly by the empirical relation formula proposed by Scott:

$$ln(L_s) = -26.8 + 28.5[lnD + 3.67]^{0.1}. \tag{20}$$

Assuming that the liquid film area is homogeneous and the gas and liquid is incompressible, the liquid phase continuity equation over the entire slug unit is:

$$u_{sl}L_u = u_{ls}H_{ls}L_s + u_{lf}H_{lf}L_f.$$

According to the above formula and $L_u = L_s + L_f$, then the length of the entire slug unit L_u can be expressed as:

$$L_u = \frac{u_{ls}H_{ls} - u_{lf}H_{lf}}{u_{sl} - u_{lf}H_{lf}} \cdot L_s \tag{21}$$

(4) liquid slug frequency

The frequency of the liquid slug refers to the number of slug units that pass through the unit time at a certain point along the pipeline. Therefore, when both the translational velocity of slug u_t and the length of slug unit L_u are known, the frequency of slug can be calculated according to the formula:

$$f_s = \frac{u_t}{L_u}. \tag{22}$$

(5) pressure drop

In addition to the gravitational pressure drop and frictional pressure drop, the acceleration pressure drop caused by the liquid entrainment at the front of the slug is also considered, and it is assumed that the acceleration pressure drop only exists in the first half of the slug. Then, according to the momentum balance, the pressure drop expression of the liquid slug is

$$\begin{aligned}\Delta p_s &= \frac{1}{2} f_m \rho_m u_m^2 \frac{\pi D L_s}{A} + \rho_m g sin\theta L_s + \rho_l H_{ls}(u_{ls} - u_{lf})(u_t - u_{ls}) \\ \rho_m &= H_{ls}\rho_l + (1 - H_{ls})\rho_g.\end{aligned} \tag{23}$$

The fanning coefficient f_m of the gas-liquid mixture in the slug and the pipe wall is still calculated according to the above methods, and the relevant parameters are as follows:

$$Re_m = \frac{\rho_m u_m D}{\mu_m} \qquad \mu_m = \mu_l H_{ls} + \mu_g(1 - H_{ls}).$$

The calculation of two-phase pressure drop in the liquid film region is calculated according to the previous stratified flow method. Then, the Eqs. (12) and (13) are combined together and the shear stress term of the gas-liquid interface is eliminated, the pressure drop expression in the liquid film region is obtained as follows:

$$\Delta p_f = \frac{\tau_f S_f + \tau_g S_g}{A} L_f + \left[H_{lf}\rho_l + (1 - H_{lf})\rho_g\right] g sin\theta L_f. \tag{24}$$

Then the pressure drop of the entire slug unit can be expressed as:

$$\Delta p = \Delta p_s + \Delta p_f. \tag{25}$$

3.2 Evaluation of Slug Tracking Model

This model can be used to express the transient flow characteristics of the slug flow. The undulating two-phase flow device is used to measure the pressure with the different time in the ascending section of the slug flow, then compare it with the predicted result of the model. The output results under specific pipeline parameters and two-phase flow rate are as follows.

It can be seen from Fig. 5 that as certain gas liquid flow rate, the pressure of the slug flow in ascending section presents periodic fluctuation changes over time, and this pressure fluctuation is also consistent with the transient and intermittent characteristics of slug flow. This is because when the slug passes the measurement point, the high liquid holdup and large liquid flow rate lead to the high pressure. However, when the liquid film passes through, the transient pressure is small due to the low liquid holdup in the liquid film area and the small liquid flow rate. As the liquid slug and the liquid

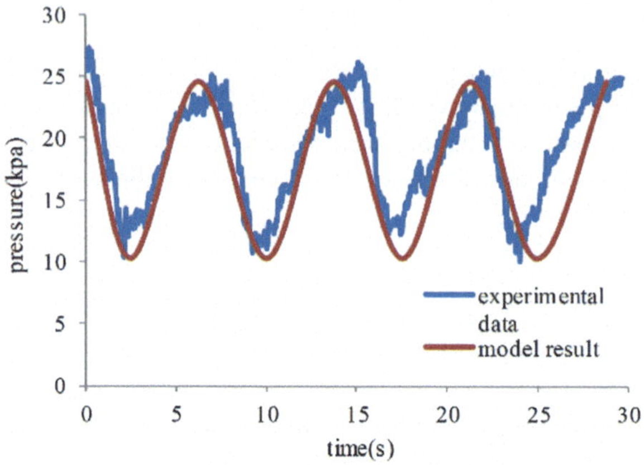

Fig. 5 Model calculation and measured pressure comparison chart (Experimental measurements Pmax = 27.35 kPa, model calculation Pmax = 24.63 kPa, model error −9.95%.)

film flows over and over, the instantaneous pressure at one point on the pipe exhibits periodic fluctuations.

4 Conclusion

(1) The pressure drop of continuous undulating pipeline can be calculated through the descending section-ascending section-bottom/top elbow sections respectively. Based on the pressure drop of each undulation unit, and then the two-phase pressure drop of the entire undulation pipeline can be obtained. In this paper, the prediction results of the continuous undulating wellbore pressure drop model are compared with those calculated by PiPesim with the same parameters, it is found that although the local pressure drop caused by a single elbow pipe accounts for a relatively low proportion of the total pressure drop, the total local pressure drop loss should not be ignored.

(2) The slug can be divided into liquid slug region and liquid film region. The liquid slug region can be treated as the bubble flow, and the liquid film region can be treated as stratified flow or wavy flow. In addition, during the calculation of the pressure drop, the acceleration pressure drop at the front of the liquid slug entrainment should not be ignored.

(3) The instantaneous pressure of the ascending section at a specific position will exhibit periodic fluctuations over time, this is because when the slug passes the measurement point, the high liquid holdup and large liquid flow rate lead to the high pressure. However, when the liquid film passes through, the transient pressure is small due to the low liquid holdup in the liquid film area and the small liquid flow rate. As the liquid slug and the liquid film flow over and over, the instantaneous pressure at one point exhibits periodic fluctuations.

Acknowledgements. This work was supported by the National Natural Science Foundation of China (Grant No. 51574256) and National Science & Technology Major Project (Grant No.2016ZX05072).

References

1. Rothe, P.H., Crowley, C.J.: Insignificant terrain effects in multiphase pipelines. Creare Incorporated (1986)
2. Scott, S.L., Kouba, G.E.: Advances in slug flow characterization for horizontal and slightly inclined pipelines. SPE-20628-MS (1990)
3. Zheng, G.H., Brill, J.P., Taitel, Y.: Slug flow behavior in a hilly terrain pipeline. Int. J. Multiph. Flow **20**(1), 63–79 (1994)
4. Zheng, G.H., Brill, J.P., Shoham, O.: An experimental study of two-phase slug flow in hilly terrain pipelines. SPE Prod. Facil. **10**(4), 233–239 (1995)
5. Dehenau, V., Raithby, G.D.: A study of terrain-induced slugging in two-phase flow pipelines. Int. J. Multiph. Flow **21**(3), 365–379 (1995)
6. Zhang, H.Q., Al-Safran, E., Jayawardena, S. et al.: Modeling of slug dissipation and generation in a hilly-terrain pipeline. In: 2nd North American Conference on Multiphase Technology, pp. 161–168 (2000)
7. Zhang, H.Q., Wang, Q., Brill, J.P., et al.: Unified model for gas-liquid pipe flow via slug dynamics. ASME, Energy Resour. Technol. **125**(4), 266–283 (2003)
8. Al-Safran, E., Sarica, C., Zhang, H.Q., et al.: Investigation of slug flow characteristics in the valley of a hilly-terrain pipeline. Int. J. Multiph. Flow **31**(3), 337–357 (2005)
9. Al-Safran, E., Sarica, C., Zhang, H.Q., et al.: Mechanistic/Probabilistic modeling of slug initiation in a lower elbow of a hilly-terrain pipeline. SPE Prod. Oper. **23**(1), 88–99 (2008)
10. Al-Safran, E., Gokcal, B., Sarica, C.: Investigation and prediction of high-viscosity liquid effect on two-phase slug length in horizontal pipelines. SPE Prod. Oper. **28**(3), 296–305 (2013)
11. Al-Safran, E.: Prediction of slug liquid holdup in horizontal pipes. J. Energy Res. Technol. **131**(2), 1–8 (2009)
12. ZHANG Qi, Principle and design of oil production engineering[M]. Dongying Shandong; China University of Petroleum Press,2004: 23 ~ 32
13. Chisholm, D.: Two-phase flow in bends. Int. J. Multiph. Flow **6**(4), 363–367 (1980)
14. Gomez, L.E., Schmidt, Z., Chokshi, R.N.: Unified mechanistic model for steady-state two phase flow: horizontal to vertical upward flow. SPE J. **5**(3), 339–350 (2000)

Chemical Mechanism in the Fluid Loss Additive Modified Oil Well Cement System

Di Chen and Jin-tang Guo(✉)

School of Chemical Engineering and Technology, Tianjin University, Tianjin, China
bettychgch@163.com, jtguo@tju.edu.cn

Abstract. Fluid loss additive (FLA) is an important chemical agent added into oil well cement to prevent cement slurry from losing water. In this work, chemical mechanism between FLA and oil well cement was investigated by multiple technologies. Compared the neat cement and the cement slurry with FLA, it is found that FLA prolongs the induction period of cement hydration, which is proved by the data of compressive strength, hydration kinetics and hydration products. The Ca^{2+} concentration, pH values and total organic carbon of cement pore solution were measured. The results show that FLA adsorbed on cement surface and the complexation occurred between Ca^{2+} and −COOH groups to build Ca^{2+} bridges between FLA and cement. Meanwhile, the fluid loss control ability of FLA is also attributed to the complexation.

Keywords: Oil well cement · Fluid loss additive · Adsorption · Complexation

1 Introduction

Polymer-based additive is one of chemical agents added into oil well cement to improve the performance of the cement slurry and to ensure cementing quality. Particularly, the polymer fluid loss additive is an important one to prevent cement slurry from losing water [1]. The most widely used for fluid loss additives are anionic copolymers synthesized by radical copolymerization of sulfonic acid and carboxylic acid monomers.

J. Lin (ed.), *Proceedings of the International Petroleum and Petrochemical Technology Conference 2019*, pp. 204–212, 2020.
https://doi.org/10.1007/978-981-15-0860-8_16

In the past years, works in the field of cement and concrete science focused on the interactions between cement and additives. Generally, it is believed that hydration reaction made the surface of cement particles to show positively charged, once cement mixed with water. The polymer with anionic groups could adsorb on the surface of cement with the positive ions through electrostatic interaction [2]. For another, the mechanism of fluid loss additive is achieved by improving the microstructure of filter cake [3]. So far, the adsorption and improving the microstructure are the main mechanisms which have been widely accepted.

In this work, the effects of FLA on the hydration of oil well cement were analyzed by compressive strength test, isothermal calorimeter and X-ray powder diffraction (XRD). The microstructure of set cement was observed by Scanning Electron Microscope (SEM). The influence of FLA on the pore solution was analyzed via Ca^{2+} concentration, pH values and total organic carbon. These data confirmed that the complexation occurred between Ca^{2+} and –COOH groups to build Ca^{2+} bridges between the FLA and cement.

2 Experimental

2.1 Materials

Class G oil well cement in this paper was provided by Jiahua Special Cement Co., Ltd (Sichuan, China). The chemical compositions of cement were measured by X-ray fluorescence spectrometry (S4 Pioneer, Bruker, Germany), which were given in Table 1.

Fluid loss additive (FLA) was obtained by free radical polymerization according to literature [4]. FLA is the terpolymer of 2-arcylamido-2-methylpropane sulfonic acid (AMPS), itaconic acid (IA) and N, N-dimethylacrylamide (NNDMA) with a solid content of 25 wt%.

Table 1 Chemical compositions of Class G oil well cement (wt%)

CaO	SiO_2	Fe_2O_3	Al_2O_3	MgO	K_2O	SO_3	Loss on ignition
71.3	14.1	6.3	2.0	0.8	0.8	2.9	1.6

2.2 Characterization of FLA

The copolymer solution was purified by dialysis against deionized water. The purified sample was freeze-dried to be used for chemical structures and performance. The purified sample was analyzed by Bruker Vector 22 Fourier infrared spectrometer (Germany) within a spectral range from 4000 to 400 cm^{-1}. Model TGA-Q500 thermal gravimetric analyzer (Shimadz, Japan) was used to investigate the thermal stability of FLA. The heating rate is 10 °C/min and the temperature range is from room temperature to 600 °C.

2.3 Preparation and Characterization of Cement Slurry

Cement slurries were prepared according to API Recommended Practice 10B issued by American Petroleum Institute (API) [5]. The water/cement (W/C) ratio of the cement paste was fixed at 0.44 and FLA content in cement paste was 3% by weight of cement (bwoc).

High-pressure, high-temperature fluid-loss apparatus (TG-71 type, Shenyang taige oil equipment Corporation, China) was used for the static fluid loss test of oil well cement slurries. Automatic pressure testing machine (Model 4207, Chandler Engineering, USA) was used for the compressive strength test at 60 °C. An automatic cement hydration calorimeter (YT12959-08, Wuhan Yite Instrument, China) was used to record the heat evolution within 72 h. Crystal structures of hardened cement pastes were investigated by X-ray powder diffraction (X'Pert Pro, PANalytical, Holland). Morphology was observed by Field Emission Scanning Electron Microscope (Hitachi S-4800, Japan).

2.4 Characterization of Cement Pore Solution

Adsorbed amount of FLA on oil well cement surface was determined by total organic carbon analyzer (Shimadzu, Japan). The filtrate was obtained by static fluid loss test and then diluted with deionized water for TOC analysis. The amount of FLA adsorbed on oil well cement was calculated from the difference between the amount of organic carbon in liquid before and after mixing. EDTA titration is used to measure the amount of Ca^{2+} in the pore solution, which can prove the chelation of Ca^{2+} [6]. Meanwhile, the pH value of the pore solutions at different hydration time was measured by pH meter (FE20, Mettler Toledo, Switzerland) [7].

3 Results and Discussion

3.1 Structure and Property of Copolymer FLA

FTIR analysis. Figure 1 represents the FTIR spectrum of FLA. The peak observed in the spectrum of FLA at 3440 cm^{-1} is attributed to the N–H of AMPS. The peaks at 2972 and 2931 cm^{-1} are associated to the C–H stretching vibration. The peak at 1659 cm^{-1} is attributed to –C=O of amide and carboxyl, the peak at 1218 cm^{-1} is the C–N group of amide, and the peak at 1193 cm^{-1} is C–O of carboxyl. The peak at 1044 cm^{-1} is attributed to –S=O group of AMPS. It shows that the product is the copolymer of AMPS, NNDMA and IA.

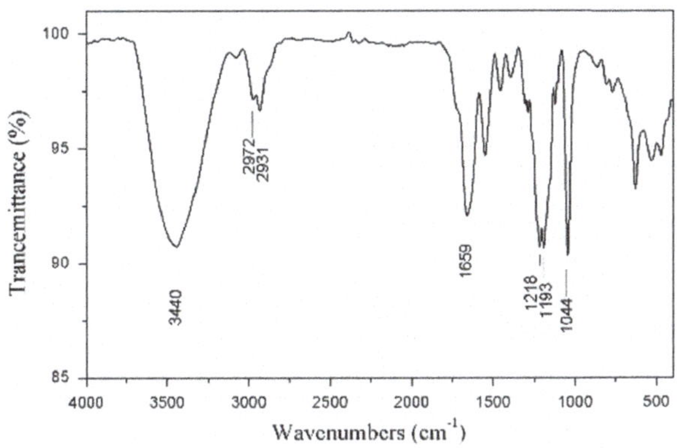

Fig. 1 FTIR spectrum of FLA

TGA analysis. Thermo Gravimetric Analysis (TGA) is widely used for characterizing the thermal stability of copolymers [8]. Figure 2 shows the TGA results of FLA which could be seen three weight loss stages. The first weight loss stage is from 60 to 153 °C with a weight loss of 3.48%, which is because the evaporation of H_2O coordinated with polymeric chain via hydrogen bonds [8]. Then, a striking weight loss occurred at the temperature up to 300 °C, this is the second weight loss stage with loss of 36.4%. The last weight loss stage with loss of 17.2% occurred from 335 to 500 °C. Thus, FLA can maintain stable structure below 300 °C and have good thermal resistance.

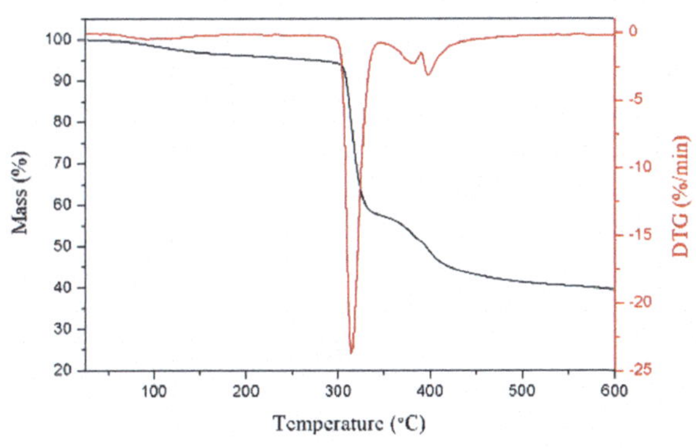

Fig. 2 TGA curves of FLA

3.2 Effects of FLA on Hydration Process

Based on the previous investigation [8] and application in oil well, the fluid loss of cement paste with FLA can be controlled within 50 ml. The static fluid loss test was repeated again to ensure the stability of the FLA. Furthermore, the compressive

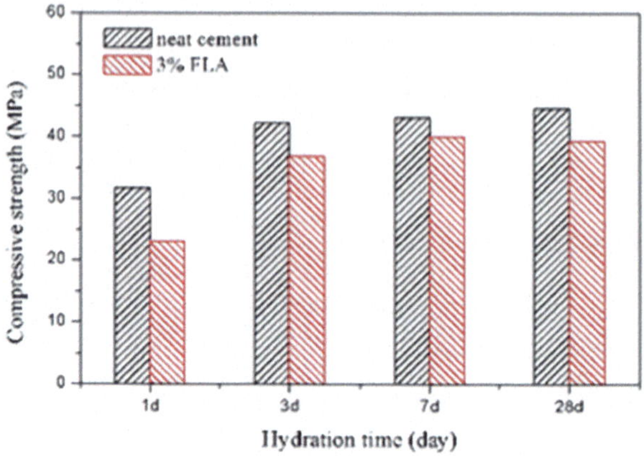

Fig. 3 Compressive strength of set cement at 60 °C

strength, hydration kinetics and hydration products were used to analyze the effects of FLA on cement hydration process.

Compressive strength. The development of compressive strength with time was measured, in order to evaluate the effect of FLA on the mechanical property of hardened cement. Figure 3 shows the compressive strength within 1, 3, 7 and 28 days at 60 °C. The oil well cement samples without FLA were treated as the neat cement. Compared with the neat cement, the compressive strength of the set cement with 3% bwoc FLA reduced due to the presence of carboxyl groups in copolymer [9].

Hydration kinetics. Figure 4 shows the heat evolution rates curves and cumulative heat curves of neat cement slurry and cement slurry with 3% bwoc FLA at 40 °C. From Fig. 4a, it can be seen that FLA prolonged the induction period of cement hydration and reduced the hydration rate in the acceleration stage [10]. At the same time, FLA decreased the total heat during curing ages, which is shown in Fig. 4b. These phenomena exactly explained why FLA reduced the compressive strength of the set cement. The anionic groups on FLA combined with Ca^{2+} from cement to prolong the cement hydration process.

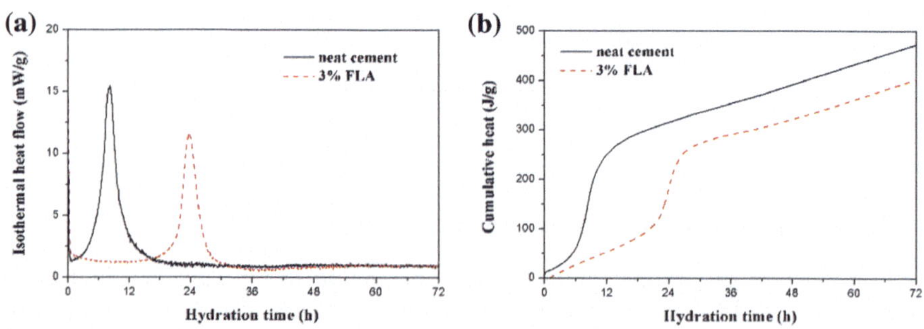

Fig. 4 The heat evolution rates (**a**) and cumulative heat (**b**) of the neat cement slurry and cement slurry with 3% bwoc FLA

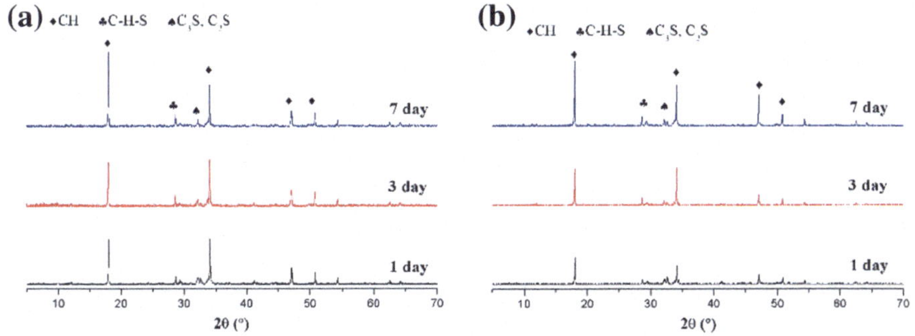

Fig. 5 XRD patterns of set cement for 1, 3 and 7 days at 60 °C, **a** neat cement, **b** set cement with 3% bwoc FLA

Hydration products. Figure 5 gives the mineral composition of set cement with 1, 3 and 7 days at 60 °C to analyze the hydration process. The mainly mineral compositions include portlandite (CH) with the peaks at 18.06°, 34.10° and 47.12°, calcium silicate hydrate (C–S–H) and unhydrated minerals (C_3S and C_2S) [11]. The precipitates of portlandite (CH) formed at the acceleration stage of cement hydration [8]. From the intensity of CH in Fig. 5a and b, the FLA put off the form of CH crystals due to the chelation with Ca^{2+}. This result is in accordance with hydration kinetics in Fig. 4.

3.3 Microstructure Analysis by SEM

SEM is a direct technology to investigate the microstructure of materials. In this section, SEM was used to observe the pore structure of set cement. As shown in Fig. 6a, lots of pores existed in the neat cement. However, the set cement with 3% bwoc FLA in Fig. 6b shows less pore structure. The hydration product CH could be clearly observed from the image. CH is scattered distributed in cement which is marked by red regions in Fig. 6a. However, it can be seen a large area of CH in the field of vision in Fig. 6b. The large area of CH is more likely to break the cement stone due to poor continuity of the C–S–H structure. Thus, FLA is of benefit to the formation of CH, but not good for compressive strength.

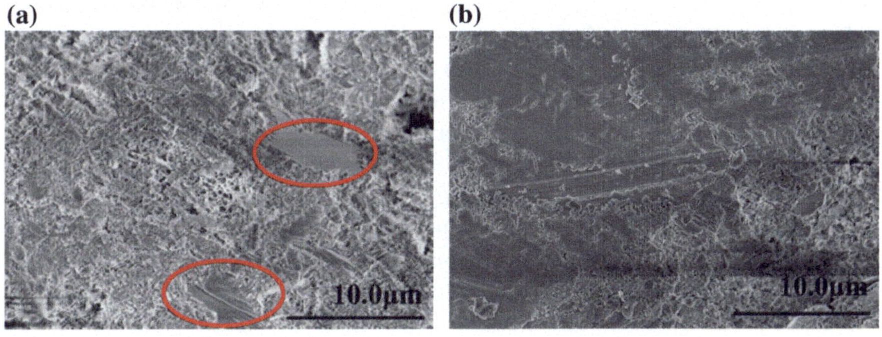

Fig. 6 SEM images of hydration cement cured for 3 days at 60 °C, **a** neat cement, **b** set cement with 3% bwoc FLA

3.4 Effects of FLA on the Pore Solution

Adsorption behavior of FLA on cement surface. The filtrate was obtained by static fluid loss test to use for TOC analysis. The adsorbed amount was calculated from the difference between the amount of organic carbon in liquid before and after mixing. Figure 7 illustrates the adsorbed amount of FLA on the surface of cement with time. It can be clearly seen that the equilibrium adsorbed amount decreases with the temperature, but the equilibrium rate increases. The molecular motion of FLA increased at higher temperature, which leads to desorption, as well as accelerating reaction rate. Usually, the surface of cement particles shows positively charged, and polymers have anionic groups. The adsorption of polymer onto cement surface occurred by electrostatic interaction [2]. Thus, the adsorption behavior is due to the chelation with Ca^{2+}.

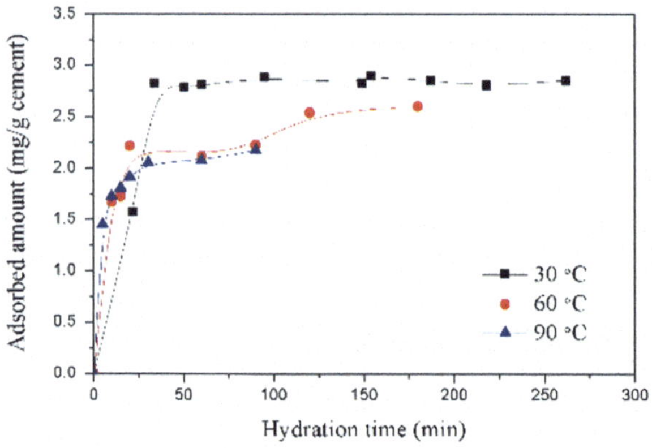

Fig. 7 The adsorbed amount of FLA on the surface of cement with time

Ca^{2+} concentration and pH values of pore solutions. The effect of copolymer on Ca^{2+} concentration in pore solutions was investigated in this section. The filtrate of cement slurry was obtained by centrifugation and used for Ca^{2+} concentration test. Figure 8a illustrates the effect of copolymer on Ca^{2+} concentrations in cement pore solutions at 30 °C. FLA makes the Ca^{2+} concentrations in cement pore solutions increase rapidly. The copolymer complexes calcium ions and inhibits the formation of CH. The nucleus formation of CH is dependent on the calcium ion concentration in the pore solution [8]. Thus, it is considered that –COOH groups on FLA are complexed with Ca^{2+} from cement and retained Ca^{2+} in the solution.

Figure 8b shows the effect of copolymer on pH values of cement pore solutions at 30 °C. Whatever the cement pore solution with and without FLA, the pH value increased during the time from 0 to 30 min. The pore solutions become alkali-rich due to the fast dissolving of some mineral components when the water mixed with cement [12]. At the same time, an intense heat is quickly released [10], which can be observed in Fig. 4a. Then, the pH value of the solution reduces and is close to equilibrium. However, it is shown that FLA makes the pH value of the solution reduces from

Fig. 8b. FLA containing the –COOH and $–SO_3H$ would lead the pH value to decrease. For another, –COOH groups are complexed with Ca^{2+} to produce $Ca(HCOO)_2$ which is slightly alkaline, while the production of $Ca(OH)_2$ in the neat cement shows a higher pH value [12].

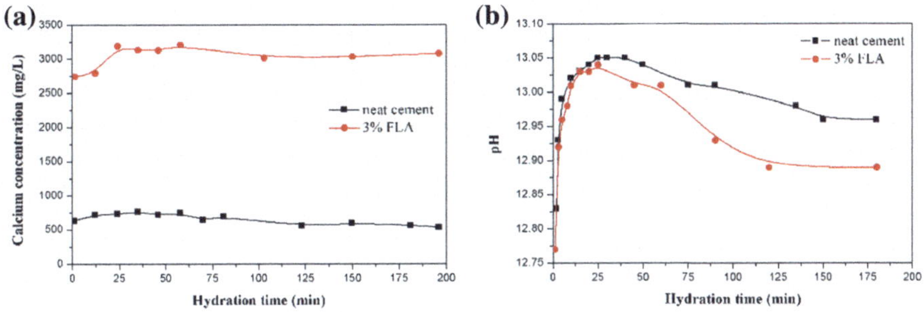

Fig. 8 The effect of copolymer on **a** Ca^{2+} concentrations and **b** pH values of cement pore solutions

4 Conclusions

In this work, the interactions between FLA and oil well cement were investigated through different testing methods. FLA containing –COOH groups could prolong the induction period of cement hydration, which is induced by the complexation of Ca^{2+} and –COOH groups. Meanwhile, the water loss control ability of FLA is also attributed to the complexation. FLA adsorbed on cement surface and the complexation occurred between Ca^{2+} and –COOH groups to build Ca^{2+} bridges between FLA and cement.

References

1. Xia, X., Feng, Y., Guo, J., Liu, S., Jin, J., Yu, Y.: Zwitterionic copolymer for controlling fluid loss in Oil well cementing: Preparation, characterization, and working mechanism. Polym. Eng. Sci. **57**, 78–88 (2017)
2. Zhong, S., Han, D., Li, J.: Some aspects about adsorption of polymer on cement grain. Adv. Mater. Res. **687**, 341–346 (2013)
3. Bu, Y., Liu, H., Nazar, A., He, Y., Song, W.: Amphoteric ion polymer as fluid loss additive for phosphoaluminate cement in the presence of sodium hexametaphosphate. J. Nat. Gas Sci. Eng. **31**, 474–480 (2016)
4. Guo, J., Lu, H., Liu, S., Jin, J., Yu, Y.: The novel fluid loss additive HTF-200C for oil field cementing. Petrol. Explor. Dev. **39**, 385–390 (2012)
5. American Petroleum Institute: ANSI/API Recommended Practice 10B-2, Recommended Practice for Testing Well Cements, 1st edn. American Petroleum Institute, Washington (2005)
6. Liu, H., Bu, Y., Sanjayan, J., Shen, Z.: Effects of chitosan treatment on strength and thickening properties of oil well cement. Constr. Build. Mater. **75**, 404–414 (2015)

7. Wang, M., Wang, R., Zheng, S., Farhan, S., Yao, H., Jiang, H.: Research on the chemical mechanism in the polyacrylate latex modified cement system. Cem. Concr. Res. **76**, 62–69 (2015)
8. Xia, X., Guo, J., Chen, D., Feng, Y., Yu, Y., Jin, J.: Hydrophobic associated copolymer as a wide temperature range synthetic cement retarder and its effect on cement hydration. J. Appl. Polym. Sci. **134**, 45242 (2017)
9. Cao, L., Guo, J., Tian, J., Xu, Y., Hu, M., Guo, C., Wang, M., Fan, J.: Synthesis, characterization and working mechanism of a novel sustained-release-type fluid loss additive for seawater cement slurry. J. Colloid Interface Sci. **524**, 434–444 (2018)
10. Kong, X., Emmerling, S., Pakusch, J., Rueckel, M., Nieberle, J.: Retardation effect of styrene-acrylate copolymer latexes on cement hydration. Cem. Concr. Res. **75**, 23–41 (2015)
11. Ma, S., Li, W., Zhang, S., Hu, Y., Shen, X.: Study on the hydration and microstructure of Portland cement containing diethanol-isopropanolamine. Cem. Concr. Res. **67**, 122–130 (2015)
12. Yoshioka, K., Tazawa, E.I., Kawai, K., Enohata, T.: Adsorption characteristics of superplasticizers on cement component minerals. Cem. Concr. Res. **32**, 1507–1513 (2002)

Quantitative Characterization of Bitumen in the Complex Carbonate of Sichuan Basin: A Case Study

Gang Wei[1(✉)], Zuo-an Zhao[2], Yu-ran Yang[2], Li-chuan Deng[1], and Xian-lei Zeng[1]

[1] Baker Hughes, a GE Company, Houston, USA
{Gang.wei,Lichuan.deng,Xianlei.Zeng}@bhge.com
[2] Southwest Oil and Gas Field Company, CNPC, Chengdu, China
{Zhaozuoan,Yangyuran}@petrochina.com.cn

Abstract. In Sichuan Basin, Southwest of China, the occurrence of bitumen has been reported in the deep-buried carbonate gas field, where the reservoir spaces are mainly dissolved pores and vugs. The bitumen occupies the pore volume and blocks the flow channel of fluid, adversely impacted on the reservoir performance. Hence, identification and quantification of bitumen are critical for the reservoir evaluation. However, lacking of contrast from conventional logs between bitumen and other fluids makes it difficult to differentiate them. Fortunately, with the invention of some hi-end logging tools, the identification of bitumen is becoming reliable, even the quantification is becoming possible. In this case, nuclear magnetic resonance (NMR) and neutron spectroscopy tools are used, originally to acquire the accurate porosity and mineralogy. The NMR response of bitumen is dominated by short transverse relaxation times (T2), providing an opportunity to identify bitumen by porosity deficit. With the application of 2D NMR, the saturation and fluid density could be determined, which are important inputs of bitumen quantification. High-energy pulse neutron tool measures the main element of bitumen, such as carbon (C) and sulfur (S) from collision of fast neutron, and calculates the grain density, providing another crucial parameter to calculate the pore volume occupied by bitumen. This case study demonstrates the bitumen volume derived from the combination of NMR, neutron spectroscopy and conventional logs. The presence of bitumen

J. Lin (ed.), *Proceedings of the International Petroleum and Petrochemical Technology Conference 2019*, pp. 213–220, 2020.
https://doi.org/10.1007/978-981-15-0860-8_17

is validated by cores. Based on the bitumen analysis, engineer could optimize their completion and acid fracturing strategy to improve the reservoir performance; meanwhile geologist could further study the evolution progress of reservoir. This method applies to those conventional reservoirs that suffer the harm of bitumen, and is also suitable in the heavy oil evaluation.

Keywords: Bitumen · Nuclear magnetic resonance · High-energy pulse neutron logging

1 Introduction

Located in Southwest of China, Sichuan Basin is well known by petroleum industry for its very thick carbonate formation and the huge gas reserve. The methane split from oil as a product of thermal decomposition, at the same time the heavy molecular composition left bituminized. That bitumen distributes in pores, vugs and fractures as observed from drilling cores, and it is extremely viscous and immovable. DENGYING IV, one of those reservoir formations, has a very low porosity due to buried more than 5000-meter-deep; each well in this field need to be acid fractured for stimulation. However, the bitumen occupies the pore volume and blocks the flow channel, reducing the production rate and making stimulation less effective. The quantitative characterization of bitumen is therefore becoming crucial when it is present in significant amount.

NMR porosity deficit, as a response to short transverse relaxation times (T2) of Bitumen, has been discussed by many Petrophysicists. In this work, we will integrate NMR with neutron spectroscopy and conventional log, and introduce a more accurate approach for bitumen volume estimation.

2 NMR Porosity Deficit of Bitumen

NMR logging tool measures the hydrogen in the rock by detecting the decay of signal amplitude. In reservoir system, hydrogen is mostly present in water and hydrocarbons. Therefore, after calibration with water, we are able to convert the signal amplitude to apparent porosity. Bitumen, as one kind of hydrocarbon, contains hydrogens should be detected NMR. However, the signal decay of bitumen is very fast, predominately determined by its extreme viscosity. In this field, the NMR response of bitumen is more like rock matrix rather fluid and undetectable by NMR tool due to hardware limitation. As a result, the porosity volume filled by bitumen will be missing in NMR porosity (see Fig. 1).

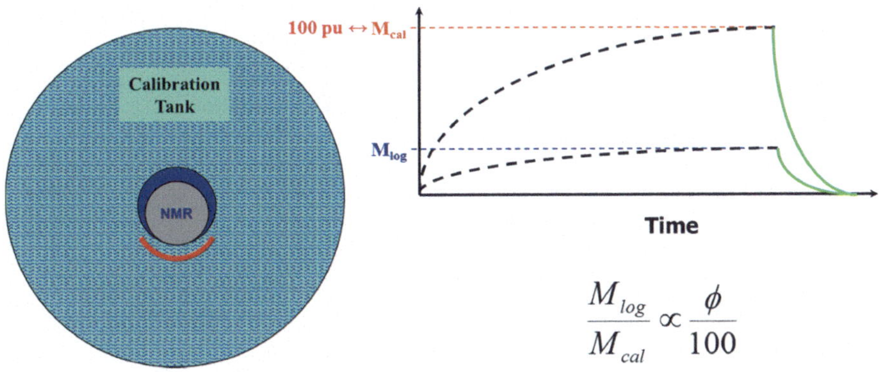

$$\frac{M_{log}}{M_{cal}} \propto \frac{\phi}{100}$$

Fig. 1 NMR measurement principle for bitumen

The other way around, conventional porosity log such as triple combo, respond to all the matrix and fluids within the sensitive volume, including the bitumen filled pores. Compare to triple combo log, the NMR porosity will be smaller when bitumen is present, in other words, NMR porosity deficit will appear. In some cases, we use the difference between NMR and triple combo approximately as the bitumen volume. Nevertheless, the triple combo response to bitumen is different from water and oil, the difference makes this quantification method not accurate enough.

3 Neutron Spectroscopy Response to Bitumen

Neutron spectroscopy tool emits high-energy neutron into formation from a pulse neutron source. As the fast neutrons are passing through the formation, they present two types of interaction with atoms of different elements, namely inelastic collision and capture process. These interactions produce gamma photons with energy characteristic for these elements involved. After recorded by BGO sensor then analyzed with specialized software, those gamma ray spectrums reveal the elements constituting the rock as well as their weight fraction.

Carbon and sulfur are two main composition elements of bitumen, and detectable for neutron spectroscopy tool. In this discussed carbonate formation, carbon also present in rock matrix, making it difficult to tie the carbon from log directly to the presence of bitumen. Firstly, we have to perform geochemistry analysis with all the elements from the log, calculating the mineralogy and the carbon needed to constitute the carbonates. After deducting the carbonate carbon from the total logged carbon, the remaining or excess carbon will represent hydrocarbon filled in pores. By utilizing excess carbon and sulfur together, the identification will be much more reliable (see Fig. 2).

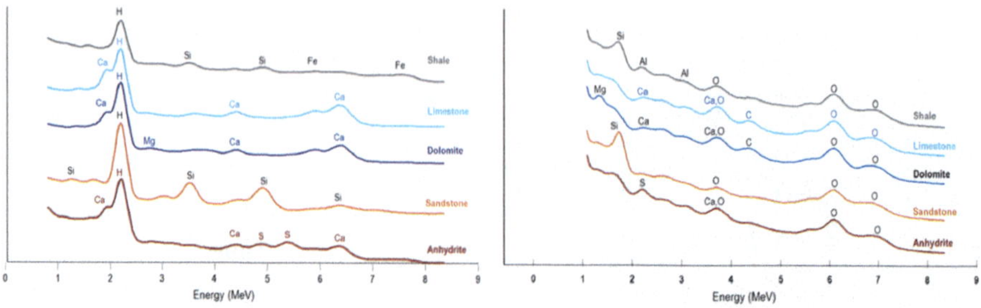

Fig. 2 Neutron Spectroscopy Response to different elements

Figure 3 shows two examples in this discussed field using carbon and sulfur to identify bitumen filled zone. In these two wells, both drilling core and neutron spectroscopy are available along with the slam log. It is very clear that excess carbon and sulfur always show up simultaneously on the log, and with the appearance of them, the black bitumen are visible from the cores. In the meantime, there is no clue of presence of bitumen from the slam log.

These two examples illustrate the success of bitumen identification in a comparative way between neutron spectroscopy log and core, and prove that this methodology is reliable.

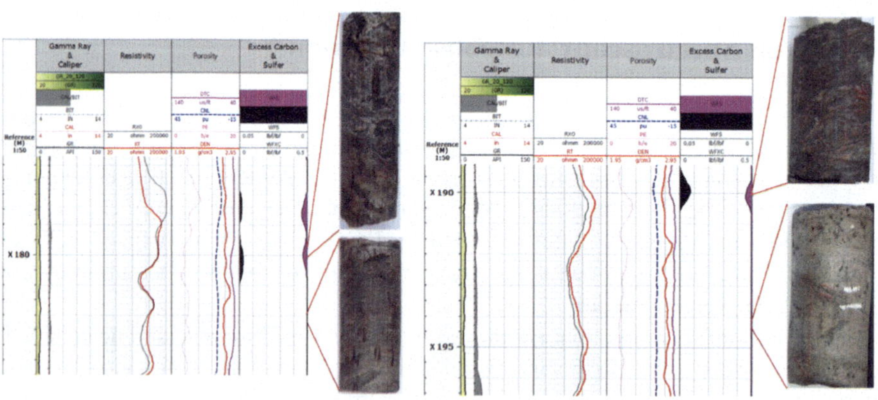

Fig. 3 Log processed result and bitumen in the core fracture

At this step, the rock matrix density, or grain density (ρ_{Grain}) is also calculated based on minerals of rock and their weight fraction. It is a very important parameter of volumetric model and we will discuss later.

4 Fit-for-Purpose Petrophysical Volumetric Model

In previous section, we discussed a quick but qualitative identification for reservoir bitumen. In this section, adding bitumen into the conventional volumetric model, we will use a revised model to quantify bitumen volume and calculate the corrected total porosity.

Density log, or neutron and acoustic, will respond to the rock matrix and entire pore inside the sensitive volume, no matter those pore are fluid filled or bitumen filled (see Fig. 4), it can be modelled using Eq. (1). The matrix density (ρ_{Grain}) is calculated with mineralogy analysis in last section. Bitumen density $(\rho_{Bitumen})$ is from laboratory analysis or field experience.

$$\rho_{bulk} = \rho_{Bitumen} \cdot \emptyset_{Bitumen} + \rho_{Fluid} \cdot \emptyset_{Fluid} + (1 - \emptyset_{Bitumen} - \emptyset_{Fluid}) \cdot \rho_{Grain} \quad (1)$$

NMR log will only see the fluid filled pore; it does not detect the bitumen, as Eq. (2).

$$\emptyset_{Fluid} = MPHS_{Total} \quad (2)$$

The fluid density derives from Eq. (3) using saturation of fluids in pore, which are water and gas in discussed case. Given gas composition, formation pressure and temperature, it is simple to calculate the gas density (ρ_{Gas}).

$$\rho_{Fluid} = \rho_{Gas} \cdot S_{Gas} + \rho_{Water} \cdot (1 - S_{Gas}) \quad (3)$$

To solve the bitumen volume, fluid filled porosity, saturations are crucial inputs, and we will discuss separately in next section.

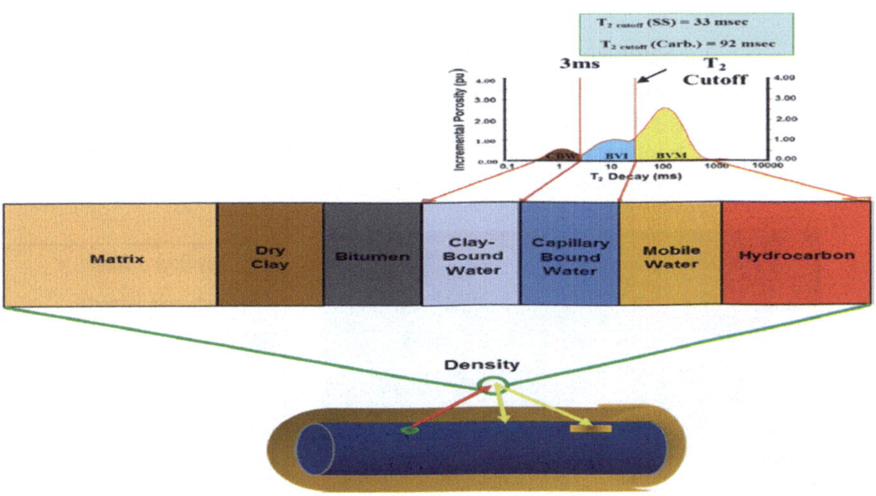

Fig. 4 T2 distribution based NMR interpretation principle

5 Gas Saturation and HI Corrected Porosity with 2D NMR

As we already discussed, NMR logging actually measures the hydrogen in rock and gives apparent porosity with calibration to pure water. The apparent porosity is therefore subject to the proton density of a reservoir fluid compared to pure water at standard temperature and pressure, i.e., hydrogen index. The hydrogen index (HI) of gas is much smaller than water, and varies with its composition, pressure and temperature. As a result, the apparent porosity in this gas field is underestimated.

To make HI correction for apparent NMR porosity, the HI under reservoir condition and gas saturation are necessary. The former is very easy to calculate with known gas composition, temperature and pressure, but the latter is more complicate to solve.

In traditional way, log analyst use Archie equation to calculate saturation. It is an empirical relationship between resistivity and porosity log of a rock. However, in this kind of gas reservoir, the NMR porosity is underestimated and density porosity is inaccurate with the presence of bitumen, the true total porosity is yet to solve. The dual lateral resistivity log has a very deep depth of investigation, and measures the uninvaded formation. Therefore, the resistivity-derived saturation is not suitable for the HI correction of NMR log, which has a very shallow depth of investigation and only measures the flushed formation.

2D NMR (T1/T2 vs. T2) provides us a mean to separate the apparent gas porosity from the total apparent porosity. Three different T2 relaxation mechanisms, which are bulk, surface and diffusion relaxation, affect T2 measurements made on porous media as Eq. (4).

$$\frac{1}{T_{2,app}} = \frac{1}{T_{2,Bulk}} + \frac{1}{T_{2,Surface}} + \frac{1}{T_{2,Diffusion}} \tag{4}$$

Only two of them control the T1 measurements with the exception of diffusion as Eq. (5).

$$\frac{1}{T_1} = \frac{1}{T_{1,Bulk}} + \frac{1}{T_{1,Surface}} \tag{5}$$

Gas has a very high diffusion coefficient compared to other reservoir fluid, say water and oil. This high diffusion does not affect the T1 properties of methane, but has a significant shortening effect on the apparent T2 of methane. The $T_1/T_{2,app}$ ratio of methane will be much bigger than water and oil, making it easy to separate the gas from

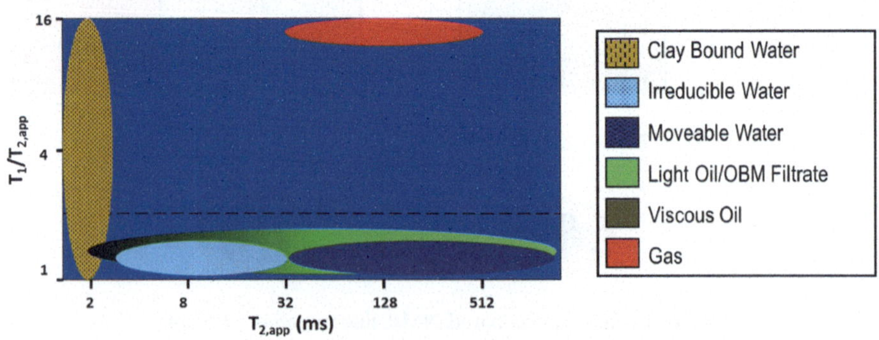

Fig. 5 2D NMR interpretation principle

other pore fluids on 2D NMR cross plot (Hursan et al. 2005) (see Fig. 5). Then it is possible to derive the HI corrected porosity and gas saturation from Eqs. 6 and 7.

$$MPHS_{Total} = MPHS_{Total,app} - MPHS_{Gas,app} + \frac{MPHS_{Gas,app}}{HI_{Gas}} \tag{6}$$

$$S_{Gas} = \frac{MPHS_{Gas,app}}{HI^{*}_{Gas}MPHS_{Total}} \tag{7}$$

Plugging into Eqs. (1), (2) and (3), the bitumen filled pore volume is finally solved out.

The calculated result are plotted in Fig. 6. The fifth track shows the excess carbon existing and the sixth track shows the bitumen contents.

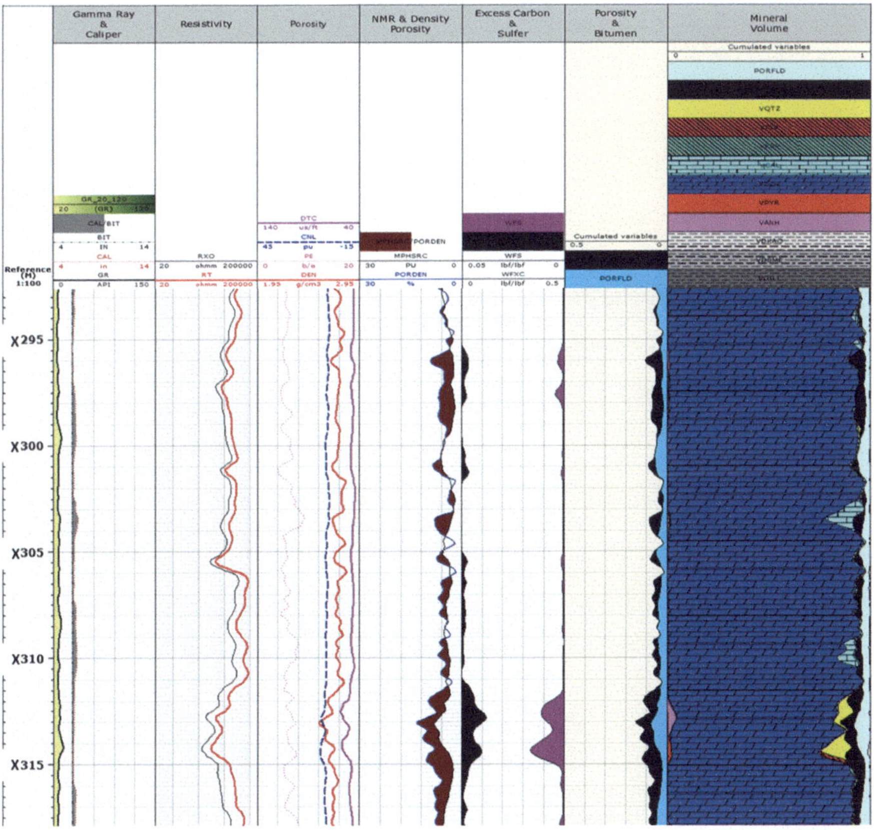

Fig. 6 A processed result shows some bitumen contents

Acknowledgements. The project is supported by Southwest Oil and Gas Field Company, CNPC(2) and Baker Hughes, a GE Company(1). The authors thank the management of South West oil and gas filed, CNPC and Baker Hughes, a GE Company(BHGE) for the permission to publish this paper.

Reference

Hursan, G., Chen, S., Murphy, E.: New NMR two-dimensional inversion of T1/T2APP vs. T2APP method for gas well petrophysical interpretation. In: SPWLA 46th Annual Logging Symposium (2005)

Integral Potential Exploitation Technology in Fault Zone of High Water-Cut Oilfield-Taking DQSB in Daqing Oilfield as an Example

Jun-hui Bai(✉)

The Third Oil Production Factory, Daqing Oil Field Limited Company, Daqing, Heilongjiang, China
jhbai@petrochina.com.cn

Abstract. Old oilfields in eastern China have generally entered the stage of high water cut, and it is increasingly difficult to tap their potential. At present, there are still three problems in excavating potential of remaining oil in old oilfields: low accuracy of description of small faults, unclear understanding of potential of fault edge and the means of tapping the remaining oil in fault area are limit, which restrict accurate tapping potential. Taking DQSB development area of Daqing Oilfield as an example, the paper proposes to use well and seismic data to cooperate with genetic analysis to realize fine characterization of various faults. Through the analysis of fault genesis and the elongation of fault spacing below 5 m, the fine interpretation of fault points is realized. On the basis of the fine understanding of fault location, the causes of residual oil at fault edge are identified by reservoir engineering method from three aspects of structure, reservoir and well pattern. Then, according to the law of residual oil enrichment, three models of overall tapping potential in fault area are put forward and optimized. Drilling well type, measures and other means to improve the relationship between injection and production in fault areas, realizing the efficient tapping of remaining oil potential. After implementation, the average daily oil increase of single well in fault area is more than 7 times that of the surrounding old wells, and the development benefit is obvious, thus forming a set of integrated potential tapping mode in fault area.

Keywords: Tapping the potential of remaining oil · Fine structure description · Efficient well · Fault · Tapping mode

J. Lin (ed.), *Proceedings of the International Petroleum and Petrochemical Technology Conference 2019*, pp. 221–232, 2020.
https://doi.org/10.1007/978-981-15-0860-8_18

1 Introduction

According to the energy structure and reservoir distribution of our country, the energy strategy of stabilizing the east and developing the west is put forward, and the oil field in the east is still an important part of the petroleum in China. Eastern onshore reservoirs are characterized by relatively developed faults, rapid lithologic changes, strong heterogeneity of reservoirs, complex distribution relationships between remaining oil and water, etc. After long-term water flooding development, the eastern oilfields have basically entered the period of high water cut development. It is becoming more and more difficult to exploit the potential of the remaining oil in the conventional way, and the fault area has become one of the main targets of the remaining oil in the area of relative abundance of remaining oil and the potential of the remaining oil in the future. The remaining oil distribution and potential tapping techniques of different types of reservoirs have been studied by Xuanna et al. in China [1]. According to the characteristics of fracture zone in outcrop and a lot of logging data, Jin Qiang et al. studied the influence of fault zone on the development of remaining oil in fault block oilfield [2]. According to the geological characteristics of Yong 3 fault block in Xin Oilfield, Houfudong et al. studied the fault scale of Yong 3 fault block by using the three dimensional seismic interpretation of ant tracing technique and the analysis of local tectonic stress field. The influence of low order faults such as sealing and combination patterns on water injection development [3]. Yan Lin and Zhang Hui have successively used logging data combined with three-dimensional seismic data, based on the fine description of reservoirs such as fine three-dimensional geological models and the analysis of the relationship between injection and production at the fault edge. A study on the potential and distribution of residual oil in the channel sand body of thick reservoir along the fault edge has been carried out [4–7].

The DQSB fault area is located in the northern part of Daqing Changyuan with an area of 34.95 km^2, and 49 developed faults, all of which are normal faults with an extension length of 0.2–2.9 km and a fault distance of 0.8–93.2 m. Having been developed more than 50 years, the comprehensive water cut has reached 94%. After the reservoir experienced water flooding, polymer flooding and ternary flooding, the remaining oil has been highly dispersed and distributed in the fault areas with low well pattern control degree and the narrow channel sand body between wells [2–8]. In order to accurately tap the potential of residual oil, this paper focuses on how to improve the accuracy of fault description and how to determine the distribution of residual potential along the fault edge in view of the development of unrecoverable reserves at the fault edge, and how to improve the accuracy of the fault description in the study area. How to enrich the methods of tapping potential in fault area is studied in three aspects. Through combining well and earthquake, analyzing the cause of remaining oil and cooperating with multiple means to excavate potential, we have formed a set of model of tapping potential as a whole in fault area, and have seen obvious development effect. It can be used for reference in the development of similar block fault area.

2 Fine Characterization of Faults

2.1 Analysis of Genesis Characteristics of Faults

The faults in the zone can be divided into four sets of fault systems according to the fault horizon and scale (Fig. 1). They are the active faults of T2-T06 depression and inversion period, the early active faults of T2-S depression, the active faults of G-T06 depression and the active faults of G-S depression. During the depression period, the stress field was mainly extensional and extensional in the east-west direction, forming 9 T2-S NNW strike faults. The inversion fault is subjected to the deformation field of left spin compression and torsion, forming 5 normal faults of G-T06 near NW direction. There are 8 T2-T06 faults, which were formed in the basin depression period and were reformed by compression and torsional deformation during the reverse period, forming the NNW strike Yanlie fault zone, with fault terrace, "Y" type, graben and barrier as the main assemblage model. There are 27 active G-S faults in the middle and late period of depression, which are small in scale and have the characteristics of two tectonic movements, but the strike is not obvious.

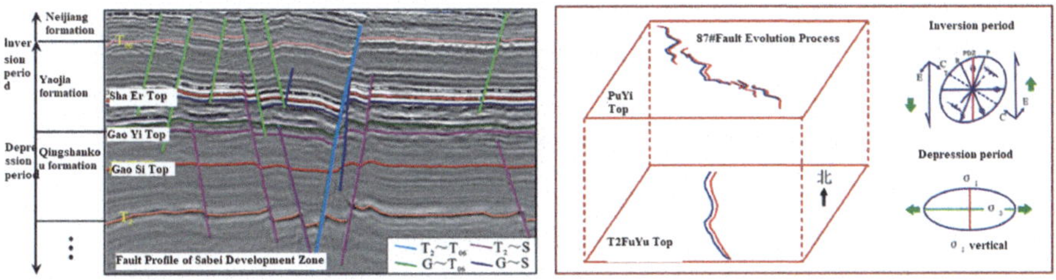

Fig. 1 Four sets of fault systems and their origin explanation in the study area

2.2 Fine Identification of Faults

In order to improve the accuracy of fault interpretation, the extension position of faults below 5 m is analyzed. By using the method of determining the location of seismic and imaging logging data and contrasting the fault distance between adjacent well curves, 12 fracture points with fault spacing less than 2 m are newly identified and 11 fault point positions are adjusted. The following figure (Fig. 2) is taken as an example. Through the accurate matching of well seismic data in three dimensional space, it is found that a well in the fault area passes through the edge of the fault but has no fracture point interpretation. The section is composed of multiple well fracture points and seismic interpretation data, which is relatively reliable. Therefore, the well seismic data can be implemented again. Firstly, the fracture point is predicted by seismic data, then the near well is contrasted in the area, and the small fracture point with 1.6 m fault distance is found in the well. Example 2 is the image of imaging logging data. The vertical resolution of XRMI imaging logging is 5 mm. The weak variation of formation dip angle can be identified by using resistivity imaging in one week of annular hole, and it is obvious that the formation dip angle of this well increases suddenly. From 5°

to 35°, the location is compared with the multi-well logging curve data, and it is determined that there exists a small fracture point of 0.8 m. Based on the fracture point, the extension range of the fault is carried out.

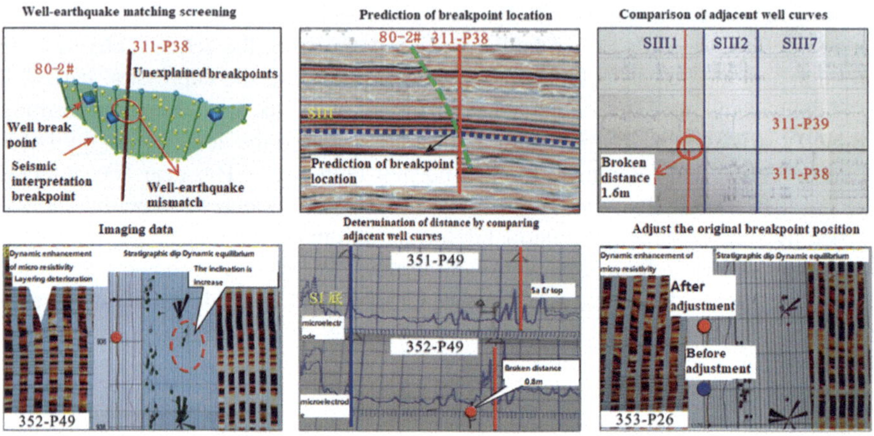

Fig. 2 Interpretation of well earthquakes combined with small breakpoints

2.3 Growth Characteristics of Large Faults

For the large fault of T2-T06, the scale is the largest and the structure is more complicated. According to the characteristic of segment growth of large fault with low value region, by drawing the fault distance position curve of 9 big faults (Fig. 3), 6 fault segments and 9 growth points are identified. It is recognized that the large faults in the study area have the characteristics of segmental growth and are formed by the lateral overlay of several small faults [9–12]. For large faults with little change in fault spacing and unclear segmental features, taking 87# fault as an example, according to the results of fine interpretation of fracture points, several wells and multiple breakpoints in the area are combined to form four segmented faults, and the fault spacing and position curves are redrawn. It can be seen that the middle fault of each segment is large, the two ends are small, and the change is reasonable. The fault spacing of ① and ② faults has obvious characteristics of growing faults, while ③ and ④ fault have the opposite change, which indicates that the two faults have different origin and are formed in late period. From the position of each small fault distance and multiple fracture points of a well, three sub-growth points can be identified. A large fault is divided into four relatively independent small faults. The combination ratio of major fault breakpoints is further increased to 97.97%, and the location of the three dimensional distribution of faults is clearer. There are obvious differences in width of fault zones in different locations, which is of great significance for the determination of safe distance between high efficiency wells and faults, and is beneficial to the development and deployment of wells along the fault side.

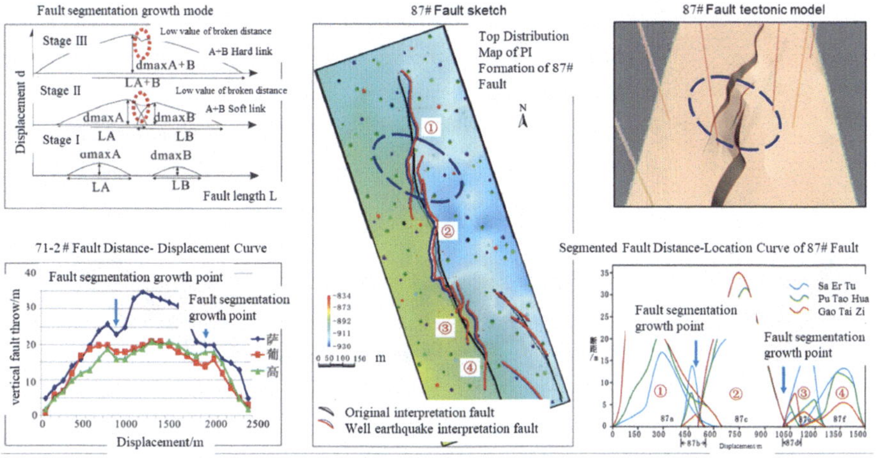

Fig. 3 Fine explanation of the origin of large faults

3 Three Types of Potential Recognition of Fault Edge

The accuracy of fault interpretation and the location of fault are improved, so that the area that could not be developed in the past is further clarified. Through the comprehensive analysis of three aspects of potential along the fault edge, the potential of fine recognition in the small well area is analyzed, which provides a reliable basis for accurate development.

3.1 Identification of Favorable Traps Controlled by Tectonic Factors

Generally, high efficiency wells at the fault edge are only deployed in a single reverse fault footwall in a fault circle type. Although different scholars have confirmed that the reverse fault footwall is more favorable for oil and gas accumulation [13, 14], the number of this type of fault circle is, after all, limited. It is difficult to find such broken circle in the study area. After extensive investigation [15–18], and through the classification of fault and stratigraphic tendency and the analysis of fault patterns in the study area, it is recognized that there are six types of fault circle in the study area (Fig. 4): reverse fault footwall, roof fault two plates, Fault-hydrodynamic reservoirs, fault terrace blocks, barrier blocks and graben fault blocks can be divided into two types: single fault and multiple fault combinations. Through lateral sealing of faults, oil and gas are blocked [14, 15], and the scope of well selection in the study area is greatly increased.

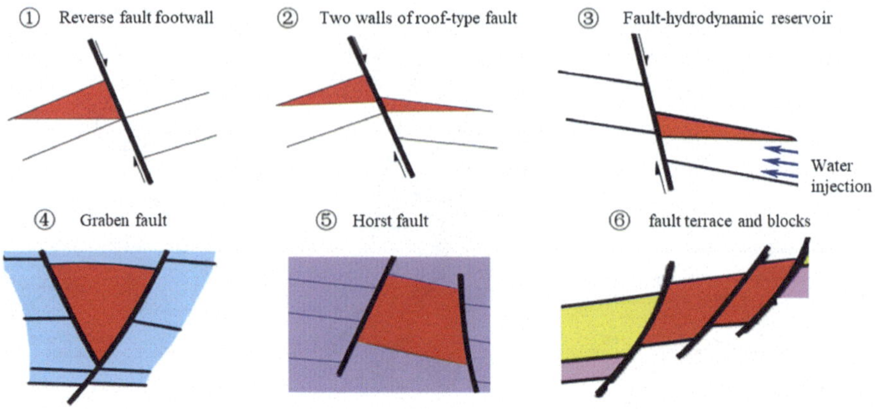

Fig. 4 Six favorable fault circle types in the fault edge of the study area

3.2 Identification of the Enrichment Sites of the Special Sand Bodies at the Edge of the Fault

According to the above analysis (Fig. 3), the faults in the study area have the characteristics of growing faults, and according to the sand control model at the fault edge (Fig. 5), the footwall of the growth normal fault is more enriched than the sand body near the upper plate. In addition, due to the occurrence of segmental fault growth at the same time in the study area, the fault transition slope is located in a local low-potential area, which results in rapid unloading of sediments and strong hydrodynamic force, resulting in a higher sand to soil ratio and more enrichment of sand bodies. Therefore, it is easier to find thick sand bodies which are favorable to the enrichment of remaining oil at the small fault spacing of the downplane, which can not be predicted effectively only by drawing the equivalent map of the thickness of the drilling wells.

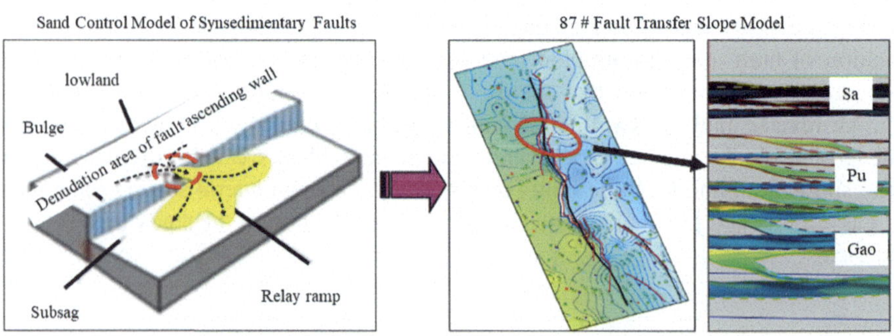

Fig. 5 Enrichment law and location of sand bodies at fault edge

3.3 Injection-Production Imperfect Area Controlled by Well Pattern Factors

The original well pattern in the study area is cut by fault, which results in the existence of residual oil in different locations. This kind of remaining oil is formed by the original area well pattern, which is different from single layer and layer remaining oil type. There are 6 sets of well pattern and 8 sets of layers in the study area. Considering the remaining reserves and the degree of clarity of the well pattern, the primary infill layer system of the high platform subunit of PuII-Gaotazi is selected and the first kind of reservoir polymer flooding system of the group PuI is exploited. There are three sets of layers of polymer flooding in the second class oil reservoir of SaII10- and SaIII formation. The three sets of layers are river course, the main body is thick layer sand, and the well pattern is clearer. From the analysis of fault and actual well pattern, four kinds of potential areas (Fig. 6) are identified in the study area [19, 20]: blank well area, injection-production distributary line location, injection and no mining area, fault block area, when the well spacing reaches above the original well spacing, If it is regarded as a favorable area for remaining oil to be exploited, the potential can be tapped. The selection range of potential area is greatly increased through the arrangement of high efficiency well stratification system in fault area [21, 22].

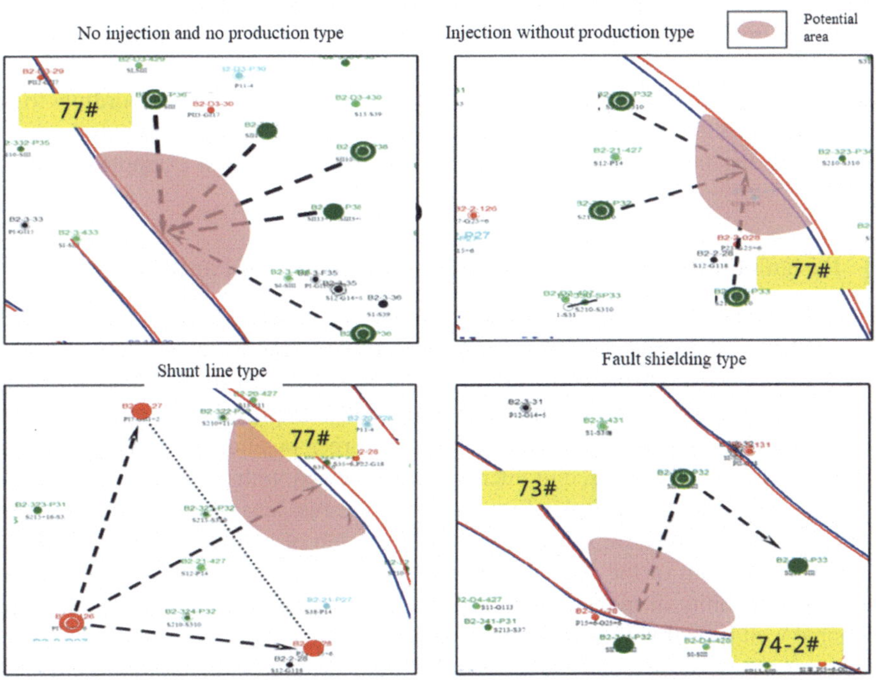

Fig. 6 Area of fault area well pattern four types of remaining oil potential areas

4 Three Models of Whole Tapping Potential in Fault Area

Based on the results of fine interpretation of faults, considering the structural characteristics, reservoir development, remaining oil distribution and well pattern of strata, 22 potential areas have been found in the study area, 139 new wells have been drilled and 29 measures have been taken to tap potential. The cumulative production of oil is 29.40×10^4 t, which forms three models of tapping potential in the fault area (Table 1).

Mode 1: for the small dip angle of fault, imperfect injection and production of the side of the fault, drilling and tapping of directional wells with large displacement in the local rich area of remaining oil, 13 high efficiency wells have been deployed in the study area, and 9 wells have been put into operation at present, and the daily production of single well is 19.0 t at the initial stage. The water content is 39.9%, and the cumulative oil production is 2.44×10^4 t for two years.

Mode two: for the fault cognition change, the injection-production system adjustment is carried out in the imperfect injection-production area. Through the combination of the supplementary drilling straight well and the old well transfer injection combined with tapping the potential remaining oil, the injection-production system adjustment in 3 blocks has been carried out, totally 126 oil-water wells have been drilled. Four years cumulative oil production 25.69×10^4 t.

Mode 3: for the change of understanding of faults, the perfection of well pattern, the local remaining oil rich area, the optimization measures, the combination of hole filling, fracturing and water injection well scheme adjustment are adopted to tap potential, 29 oil wells are implemented, and the daily oil is 153.2 t. The cumulative oil increment is 1.27×10^4 t.

According to the potential analysis of the specific well area, according to the distribution of different remaining oil along the fault edge, the potential tapping techniques in the fault area are described by a single well design example as follows (Fig. 7).

4.1 Design of a Directional Well with High Inclination of Parallel Section

The applicable conditions are as follows: (1) there is no filling hole in the side of the fault, (2) there is imperfect injection and production in more than two sets of layers.

Trajectory design: (1) parallel section, distance keeping 20–40 m; (2) track azimuth according to ground well arrangement condition, collision prevention requirement with old well, multiple sets determined by optimum target position of layer system and maximum drilling thickness requirement, can be skewed with fault strike.

Taking well 1 as an example (Fig. 7a), there are two sets of wells with imperfect injection-production pattern. Because the drilling depth of an old well in the fault side is only at the bottom of the II formation and is too close to the original well pattern, there is no possibility of making up a hole too far from the fault. A large displacement directional well is deployed in this area. Considering the requirement of 50 m radius of drilling operation surface and 30 m distance from the old well, the optimum target position of two sets of well pattern is designed. The trajectory of the well is 340 m and the parallel section is designed. The azimuth and fault strike are oblique and 40 m away from the fault plane.

Table 1 Three models of tapping potential in fault area

Model	Application condition	Potential tapping method	Schema chart
Model one	The dip angle of fault is small, the injection and production of fault edge is not perfect, and the remaining oil is enriched	Drilling high efficiency wells: long reach directional wells, horizontal wells, straight wells	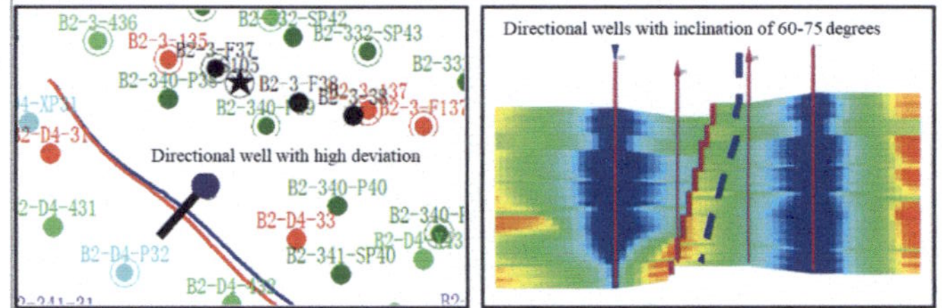
Model two	Fault cognition changes and injection-production imperfect	Implementation of Injection-production system Adjustment: combination of supplemental drilling straight well and turn injection	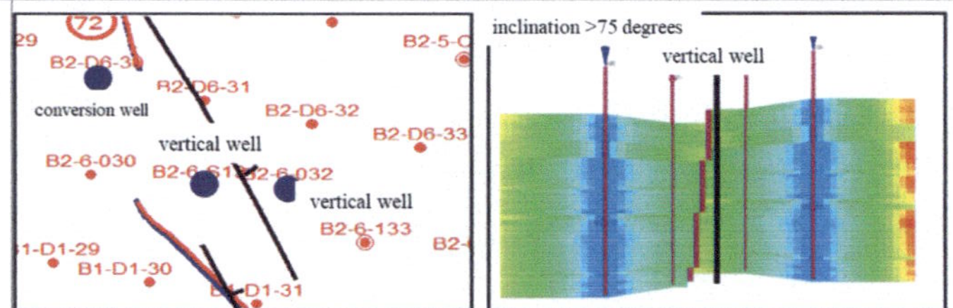
Model three	Fault recognition change, well pattern perfect, local remaining oil enriched	Optimization measures: combination of hole filling, fracturing and water injection well scheme adjustment	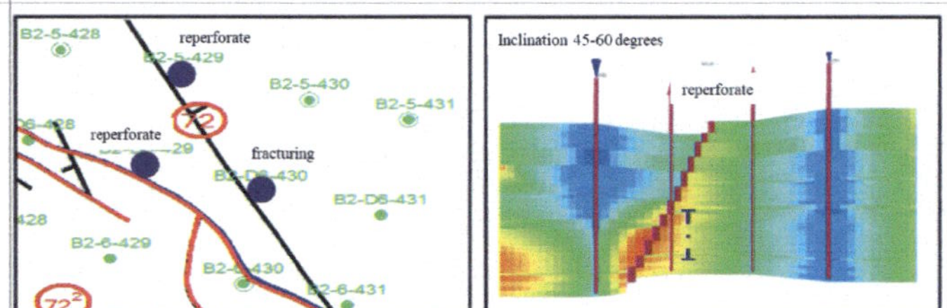

4.2 Design of High Efficiency Straight Wells with Cross Section

The applicable conditions are as follows: (1) there is no hole in the edge of the fault to make use of the wells; (2) there are imperfect injection-production parts in the single set of layers or imperfect injection-production parts in the complex fault block area.

The trajectory design is as follows: (1) it is necessary to ensure that the production series is not broken in the cross section straight well; (2) the distance between the straight well and the section in the production layer is controlled within 100 m; (3) the vertical well can be adjusted to a general directional well considering the condition of surface distribution.

Taking well 2 as an example (Fig. 7b), there is residual oil in this area due to fault occlusion, because there is only one set of potential layers, such as large investment in deploying extended displacement wells, and the safe distance between well layout and section is not easily controlled in cross fault blocks. Therefore, the optimal selection of vertical wells is designed to drill into a fault at the top of the target layer, without losing the target layer and keeping the bottom within 100 m, so as to ensure the exploitation of remaining oil in the edge of the fault.

4.3 Design of Hole Mending in Old Wells

The applicable conditions are as follows: (1) there are perforated oil wells; (2) the injection and production systems are imperfect and unbroken; (3) the thick layers have been opened or the original wells have reached the upper limit of water cut.

Trajectory design: the distance between the old well and the cross section is controlled within 100 m.

Taking well 3 as an example (Fig. 7c), the three sets of wells in this area have imperfect injection-production pattern, the crude oil has been shot into thick layer, and the target layer is not broken and is close to the fault, so it can be used to make up the hole and tap the potential. According to the effect of pore filling in group I, the recognition of remaining oil in this area is confirmed.

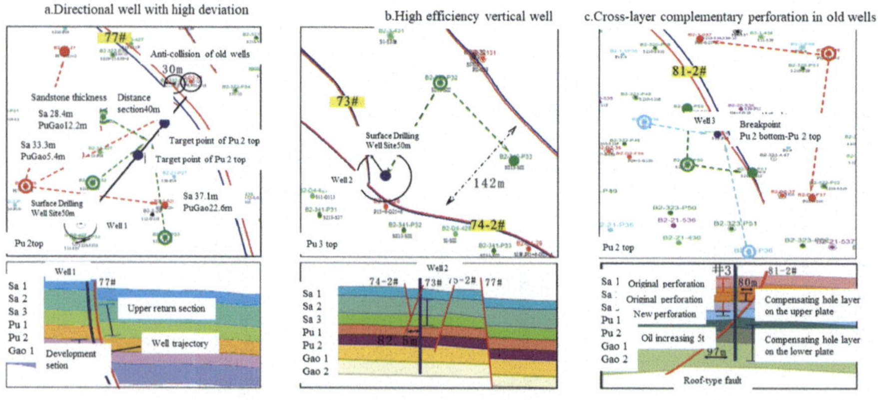

Fig. 7 Design example of high efficiency well at fault edge

The three methods mentioned above are designed for different potential types and adapt to different situations. In addition, from the economic point of view, the old well is the least expensive, and the most expensive to drill the large slope well. But because the potential of no-well gap area is greater and the remaining oil can only rely on the deployment of new wells to tap potential, the principle of preferential deployment of high-efficiency wells is adopted in general.

5 Conclusion

(1) From fault origin analysis, well shock combined with small fault point identification, fault growth characteristics analysis, the fine description of faults in the study area is realized in three steps, and the combination rate of fracture points in the study area reaches 97.97%, which provides the basis for tapping the remaining oil in the edge of the fault.
(2) The cause of formation of residual oil in fault edge is analyzed from three aspects of structure, reservoir and well pattern, and the range of potential area of fault edge is expanded, 6 types of fault circle, small fault space of downplane of fault and 4 kinds of injection-production relation are more favorable for remaining oil enrichment in fault area.
(3) By optimizing the selection of potential targets, the way of tapping potential, and the deployment of high efficiency wells along the fault edge, the situation that there is no well cloth on the edge of the fault in the study area and that the remaining oil is unable to tap the potential meticulously can be changed, thus realizing the development of the fault-distinguishing step and the orderly development. On this basis, three models of tapping potential in the fault edge are summarized and formed, which is helpful for the similar blocks to use for reference.

Acknowledgements. The project is supported by the National Special Project (No. 201605054).

References

1. Xuanna, Mulongxin, Qiuyinan. Distribution of reserves and movable remaining oil in different types of reservoirs in China. Petrol. Explor. Devel. **25**(5), 41–44 (1998)
2. Jin, Q., Zhou, J., Wang, D., et al.: Identification of fault fracture zone and its application in fault block oilfield development. Acta Petrol. Sinica **33**(1), 82–89 (2012)
3. Hou, F., Shaochun, Y., Lu, Z., et al.: Characteristics of low order faults in Yong3 fault block of Dongxin oilfield and its influence on water injection development. Fault block Oil Gas Field **22**(04), 488–491 (2015)
4. Zhang, Y.: Study on Reservoir description and residual oil distribution in the later stage of development of he 31 fault block in complex fault block reservoir. Drill. Prod. Technol. **33** (03), 75–76 (2010)
5. Fu, Y., Feng, M., Liu, H., et al.: Analysis of remaining oil distribution and control factors in high water cut oilfield. Drill. Prod. Technol. **33**(06), 61–63 (2010)

6. Yan, L., Sun, W., Zhu, Y., et al.: Distribution of remaining oil in Sanjianfang formation reservoir of Qiulin oilfield. J. Northwest Univ. (Nat. Sci. Ed.) **35**(02), 200–206 (2005)
7. Zhang, H.: Study on Residual Oil Potential of Fault Edge of North West Block in Lamadian Oilfield, pp. 17–33. Zhejiang University (2010)
8. Li, X., Song, B., Jiang, Y., et al.: Optimization design of high efficiency wells near faults in super high water cut old oil fields. Petrol. Geol. Oilfield Dev. Daqing **34**(01), 56–58 (2015)
9. Zhao, Junwei, Xu, Huaimin, Xu, Chaohui, et al.: Control mechanism of development fluid potential on remaining oil distribution in medium and high permeability reservoirs. J. China Univ. Min. Technol. **03**, 535–543 (2016)
10. Wang, Duanyang: Re-recognition of structural faults based on 3D seismic interpretation. Oil-Gasfield Surf. Eng. **32**(5), 11–12 (2013)
11. Scholz, C.H., Dawers, N.H., Yu, J.Z., et al.: Fault growth and fault scaling laws: preliminary results. J. Geo-phys. Res. **98**, 21951–21961 (1993)
12. Childs, C., Manzocchi, T., Walsh, J.J., et al.: A geometric model of fault zone and fault rock thickness variations. J. Struct. Geol. **31**, 117–127 (2009)
13. Fu, X., Chen, Z., Yan, B., et al.: Analysis of main controlling factors for hydrocarbon accumulation incentral rift zones of the Hailar-Tamtsag Basin using a fault-caprock dual control mode. Sci. China Earth Sci. **56**(8), 1357–1370 (2013)
14. Peacock, D.C.P., Sanderson, D.J.: Geometry and development of relay ramp sin normal fault systems. American Assoc. Petrol. Geol. Bull. **78**(2), 147–165 (1994)
15. Weidaning, F.G.: The reverse fault footwall is easier to enrich the oil and gas mechanism than the upward side of the forward fault. J. Jilin Univ. Geosci. **46**(3), 702–710 (2016)
16. Fu, G., Lang, Y., Hu, X.: Study on the difference between reverse and forward fault lateral seal oil and gas. Lithol. Reservoirs **26**(6), 28–33 (2014)
17. Zhang, S.M., Wang, Z.: Analysis of oil and gas accumulation conditions and main controlling factors of oil and gas enrichment in fault sand configuration. Sci. Technol. Eng. **14**(20), 198–204 (2014)
18. Fu, G., Chen, J.: Comprehensive evaluation of oil-gas accumulation source-fault-potential control in fault traps: a case study of the southern section of the main fault depression zone in the central part of the Haita Basin. Petrol. Geol. Recovery Effi. **N22**(6), 1–6 (2015)
19. Micarelli, L., Benedicto, A.C., Wibberley, A.J.: Structural evolution and permeability of normal fault zones in highly porous carbonate rocks. J. Struct. Geol. **28**, 1214–1227 (2006)
20. Fu, Xiaofei, Li, Wenlong, Lu, Yanfang, et al.: Lateral sealing of faults and control of oil-water relationship in fault zone. Geol. Rev. **57**(3), 387–397 (2011)
21. Duan, Yajun, Li, Baichuan, Liu, Xiao, et al.: Effects of faults on reservoir development. Technol. Innov. Appl. **10**(19), 155 (2016)
22. Zhang, Qing: Study on the application of optimal deployment technology of well pattern in stratified system development A well area in block E as an example. J. Yangtze Univ. (Nat. Sci. Ed.) **13**(1), 68–71 (2016)

Application of Ion-Responsive Hydrogel in Self-healing of Oil Well Cement Sheath

Miao-miao Hu and Jin-tang Guo(✉)

Department of Polymer Science and Engineering, School of Chemical Engineering and Technology, Tianjin University, Tianjin, China
m15822890856@163.com, jtguo@tju.edu.cn

Abstract. Oil well cement (OWC) performs multi-functions, including providing zonal isolation along the well, providing mechanical support. Cracks due to the intrinsic fragile property, shrinkage and external loads commonly reduce the functionality and durability cement sheath. This paper proposed an innovative ion-responsive hydrogel to achieve self-healing of oil well cement and control the reduction of compressive strength. We prepared the hydrogel by simply crosslinking with Ca^{2+}, and graphene oxide was added to reinforce the material. The ion-responsive behavior of hydrogel was measured, and results showed that higher swelling potentials could be obtained at lower Ca^{2+} concentrations, which was consistent with what was expected with the formation of a less dense network. This interesting dependence on Ca^{2+} concentration was supposed to avoid the formation of large macro-pores during the mixing process and result in increased swelling when cracks occur and water enters the crevices.

Keywords: Ion-responsive · Graphene oxide · Sodium alginate · Self-healing

1 Introduction

The oil and gas exploration has faced increasing challenges, the complicated operational scenarios require cement sheath with higher durability and some new functions [1]. Cracks in oil well cement sheath caused by internal structural shrinkage and external loads should be addressed, since the integrity and durability of cement sheath are important to ensure the proceeding of exploration [2]. The occurrence of cracks or micro-gaps of cement sheath will lead to serious consequences: (1) reduction of the productivity resulting from the leakage of oil/gas fluid; (2) corrosion of steel pipe due to the invasion of aggressive species such as saline fluid, CO_2 and H_2S [3]. Considering the extremely difficult environment of the cement sheath, manual repairing will lead to a higher cost and be impossible in some situations.

Self-healing of cement and concrete has been studied for a long period, and many creative materials have been reported such as shape memory materials [4], microcapsules [5], bacteria [6] and activated chemicals of cement [7]. However, these methods reported in building materials are not so efficient in case of oil well cement. Recently, the self-healing triggered by the leakage of aqueous or oily fluids attracts much

J. Lin (ed.), *Proceedings of the International Petroleum and Petrochemical Technology Conference 2019*, pp. 233–239, 2020.
https://doi.org/10.1007/978-981-15-0860-8_19

attention. Lu [8] prepared a creative oil swellable polymer and the self-healing by absorbing of oil was achieved in cement sheath. The cracks in polymers modified cement composites can be sealed by swelling, while the binding force between matrix and hydrogel is too low to prevent the secondary crack.

As a biopolymer, alginate(Alg) can be isolated from brown seaweeds. CaAlg is a water-absorbant hydrogel, and has been used as a biomaterials, its applications in scaffolds for tissue engineering, delivery vehicles for drugs, and biosensing devices, have been widely reported [9]. It can be prepared through ionic crosslinking via calcium bridges between the guluronic acid residues in alginate [10]. This paper endeavours to achieve limited reduction in compressive strength and efficient self-healing by the Ca^{2+}-sensitive property of CaAlg. Besides the ion-responsive behavior, the unique synthesis procedure make it contain many reacting Ca ions, which is supposed to react in self-healing process and accelerate the formation of healing crystals. In this paper, graphene oxide was used to reinforce the CaAlg. The ion-sensitive properties and self-healing potential were studied.

2 Materials and Experiment

2.1 Materials

Graphene oxide was purchased from Tanfeng Tech. Inc., (Su Zhou). Sodium alginate, calcium chloride ($CaCl_2$) and sodium silicate were obtained from Jiangtian Chemical Co., Ltd.

2.2 Preparation of GO/CaAlg

GO was firstly dispersed in water, and then sodium alginate was added at 8000 rmp stirring to obtain the unique GO/Alg solution. The black solution was dropped into $CaCl_2$ solution to achieve crosslinking, the obtained beads (GO/CaAlg) were then freeze-dried and sieved through 70 mesh for further characterizations and studies.

2.3 Characterization

The reaction between GO and NaAlg was characterized by FTIR, and the morphology of formed GO/CaAlg was measured by SEM. The precipitation of calcium silicate hydrate (C–S–H) on hydrogel was characterized by SEM-EDS. The swelling capacity of GO/CaAlg in solutions at various Ca concentration was determined using Eq. (1):

$$\text{Swelling capacity} = (m_o - m_{filter})/m_{GO/CaAlg} \quad (1)$$

m_o: the mass of the water initially added (g); m_{filter}: the mass of the water that passed through the filter (g); $m_{GO/CaAlg}$: the initial mass of GO/CaAlg.

3 Results and Discussion

3.1 Intereaction Between GO and Alg

FTIR spectra of Alg and GO/Alg are shown in Fig. 1. In Alg sample, the peak at 1022 cm^{-1} is attributed to the stretching vibration of –C–O–C–, and peaks at 1624 cm^{-1} and 1428 cm^{-1} are assigned to the symmetric and asymmetric COO– stretching vibration of carboxylate group, respectively [11]. The addition of GO has no obvious change in these peaks, however the peak at 3421 cm^{-1}, corresponding to –OH stretching vibration, broadens and shifts to smaller wavelengths (3413 cm^{-1}). Thus the interaction of Alg and GO through intermolecular hydrogen bonds, and the reaction mechanism of GO and Alg can be schematically shown in Fig. 2. The strong hydrogen bonds between GO and Alg demonstrate excellent adhesion between the matrix and reinforcement, which can ensure uniform dispersion and certain degree of alignment of GO within matrix.

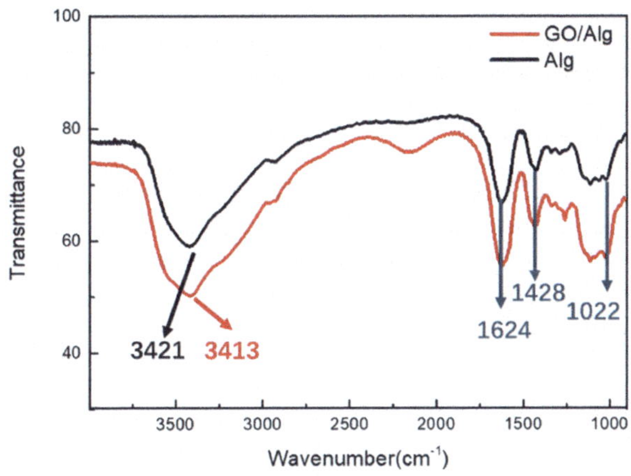

Fig. 1 FTIR spectra of Alg and GO/Alg

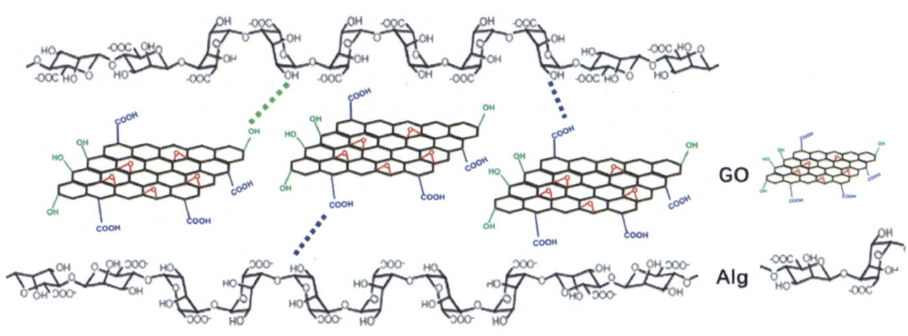

Fig. 2 Reaction mechanism of GO and Alg

3.2 Morphology of GO/CaAlg

As shown in our previous study [12], the surface of crosslinked CaAlg are homogeneous and smooth, and its structure for water storage was obvious. In Fig. 3a, there were also few GO sheets dragged out and the surface of hydrogel are obviously different from pure CaAlg. The addition of GO results in a rather rough surface, an ordered striped structure is observed along the stirring direction. The rough surface of GO/Alg will facilitate the precipitation of C–S–H, thus accelerates the self-healing process of cement.

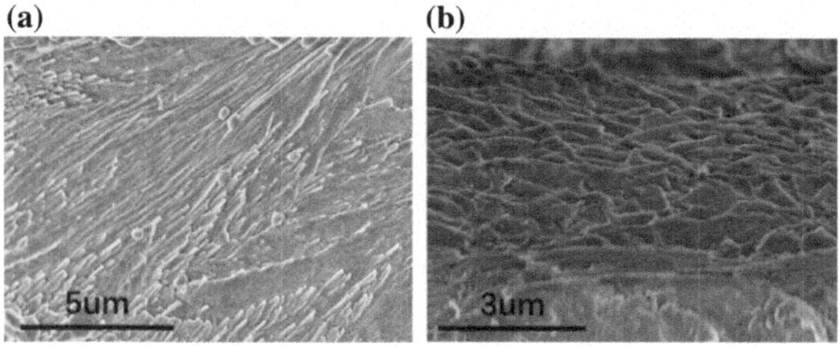

Fig. 3 SEM images of GO/CaAlg surface (**a**) and fracture section (**b**)

Alginate is water-soluble and can form hydrogels by ionic crosslinking via calcium bridges between the guluronic acid residues on adjacent chains. Figure 3b shows the fracture section of GO/CaAlg, the water-storage structure is similar to the pure CaAlg prepared in our previous study. Thus, GO/CaAlg will swelling when cracks occur and water enters this crevices.

3.3 Swelling Capacity of CaAlg

To achieve the equilibration of absorbance in solutions with various Ca^{2+} concentration, GO/CaAlg hydrogels are incubated for 24 h until the measurement. The ion-responsive behavior is shown in Fig. 4. The swelling capacity of CaAlg reduces with the increasing concentrations of Ca^{2+} in solution. During cement mixing process, the concentration of Ca^{2+} are increasing with the dissolving of cement, the gel will further complex with the Ca^{2+} from dissolution of cement. The dense network formed is helpful to avoid the formation of large maropores during the mixing of the cement. Thus, the compressive reduction will decrease when compared to other superabsorbent polymers reinforced cement materials. Moreover, the highest swelling in deionized water demonstrates the swelling capacity of GO/CaAlg when cracks occur and water enters. The self-healing potential of the hydrogel has been demonstrated in this section.

3.4 Precipitation Capacity of Calcium Silicate Hydrate (C–S–H) on GO/CaAlg

Besides swelling to seal cracks, the precipitation capacity of self-healing products is more important, which can improve the binding force between hydrogel and cement

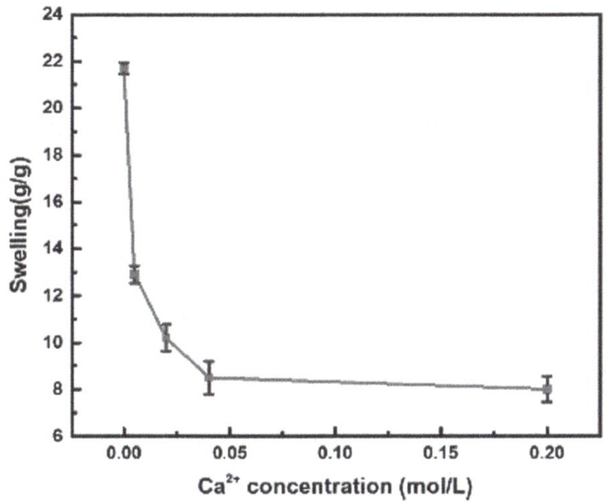

Fig. 4 Swelling capacity measurements in aqueous solutions with a varying Ca^{2+} concentration

matrix. We used the precipitation of C–S–H in sodium silicate solution to define the healing capacity of GO/CaAlg. The Ca^{2+} ions containing in GO/CaAlg react with sodium silicate, the formed C–S–H can enhance the interface bonding and avoid the formation of secondary cracks [13]. This kind of interesting combination of organic and inorganic phases has significantly better mechanical properties due to the sacrificial organic phase as reported in other studies [14].

As shown in Fig. 5a, the surface of GO/CaAlg was covered by the formed self-healing products in sodium silicate system, the corresponding element maps

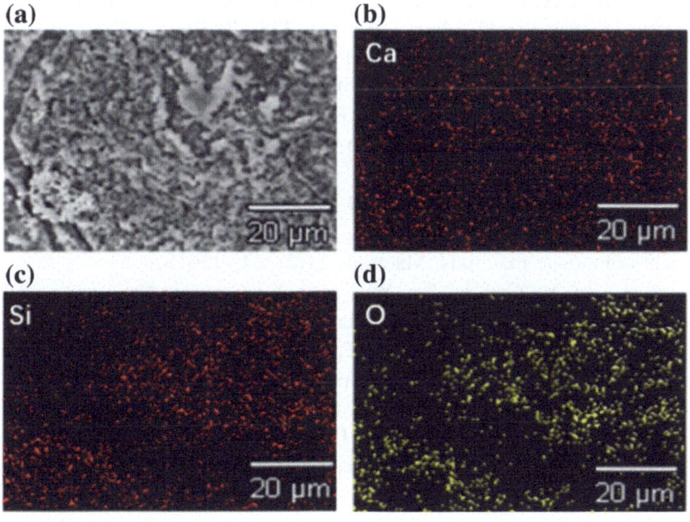

Fig. 5 Morphology and distribution of precipitated C–S–H on GO/CaAlg (**a**), and corresponding element maps (Ca, Si and O)

demonstrate the formation of C–S–H. The effects of alignment of GO in CaAlg hydrogel are not obvious in our research, and further investigations are needed to explore this interesting phenomenon.

4 Conclusion

(1) The strong hydrogen bonds between GO and Alg demonstrate excellent adhesion between the matrix and reinforcement, which can ensure uniform dispersion and certain degree of alignment of GO within CaAlg.
(2) The ion-responsive behavior of GO/CaAlg by crosslinking reaction with Ca^{2+} allows a smaller reduction in the compressive strength and achieves efficient self-sealing of cracks in oil well cement sheath.
(3) In addition to the sealing effects on cracks, the Ca^{2+} containing in hydrogel leads to the precipitation of C–S–H, which enhances the interface binding force between hydrogel and the cement matrix in healed areas.
(4) The interesting combination of organic and inorganic phases(GO/CaAlg and C–S–H) in self-healed areas results in more excellent mechanical properties in healed area due to the sacrificial organic phase, which still need further investigation.

Acknowledgements. The project is supported by National Nature Science foundation of China (Grant No. 51874210).

References

1. de Costa, B.L.S., de Freitas, J.C.O., de Melo, D.M.A., da Araujo, R.G.S., de Oliveira, Y.H., Simão, C.A.: Evaluation of density influence on resistance to carbonation process in oil well cement slurries. Constr. Build. Mater. **197**, 331–338 (2019)
2. Bu, Y., Du, J., Guo, S., Liu, H., Huang, C.: Properties of oil well cement with high dosage of metakaolin. Constr. Build. Mater. **112**, 39–48 (2016)
3. Zhang, J.F., Yang, J.L., Liu, K., Wang, B., Hou, R.X.: Carbon dioxide corrosion and corrosion prevention of oil well cement paste matrix in deep wells. Appl. Mech. Mater. **692**, 433–438 (2014)
4. Kim, E.H., Lee, H., Kim, J.H., Bae, S.M., Hwang, H., Yang, H.: Electrical/mechanical monitoring of shape memory alloy reinforcing fibers obtained by pullout tests in SMA/cement composite materials. Materials **11**(2), 315 (2018)
5. Hu, M., Guo, J., Yu, Y., Cao, L., Xu, Y.: Research advances of microencapsulation and its prospects in the petroleum industry. Materials **10**(4), 369 (2017)
6. Charpe, A.U., Latkar, M.V., Chakrabarti, T.: Microbially assisted cementation—a biotechnological approach to improve mechanical properties of cement. Constr. Build. Mater. **135**, 472–476 (2017)
7. Nguyễn, H.H., Choi, J.I., Song, K.I., Song, J.K., Huh, J., Bang, Y.L.: Self-healing properties of cement-based and alkali-activated slag-based fiber-reinforced composites. Constr. Build. Mater. **165**, 801–811 (2018)

8. Lu, Z., Kong, X., Yang, R., Zhang, Y., Jiang, L., Wang, Z.: Oil swellable polymer modified cement paste: expansion and crack healing upon oil absorption. Constr. Build. Mater. **114**, 98–108 (2016)
9. Ionita, M., Pandele, M.A., Iovu, H.: Sodium alginate/graphene oxide composite films with enhanced thermal and mechanical properties. Carbohyd. Polym. **94**, 339–344 (2013)
10. Golafshan, N., Kharaziha, M., Fathi, M.: Tough and conductive hybrid graphene-PVA: Alginate fibrous scaffolds for engineering neural construct. Carbon **111**, 752–763 (2017)
11. He, Y., Zhang, N., Gong, Q., Qiu, H., Wang, W.: Alginate/graphene oxide fibers with enhanced mechanical strength prepared by wet spinning. Carbohyd. Polym. **88**, 1100–1108 (2012)
12. Hu, M., Guo, J., Du, J., Liu, Z., Li, P., Ren, X., Feng, Y.: Development of Ca^{2+}-based, ion-responsive superabsorbent hydrogel for cement applications: Self-healing and compressive strength. J. Colloid Interface Sci. **538**, 397–403 (2019)
13. Tang, Z., Kotov, N.A., Magonov, S., Ozturk, B.: Nanostructured artificial nacre. Nat. Mater. **2**(6), 413–418 (2003)
14. Mredha, M.T.I., Guo, Y.Z., Nonoyama, T., Nakajima, T., Kurokawa, T., Gong, J.P.: A facile method to fabricate anisotropic hydrogels with perfectly aligned hierarchical fibrous structures. Advanced Materials **30**(9), 1704937 (2018)

Research and Application of New Reservoir Reconstruction Technology for Fractured Bottom Water Reservoir

Mingwei Kong[1(✉)], Z. Peter Wang[2,3], Rongli Nan[1], Wenna Zhang[3], Wanbin Wang[1], and Fuhe Wang[3]

[1] XinJiang Oilfield Production Technology Research Institute, Xinjiang, China
{kongmingwei, nanrl, wwanbin}@petrochina.com.cn
[2] IUT Group, LLC, Beijing, China
zwang@iutgroup.com
[3] Beijing Huaxiaxiaoneng Petrotec, Ltd, Beijing, China
dorice0113@sina.com, huaxiaxiaoneng2012@126.com

Abstract. Igneous rocks in a block of western Xinjiang, China, are fractured reservoirs in exploration phase currently, and most of those oil reservoirs are found in massive bulk format with water at the bottom. It is difficult to control the height of fractures during fracturing, and the oil and water are easy to flow out after fracturing. In addition, natural fractures are developed, so structure control measures are difficult, resulting in a small scale of fracturing reconstruction, not up to the expected effect of fracturing stimulation; In order to quicken the pace of exploration and development in this area, and make the resource amount into the recoverable reserves as soon as possible, it is necessary to carry out the study on the fracturing technology of igneous rocks fractured reservoir in this area, and form targeted applicable technology, so as to improve the success rate of igneous reservoir fracturing and the output per well. A unique diagnostic method for igneous rock fractured reservoirs was developed by testing the combination of fracturing interpretation parameters and interpretation curves (by introducing generalized G function analysis method). The dynamic gel plug controlled seam technology is adopted, and the integrated thinking of geological comprehensive understanding and process technology is introduced. Through the optimization design and the application of new materials, the well area will achieve effective oil increase and water control; The numerical simulation method helped optimizing the fracturing engineering parameters, which can greatly reduce the difficulty during fracturing job implementation causing by geological complexities, and provides new technical support for improving the fracturing success rate and reconstruction scale of the well stimulated volume.

Keywords: Fractured bottom-water reservoir · Igneous rock · Controlling water and increasing oil · Testing fracturing · Dynamic gel plug controlling fracture height technology

With the deepening of exploration, the fractured reservoir mainly composed of Igneous rocks in a block in western China has been discovered, at the bottom are mostly Oil-water Layer. It is difficult to control fracture height in fracturing construction, oil and water comes out easily after fracture, combined with the natural fracture development

J. Lin (ed.), *Proceedings of the International Petroleum and Petrochemical Technology Conference 2019*, pp. 240–253, 2020.
https://doi.org/10.1007/978-981-15-0860-8_20

and construction control measures is difficult. As a result, the fracturing reconstruction scale is too small to achieve the expected fracturing stimulation effect. In order to accelerate the pace of exploration and development in the area, it is necessary to adopt effective stimulation and reconstruction technology to make it valuable for industrial exploitation [1, 2], so that the amount of resources can become active reserves as soon as possible. Therefore, it is necessary to study the fracturing technology of Igneous fractured reservoir in this block, so as to form targeted applicable technology and improve the success rate of Igneous reservoir fracturing and single well output.

1 Geological Background

The Igneous reservoir in the area is generally divided into two parts. The southwest block is JL Well 10 area and the northeast block is JL Well 2 well area. At present, 5 carboniferous Igneous traps with an area of 98.85 km [2] have been implemented in ZhongGuai uplift, with estimated oil reserves of nearly 100 million tons. The favorable target of Igneous rocks in the JiaMuHe formation of the Permian is 47.4 km [2], with the predicted oil reserves of 60 million tons.

The Igneous rock structure in JL Well 10 area has a large range. The distance between JL 101 well and JL 10 well is 2105 m with a height difference of 280 m. Carboniferous reservoirs are mainly composed of Igneous rock, the reservoir properties and reservoir potential and reservoir distribution depends on the characteristics of lithology and lithofacies and reservoir space to a certain extent. According to the drilling, well logging, mud logging, wafer identification, and core description data, the end of the turn in the analysis found that constitute a bump carboniferous Igneous rock strata of rock type is more, the basalt, andesite, Igneous breccia and tuff are common.

The average porosity of Igneous breccia and andesite is relatively high, which is 10.6% and 8.87% respectively, followed by tuff and basalt, which is about 5%–8% respectively. Permeability, the carboniferous Igneous reservoir permeability variation range is larger, about 80% of the sample permeability distribution range of 0.1×10^{-3}–1×10^{-3} μm^2. Carboniferous Igneous rocks are generally of poor physical properties, with strong heterogeneity and mainly low porosity, low permeability or ultra-low permeability reservoirs. Among them, Igneous breccia and andesite have the best physical properties and are the most favorable reservoir rock facies of carboniferous, followed by tuff and basalt.

2 Fracture Development Characteristics

Fracture is the main reservoir space of Igneous rock reservoir, the main channel of fluid migration and oil and gas migration. Fractures in Igneous reservoirs include tectonic fractures, explosion fractures, condensation shrinkage fractures, dissolution fractures, and so on. Structural fractures and dissolution fractures are the major reformations. According to the analysis of fracture distribution characteristics based on seismic data, fracture activity has a certain control effect on fracture development. The degree of fracture development is relatively high near the fault zone and relatively low far from the fault zone [3].

According to the observation of collected cores and the analysis of cast thin sections and fluorescence thin sections, the pore types of carboniferous reservoirs in the area are mainly corrosion pores, pores and micro-fractures. The micro-resistivity log (FMI) data show that fractures are well developed in the carboniferous reservoir. The fractures of lava are relatively developed compared with Igneous breccia. The carboniferous system of JL Well 10 area is generally located in the eruptive facies area, in which Igneous breccia is loose and lava has a large number of matrix pores such as almond pores. In addition, in the later stage of leaching, modification and fracture communication, Igneous breccia and lava can become good Igneous rock reservoirs.

Therefore, the carboniferous reservoir in the area is a fracture-porosity dual medium reservoir. How to fracture such fractured Igneous reservoirs is a world-class problem, and if the fracturing of such reservoirs is successful, it will have an important impact on Igneous exploration and development [4] (Fig. 1).

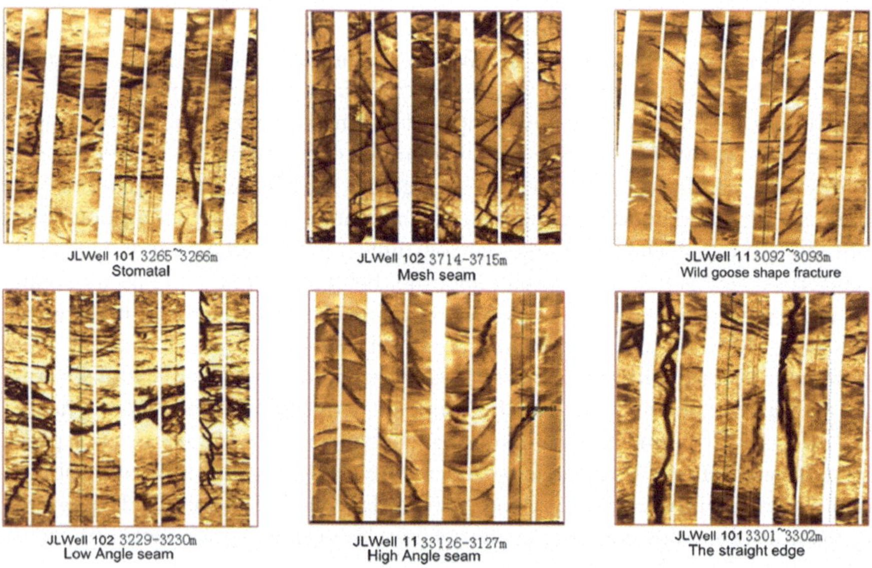

Fig. 1 Fracture development in JLWell 10 area

3 Rock Mechanics Characteristics

According to the rock mechanics parameters of JL Well 10, the minimum horizontal principal stress of the target zone is about 57 MPa, and the difference between zones is not obvious. Porosity, reticular fractures and oblique fractures are developed in the measure interval, and the longitudinal fracture height is not easy to control. It is estimated that the formation fracture pressure is 67 MPa and the formation fracture pressure gradient is 0.0202 MPa/m. The rock mechanics parameter interpretation of JL

Well 101 explains that the minimum horizontal principal stress of the target section is about 50–53 MPa, which is less different from the stress of the adjacent layer (there is a weak stress layer in the lower layer, which has a weak blocking effect on fracture extension). At the same time, there are parallel fractures, irregular reticular fractures, conjugate fractures, single micro fractures and straight fractures in the measure interval, and the longitudinal extension of fractures is not easy to control. It is expected that the formation fracture pressure is 63 MPa and the formation fracture pressure gradient is 0.0197 MPa/m.

According to the interpretation results of in situ stress of JL Well 10 and JL Well 101, the static poisson ratio of carboniferous Igneous rock reservoir in JL Well 10 is 0.17–0.27, young's modulus is 33,000–39,000 MPa, and the formation fracture pressure gradient is 0.0197–0.0202 MPa/m. Poisson's ratio is low to medium, and young's modulus is medium to high. The stress difference between layers is 0–3 MPa. In view of the fact that it is difficult to crack Igneous reservoir, it is necessary to study the rock mechanics and mineralogical characteristics of the reservoir, find out the causes of reservoir hydraulic fracturing failure, and propose corresponding measures [5].

In addition, the ground stress difference is a very important factor affecting the crack height. When the stress difference is small, a small increase in the stress difference will cause a sharp decrease in the crack height. With the further increase of the stress difference in the ground, the effect gradually flattens out. For the target well area, the stress in most of the Wells can hardly prevent the fracture height from extending. Some Wells have relatively high stress difference in the interval, and due to the influence of natural fracture development, the effectiveness of the stress interval is not high.

In terms of petrophysical properties, the ductility and plasticity of the rock as well as the slippage between layers have a great influence on the vertical extension of fractures. The ductility of the pay zone is shown as young's modulus. The higher the young's modulus is, the higher the crack height will be. Young's modulus affects the growth of crack height mainly in the following two ways: the crack reaches the contact surface of two different substances, and the difference in the contact surface can slow down the growth of crack height; When the young's modulus of the interval is larger than that of the oil layer, the young's modulus can prevent the growth of fractures.

4 New Technology for Reservoir Reconstruction of Fractured Bottom Water Reservoir

4.1 Diagnosis and Analysis Method of Generalized G Function

Ever before more complex reservoir sand fracturing construction, the first test fracturing, and then the obtained data is processed and analyzed, the analysis of small fracturing test data can help engineers to understand the development of fractures in the formation, and at the same time to grasp whether there is natural fracture, filtration loss and stress shielding in the formation.

The introduction of G function curve analysis method and typical curve method into the analysis of mini fracturing test data by Stimplan fracturing software can make the analysis results more accurate and reliable. This section aims at the complex situation of Igneous fractured reservoir and the same layer of oil and water in the area. On the basis of testing fracturing, the combination application of fracturing collection and interpretation parameters and interpretation curves can be used to identify and diagnose Igneous fractured reservoir (Fig. 2).

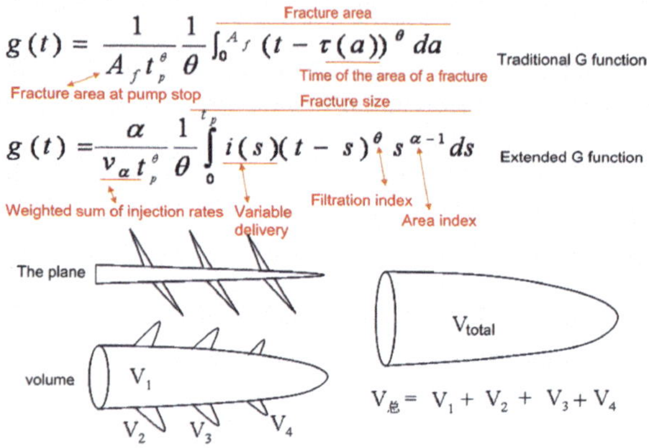

Fig. 2 The extended G function model

The basic idea of "G function" applied to fracturing pressure analysis is developed by Nolte [6] first, and it has been used in the fracturing world ever since and continues to this day. In 2002, the physical model and mathematical model of Igneous rock fracturing were preliminarily established, and the evaluation parameter standard and "G function" characteristic chart for diagnosing the difficulties in Igneous rock fracturing were formed [7]. However, with the deepening of exploration and development, the new situation requires the study on the interpretation of new theoretical models by testing fracturing parameters of Igneous reservoir [8].

An overview of generalized G function analysis methods

Under the condition of constant displacement, the fracture area generated has a relationship with time:

$$\frac{a}{A_f} = \left(\frac{\tau(a)}{t_p}\right)^{\alpha} \tag{1}$$

where, a is the area generated at $\tau(a)$ time, A_f is the fracture area at shut-in time, t_p is the injection time,α is the area index, generally between 1/2 and 1. Under the above assumptions, the traditional 'g' f can be written as:

$$g(t) = \left(\frac{t}{t_p}\right)^{\alpha+\theta} \frac{\alpha}{\theta} B\left(\alpha, \theta+1, t_p/t\right) \tag{2}$$

where, θ is the filtration index (θ = 1/2 for the general case), B is the equation of infinite approximation allowed by mathematical calculation (incomplete Beta function):

$$B(a,b,x) = \int_0^x s^{a-1}(1-s)^{b-1}ds \tag{3}$$

When the hypothesis is satisfied, such as the displacement of XS_A well is not constant, the above equation cannot be used to calculate and explain the pressure drop. Therefore, the report discusses whether natural fracture pressure changes are hidden when the injected displacement changes, and whether there are new functional forms that can solve this defect or provide suggestions for new related engineering injection displacement.

When displacement changes, the fracture area cannot be simply stated as above. We are trying to find a new G function (called generalized G function) to replace the traditional G function. It is assumed that the displacement is unchanged in an infinitely short time period (s_{n-1}, s_n), and is i_n, where:

$$0 = s_0 < s_1 < s_2 < \cdots < s_N = t_p \tag{4}$$

We can assume that the displacement is constant in each infinite short period, then the fracture area in this period conforms to the traditional G function area formula, so that the 'G' function can be deduced into the form suitable for variable displacement as follows:

$$g(t) = \frac{\alpha}{V_\alpha \theta}\left(\frac{t}{t_p}\right)^\theta t^\alpha \sum_{n=1}^{N} i_n[B(\alpha, \theta+1, s_n/t) - B(\alpha, \theta+1, s_{n-1}/t)] \tag{5}$$

where, B equation has been defined above; V is general injection volume.

$$V_\alpha = \sum_{n=1}^{N} i_n\left(s_n^\alpha - s_{n-1}^\alpha\right) \tag{6}$$

In this way, the new equation can be applied to variable displacement in theory.

The physical meaning of the relevant variables is as follows,

s_f: Fracture stiffness; C_L: Filtration coefficient; V(t): Crack volume at time t; A_f: Single area of crack wall; L(t): The volume of fluid loss; r_p: The expression of seepage area; S_p: Initial filtration loss coefficient; t: Pump until fracture closure time; t_p: Pump injection time; $\bar{\varpi}$: Mean crack width; θ: Fluid loss index; α: Area index; ξ: Area index to crack ratio; s: Each unit during the pumping period; $\hat{a}(s)$: The area at s time point; i(s): Delivery at s time point.

Instance Analysis. JL2006 well test fracturing simulation analysis shows that the test reservoir is located in Igneous rock of JiaMuHe formation, well section 3710.0–3725.0 m, with a perforated thickness of 8 m, lithology of andestic cementite, rock density of 2.44–2.48 g/cm^3, and logging interpretation as oil layer, with average porosity of 11.01–15.50% and permeability of 3.5 mD. Formation temperature is 95.0 °C and the formation pressure is 46.68 MPa.

According to G function interpretation curve, closure pressure is 58.2 MPa and liquid efficiency is 48.6%. In the stage of slickwater test, the net pressure did not fluctuate, indicating that the crack height did not break through the barrier, the crack height had certain control conditions, and the G function curve showed multiple cracks. However, there is no friction near the well fracture from the calculation of shut-in pressure. By this method, the friction near the well, the shut-in pressure gradient, the fluid efficiency and the rock compressibility index are determined, which are obviously different from the previous test fracturing interpretation characteristics of other Igneous rocks. Based on this, the treatment process and measures can improve the success rate of fracturing (Fig. 3).

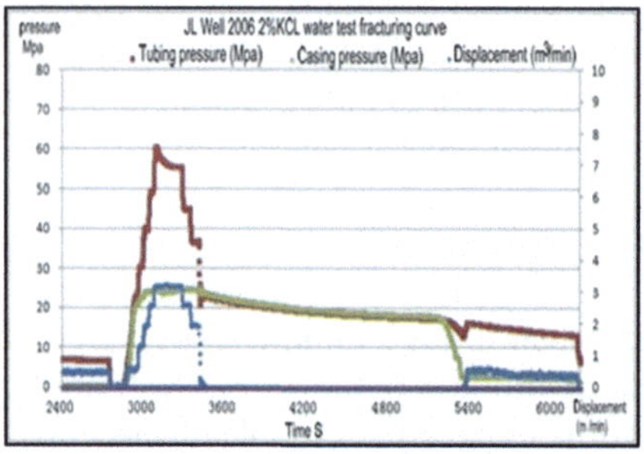

Slick water test fracturing acquisition curve

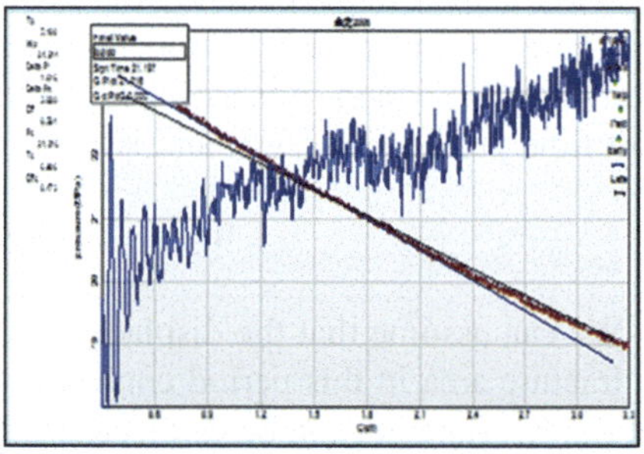

The G function interpretation curve

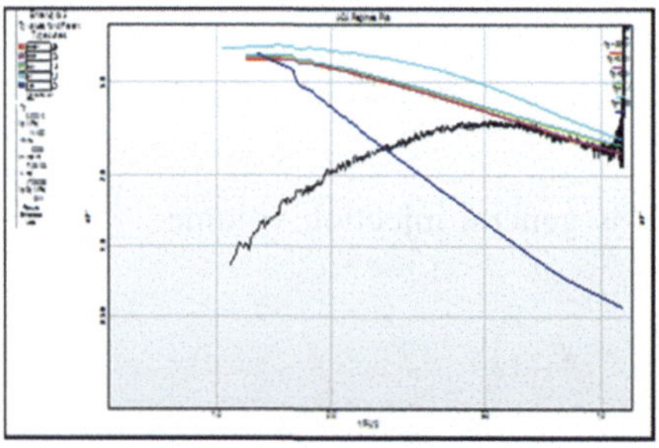

Typical curve fitting (effective K: 1.7mD)

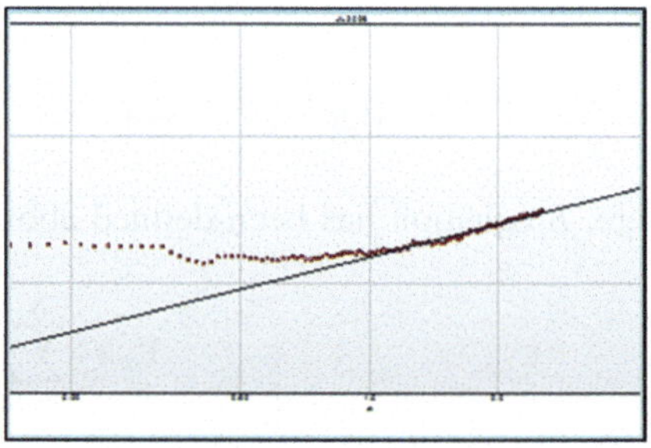

The net pressure interpretation curve

Fig. 3 Analysis and interpretation of fracturing test in Well JL2006

In a word, if the four information indicators are pointing in the same direction or in the same coordinate quadrant, it can be considered that we have basically mastered the characteristics that should be displayed in fracturing of such reservoirs. On this basis, it will also be targeted to propose more targeted control measures.

4.2 Dynamic Gel Plug Control Fracture Height Technology

When the oil/gas layer is very thin or the minimum horizontal principal stress difference between the producing layer and the shielding layer is small, the pressed fracture is easy to enter the shielding layer, which not only prevents the fracture from extending in the horizontal direction as required, but also causes flooding when there is bottom water. Fracture control fracture height technology has been widely used in oilfields. There are generally two common technical methods: One is to control the height of crack by controlling the displacement of construction and reducing the viscosity of working fluid. The other is to use the floating agent and sinking agent to establish artificial shielding to achieve the goal of controlling the height of the crack.

In view of the complex situation of Igneous fractured reservoir and oil-water layer in the area, if conventional methods are adopted to control the flow rate and reduce the liquid viscosity, effective artificial fractures cannot be formed for the fractured reservoir. If floating agent and sinking agent method are adopted, the oil-water layer reservoir can only control the fracture height but cannot achieve water control. Obviously, the two methods mentioned above are not suitable for the fracture height control technology in Igneous reservoirs in this region. Therefore, a new method, namely dynamic gel plug control fracture height technology, should be introduced to solve the above technical problems.

Principle of dynamic gel plug. The method is to inject a certain amount of liquid gel plug at the front end of the pad fluid. The viscosity of the liquid gel plug is large, the fluidity is poor, the gel plug is filled in the upper part and the lower part of the fracture, which improves the singularity of the crack tip, thus achieving the purpose of controlling the crack height. Using the characteristics of low viscosity and good fluidity of the fracturing base fluid, the secondary fracture is made to form a good fracture shape near the well, so that the fracture can be extended forward normally and the reservoir can be effectively reformed. This technique is suitable for the target zone with weak stress shielding, far distance from gas-water interface and natural fractures (Fig. 4).

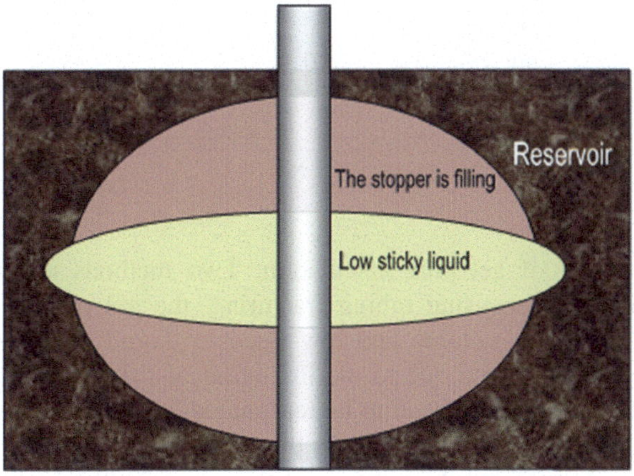

Fig. 4 Schematic diagram of dynamic gel plug

Firstly, differential injection and the pressure decline test after shut-in under different injection rate is carried out. According to the characteristics of G function curve and filter loss, the fracture height extension is analyzed, which provides the basis for the optimization of gel plug dosage and measure method. Secondly, injecting the gel plug with the critical displacement of perforating layer; then, the low-viscosity fracturing fluid is used for extruding the gel plug to a certain crack depth, and the two liquid viscosity difference is utilized, and the subsequent sand-carrying liquid is formed into the water inlet and the sand-adding fracturing construction is conducted.

Calculation and optimization of the quantity of dynamic gel plug.

① Gel plug injection process; The injection process is as follows: calculate the fracturing string volume → small displacement gel plug → small displacement squeeze into low viscosity liquid (full wellbore volume after squeeze) → large displacement injection gel plug (add 20% on the base of the proposed main fracturing displacement to make the gel plug full of cracks) → replace low viscosity liquid with small displacement → shut-in (15–30 min); For example, the fracturing string volume of a well is 15 m^3, and the construction procedure of dynamic gel plug is shown in Table 1.

Table 1 Construction procedure table of dynamic gel plug treatment for XX reservoir in well XX

Step	Construction time (min)		Production processes	Displacement (m^3/min)	Fracturing fluid consumption (m^3)	
	Stage	Accumulate			The dosage	The cumulative
1	10	10	Gel plug	1	10	10
2	5	15	For the base fluid	1	5	15
3	2.5	17.5	Injection plastic plug	4	10	25
4	6.3	23.8	Injection base	0.8	5	30
5			Stop the pump 20 min			

② The amount of gel plug calculation; Two methods are usually adopted. Method 1: when using tubing fracturing, the gel plug is 65–70% of the fracturing string volume and low-viscosity liquid is 35–30% of the fracturing string volume due to the limitation of the fracturing string volume. This ratio has been proved to be reasonable in small and medium fracturing scale (pad fluid consumption less than 150 m^3) by field multi-well practice. Method 2: in case of casing fracturing, due to the increase of wellbore volume, according to field experience, the amount of gel plug is 8–12% of the amount of pad fluid. In this way, dynamic gel plug method can still be used to control fractures in Wells with larger fracturing scale.

③ Software numerical simulation of dynamic gel plug, see Sect. 4.3.

Indoor evaluation of gel plug. Indoor evaluation is carried out for the designed gel plug, and the technical performance of the gel plug is evaluated mainly according to the industry standard of fracturing fluid and the conditions of simulated formation temperature and construction time.

① Technical properties of gel plug;

A: Shear at 170 s^{-1}, the initial viscosity is greater than 1000 mPa s, and the terminal viscosity is >250 mPa s after 90–180 min.

B: After the construction, the gel plug will be degraded and broken within 2 h with a viscosity of <10 mPa s, so as to close the cracks in the filling part of the gel plug (Table 2).

Table 2 Gel plug performance table

Shear viscosity mPa s					
Initial	10 min	30 min	60 min	120 min	180 min
≥ 1000	≥ 900	≥ 500	≥ 300	≥ 250	≥ 200

C: Gel plug heat-resistance is 90 and 100 °C.

D: Evaluation project: Temperature and shearing resistance, viscoelasticity, static filtration, dynamic filtration and gel breaking performance;

E: Performance table of gel plug

If the performance adjustment of gel plug is not up to the standard or the stability or shear resistance of gel plug is improved, the fiber can be added to improve its performance index.

② Gel plug configuration: Based on the existing fracturing fluid formulations in this area, the ratio of thickener of conventional fracturing fluid is 0.35–0.4%, the ratio of thickener of gel plug is 0.7–0.8%, and the hinge ratio is determined by indoor evaluation (Table 3).

③ Indoor evaluation results

Table 3 Indoor evaluation results

Project	Shear viscosity mPa s/100 °C, 170 s^{-1}					
	The initial	10 min	30 min	60 min	120 min	180 min
Standard	≥ 1000	≥ 900	≥ 500	≥ 300	≥ 250	≥ 200
Indoor	>2000 normal temperature	≥ 2000	≥ 2000	≥ 600 (100 °C)	≥ 400 (100 °C)	
The experimental data	≥ 1000(100 °C)	≥ 1000 (100 °C)	≥ 600 (100 °C)			

From the test and evaluation results of indoor formula adjustment, its performance reaches the application standard of gel plug, and the use of fiber is not considered.

Notes for dynamic gel plug application. How to effectively use dynamic gel plug for fracture control, should be done in the following aspects.

① Perforation section requirements; Dynamic gel plug is a control method for fracture height, if the perforating hole is too big, when low viscosity liquid into the perforation and fracture, fracture height on the rest of the gel plug is filling area decreased significantly, greatly reduces the vertical of the gel plug is in meet the demand of fracturing, on the basis of the choice of bearing strength and resulting in a decline in the effectiveness of the gel plug, perforation thickness control within 20% of the fracture opening, which can effectively improve the gel is bearing strength and the effectiveness of the gel plug.

② The design of the amount of gel plug is carried out under the condition of the wellbore volume. When the design amount exceeds the wellbore volume, the excess gel plug cannot meet the technical requirements of the process conditions. Therefore, when tubing fracturing, the tubing depth can be appropriately adjusted or the tubing diameter can be increased to meet the design requirements of the plug.

4.3 Numerical Simulation of Dynamic Gel Plug Fracturing Effect—Evaluation of Fracturing Process Effect

Finally, fracturing optimization simulation software was used to simulate the fracturing effect of liquid gel plug numerically, and construction parameters were optimized to achieve the effect of controlling fracture height. The simulation software adopts StimPlan full 3D fracturing design and analysis software for simulation. The simulation conditions are three kinds of liquid viscosity: clear water 1 mPa s, base fluid 50 mPa s, and gel plug initial 1000 mPa s [9].

At the rate of 1.0 m^3/min, 15 m^3 water in the original pipe string is injected into the formation. At this time, the dynamic fracture width is within the range of 0.1–0.4 cm, with a fracture height of 7 m and a half fracture length of 110 m. The fracture geometry is the half-length crack is long, fracture height controllable and narrow in fracture width. As shown in Fig. 5.

With a displacement of 4.0 m^3/min and a gel plug of 10.0 m^3, the dynamic fracture width is more than 0.1 cm, the height of the fracture is 33 m, and the length of the half fracture is 56 m. When the dynamic fracture width more than 0.6 cm, the half fracture length is 20 m, the fracture height is 17 m, and the maximum fracture width near the well is 0.9 cm. The geometrical dimensions of fractures show that the half-length of fracture becomes shorter, the height increases, and the width of fractures increases. As shown in Fig. 6.

With a displacement of 0.8 m^3/min and a base fluid (low-viscosity liquid) of 5.0 m^3 injected, the dynamic fracture width of 0.1 cm at this time is 30 m in fracture height and 65 m in fracture half length; among which the dynamic fracture width is 0.6 cm, 2 m in fracture height and 25 m in fracture half length. The geometrical dimensions of

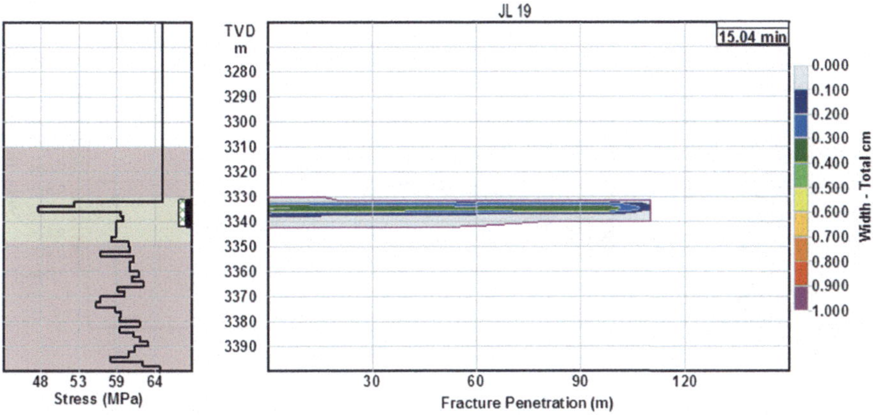

Fig. 5 Fracture geometry after water injection into the formation

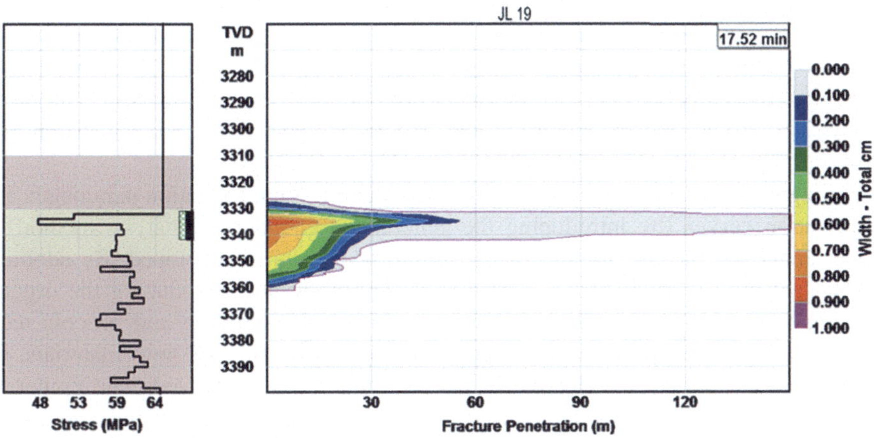

Fig. 6 Fracture geometry after injection of gel plug into the formation

the fractures changed, showing that the half-length of the fracture increased, the height of the fracture decreased, and the width of the fracture decreased compared with Fig. 6, indicating that the low-viscosity liquid reached the required radius for dredging. As shown in Fig. 7.

From the simulation of the above dynamic plug injection process, it is confirmed that the influence and change of liquid viscosity and displacement on the fracture morphology. The larger the displacement is, the higher the liquid viscosity is, the higher the fracture height will be and the shorter the fracture length will be. The applicability of dynamic gel plug technology for this type of reservoir reconstruction is also proved.

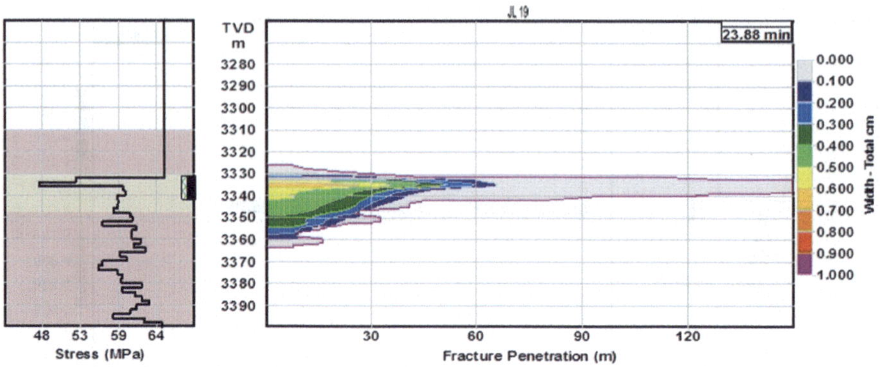

Fig. 7 Fracture geometry after injection of low-viscosity liquid into the formation

5 Conclusion

The Igneous reservoir in a block in western China is controlled by tectonic structure of the water reservoir and lithologic reservoir, the implementation of JL 10 wellblock technology difficulty is effective in increasing oil and controlling water problem, improve the success rate of fracturing and transform scale, through the research proposed the improvement direction of the fracturing technology.

Through combination application of testing fracturing interpretation parameters and interpretation curves (by introducing the generalized G function analysis method), a unique diagnostic method for Igneous fractured reservoirs is developed. By adopting dynamic gel plug fracture height controlling technology and introducing the organic combination thinking of geological comprehensive understanding and process technology, through the optimization design and the application of new materials, oil increase and water control effective in JL well area will be realized. The numerical simulation method optimizes the fracturing construction parameters, which can greatly reduce the influence of complex working conditions caused by multiple fractures on the construction, provide new technical support for improving the fracturing success rate and reconstruction scale of the well area, and put forward new ideas for the reconstruction technology of this type of reservoir.

References

1. Zhang, X., Feng, C., Xie, J., et al.: Stimulation Technology for Deep Tight Gas Reservoirs. In: Papers on the 40 years discovery of Daqing oilfield. petroleum industry press, Beijing (1999)
2. Liu, H., Yan, J., et al.: New fracturing technology for deep igneous gas reservoirs in SongLiao basin. Daqing Petrol. Geol. Dev. (2003)
3. Qu, Y., Shu, P., Wang, Q., et al.: Characteristics of igneous reservoir in XingCheng gas field. Nat. Gas Explor. Dev. **9**, 17–20 (2006)

4. Feng, C., Xie, C., Zhan, Y., et al.: Fracturing technology test of deep fractured Igneous gas reservoir in Daqing. Nat. Gas Ind. **6**, 108–110 (2006)
5. Li, Feng, Xianjun, Wang, Xuanfu, Wang, et al.: Mineralogical and rock mechanical characteristics of tuff Igneous reservoir in Hailaer basin. Petrol. Geol. **2**, 39–42 (2006)
6. Nolte, K.G.: Determination of fracture parameters from fracturing pressure decline. In: Paper 8341-MS presented at the SPE Annual Technical Conference and Exhibition, 23–26 Sept 1979, Las Vegas, Nevada, USA, New York, SPE (1979)
7. Yongping, Zhang, Shicheng, Zhang, Xiufen, Wei: Research and application of diagnostic characteristic parameters of fracturing test for fractured Igneous gas reservoir. Daqing Petrol. Geol. Dev. **27**(2), 91–93 (2008)
8. Yuguang, Zhang, et al.: Application of extended "G function" model in the interpretation of fracturing parameters in igneous rock testing. Dev. Eng. **4**, 61–65 (2013)
9. Xiuling, Han, Xianyou, Yang, Yang, Shi, et al.: Study on construction parameter optimization for fracturing development of ultra-deep fractured reservoir. Fractur. Acid. Technol. **8**, 1–6 (2014)

Wax Control in Paraffinic Crudes: Investigating the Effectiveness of Novel Non-ionic Surfactants as Wax Inhibitor/Dispersant

Khatere Sokhanvarian(✉), Alpha Diarra, Jorge M. Fernandez, and Cornell Stanciu

Sasol Performance Chemicals, Houston, USA
kate.sokhanvarian@us.sasol.com

Abstract. Wax and paraffin precipitation is a major problem around the world, costing the petroleum industry billions of dollars yearly. As temperature drops below the Wax Appearance or Wax Precipitation Temperature (WAT/WPT) of crudes, paraffin starts to precipitate out and restrict or block the effective flow. There are different methods, such as mechanical and chemical remediation to deal with wax issues. Among the latter ones, the use of surfactants is favorably looked upon since they are small molecules with surface activity properties. This study aims to introduce novel aliphatic non-ionic surfactants with different chain length and degree of ethoxylation. In addition to chain length, the impact of branching on the hydrophobic part of the surfactants was also studied. A waxy crude oil from Brazil was characterized through determining its carbon distribution, WAT, viscosity and density based on industry standard methods. Several surfactants with different combinations of chain length/ethoxylation number were then selected for screening. The performance of surfactants was evaluated based on data obtained from treated crude versus the control sample through different experiments. Rheology studies were conducted at 50 to −10 °C and at shear rates of 5 and 300 s^{-1}. The cold finger instrument was utilized to determine paraffin content of the untreated and treated crude. Finally, the paraffin crystal size was analyzed through microscopic studies. The results showed that shear rate can affect the wax treatment outcome as well as the effective concentration of surfactant. Therefore, it is important to assess the rheology at high and low shear rates. Some surfactants in the present study performed great at both low and high shear rates and were able to reduce the viscosity by 80% at temperatures well below WAT of the crude oil. The microscopy results confirmed that wax crystals were reduced in size and were more dispersed after treating the crude with these surfactants. The current study addresses the wax precipitation/deposition challenges of heavy crudes and proposes mitigating them through the use of some new non-aromatic non-ionic surfactants. The chemistries and findings of this research help the oil and gas industry to save money and time by mitigating flow assurance problems.

Keywords: Non-ionic surfactant · Flow assurance · Paraffin dispersant · Rheology modifiers

J. Lin (ed.), *Proceedings of the International Petroleum and Petrochemical Technology Conference 2019*, pp. 254–268, 2020.
https://doi.org/10.1007/978-981-15-0860-8_21

1 Introduction

Crude oil is a commodity that is being actively traded and affects the economy and politics of the whole world. Crude oils are categorized and characterized based on physical and chemical properties. Asphaltene and resins are important components of the complex dispersive system of oil. Asphaltene in a colloidal state can be the attraction and precipitation focal points for the resins molecules that are dissolved in the oil (Can attract the resins and act as nucleation sites for them). Asphaltene-resin complexes can react with liquid hydrocarbons and these tiny aggregates can coagulate with Asphaltene-resin complexes below the crystallization temperature of hard paraffin. Therefore, Asphaltene-resin-paraffin complexes are created (Turbakov and Riabokon 2014; Kumar et al. 2016). Waxes are formed from living and fossilized sources or through a synthetic process and they are polydisperse association of polymethylene chain compounds. Although the petroleum wax that is separated in the refineries is very useful in other consumer products, it is very problematic in some petroleum production operations (Jennings and Weispfennig 2006). In particular, crude oils with high concentrations of high molecular weight paraffin are more susceptible to wax problems (Jennings and Weispfennig 2006). There is a significant difference in chemical composition of waxes; however, thermo plasticity appears to be their universal physical characteristic (Struchkov and Rogachev 2017). Wax is soluble in oil reservoirs under high pressure high temperature (HPHT) conditions. As the temperature is decreased below the wax appearance temperature (WAT), the wax crystallization process is started.

Organic deposition is mainly the result of continuous three-dimensional macromolecular structure formation that is stabilized by Asphaltene and resin in the crude oil (Struchkov and Rogachev 2017). The precipitated wax crystal can deposit on the internal surface of pipeline throughout different phases of production either in offshore or onshore environment. Furthermore, wax deposition during oil flow causes reduction in effective flow rate or can completely block the flow (Labes-Carrier et al. 2002; Harun et al. 2016).

There are three different methods to deal with wax issues such as mechanical, thermal, and chemical, which each have their own advantages and drawbacks. Traditional treatments such as hot oiling and hot watering still are used to address wax deposition and precipitation that is a significant problem for oil industry. The biggest advantages of these methods are simplicity, cost, and immediate results. However, one of the important considerations in hot oiling/watering treatment is their effectiveness since the heat capacity of injected fluid is much lower than the heat capacity of the well, so at higher depths this practice is questionable (White et al. 2017). Moreover, a lot of hot oiling treatments are conducted using the same oil caused the wax deposition to reoccur and this can be problematic since the wax carrying capacity of this oil is low or it wouldn't precipitate wax (Becker 2000). Formation damage due to different reasons can also happen during hot oiling treatments (Barker 1989).

Hot watering and hot watering surfactant were introduced as an alternative to hot oiling, however, plain hot water treatments cannot provide the solvency that hot oiling offers. Hot watering in the presence of surfactants usually produce large water wet particles and as a result more demulsifiers are required to address water wet waxes. Aromatic and aliphatic solvents can also be used as a method to address the wax problem; however, this method is used without the benefit of using heat. Molecular level separation can be achieved through using the solvents with high wax carrying capacity (Becker 2000).

Mechanical means are the oldest techniques to remove wax, which are economical. The main disadvantages with these methods are perforation plugging and wireline scrapers being stuck in wells during post-cleaning stages (White et al. 2017).

Chemical methods can be very effective since they work at molecular level. Wax inhibitors (WI) can be added to oils to diminish oil transportation issues, and they categorize into wax crystal modifiers, dispersants, and pour point depressants (PPD) (White et al. 2017). The crystal growth of waxes can be reduced through the use of WI and as a result the smaller crystals allow larger free space of the liquid fraction of crude oil to flow freely (Halim et al. 2011). One of the disadvantages of PPDs and polymeric WIs is that their performance is poor at temperatures below WAT under higher shear rates. The latter can be due to the breakage of crystal-crystal bond under high shear rate and lack of its reforming (Daraboina et al. 2016). Another problem is with the degradation of large molecules of PPDs so smaller molecules are favorable. On the other hand, non-ionic surfactants are becoming more important since they don't pose any incompatibility issues with other additives and surfactants and they are soluble in water and many organic solvents. Non-ionic surfactants are better choices than ionic surfactants due to some advantages such as more degradability and low toxicity. The most typical and used non-ionic surfactants are ethoxylated compounds due to availability of many hydrophobes and lower cost (El-Shamy et al. 2011). Therefore, the objectives of the present study were to: (1) Investigate the effect of non-ionic surfactants as rheology modifiers (2) Study the impact of hydrophobe structure, number of EO and PO in non-ionic surfactants as wax inhibitor/dispersant, (3) Assess the performance of these non-ionic surfactants through cold finger experiments.

To investigate the deliverability of these objectives, laboratory tests were conducted to study: steady shear viscosity measurements as crude was cooled down below its WAT, microscopic studies of untreated and treated crude, and wax deposition rate of treated and untreated crude through cold finger.

2 Materials and Experimental Studies

2.1 Crude Oil Samples

Oil samples were received from Lagoa do Paulo field, and they were heated and homogenized. Saturates, Aromatic, Resin, Asphaltene (SARA) analysis and HTGC were conducted to determine crude oil properties and carbon distribution based on ASTM D2007-11 and ASTM D5442-93, respectively. Tables 1 and 2 show the crude oil characteristics. Wax appearance temperature of the crude oil was determined based on the viscosity deflection method (Fig. 1). High shear forces are required for better flow of crude oil through pipelines based on the surface tension of 29 m N m^{-1}. It can be seen that this crude is high in saturates (Fig. 2).

Table 1 Characteristics of crude oil used in the present study. The crude oil is high in saturates

Field	Paraffin% (HTGC ≥ C18)	Saturates	Aromatics	Resin	Asphaltene	API Gravity	WAT, °C
Lagoa do Paulo	31	64.8	19.05	15.82	0.34	30.5	37

Table 2 Physical properties of crude oil used in this study

Field	Surface tension, mN/m @50 °C	Density, g/cm^3 @ 20 °C	Density, g/cm^3 @ 60 °C	Water content, wt%
Lagoa do Paulo	29.7	0.8735	0.8425	12

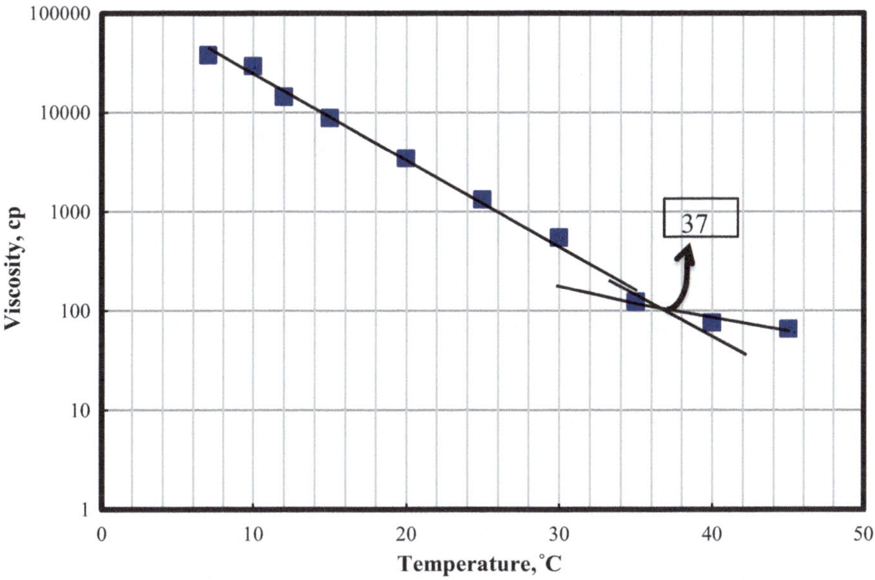

Fig. 1 WAT of crude oil from Lagoa do Paulo field based on the viscosity deflection method

2.2 Non-ionic Surfactants

Different non-ionic surfactants were used in the present study with varying hydrophobe chain length, structure, and EO/PO (Table 3). The physical properties of selected surfactants are shown in Table 4. The fact that some of these surfactants have a very low pour point is very favorable since it eases field application under cold temperature conditions.

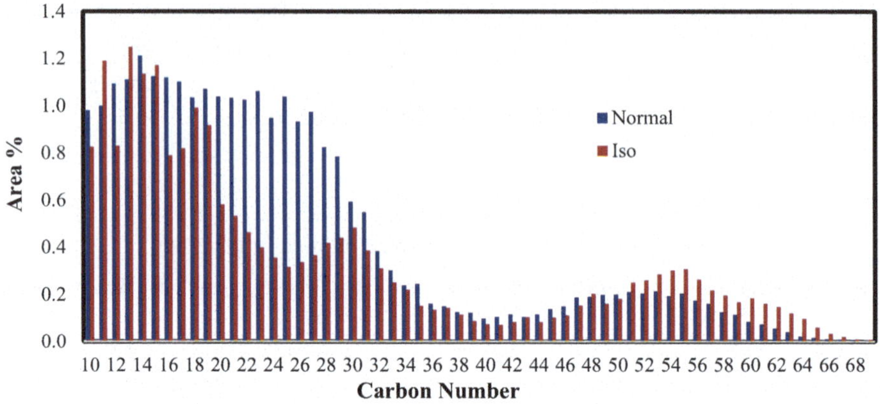

Fig. 2 Carbon number distribution analysis based on HTGC

Table 3 Different type of non-ionic surfactants used in the study. Short chain length hydrophobes were also tested but were not effective as wax inhibitors/dispersants

Surfactant name	Hydrophobe length	Hydrophobe structure	EO#	PO#
TERRAVIS N	long	linear	Low[a]	High[c]
TERRAVIS N'	long	linear	Low	High
TERRAVIS M	medium	branched	Medium[b] Medium Medium	0
TERRAVIS S	medium	semi-branched	Medium Medium Medium	0

[a]1–3; [b]4–15; [c]16–30

Table 4 Physical properties of non-ionic surfactants used in the study

Surfactant name	pH@ 23 °C	Density, g/cm^3 @ 23 °C	Pour point, °C
TERRAVIS N	6.22	1.02	10
TERRAVIS N'	6.84	1.02	10
TERRAVIS M4	5.1	1.01	−25
TERRAVIS S4	5.2	1.01	−5

2.3 Viscosity Measurements

A coaxial cylinder, rotational viscometer with R1/B1 combination and ±0.5% of torque span or better was used to assess the viscosity profile of untreated and treated crude oil. Refrigerated and Heating Circulator for heating and cooling purposes was used in rheology measurements with a temperature stability of ±0.01 °C. The oil sample was heated to 50 °C and then cooled down to −10 °C at a cooling rate of 1.2 °C/min.

The viscosity measurements for each sample were conducted at shear rates of 5 and 300 s^{-1}. The viscosity profile of treated crude oil was compared against the viscosity of the control sample (untreated crude).

2.4 Optical Microscopy

A small aliquot of sample was placed onto a microscope slide then a cover slip was placed on top. Images were collected utilizing a 10×, 20×, and 50× objective lenses. The data were collected on an Olympus BX51 research grade optical compound microscope with a ColorView IIIu or CoolSNAP digital color charged-coupled device (CCD) camera. All the images were collected in the bright field reflectance mode with polarized light.

2.5 Cold Finger

The cold finger consists of a closed stainless steel tube (straight, diameter 12 mm, length 72 mm), inside which flows a cooling fluid with temperature range accuracy of ±0.2 °C. The cold finger is submerged inside the oil sample that is heated and homogenized with the aid of a stirrer. The stirring rate was set at 500 RPM throughout the test. When the cold finger is cooled below the WAT of the crude, wax starts to deposit on the cold finger surface and as a result the cold finger mimics the oil pipeline's wall. The instrument is a multi-cold finger and six cold finger tests can be run simultaneously and each one is temperature controlled through valves allowing the specific amount of cooling fluid to flow through each cold finger.

3 Results and Discussion

3.1 Effect of Non-ionic Surfactants Structures on Their Performance as Wax Inhibitors/Dispersants

Nine different surfactants were tested in this study, and their characteristics are summarized in Table 3. The crude oil was treated with 500 ppm of surfactant above the WAT of the crude oil and then the viscosity was measured as the sample was cooled down to temperatures well below the WAT. It can be seen from Figs. 3 and 4 that TERRAVIS M4 outperformed TERRAVIS M7 and M9, however, TERRAVIS S4, S7, and S9 almost performed the same. Figure 5 shows the effect of each surfactant on the viscosity of crude oil as it was cooled down to 10 °C. All of the surfactants were effective to decrease the viscosity of crude oil at the applied shear rate of 5 s^{-1} and temperature ramp of 40 to 10 °C. Moreover, based on the comparison between the surfactants with medium chain length hydrophobe (TERRAVIS M4 and TERRAVIS S4) TERRAVIS M4 performed better and this can be due to the degree of branching of hydrophobe. They both have the same chain length of hydrophobe and the same number of EO; however, TERRAVIS M4 is highly branched, whereas TERRAVIS S4 is semi-branched.

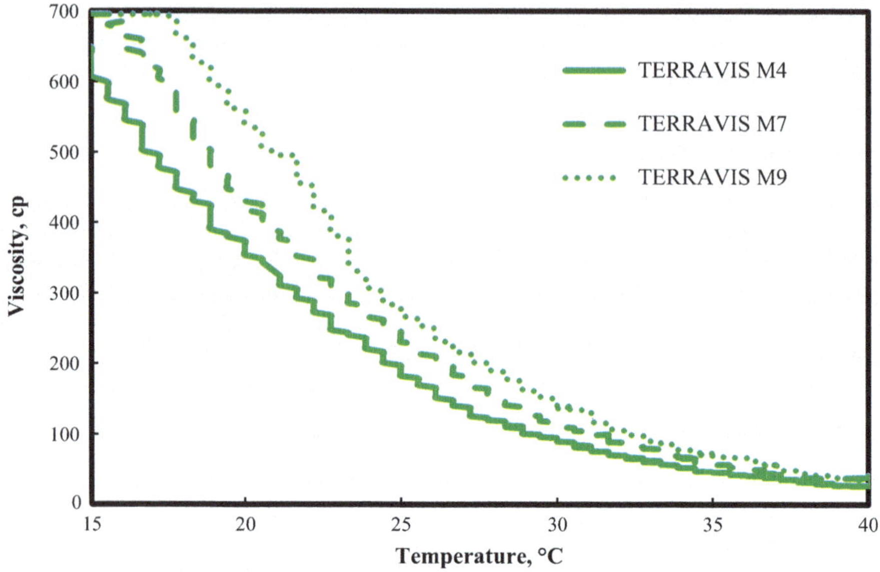

Fig. 3 Effect of number of EO on the performance of TERRAVIS M as wax inhibitor/dispersant at 300 s^{-1}

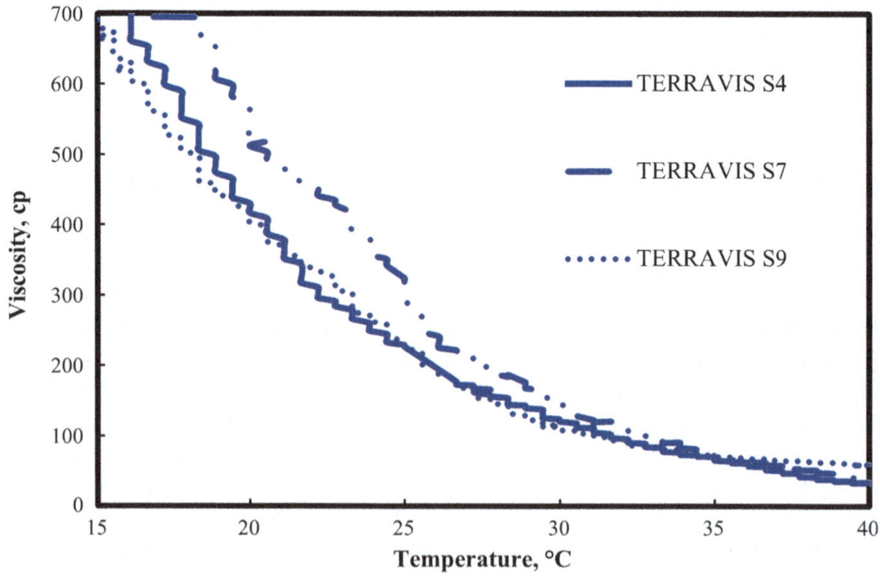

Fig. 4 Effect of number of EO on the performance of TERRAVIS S as wax inhibitor/dispersant at 300 s^{-1}

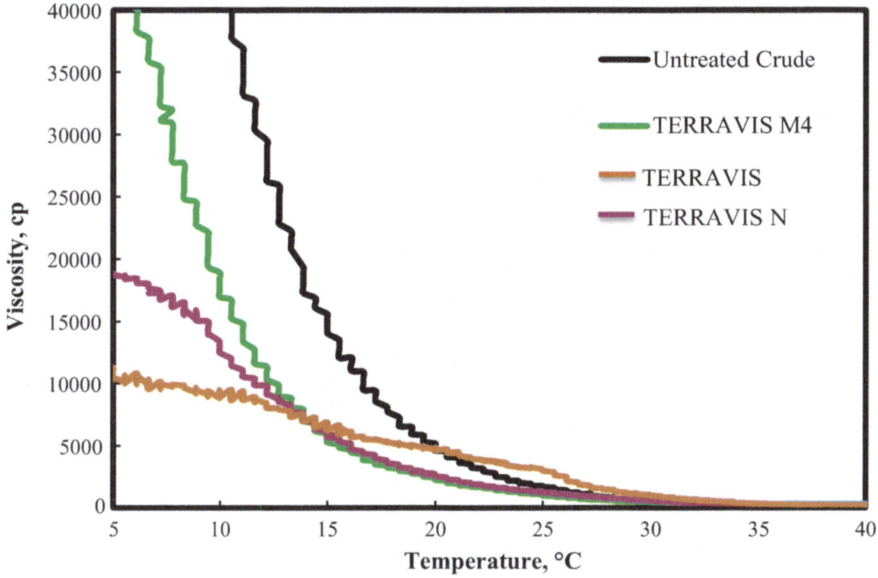

Fig. 5 Viscosity as a function of temperature at shear rate of 5 s^{-1}

Any fluid in the oil and gas industry is exposed to different shear rates during different stages of any treatment. For example high shear rates are applied to waxy crudes during long distance transportations whereas low shear rates inside the oil pipelines is existed during long-term shutdowns. Therefore, it is crucial to assess the performance of any treatment fluid under both high and low shear rates. Based on the results of the present study, the effective concentration of surfactant to reduce the crude oil viscosity may change under high shear environment. The viscosity of the crude oil after treating it with 500 ppm of each surfactant was measured at 300 s^{-1} (Fig. 6). TERRAVIS M4 and TERRAVIS S4 were not able to inhibit/disperse wax crystals in the crude oil at 300 s^{-1}. TERRAVIS N outperformed all the other surfactants and TERRAVIS N' also was able to reduce the viscosity of the crude compared to untreated crude oil. A better performance of TERRAVIS N compared to TERRAVIS N' due to the fact that TERRAVIS N has less linear hydrophobe since all the other factors (hydrophobe chain length, number of PO and EO) are the same for both surfactants. A possible explanation is that the side chain or branching can cause more hindrance effect so the crystal growth would be retarded more effectively (Daraboina et al. 2016).

Poor performance of TERRAVIS M4 and S4 (at 500 ppm dosage) at high shear rate can be attributed to higher heat transfer rate and increased rate of molecular collisions. Moreover, wax crystals can break under shear and these fragments can serve as a new nucleation sites for future crystal growth (Blake and Marangoni 2015). Higher

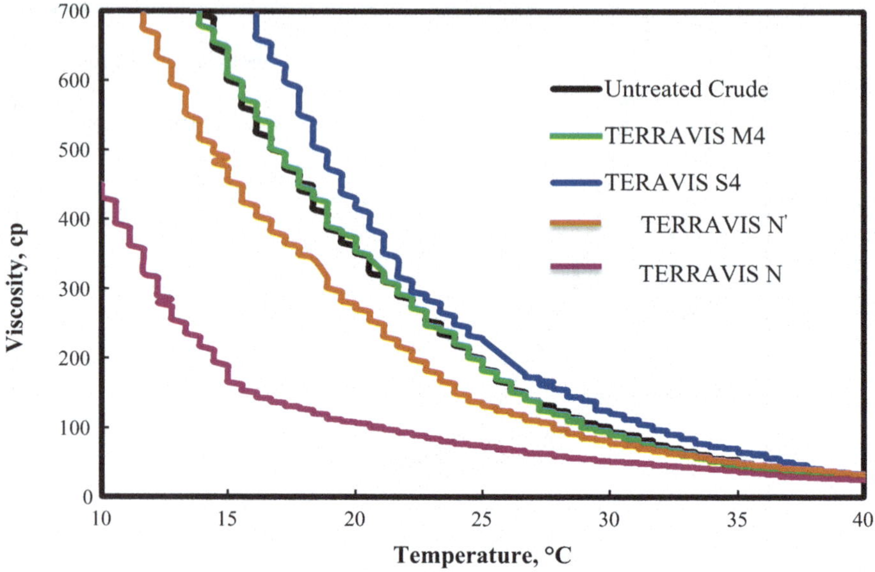

Fig. 6 Viscosity as a function of temperature at shear rate of 300 s^{-1}

shear rate and faster cooling rate also can decrease the interaction time between the treatment chemical and wax crystals. Moreover, for an increased porosity lattice, such as the one TERRAVIS N would form, the entrapped liquid oil would still be able to flow through the pores, especially at higher shear, which supports again the observed behavior of TERRAVIS N and TERRAVIS M4 at higher shear rates.

To investigate the effect of concentration of TERRAVIS M4 on its performance at shear rate of 300 s^{-1}, concentrations of 1000 and 1500 ppm were examined. Furthermore, it was found that 1500 ppm of TERRAVIS M4 is able to reduce the viscosity of crude oil compared to untreated crude when the temperature is decreased from 50 to 10 °C at a fixed shear rate of 300 s^{-1}. However, 500 ppm of TERRAVIS N still outperformed 1500 ppm of TERRAVIS M4 (Fig. 7).

3.2 Microscopic and Particle Size Studies

The optical polarized microscope can recognize the crystalline areas inside the sample due to the fact that they have different refractive indexes compared to the surrounding hydrocarbon environment. The wax crystals can be seen in Fig. 8a–c. It is obvious that the wax crystals in untreated crude are denser (Fig. 8a), which triggers agglomeration and as a result, higher viscosity of crude as it cools down. Treated crude using TERRAVIS N leads to less nuclei number density and different wax crystals sizes (Fig. 8b) that can be the reason behind its better performance compared to other surfactants regarding the ability to reduce the viscosity of the crude oil. In the case of

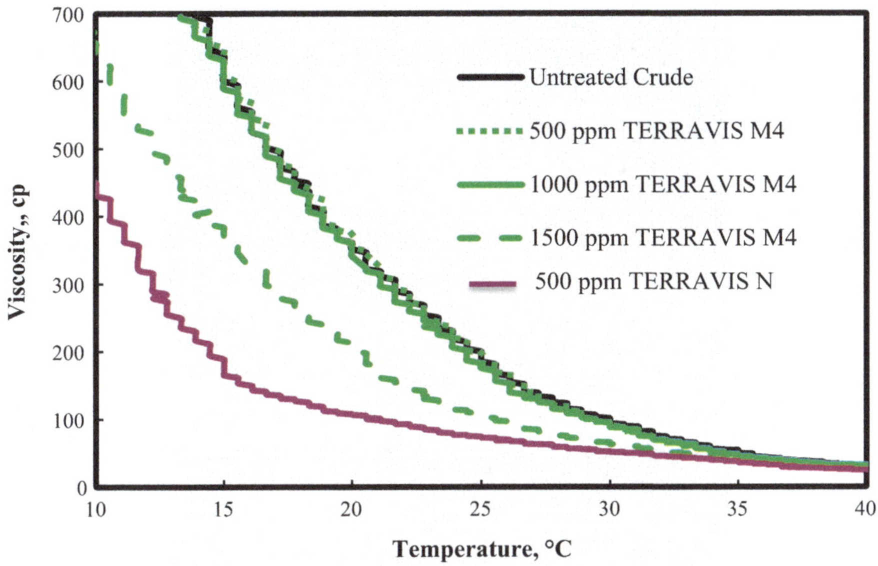

Fig. 7 Effective concentration of TERRAVIS M4 is increased at shear rate of 300 s^{-1}

TERRAVIS M4, the wax crystals are significantly smaller (Fig. 8c) than the wax crystals in the presence of TERRAVIS N, but higher density of nuclei is observed.

The results of this section are in agreement with the findings from the rheology section, which showed that TERRAVIS N outperformed TERRAVIS M4 in reducing the crude oil viscosity as it was exposed to temperatures well below WAT.

3.3 Cold Finger Studies

The wax inhibition studies were conducted to assess the performance of each surfactant. The samples were heated to 55 °C and then the fingers were cooled down to 25 °C. Figure 9 shows the wax deposition at the end of the tests. TERRAVIS M4 could reduce the wax deposition by 30%, whereas TERRAVIS N did not appear to decrease the wax deposition although its performance was better than the other surfactants in regards to reducing the viscosity of crude and producing less wax crystals based on microscopic studies. Second round of cold finger experiments was run where the samples were cooled down from 55 to 18 °C. The wax deposit on the cold finger wax and crude oil entrained within the deposit matrix (Jennings and Weispfennig 2006).

It should be noted that the reason why TERRAVIS M4 a thinner deposition compared to TERRAVIS N can be due to the wax porosity. The free space between the wax crystals, which is filled by the confined liquid oil, defines wax porosity and a high wax porosity normally results in a thicker but softer wax deposit (Labes-Carrier et al. 2002). In addition, the small crystals favored by TERRAVIS M4 are more likely to fill

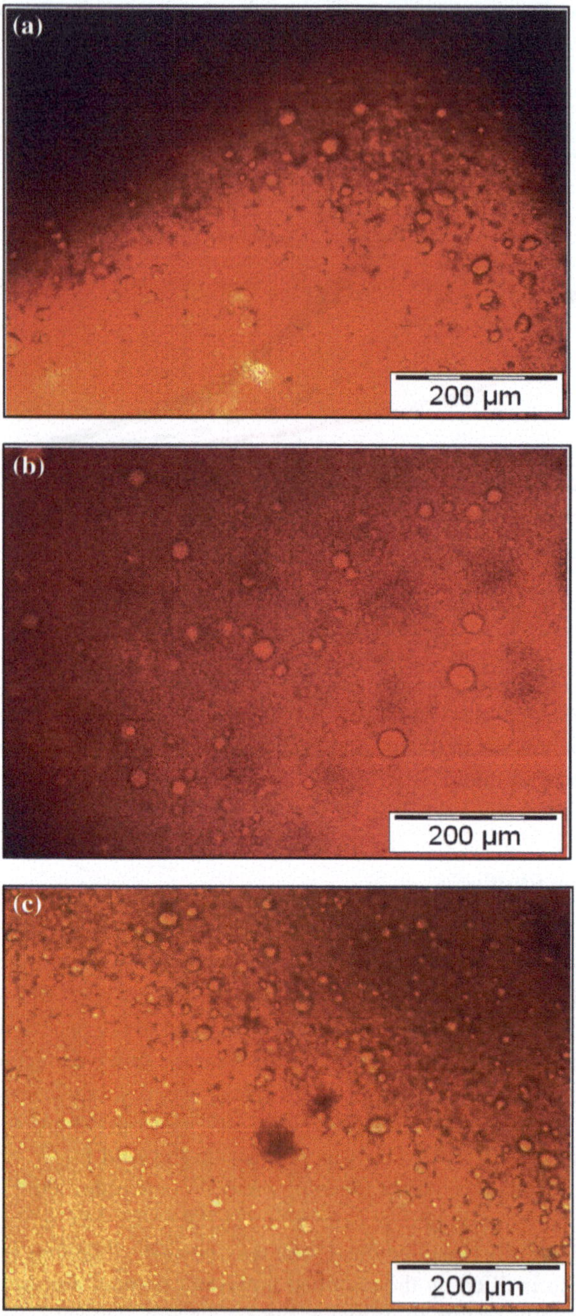

Fig. 8 Wax crystals shape and distribution at 22 °C and a magnification of 20×, **a** blank crude oil, **b** crude oil + 500 ppm of TERRAVIS N, **c** crude oil + 500 ppm of TERRAVIS M4. The wax crystals are bigger but less and more disperse (in the presence of TERRAVIS N) and smaller and denser wax crystals can be seen (in the presence of TERRAVIS M4). Both surfactants were successful to disperse/inhibit wax crystals from further growth

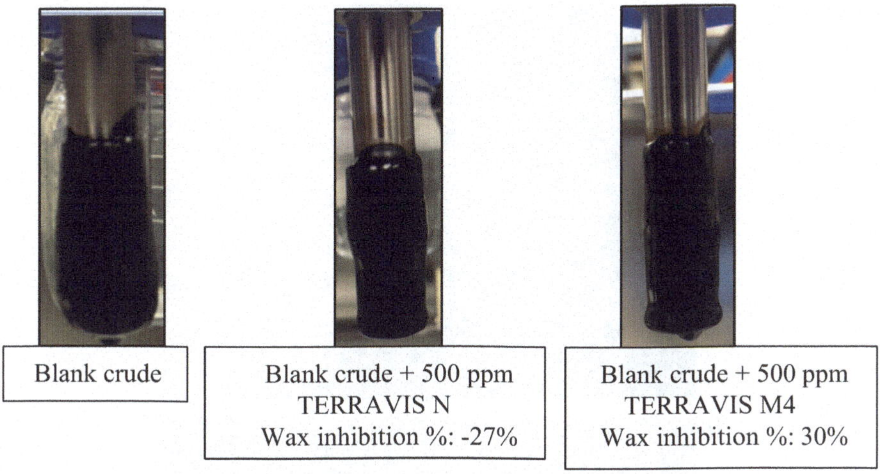

Fig. 9 Wax deposition from the cold finger experiments with a cooling rate of 1 °C from 55 to 25 °C. TERRAVIS M4 outperformed TERRAVIS N

the pore spaces inside the deposit, leading to a higher degree of compaction and a thinner deposit. However, larger crystals can buildup on the deposit surface and create a thicker deposit (Daraboina et al. 2016).

The wax deposit on the fingers was analyzed using optical microscope (Fig. 10). Figure 10a shows a continuous network of wax crystals. In the presence of 500 ppm of TERRAVIS N and TERRAVIS M4 (Fig. 10 a, b), the wax crystals became dispersed and the continuous network is not existed anymore.

4 Conclusions

Different non-ionic surfactants were used as wax inhibitors/dispersants for high paraffinic crude. Several techniques and experiments were run to assess the performance of each surfactant. The conclusions below are based on the results and findings of the present study:

1. Medium and long chain non-ionic surfactants in this study are capable of reducing the precipitation and deposition of wax.
2. Two surfactants outperformed the other ones, one with low HLB and the other one with high HLB. The latter is advantageous since some surfactants are only effective in the presence of water.
3. The hydrophobe with higher degree of branching and more EO (smaller and more polar ones) can be considered more as a wax inhibitor rather than dispersant due to the hindrance effect and polar-non-polar repulsive interactions. Longer chain hydrophobes are effective more as dispersant.

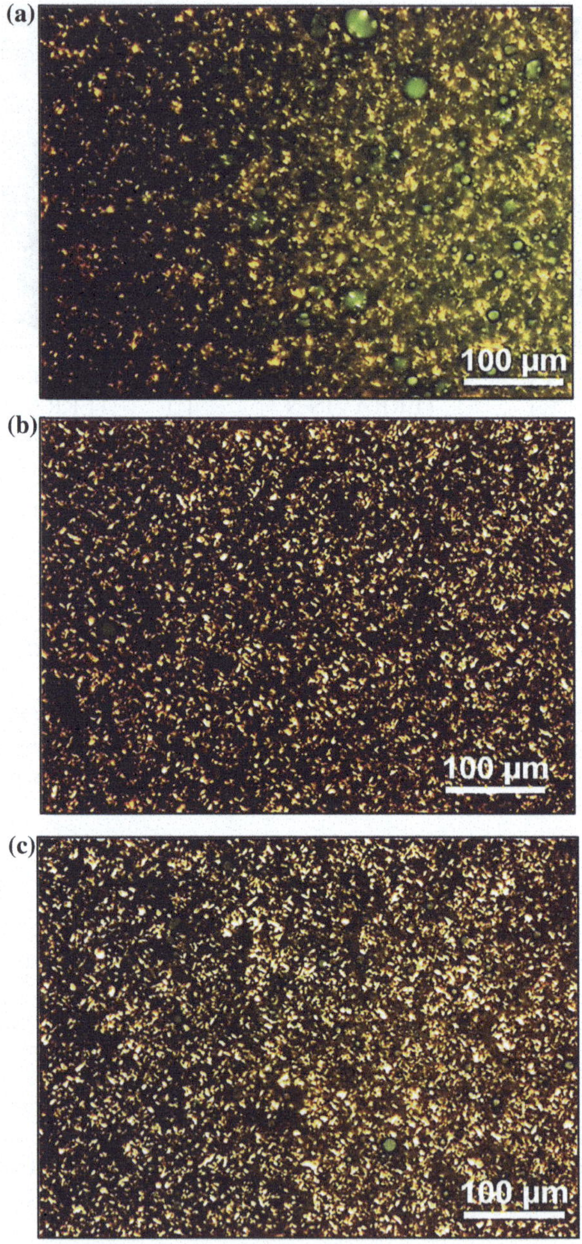

Fig. 10 Wax crystals shape and distribution at 22 °C and a magnification of 20×, **a** blank crude oil, **b** crude oil + 500 ppm of TERRAVIS N, **c** crude oil + 500 ppm of TERRAVIS M4. The wax crystals are more disperse (in the presence of TERRAVIS N) and smaller and denser wax crystals can be seen (in the presence of TERRAVIS M4) compared to blank crude. Both surfactants were successful to disperse/inhibit wax crystals from further growth

This study shows that despite the complex nature of the possible interactions occurring between rheology modifiers or pour point depressants and different types of crudes with a wide SARA distribution, a good understanding of a few basic facts about the latter ones allows for a good choice of surfactants and thus for the operational success sought for by the service companies or operators dealing with paraffin/asphaltene deposits of heavy and/or high paraffinic oils.

Acknowledgements. The authors thank Sasol Performance Chemicals for permission to present and publish this paper. Riaan Bekker and Rebecca Sanders are acknowledged for their analytical support.

References

Barker, K.M.: Formation damage related to hot oiling. SPE Prod. Eng. **4**(4): 371–375 (1989). SPE-16230-PA. https://doi.org/10.2118/16230-PA

Becker, J.R.: Oilfield paraffin treatments: hot oil and hot water compared to crystal modifiers. In: Presented at the SPE Annual Technical Conference and Exhibition, 1–4 October, Dallas, Texas. SPE-63123-MS. https://doi.org/10.2118/63123-MS (2000)

Blake, A.I., Marangoni, A.G.: The effect of shear on the microstructure and oil binding capacity of wax crystal networks. Food Biophys. **10**(04), 403–415 (2015). https://doi.org/10.1007/s11483-015-9398-z

Daraboina, N., Soedarmo, A., Sarica, C.: Microscopic study of wax inhibition mechanism. In: Presented at the Offshore Technology Conference, 2–5 May, Houston, Texas. SPE-26973-MS. https://doi.org/10.4043/26973-MS (2016)

El-Shamy, O.A.A., Khid, T.T., Doheim, M.M.: Effect of ethoxlate chain length on the pour point depressant of middle distillate fuel oil. J. Disp. Sci. Technol. **32**(05), 654–658 (2011). https://doi.org/10.1080/01932691003800023

Halim, N., Ali, S., Nadeem, M., et al.: Synthesis of wax inhibitor and assessment of squeeze technique application for malaysian waxy crude. In: Presented at the SPE Asia Pacific Oil and Gas Conference and Exhibition, 20–22 September, Jakarta, Indonesia. SPE-142288-MS. https://doi.org/10.2118/142288-MS (2011)

Harun, A., Ab Lah, N.K.I.N., Husin, H., et al.: An overview of wax crystallization, deposition mechanism and effect of temperature and shear. In: Presented at the 2016 International Conference on Industrial Engineering, Management Science and Application (ICIMSA) (2016)

Jennings, D.W., Weispfennig, K.: Effect of shear on the performance of paraffin inhibitors: coldfinger investigation with Gulf of Mexico crude oils. Energy Fuels **20**(06), 2457–2464 (2006)

Kumar, R., Banerjee, S., Mandal, A., et al.: Improvement in transportability of indian heavy crude oil using novel surfactant. Indian J. Chem. Technol. **23**(04), 262–270 (2016)

Labes-Carrier, C., Ronningsen, H.P., Kolnes, J., et al.: Wax deposition in north sea gas condensate and oil systems: comparison between operational experience and model prediction. In: Presented at the SPE Annual Technical Conference and Exhibition, 29 Sept 2 October, San Antonio, Texas. SPE-77573-MS. https://doi.org/10.2118/77573-MS (2002)

Struchkov, I.A., Rogachev, M.K.: Risk of wax precipitation in oil well. Nat. Resour. Res. **26**(1), 67–73 (2017)

Turbakov, M.S., Riabokon, E.P.: Improving of cleaning efficiency of oil pipeline from paraffin. In: Presented at the SPE Russian Oil and Gas Exploration and Production Technical Conference and Exhibition, 14–16 Oct, Moscow, Russia. SPE-171295-MS. https://doi.org/10.2118/171295-MS (2014)

White, M., Pierce, K., Acharya, T.: A review of wax-formation/mitigation technologies in the petroleum industry. SPE Prod. Oper. **33**(03): 476–485. SPE-189447-PA. https://doi.org/10.2118/189447-PA (2017)

New Non-aromatic Non-ionic Surfactants to Form a Stabilized Emulsified Acid for Efficient HP/HT Stimulation

Khatere Sokhanvarian[1(✉)], Cornell Stanciu[1], Jorge M. Fernandez[1], Ahmed Ibrahim[2], and Hisham A. Nasr-El-Din[2]

[1] Sasol Performance Chemicals, Houston, USA
kate.sokhanvarian@us.sasol.com
[2] Texas A&M University, College Station, USA

Abstract. Matrix acidizing has been used for permeability and productivity enhancement purposes in oil and gas wells. Hydrochloric acid has been always a first choice due to so many advantages that it can offer. However, HCl in high pressure/high-temperature (HP/HT) wells is a concern because of its high reactivity resulting in face dissolution, high corrosion rates, and high corrosion inhibition costs. There are several alternatives to HCl, among them emulsified acid is a favorable choice due to inherent corrosion inhibition, deeper penetration into the reservoir, less asphaltene/sludge problems, and better acid distribution due to its higher viscosity. Furthermore, the success of the latter system is dependent upon the stability of the emulsion especially at high temperatures. The emulsified acid must be stable until it is properly placed and it also should be compatible with other additives in an acidizing package. This study presents the development of a stable emulsified acid at 300 °F (150 °C) through investigating a novel aliphatic non-ionic surfactants. This paper introduces some new non-aromatic non-ionic surfactants to form an emulsified acid for HP/HT wells where the conventional acidizing systems face some shortcomings. The type and quality of the emulsified acid was assessed through conductivity measurements and drop test. Thermal stability of the system was monitored as a function of time through the use of pressure tubes and a preheated oil bath at 300 °F (150 ° C). The viscosity of the emulsified acid was tested as a function of shear rate (0.1–1000 s^{-1}). The microscopy study was used to investigate the droplets size and shape of the emulsified acid that was formed. Coreflood study was conducted to determine the performance of the newly developed stable emulsified acid in creating wormholes and reaching break through. Inductively Coupled Plasma (ICP) was used to determine dissolved cations. Superior stimulation results were achieved at 300 °F with the newly developed emulsified acid system. This study will assist in developing a more cost-effective and efficient design of acid treatments through introducing the development of a new and effective aliphatic non-ionic surfactants, which lead to the formation of a stable emulsified acid. This new emulsified acid system efficiently stimulates HP/HT carbonate reservoirs.

Keywords: Non-ionic surfactant · Emulsified acid · Retarded acid · Cost-effective

J. Lin (ed.), *Proceedings of the International Petroleum and Petrochemical Technology Conference 2019*, pp. 269–283, 2020.
https://doi.org/10.1007/978-981-15-0860-8_22

1 Introduction

Matrix acidizing is a technique, in which acid is injected below the fracture pressure of formation to remove or bypass the formation damage. In principle, acid moves outward from the wellbore radially toward the formation and as the rock is dissolved long pathways are created (wormholes). The created paths can improve the flow and production of hydrocarbons from the reservoir toward the wellbore. One of the most typical acids that have been used in the industry is hydrochloric acid (HCl) due to its availability, price, and solubility of its reaction products. However, HCl is a very strong acid and its aggressiveness is increased at temperatures greater than 250 °F. The latter is one of the main limitations of HCl along with high corrosion rates. Several alternatives as acidizing systems such as organic acids and chelating agents were introduced. Organic acids should be used at high temperatures due to their slow reaction rates, and also they should be uses at limited concentration due to low solubility of their reaction products. Chelating agents is another solution as a stand-alone stimulation agent; however, their thermal stability is a concern especially at temperatures greater than 350 °F (Mahmoud et al. 2011; Sokhanvarian et al. 2013, 2016). Moreover, a new in situ generated HCl acid system was developed recently; which could be injected at lower flow rates without face dissolution for carbonate dissolution in sandstone and carbonate reservoirs (Sokhanvarian et al. 2017). Another alternative that can address the limitations of HCl is to use emulsified acid (Crowe and Miller 1974; Bergstrom and Miller 1975). Emulsified acid offers several benefits such as minimum corrosion rate, slower reaction through smaller diffusion rate, deeper acid penetration, less asphaltene, and iron issues. The presence of iron (III) in the live acid is a major concern since it can precipitate out around pH 2, and can adversely interact with acid additives, and increase the sludging tendency (Nasr-El-Din et al. 2008 SPE). Therefore, the fact that there is less iron (III) in the live acid in the case of emulsified acid is very favorable.

The acid emulsion is water in oil type emulsion (W/O) and consists of non-aqueous (continuous phase) and acid (dispersed phase) in the presence of an emulsifier, which assists in emulsion formation and stability (Al-Anzi et al. 1998; Bazin and Abdulahad 1999). The success of the acidizing treatment based on the emulsified acid is strongly dependent upon the emulsion stability within the tubular until it reaches the formation. There are many studies on the effect of emulsifier concentration, the ratio of aqueous/non-aqueous, preparation method, etc. on the stability of the acid emulsion systems. It has shown that 70:30 volume ratio of acid to diesel is the best ratio in terms of stability. Moreover, higher concentrations of emulsifier leads to smaller droplet size that is translated to a more sable emulsified acid system. The type of emulsifier is very important in determining the quality and stability of the emulsified acid.

To the best of the authors' knowledge, most of the emulsifiers that have been used previously (whenever the type of emulsifier was disclosed in literature) or in recent studies are cationic based surfactants and on very few occasions anionic or mixtures of cationics or anionic/non-ionic surfactants were used (Crow and Miller 1974; Al-Anzi et al. 1998; Nasr-El-Din et al. 2000, 2001; Al-Mutairi et al. 2007, 2008, 2009a, b; Cassidy et al. 2012; Sabhapondit et al. 2012; Sayed et al. 2012a, b; Zakaria et al. 2012; Pandya and Wadekar 2013; Sayed et al. 2013; Sidaoui and Sultan 2016; Sidaoui et al.

2016, 2017; Musa et al. 2018). It is worth to mention that non-ionic surfactants are becoming more important since they don't pose any incompatibility issues with other additives and surfactants and they are soluble in water and many organic solvents. Non-ionic surfactants are better choices than ionic surfactants due to some advantages such as more degradability and low toxicity. In addition, one of the drawbacks with common emulsifiers (cationic) used to form emulsified acid is that the corrosion products (Fe^{+2}, etc.) are ionic species that interfere in the action of the cationic surfactants (Syafii et al. 2016). The most typical and used non-ionic surfactants are ethoxylated compounds due to availability of many hydrophobes and lower cost (El-Shamy et al. 2011). Moreover, non-ionic surfactants are more suitable for acidizing purposes since they have minimum interactions with other acid additives (Nasr-El-Din et al. 2003). Therefore, some new non-aromatic non-ionic surfactants were developed and are introduced in the present study to form a stable emulsified acid for high temperature acidizing treatments. The objectives of the present study are to: (1) Investigate the ability of aliphatic non-ionic surfactants to form an acid in oil emulsion (2) Study the thermal stability and rheology of the emulsified acid, and (3) Determine the required pore volume of the emulsified acid to breakthrough at 300 °F. To investigate the deliverability of these objectives, laboratory tests were conducted to study: thermal stability of emulsified acid at 300 °F using heating oil bath, microscopic studies of acid droplets in diesel, and viscosity of the emulsified acid as a function of shear rate and temperature.

2 Materials and Experimental Studies

Diesel number 2 was used as the non-aqueous phase in forming the emulsified acid. DI-water of resistivity greater than 18.2 MΩ cm at 77 °F was used to dilute hydrochloric acid to 15 wt% in all the experiments and the HCl was ACS grade (36–37 wt%). The physical properties of diesel (non-aqueous phase) and HCl (aqueous phase) are listed in Table 1. A commercial corrosion inhibitor was provided by a service company. Different non-ionic surfactants were synthesized and used in the present study with varying hydrophobe chain length, structure, and EO/PO ratios. Among the various emulsifiers, one was chosen to further extend the study for acidizing applications purposes. The structure of synthetized non-ionic surfactant and its physical properties are shown in Tables 2 and 3, respectively. An outcrop Indiana limestone with dimensions of 1.5 in diameter and 6 in. length was used in this study.

Table 1 Physical properties of diesel and HCl used in forming W/O emulsion

1. Aqueous and non-aqueous phases	Viscosity, mPa s @ 25 °C	Density, g/cm^3 @ 25 °C
Diesel	2.17	0.866
15 wt% HCl	1.22	1.073

Table 2 The structure description of emulsifier synthesized in the present study

Surfactant name	Hydrophobe length	Hydrophobe structure	EO#	PO#
Acid emulsifier	Detergent range	linear	High	High

Table 3 Physical properties of non-ionic surfactant used in the study

Surfactant name	Viscosity, mPa s @ 50 °C	Density, g/cm^3 @ 23 °C	Pour point, °C
Acid emulsifier	79.24	0.97	−10

2.1 Conductivity and Drop Test

The type of the emulsion (O/W or W/O) was determined using the conductivity meter. If W/O emulsion is formed, the conductivity meter will read small to zero µs/cm, however if O/W emulsion is formed the conductivity will be high. Drop test (solubility) was used to verify the quality of W/O emulsion. If the placed droplet of emulsified acid disperses in water, then the external phase is water. If the placed droplet in water sinks and holds its spherical shape, the external phase is oil. The conductivity and drop tests were performed immediately after the emulsion was formed as well as at the end of thermal stability experiment.

2.2 Thermal Stability of Emulsified Acid at 300 °F

An oil bath was used to monitor the stability of acid emulsion every one hour for the total of four hours. The Ace glass pressure tubes with the rating of 150 psi at 248 °F were used. At the end of the four hours test, the conductivity was measured and the quality of the emulsified acid was determined through the drop test.

2.3 Viscosity Measurement

A coaxial cylinder, rotational viscometer with R1/B1 combination and the accuracy of ±0.5% of torque span or better was used to assess the viscosity profile of the emulsified acid. The viscometer features acid-resistant wetted material (Hastelloy C-276). The shear rate range of the instrument is 0.0038–1020 s^{-1}. The viscosity measurements for the sample were conducted at shear ramp of 1–1000 s^{-1} and temperatures up to 300 °F.

2.4 Optical Microscopy

A small aliquot of sample was placed onto a microscope slide then a cover slip was placed on top. Images were collected utilizing a 20× objective lenses. The data were collected on an Olympus BX51 research grade optical compound microscope with a ColorView IIIu or CoolSNAP digital color charged-coupled device (CCD) camera. All the images were collected in the bright field reflectance mode with polarized light.

2.5 Core Flood

Figure 1 shows a schematic of the coreflood setup. 1,800 psi overburden pressure was used, and a back pressure of 1,100 psi was applied to keep the CO_2 in solution (Fredd and Fogler 1998). The core holder was placed inside an oven to provide a uniform heating environment. The acid injection was continued until the sharp drop in pressure drop occurred (breakthrough). After the injection of half a PV of acid, the core effluent sample was collected every 0.33 PV. The core effluent samples were collected until they became colorless and pressure drop was stabilized.

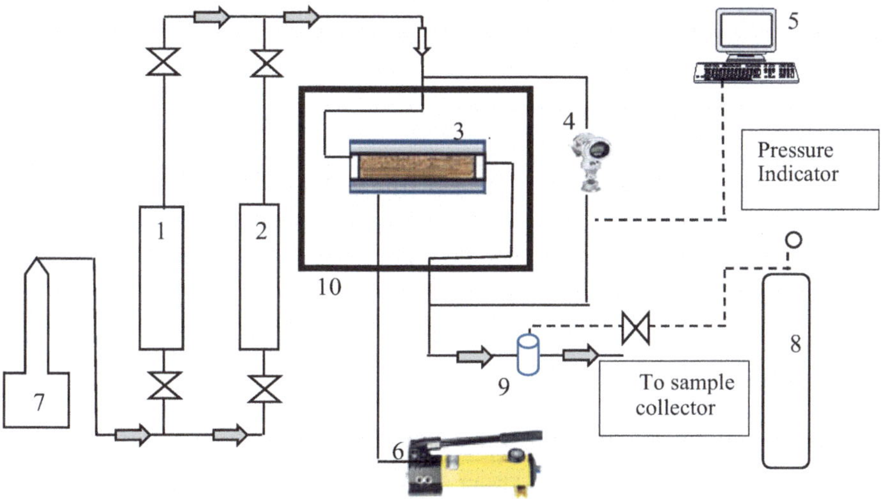

Fig. 1 Schematic for coreflood setup. 1&2 = accumulators, 3 = core holder, 4 = pressure transducer, 5 = PC recorder, 6 = hand pump for overburden pressure, 7 = syringe pump, 8 = N_2 cylinder, 9 = back pressure regulator, 10 = oven

- The core was dried in an oven at 150 °F overnight and the dried-core weight was then measured.
- The core was saturated with DI-water under vacuum for 4 h and then left submerged in the water for 24 h. The weight of saturated-core was measured. The core pore-volume (PV) was calculated as a function of the weight difference between the saturated and the dry cases and the DI-water density (0.9995 g/cm^3).
- The core was inserted in the core holder to measure the initial permeability at 75 °F. The initial permeability for the cores was found to be 5.4 and 7 md with PVs of 26.11 and 27.2, respectively.
- The core holder was put in the oven for uniform heating and the oven temperature was set to 300 °F.
- DI-water was injected at constant flow rates of 2 cm^3/min and at 300 °F for 3 h to ensure the system temperature is at 300 °F.
- The emulsified acid system was then injected and the effluent samples were collected every 0.33 PV.

- The pressure drop across the core was monitored until breakthrough, which was determined from a sudden drop in the pressure drop across the core.
- DI-water was then injected to flush the system and displace the acid out of the core.
- The effluent samples were analyzed using density and pH measurements. Inductive coupled plasma (ICP) was conducted to measure the Ca^{+2} concentrations in the effluent samples.

2.6 Inductively Coupled Plasma (ICP)

The core effluent samples were diluted 2500 time with DI-water and the dissolved calcium was measured after the calibration of the instrument using the standard solution at three different concentrations (5, 15, and 30 ppm).

2.7 Density and PH Measurements

The pH of the core effluent samples was measured using an epoxy electrode, which was calibrated with three pH standards (4, 7, and 10). A digital portable density meter was used to measure the density of the effluent samples. Density and pH data helps in evaluating the dissolution capacity and the depth of penetration of the emulsified acid system, respectively.

3 Results and Discussion

3.1 Stability of Emulsified Acid and Particle Size Analysis

One of the main requirements that should be met for emulsified acid application is the emulsion stability. Stability of an emulsified acid at the target temperature until the emulsion is reached the formation is crucial. Therefore, thermal stability of the emulsified acid was assessed at 300 °F in a static mode. The presence of any phase separation was monitored every one hour. Figure 2 shows that the aqueous phase separation (bottom layer) after four hours of heating at 300 °F is insignificant. On the other hand, the downward creaming (diesel layer at top of the emulsion) is not present, which proves the stability of the emulsion (Al-Mutairi et al. 2007). The latter confirms the suitability of the synthesized non-ionic surfactant to form a stable emulsified acid in the present study, which can translate to a deeper acid penetration with less pore volume of acid. The latter will be confirmed in the coreflood section. It is worth to mention that the conductivity of the emulsified acid was measured before and after heating at 300 °F and was close to zero µs/cm. The quality of the emulsion was examined through the drop test (Fig. 3).

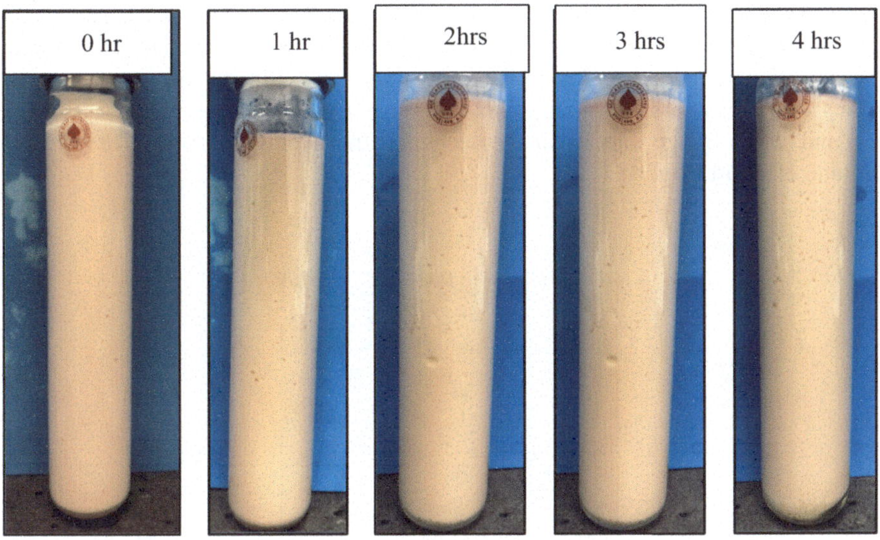

Fig. 2 Thermal stability of emulsified acid at 300 °F as a function of time

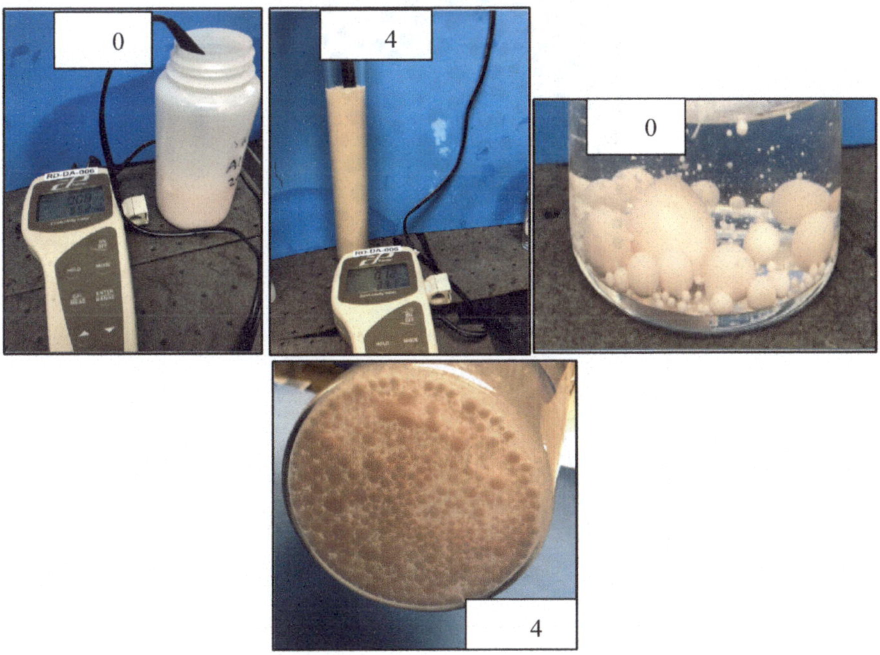

Fig. 3 Conductivity and drop test of emulsified acid confirms the type of emulsion (W/O) and the quality of the emulsion formed

Particle size plays an important role in the stability of emulsions, and smaller particle size can result in a more viscous and stable emulsions. The high thermal stability of the emulsified acid in the present study can be due to the fact that the acid droplets in diesel phase are small (Fig. 4), Image J software was used on a clarified black and white image and it can be seen that the average particle size using this technique is 2 μm (Fig. 4).

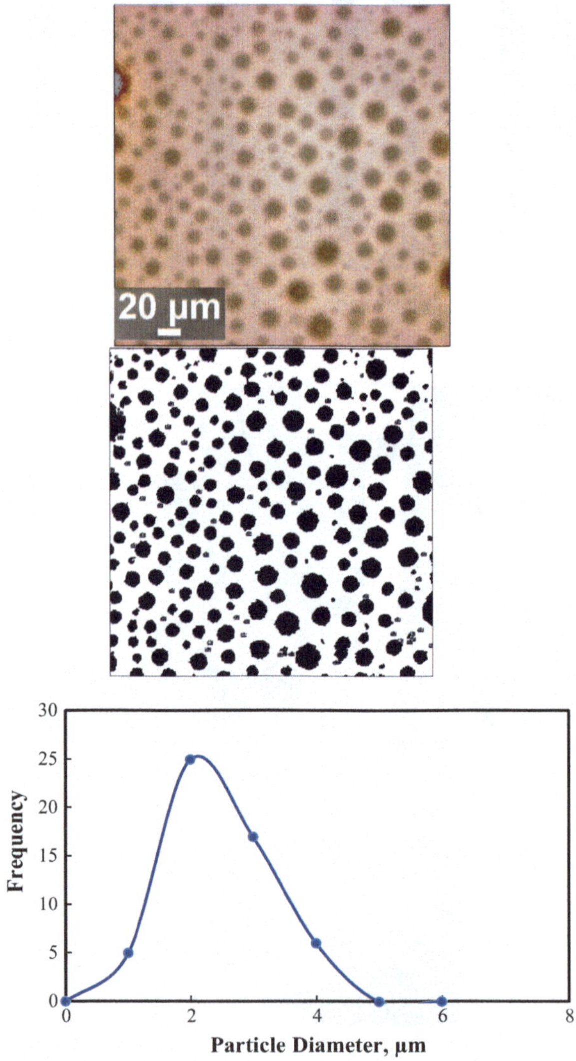

Fig. 4 Microscopic image of acid droplets dispersed in diesel at 20X magnification

3.2 Rheology of Emulsified Acid

Viscosity of the emulsified acid is an important factor that is considered as a candidate for field treatment. High initial viscosity equal and greater than 150 cp at surface is required (Bartko et al. 1992; Appicciutoli et al. 2010). Therefore, the apparent viscosity of the emulsified acid was measured at 75, 150, 200, and 300 °F at shear rates of 1 to 1000 s^{-1}, and the viscosity at 75 °F is within the acceptable range mentioned in the literature (Fig. 5). The apparent viscosity is decreased as the shear rate is increased. The latter behavior is in agreement with the trend of a non-Newtonian shear-thinning fluid, and therefore can be explained and fitted based on the power-law model shown below:

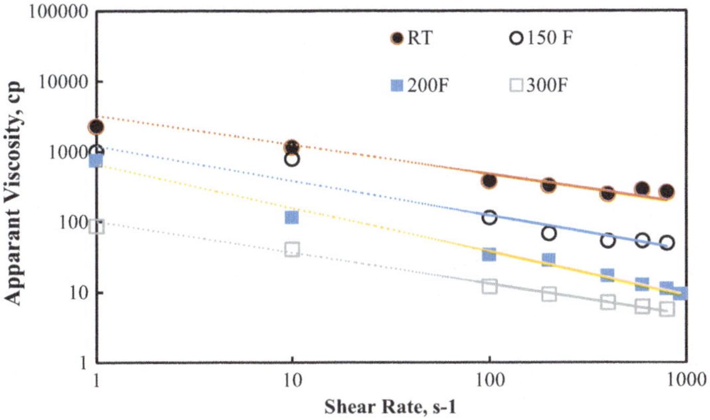

Fig. 5 Apparent viscosity profile as a function of shear rate at different temperatures up to 300 °F-The acid emulsion is a shear thinning fluid

$$\mu_a = K\gamma^{n-1}$$

where K is the Power-law consistency factor, g/cm s^{n-2}, n is the power-law index, μa is the apparent viscosity of the fluid, cp, and γ is the shear rate, s^{-1}. Table 4 shows the summary of the power-law parameters for the emulsified acid.

Table 4 Power-law parameters at different temperature up to 300 °F for the emulsified acid

Temperature, °F	K, mPa s^n	n	R^2 factor
75	3290	0.415	0.965
150	1206	0.493	0.9646
200	674	0.622	0.9952
300	100	0.44	0.9962

3.3 Corefloods Studies

The stability of emulsified acid was confirmed in the previous sections and its viscosity was shown to be acceptable. In this phase of study, two coreflood experiments were conducted to evaluate the retardation performance for the emulsified acid system at 2 cm^3/min. Figure 6 shows the pressure drop across the core for the experiments ran at 2 cm^3/min. The pressure drop stabilized at 24 psi during DI-water injection. The pressure drop was then increased once the acid reached the core interface due to the high viscosity of the emulsified acid and the generation of CO_2 through the reaction of acid with the core matrix. As the acid propagated inside the core and wormholes were created, the pressure drop across the core decreased continuously until the acid breakthrough. The cumulative injected acid volume to breakthrough (PVBT) was 0.19 PV. The fact that a small pore volume (0.19) of the emulsified acid results in breakthrough at low flow rate (2 cm^3/min) is very favorable from the field application point of view. Moreover, the stable emulsified acid decreases the need for the larger pumps and high pressure drops during acid injection (Cairns et al. 2016).

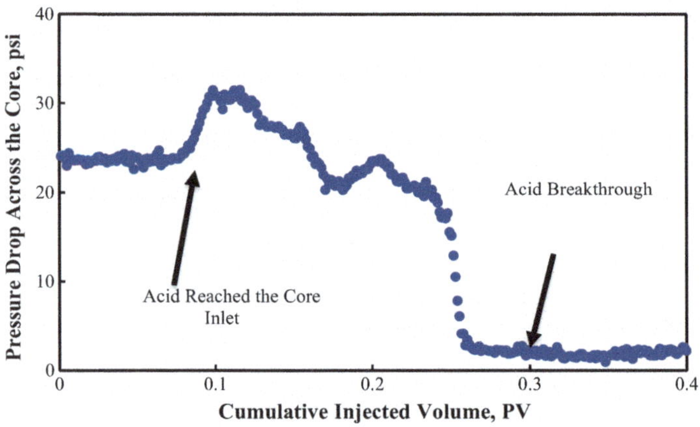

Fig. 6 Pressure drop across the core during the injection of the emulsified acid system at 2 cm^3/min and 300 °F

The pH and density measurements for the effluent samples from the coreflood experiments at 2 cm^3/min is shown in Fig. 7. The density of the effluent samples reflects the concentration of $CaCl_2$ in the solution. It can be seen that the density started at 1 g/cm^3, and then increased during the acid injection and dissolution of the core matrix. After the breakthrough and injection of DI-water, the effluent samples densities decreased to 1 g/cm^3. The weight of Ca^{+2} in the effluent samples was calculated based on the area under the curve in Fig. 8 to be 0.186 g in the case of 2 cm^3/min.

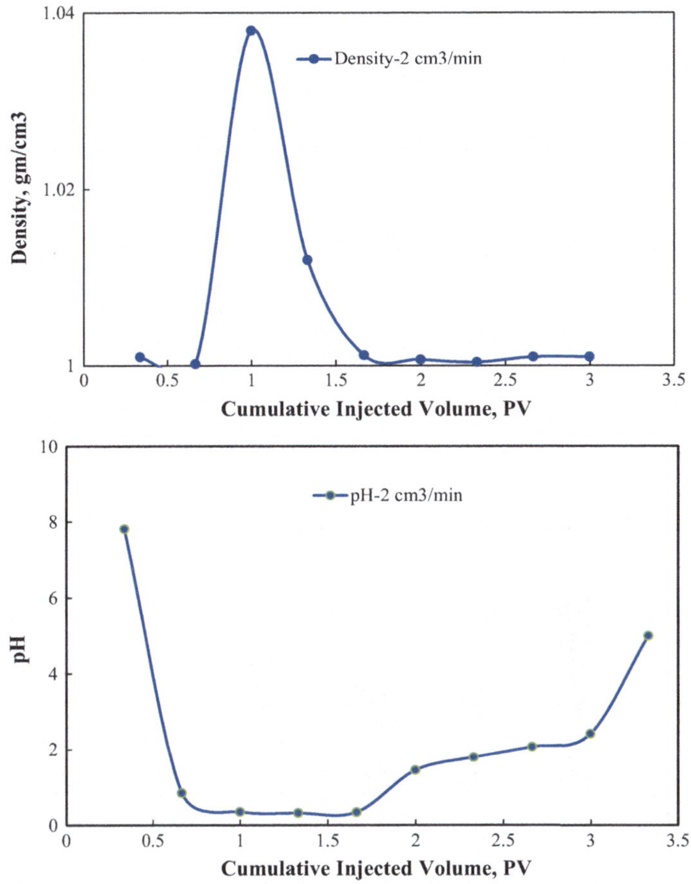

Fig. 7 pH and density measurements for the effluent samples at acid injection flow rate of 2 cm^3/min

The use of non-aromatic non-ionic surfactant introduced in the current study resulted in an emulsified acid system with a superior performance compared to the results from the literature (Sayed et al. 2014; Cairns et al. 2016). The proposed emulsified acid system needs low PVBT (0.19) compared to previous literature where the PVBT reached up to 1.36 (Fig. 9). The emulsified acid system in the present study was able to generate one dominant straight wormhole that led to low spent acid volume (dissolved 0.5 g $CaCO_3$ from the core matrix at 2 cm^3/min). By contrast, a branched wormhole was observed in the previous work by Cairns et al. (2016) and more rock matrix was dissolved (3.14 to 3.45 g $CaCO_3$ at 2 cm^3/min for the three tested fluids), which increased the required acid volume to breakthrough. It is important to consider not only PVBT but also the amount of dissolved rock since lower carbonate solubility can be representative of higher stability of the emulsified acid.

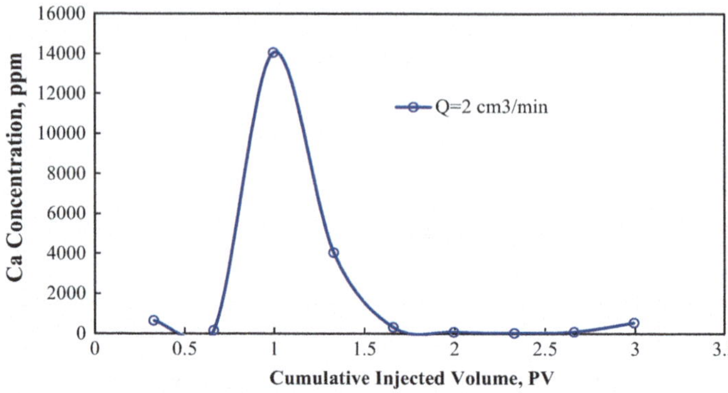

Fig. 8 Calcium concentration in the effluent samples during acid treatment on Indiana limestone cores at 2 cm^3/min

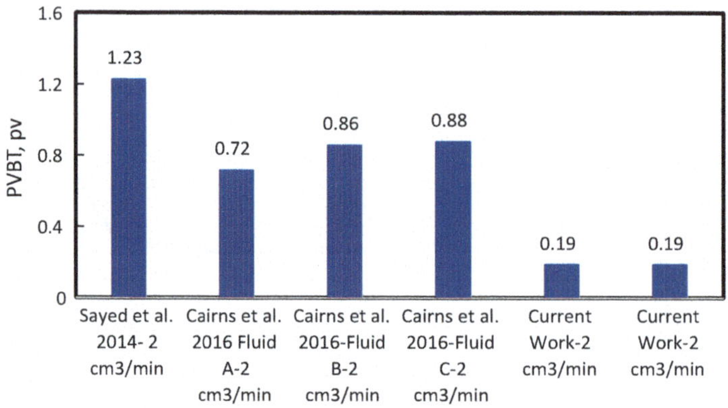

Fig. 9 Comparison between the injected acid volume to breakthrough in the current study and the previous data from the literature

4 Conclusions

A new aliphatic non-ionic surfactant was introduced as an emulsifier to form emulsified acid for high temperature acidizing purposes. Several experiments were conducted to assess the stability, rheology, and rock/fluid interactions. Based on the results of this study, the following conclusions can be drawn:

1. A non-aromatic non-ionic surfactant (detergent range) with the right level of derivatization can lead to a highly stable emulsified acid at 300 °F.
2. Initial high viscosity and average particle size of 2 μm were achieved through the use of suitable emulsifier.
3. Based on the results of the current study, the required PVBT for carbonate core is decreased compared to some recent data in literature.

The present study assists in creating a stable emulsified acid system for acidizing of carbonate formations through introducing a new non-ionic surfactant, which addresses the shortcomings associated with conventional systems. The high stability and low PVBT of the new emulsified acid system result in more efficient treatments through reducing the cost of acid and other additives. The developed emulsified acid system may be also used for acid fracturing purposes.

Acknowledgements. The authors thank Sasol Performance Chemicals for permission to present and publish this paper. Rebecca Sanders is acknowledged for her analytical support.

References

Al-Anzi, H.A., Nasr-El-Din, H.A., Mohamed, S.K.: Stimulation of tight carbonate reservoirs using acid-in-diesel emulsions: field application. In: Presented at the SPE Formation Damage Control Conference, 18–19 Feb. Lafayette, Lousiana. SPE-39418-MS. https://doi.org/10.2118/39418-MS (1998)

Al-Mutairi, S.H., Hill, A.D., Nasr-El-Din, H.A.: Fracture conductivity using emulsified acids: effects of emulsifier concentration and acid volume fraction. In: Presented at the International Petroleum Technology Conference, 3–5 Dec. Kuala Lumpar, Malaysia. IPTC-12186-MS. https://doi.org/10.2523/IPTC-12186-MS (2008)

Al-Mutairi, S.H., Hill, A.D., Nasr-El-Din, H.A.: Effect of droplet size, emulsifier concentration and acid volume fraction on the rheological properties and stability of emulsified acids. In: Presented at the SPE European Formation Damage Conference and Exhibition, Scheveningen, The Netherlands, 30 May–1 June. SPE-107741-MS. https://doi.org/10.2118/107741-MS (2007)

Al-Mutairi, S.H., Nasr-El-Din, H.A., Hill, A.D.: Effect of droplet size on the reaction kinetics of emulsified acid with calcite. SPE J. **14**(04), 606–616. SPE-112454-PA. https://doi.org/10.2118/112454-PA (2009a)

Al-Mutairi, S.H., Nasr-El-Din, H.A., Hill, A.D.: Droplet size analysis of emulsified acid. In: Presented at the SPE Saudi Arabia Section Technical Symposium, Al-Khobar, Saudi Arabia, 9–11 May. SPE-126155-MS. https://doi.org/10.2118/126155-MS (2009b)

Appicciutoli, D., Maier, R.W., Strippoli, P., et al.: Novel emulsified acid boosts production in a major carbonate oil field with asphaltene problems. In: Presented at SPE Annual Technical Conference and Exhibition, Florence, Italy, 19–22 Sep. SPE-135076-MS. https://doi.org/10.2118/135076-MS (2010)

Bartko, K.M., Conway, M.W., Krawietz, T.E., et al.: Field and laboratory experience in closed fracture acidizing the lisburne field, Prudhoe Bay, Alaska. In: Presented at SPE Annual Technical Conference and Exhibition, Washington, D.C., 4–7 Oct. SPE-24855-MS. https://doi.org/10.2118/24855-MS (1992)

Bazin, B., Abdulahad, G.: Experimental investigation of some properties of emulsified acid systems for stimulation of carbonate formations. In: Presented at the SPE Middle East Oil Show and Conference, 20–23 Feb. Bahrain. SPE-53237-MS. https://doi.org/10.2118/53237-MS (1999)

Bergstrom, J.M., Miller, B.D.: Results of acid-in-oil emulsion stimulations of carbonate formations. In: Presented at the SPE Annual Fall Meeting, 28 September-1 October, Dallas, Texas. SPE-5648-MS. https://doi.org/10.2118/5648-MS (1975)

Cairns, A.J., Al-Muntasheri, G.A., Sayed, M., et al. Targeting enhanced production through deep carbonate stimulation: stabilized acid emulsions. In: Presented at SPE International

Conference and Exhibition on Formation Damage Control, Lafayette, Louisiana, USA24–26 Feb. SPE-178967-MS. https://doi.org/10.2118/178967-MS (2016)

Cassidy, J.M., Wadekar, S., Pandya, N.K.: A unique emulsified acid system with three intensifiers for stimulation of very high temperature carbonate reservoirs. In: Presented at the SPE Kuwait International Petroleum Conference and Exhibition, Kuwait City, Kuwait, 10–12 December. SPE-163308-MS. https://doi.org/10.2118/163308-MS (2012)

Crowe, C.W., Miller, B.D.: New, low-viscosity acid-in-oil emulsions provide high degree of retardation at high temperature. In: Presented at the SPE Rocky Mountain Regional Meeting, 15–16 May, Billings, Montana. SPE-4937-MS. https://doi.org/10.2118/4937-MS (1974)

El-Shamy, O.A.A., Khid, T.T., Doheim, M.M.: Effect of ethoxlate chain length on the pour point depressant of middle distillate fuel oil. J. Disp. Sci. Technol. **32**(05), 654–658 (2011). https://doi.org/10.1080/01932691003800023

Fredd, C.N., Fogler, H.S.: Influence of transport and reaction on wormhole formation in porous media. AIChE J. **44**(9), 1933–1949 (1998)

Mahmoud, M.A., Nasr-El-Din, H.A., de Wolf, C.A., LePage, J.N., Bemelaar, J.H.: Evaluation of a new environmentally friendly chelating agent for high-temperature applications. SPE J. **16** (3), 559–574 (2011). https://doi.org/10.2118/127923-pa

Musa, A., Sultan, A., Qiu, X., et al.: A novel emulsified acid for deep wells stimulation: rheology, stability, and coreflood study. In: Presented at the SPE Kingdom of Saudi Arabia Annual Technical Symposium and Exhibition, 24–27 Apr. Dammam, Saudi Arabia. SPE-192312-MS. https://doi.org/10.2118/192312-MS (2018)

Nasr-El-Din, H.A., Al-Anzi, H.A., Mohamed, S.K.: Stimulation of water-disposal wells using acid-in-diesel emulsions: case histories. SPE Prod. Fac. **15**(03), 176–182. SPE-65069-PA. https://doi.org/10.2118/65069-PA (2000)

Nasr-El-Din, H.A., Solares, J.R., Al-Mutairi, S.H., et al.: Field application of emulsified acid-based system to stimulate deep, sour gas reservoirs in Saudi Arabia. In: Presented at SPE Annual Technical Conference and Exhibition, New Orleans, Louisiana, 30 Sep–3 Oct. SPE-471693-MS. https://doi.org/10.2118/71693-MS (2001)

Nasr-El-Din, H.A., Samuel, E., Samuel, M.: Application of a new class of surfactants in stimulation treatments. In: Presented at the SPE International Improved Oil Recovery Conference in Asia Pacific, Kuala Lumpur, Malaysia, 20–21 Oct. SPE-84898-MS. https://doi.org/10.2118/84898-MS (2003)

Nasr-El-Din, H.A., Al-Dirweesh, S., Samuel, MM.: Development and field application of a new, highly stable emulsified acid. In: Presented at SPE Annual Technical Conference and Exhibition, Denver, Clorado, USA, 21–24 Sept. SPE-115926-MS. https://doi.org/10.2118/115926-MS (2008)

North Africa Technical Conference and Exhibition, Cairo, Egypt, 20–22 Feb. SPE-150337-MS. https://doi.org/10.2118/150337-MS

North Africa Technical Conference and Exhibition, Cairo, Egypt, 20–22 Feb. SPE-152844-MS. https://doi.org/10.2118/152844-MS

Pandya, N., Wadekar, S.: A novel emulsified acid system for stimulation of very high-temperature carbonate reservoirs. In: Presented at the International Petroleum Technology Conference, 26–28 Mar. Beijing, China. IPTC-16452-MS. https://doi.org/10.2523/IPTC-16452-MS (2013)

Sabhapondit, A., Vielma, J.R., Prakash, C.: Laboratory optimization of an emulsified acid blend for stimulation of high-temperature carbonate reservoirs. In: Presented at the SPE (2012)

Sayed, M., Assem, AI., Nasr-El-Din, H.A.: Effect of presence of crude oil on the performance of emulsified acids. Presented at the SPE (2012a)

Sayed, M.A.I., Zakaria, A.S., Nasr-El-Din, H.A., et al.: Core flood study of a new emulsified acid with reservoir cores. In: Presented at the SPE International Production and Operations

Conference and Exhibition, Doha, Qatar, 14–16 May. SPE-157310-MS. https://doi.org/10.2118/157310-MS (2012b)

Sayed, M., Nasr-El-Din, H.A., Nasrabadi, H.: Reaction of emulsified acids with dolomite. J. Can. Pet. Technol. **52**(03), 164–175. SPE-151815-PA. https://doi.org/10.2118/151815-PA (2013)

Sayed, M.A., Assem, A.I., Nasr-El-Din, H.A.: Effect of oil saturation on the flow of emulsified acids in carbonate rocks. SPE Prod. Oper. **29**(01), 29–41 (2014). https://doi.org/10.2118/152844-PA

Sidaoui, Z., Sultan, A.S., Qiu, X.: Viscoelastic properties of novel emulsified acid using waste oil: effect of emulsifier concentration, mixing speed and temperature. In: Presented at the SPE Kingdom of Saudi Arabia Annual Technical Symposium and Exhibition, 25–28 Apr. Dammam, Saudi Arabia. SPE-182845-MS. https://doi.org/10.2118/182845-MS (2016)

Sidaoui, Z., Sultan, A.S., Brady, D.: A novel approach to formulation of emulsified acid using waste oil. In: Presented at the SPE Kingdom of Saudi Arabia Annual Technical Symposium and Exhibition, 24–27 April, Dammam, Saudi Arabia. SPE-188116-MS. https://doi.org/10.2118/188116-MS (2017)

Sidaoui, Z., Sultan, A.S.: Formulating a stable emulsified acid at high temperatures: stability and rheology study. In: Presented at the International Petroleum Technology Conference, 14–16 Nov. Bangkok, Thailand. IPTC-19012-MS (2016)

Sokhanvarian, K., Nasr-El-Din, H.A., de Wolf, C.A. Thermal stability of oilfield aminopolycarboxylic acids/salts. SPE Prod. Oper. **31**(01), 12–21. SPE-157426-PA. http://dx.doi.org/10.2118/157426-PA (2016)

Sokhanvarian, K., Pummarapanthu, T., Arslan, E., et al.: A new in-situ generated acid system for carbonate dissolution in sandstone and carbonate reservoirs. In: Presented at the SPE International Conference on Oilfield Chemistry, Montgomery, Texas, USA, 3–5 Apr. SPE-184506-MS. https://doi.org/10.2118/184506-MS (2017)

Sokhanvarian, K., Nasr-El-Din, H.A., de Wolf, C.A.: Thermal decomposition of chelating agents and a new mechanism of formation damage. In: Presented at the SPE European Formation Damage Conference and Exhibition, Noordwijk, The Netherlands, 5–7 June. SPE-165153-MS. https://doi.org/10.2118/165153-MS (2013)

Syafii, I., Pandya, N., Sabhapondite, A., et al.: High-temperature acidizing: advantages of inhibitor-intensifier synergy. In: Presented at the Abu Dhabi International Petroleum Exhibition and Conference, 7–10 Nov. Abu Dhabi, UAE. SPE-183034-MS. https://doi.org/10.2118/183034-MS (2016)

Zakaria, A.S., Sayed, M.A.I., Nasr-El-Din, H.A.: New Insights into propagation of emulsified acids in vuggy dolomitic rocks. SPE J. **19**(01), 150–160. SPE-163288-PA (2012). https://doi.org/10.2118/163288-PA

Research and Application of a New Plugging Agent—Oil Soluble Resin

Zhong-hai Qin[1(✉)], Wen-jie Wu[2], Zong-xian Pei[1], Wei Zhang[1], Li-li Zhang[1], Ai-jun Wei[1], Hai-jun Yan[1], Juan Xie[1], Xuan-qi Yan[1], Na Su[1], Fang Shang[1], Li-li Wei[1], Ying Liu[1], and Zhi Ma[1]

[1] Engineering Technology Research Institute of Huabei Oilfield Company, Renqiu Hebei, People's Republic of China
cyy_qzh@petrochina.com.cn

[2] China Petroleum Engineering & Construction Corp, North China Company, Renqiu Hebei, People's Republic of China

Abstract. Due to the oil-water layers crisscrossing in the oilfield exploitation, oil production engineers keep on research more and more agent with selective water-plugging/profile modifying in order to reduce water and increase oil. In this paper, researchers use the oil soluble resin as the main raw material, and develop the emulsified oil soluble resin selective plugging agent, and the oil soluble resin is emulsified and dispersed by special processing technology to form a stable oil-in-water emulsion. The emulsion can be mixed with any proportion of water, has good fluidity and easy injection into the layer. After entering the layer, the stability of the system is lost under the action of special demulsifiers, contacting rocks, and the formation temperature, the oil-soluble resin occupy the space to plug by the aggregation and mutual adhesion of the particles to form clusters, or they are attached to the rock surface. By changing the wettability of rock surface and forming mechanical blockage of water flow, the purpose of reducing water production and increasing crude oil production is achieved. The experimental results show that the breakthrough pressure of cores with different permeability can be increased by an average of more than 10 times, and the plugging rate of sandstone cores can reach 93% by using 25% emulsified oil soluble resin selective plugging agent. And the plugging rate of water displacement with 300 times pore volume of core is stable. Because the

This paper was prepared for presentation at the 2019 International Petroleum and Petrochemical Technology Conference in Beijing, China, 27–29, March, 2019.

J. Lin (ed.), *Proceedings of the International Petroleum and Petrochemical Technology Conference 2019*, pp. 284–292, 2020.
https://doi.org/10.1007/978-981-15-0860-8_23

system is insoluble in oil and water, it has self-selectivity after injection. Through the field application of six times sandstone reservoir and once limestone reservoir, the remarkable effect of increasing oil production and reducing water is obtained, which provides a new technical means for water plugging in reservoir and increasing oil production and reducing water production in single well.

Keywords: Emulsification · Oil soluble resin · Self-selectivity · Water-plugging/profile modifying agent · Practice

1 Introduction

In the late stage of oilfield development, with regard to oil-water layers crisscrossing, researchers actively explore water shut off agents with selective to achieve the purpose of increasing oil and decreasing water [1–3], like heavy activity oil [4, 5], oil base cement and so on, but these plugging agents are hard to popularize in the application because of demanding in harsh conditions [6, 7].

Researchers use the oil soluble resin as water plugging material and have been developed the emulsified oil soluble resin selective plugging agent. Because it has big differences in surface tension between the emulsified oil soluble resin and the water, they can't immiscible with each other under room or higher temperature. In another way, when it has been in high speed centrifugation 'shear' impact and other mechanical action under special conditions, it will become into the particles which diameter is 1–2 μm, then spreads into the medium containing emulsifier and stabilizer. Due to emulsifier directional adsorbing on the particles surface, it can reduce the water and interfacial tension of the material, then forms a stable dispersion in water system, which is oil-in-water emulsion. This decentralized system has better fluid property at room temperature which can be easily injected into the layer. After being injected into the layer, it will lose its stability under the effect of contacting the rock and the layer's temperature. The oil soluble resin particles gathers, occupies the space to plug by the aggregation and mutual adhesion of the particles to form clusters, or attaches to the surface of rock, changes the wettability of the rock surface, forming mechanical blockage of water, achieve the purpose of reducing water production increase production of crude oil. Due to the oil soluble resin be completely solving in oil, it can guarantee blocking gradually dissolves and eliminates in the remaining oil rich areas, and plugging is long term in the water channel.

2 The Formula Research on the Emulsified Oil Soluble Resin Selective Plugging Agent

2.1 The Dosage Selection of Emulsifier

Resin emulsifier should include lipophilic group and hydrophilic group, which is enable to adsorption on the exclusive interface of oil soluble resin and water mutually, and reduce the interfacial tension between each other.

2.1.1 The Estimate of Emulsifier Foam Volume

Using Ross-Miles method, researchers select and evaluate cationic emulsifiers and anionic emulsifiers, the results are shown in Tables 1 and 2.

Table 1 The performance of multiple cationic emulsifiers

Emulsifiers	Ion type	The state of emulsion	Foam volume (mL)
Octadecyl trimethyl ammonium chloride	Cation	Emulsion is brown and resin particles small homogeneous	10
Cetyl trimethyl ammonium bromide	Cation	Emulsion is brown and resin particles small homogeneous	20
Alkyl amide Quito	Cation	Emulsion is black and resin particles coarse uneven	2
Alkyl propylene diamine	Cation	Emulsion is black and resin particles coarse uneven	3

Table 2 The performance of multiple anionic emulsifiers

Emulsifiers	Ion type	The state of emulsion	Foam volume (mL)
Oleic acid	Anion	Emulsion is black and resin particles coarse uneven	3
Sodium dodecyl sulfate	Anion	Emulsion is brown and resin particles small homogeneous	20
Naphthenic acid aluminum soap	Anion	Emulsion is black and resin particles coarse uneven	2
Dodecyl benzene sulfonic acid sodium	Anion	Emulsion is brown and resin particles small homogeneous	10

The results show that the cationic emulsifier of octadecyl trimethyl ammonium chloride, cetyl trim ethyl ammonium bromide and the anionic emulsifier of dodecyl benzene sulfonic acid sodium have larger foam volume, and emulsion product is uniform particles after the resin emulsion, which suggest that its emulsifying performance is better.

2.1.2 Storage Stability Evaluation

Examinations of different emulsifiers emulsify effects on the stability of system.

Referring to the test method of storage stability of emulsified asphalt pavement under the industry standard, researchers investigate the emulsifying effect of cationic emulsifiers and anionic emulsifiers on the stability, the results are shown in Tables 3 and 4.

Table 3 The storage stability of cationic emulsifiers

Cationic emulsifiers	Storage stability (5d)/(%)
Octadecyl trimethyl ammonium chloride	5
Cetyl trimethyl ammonium bromide	9
Alkyl amide Quito	10
Alkyl propylene diamine	11

Table 4 The storage stability of anionic emulsifiers

Anionic emulsifiers	Storage stability(5d)/(%)
Oleic acid	10
Sodium dodecyl sulfate	8
Naphthenic acid aluminum soap	9
Dodecyl benzene sulfonic acid sodium	4

The experiment results show: the cationic emulsifier of octadecyl trimethyl ammonium chloride and the anionic emulsifier of sodium dodecyl benzene sulfonic acid have good storage stability.

2.2 The Determination of the Emulsifie's Amount

The range of emulsifier's dosage is determined by checking the storage stability. The test results are shown in Table 5.

Table 5 The effect of storage stability by the SDBS amounts

Amounts (%)	0.1	0.2	0.3	0.4	0.5	0.6	0.7	0.8
Storage stability (%)	60	36	19	13	13	13	13	13

The results show that the storage stability decreases with the increase of emulsifier dosage, and the stability of emulsified resin does not change much when the emulsifier content is more than 0.4%, which indicates that the emulsifier content in the emulsion has reached the critical micelle concentration. Therefore, the amount of emulsifier should be controlled at 0.4% and 0.5%.

2.3 The Formula of the Emulsified Oil Soluble Resin Selective Plugging Agent

The composition of the emulsified oil soluble resin selective plugging agent is obtained by optimization of experiment, selection of components and evaluation of performance.

The main agent: industrial oil soluble resin, the dosage is 10–60%; Additives: SX, the dosage is 5–10%; Control agent: SL, the dosage is 10–15%; Emulsifier: anionic surfactant, such as octadecyl trim ethyl ammonium chloride, the dosage is 0.5–1.0%; Cationic surfactant such as sodium dodecyl benzene sulfonic acid, the dosage is 0.5–1.0%; Stabilizer, hydroxymethyl cellulose (CMC), the dosage is 0.3–0.8%; Or hydroxyethyl cellulose, the dosage is 0.5–0.9%; The rest is water;

Adjuvant: special demulsifier, the dosage is 1.2–3.5%.

2.4 The Preparation of the Emulsified Oil Soluble Resin Selective Plugging Agent

The preparation of the emulsified oil soluble resin selective plugging agent is:

(1) According to the composition of emulsified oil soluble resin, researchers add raw material to reaction kettle, when the main agent is heating to basic molten state, gradually adding additive and control agent into it, and then researchers control the mixture's temperature to 110–120 °C after these completely dissolved, spare;
(2) According to the composition of oil soluble resin, researchers add emulsifier and stabilizer into the clean water, which is being stirring and heating, then researchers control the mixture' heating temperature to 60–90 °C after dissolving completely, spare;
(3) Researchers put (1) and (2) in a ratio of 1:2 into the rotating colloid mill whose rotate speed has add up to 6000–8000 r/min, oil soluble resin is cut into particles by rotor, then mixed with emulsifier aqueous solution, researchers get the oil soluble resin emulsion and can prepare it continuous;
(4) When the preparation of emulsion is in the process of natural cooling, researchers stir the mixture for 2.0–2.5 h, then get the emulsified oil soluble resin selective plugging agent;
(5) When used in the application, researchers make special auxiliary agent and oil soluble resin mixed with water, mixing uniformity.

3 The Performance Evaluation of the Emulsified Oil Soluble Resin Selective Plugging Agent

3.1 The Water Miscibility

Researchers make the emulsified oil soluble resin and water mixed together with different ratio at room temperature. Test results are shown in Table 6.

Table 6 The miscible performance of emulsified oil soluble resin with water

Resin: water	1:1	1:2	1:3	2:1	3:1
Test results	Homogeneous emulsion	Homogeneous emulsion	Homogeneous emulsion	Homogeneous emulsion	Homogeneous emulsion

The results show that the emulsified oil soluble resin plugging agent has good compatibility with water and can be mixed with water in any ratio to form a uniform emulsion.

3.2 The Oil Solubility Evaluation

At 35 °C, researchers add the different quality of emulsified oil soluble resin plugging to every piece of 500 ml light crude, each stirring interval is 0.5 h, weighing them after 6 h to calculating its solubility in the oil. Test results are shown in Table 7.

Table 7 the oil-soluble performance of emulsified oil soluble resin

Weight (g)	50	100	150	200	250
dissolution rate (%)	96.6	92.3	89.6	78.7	73.8

The test results show that the emulsified oil soluble resin has good solubility in light crude, which can make 92.3 g oil soluble resin dissolve in 100 g light crude, and the emulsified oil soluble resin completely insoluble in water. Therefore, because of this special property, it will have water plugging but not blocking the oil.

3.3 The Physical Model Test Evaluation

Physical model test results showed that 25% of the emulsified oil soluble resin can make the breakthrough pressure of different permeability core average increased by more than 10 times, the sandstone core plugging rate up to 93%, and with 300 times the core pore volume of water displacement plugging rate remained stable. Therefore, emulsified oil soluble resin has characteristics of high blocked strength, and resistance to water flushing for long time.

4 Filed Applications

4.1 The Application of Sandstone Reservoir with Bottom Water

The type of the oil reservoirs is tectonic-lithologic logical reservoir, the reservoir buried depth is 1100–1300 m, the sedimentary type is distributaries channel sediment [8, 9], the effective porosity is 18.2% on average, air permeability is 33.8 mD, due to the existence of the reservoir natural fracture and local high permeability zone that leads to fracture direction of the lord or high permeability zone water cut rising fast [10]. In view of the present situation, since June 2010, researchers combine with the reservoir geologic features, using oil soluble resin to carry out a water plugging experiment. Researchers put it in use at the bottom water reservoir sandstone mines up to six times, and the measure success rate amount to 100%, obtaining the significantly increasing oil and reducing water's result, which provides a new technical and measure to improve the single well's oil production and reduce water production [11]. The application instance well's condition and water plugging effect is as follows:

4.1.1 The Basic Situation of This Well

D66-16 is a low permeability sandstone oil well, and it puts into production in April 1998. initial daily fluid production is 5.60 m^3, in which 3.68 t is oil, the dynamic liquid level is 1113 m. Corresponding to the injection wells are two, which is located in the reservoir edge, long term injection but no effect. In May 2007, to this well measures the repeated fracturing. The daily liquid, water content increase by a large margin after fracturing measures. Liquid production is 13.14 m^3 and oil production is 0.45 t before water plugging measure, the dynamic liquid level is 529 m. The well formation pressure is 6.96 MPa, 91.0% level. The bottom of the reservoir bottom water 10.9 m thick and the bottom water inter layer is 0.4 m thick. After fracturing measures, researchers analysis the main cause of water rising is that the fracturing parameters of the relatively big perforation communicate bottom water, which cause the water cuts rising in the well. Therefore, the purpose of controlling water and increasing oil production is realized by implementing water plugging measures in the well.

4.1.2 Water Plugging Operation and Its Effect

On June 11, 2015, we injected 3 m^3 pre-treatment liquid into the well. The injection capacity is 0.2 m^3/min and the injection pressure is 0–5 MPa; 38 m^3 liquid, of which 25% is the emulsified oil soluble resin, was positive injected into the well, the injection capacity is 0.23 m^3/min and the injection pressure is 5–10 MPa; researchers inject 4 m^3 treating fluid and then positive injecting 3.5 m^3 water, the injection capacity is 0.23 m^3/min, the injection pressure is 10 MPa. At last, we close the well. After 3 days, the injection pressure is released. The pump is put into production according to the original working system. The initial effect of plugging measures is good, the daily oil production increased from 0.4 to 2.4 t, and then remained stable at 1.6 t, the water content reduces from 95.8 to 73.6%, and then remained stable at around 82%. The effective period of this measure is more than 540 days.

4.2 The Application of Horizontal Well in Limestone Reservoirs

4.2.1 The Basic Situation of This Well

RP 5 is a limestone horizontal well, which was completed on January 16 2008. This well has been acidizing unblocking operation before production. The injection pressure is 31 MPa, and construction displacement is 0.9–1.4 m^3/min. After acidizing operation, the well is put into production on April 11, 2008 by using rod pump of pumping unit. Daily fluid production is 50.6 m^3, and daily oil production is 43 t and working fluid level is 171 m. In July 2011, Φ56 mm rod pump is produced, with daily production liquid 50.17 m^3, daily oil 7.66 t, water content 84.73%, dynamic liquid level 143 m, and cumulative oil production 3.3413 × 104 t. The well is cross-section into the mountain, and belongs to the high structural position. In March 2010 by tuning parameter lifts, the daily fluid production is about 23 m^3, which has a certain effect on increasing oil production, but water content rising speed up. The water of liquid from this well is 84.73% at present, which is the best opportunity to limestone reservoir operating water shutoff, therefore we decided to use oil soluble resin to implement water plugging measures.

4.2.2 Water Plugging Operation and Its Effect

On August 20, 2016, researchers inject 160 m^3 the emulsified oil soluble resin selective plugging agent into the well. The injection capacity is 0.26 m^3/min and the injection pressure is 2–9 MPa. After 3 days of shutoff the well, the pump is put into the well to production according to the original working system, the initial effect of water shutoff measures is good: the daily oil production increases from 7.6 t to 17 t, the water content decreases from 84.73 to 64.4%, and continued to be effective.

5 Conclusion

(1) That researchers develop the emulsified oil soluble resin selective plugging agent which uses oil soluble resin as the main raw material, has the unique property of oil-soluble water insoluble in comparison with the conventional plugging agent. It can be mixed with water in any proportion to form a uniform emulsion, which adapts to different plugging targets.
(2) This system is not sensitive to temperature, so it can adapt to a wide range temperature of oil well water plugging and water well profile control.
(3) The pilot test result shows that the application of this system can't only gain significantly effect on increasing oil and reducing water, but also provide a new technology for water plugging to different lithologic reservoir layer and improving the single well's oil and reducing water production.

References

1. Wei, FaLin, Liu, YuZhang, et al.: Advancement of research and application of selective water shutoff agents. Oilfield Chem. **24**(1), P93–P96 (2007)
2. Xie, Shuixiang, Li, Kehua, et al.: Study of selective water plugging technology. Adv. Fine Petrochem. **4**(1), P12–P14 (2003)
3. Li, KeHua, Zhao, MingKui, et al.: Study on JHXD oil soluble polymer as selective water shutoff agent. Sci. Technol. Chem. Ind. **9**(4), P29–P30 (2001)
4. Ma, QingZhuang, Wen, ShuXin, et al.: Study and application of new type selective blocking agent for heavy oil. PGRE **9**(2), P74–P75 (2002)
5. ZhaoXuan, Qu, Wang, GuiXun: Lab tests of the selective plugging agents for emulsified thickened oil recovery. Petrol. Drill. Tech. **31**(4), P56–P58 (2003)

6. Shuhua, Ma.: Study on physical and chemical properties of emulsified heavy oil shutoff agent. Inner Mongolia Petrochem. Ind. **14**(13), P85–P86 (2003)
7. Chen, F., Wei D., Yu, X., et al.: Reservoir forming mechanism of Yanchang formation of upper triassic in northwest ordos basin. Core Oil Gas Reserv. **22**(1), 43–47 (2010)
8. Li, Y., Zhang, M., Wang, X., et al.: Reservoir formation mechanism of Yanchang formation of upper triassic in northwestern ordos basin. Core Oil Gas Reserv. **22**(2), 32–36 (2010)
9. Wang, Q., Liu, H., Cao, L.: Study on water flooding of horizontal wells in heterogeneous bottom water reservoir. Core Oil Gas Reserv. **22**(1), 122–125 (2010)
10. Jinquan, Li: Study on selective water shutoff of modified paraffin wax and its field application. Special Oil Gas Reserv. **14**(4), P78–P80 (2007)
11. Yuan, Chen: Technology of deep water shutoff in oil well to improve development effect. Special Oil Gas Reserv. **15**(6), P79–P82 (2008)

Research and Application of High Efficiency Well Flushing Fluid System

Na Su[1(✉)], Nai-xu Gao[1], Run-zhe Wang[1], Ai-jun Wei[1], Li-li Wei[1], Bo Zhao[2], Zhi Ma[1], Guan Wang[1], Fang Shang[1], Zhong-hai Qin[1], and Ying Liu[1]

[1] Engineering Technology Research Institute of Huabei Oilfield Company, Renqiu Hebei, People's Republic of China
cyy_suna@petrochina.com.cn

[2] Petrochina North China Petrochemical Company, Renqiu, Hebei, People's Republic of China

Abstract. Well workover is an indispensable production link in oilfield development. Hot washing is an important measure to maintain normal production of oil wells before well workover. Based on the principle of "source control" of pollution prevention and control in oilfield production, this paper guarantees that crude oil does not fall to the ground, adopts the way of adding chemicals to wash wells, takes the wellbore production string as the starting point of prevention and control, and uses the unique properties of surfactants, develops a new type of well-washing fluid system, which can effectively reduce the field dispensing temperature. The system can wash oil more than 90% at 65 °C, and the concentration of chemicals is 0.6%. The interfacial tension between oil and water decreases to 10–4 MN/m. Effectively avoid oil pollution from wellbore when workover operation, resulting in surface and environmental pollution. The expected goal has been achieved through well washing experiments of more than 10 oil wells before workover operation.

Keywords: Workover operation · Flushing fluid · Surfactant

J. Lin (ed.), *Proceedings of the International Petroleum and Petrochemical Technology Conference 2019*, pp. 293–298, 2020.
https://doi.org/10.1007/978-981-15-0860-8_24

1 Introduction

Huabei Oilfield, heavy oil wells and heavy wax deposited wells are often used to add medicine or wash wells with hot water above 80 °C to maintain the normal production of oil wells before operation. Conventional flushing fluids are strongly alkaline, and the loss of flushing fluids during operation will cause damage to reservoirs [1]. Simply relying on hot water to wash wells, the effect of well washing is poor, and the risk factor is high and the operation cost is high in the operation process.

The key to oil and water pollution control lies in the wellbore, and all kinds of pollutants will effectively avoid the subsequent tedious work [2]. In order to reduce the hot washing temperature, reduce maintenance costs and effectively protect the surrounding environment of the well site, a highly efficient water-based well washing fluid system was developed, which can effectively clean the oil contamination adhered to the tubing rod with small dosage and non-toxicity. It has been applied in the oil production plant of North China Oilfield and achieved good results.

2 Study on Well Flushing Fluid System

2.1 Scale Sample Analysis

Scale analysis experiments were carried out on wellbore of oil well and surface oil scale of import and export. According to the analysis of scale samples, crude oil is the main component of scale samples, and the content of metal ions in scale samples is more than 65%. The iron content in well C12 is higher than that in well C12.

2.2 Indoor Test Design

The construction of surfactant well-washing fluid system is mainly based on the high interfacial activity of surfactant, the easy formation of active water film and polar groups of surfactant, and the stronger hydrogen bond formed by the interaction between resin and asphaltene in oil scale, and the weakening of the association mechanism between resin and asphaltene [3].

2.2.1 Screening of Main Agent for flushing Fluid

The main agent of flushing fluid should have strong permeability, solubilization, emulsification and good compatibility with formation and other fluids [4]. On the premise of screening a single main agent and taking oil washing efficiency as an evaluation index, laboratory experiments were carried out on five kinds of surfactant, and a single surfactant with better oil washing effect was selected as the main agent.

Fluid oil washing efficiency experiment was designed in the laboratory to simulate flow well washing in the field. Oil washing efficiency was calculated according to the following formula: Viscosity reducer formula is as follows:

$$\text{Displacement Efficiency} = \frac{W_1 - W_2}{W_1 - W_0} \times 100\%$$

displacement efficiency =

Among them:

W_0 steel sheet quality;

W_1 The quality of steel sheets dried after smearing the sampled crude oil;

W_2 The quality of steel sheets dried naturally after oil washing.

Field oil samples from well Cha 12-122 were selected for laboratory test. The mass concentration of five surfactants (alkyl phenol polyoxyethylene ether, octyl phenyl polyoxyethylene ether, sorbitol monooleate polyoxyethylene ether, sodium dodecyl sulfate and sodium dodecyl benzene sulfonate) was controlled at 0.6–1%, and 500 ml aqueous solution was prepared. The softening point of asphalt in crude oil is 47.5–60 °C, so the oil washing experiment is carried out at 60 °C.

The results show that the oil washing ability of octylphenyl polyoxyethylene ether is better than that of anionic surfactant. However, its static stability is poor. Adding a small amount of anionic emulsifier to non-ionic emulsifier can significantly improve its static stability.

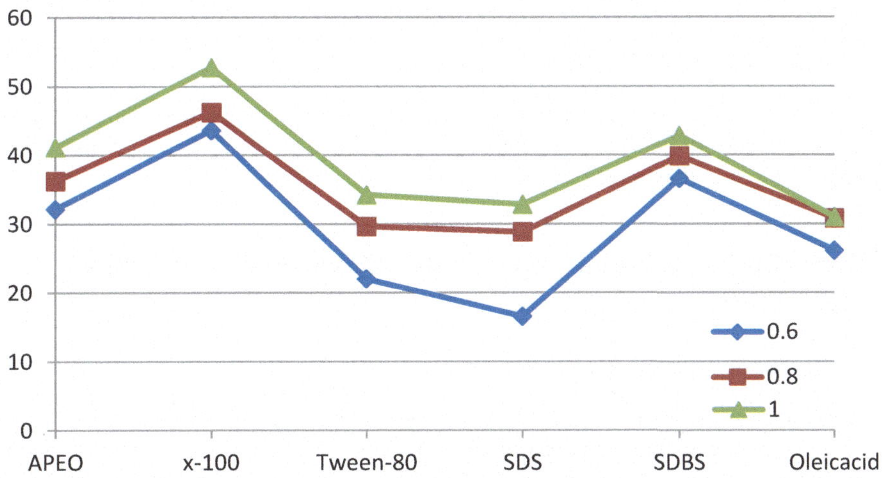

Different Surfactant Displacing Oil Efficiency

2.2.2 Compound of Main Agent for Well Flushing Fluid

There are synergistic effects among surfactants, which can obtain better performance after compounding, reduce the cost of using surfactants and reduce the environmental pollution caused by surfactants [5]. In the experiment, surfactants were mixed in different proportions, the temperature was 60 °C, the concentration of the agent was 0.6%, and the oil washing experiment was carried out.

The results of oil washing experiment are as follows: when octylphenyl polyoxyethylene ether: sodium dodecyl sulfate = 1:2, 1:3, 1:4, 2:1, 3:1, 4:1 and 1:5, the well washing efficiency is 67.27%, 64.29%, 80.82%, 82.53%, 71.67%, 27.27%, 62.71%, respectively. When octyl phenyl polyoxyethylene ether: sodium dodecyl sulfate (mass ratio) = 1:4, the reagent has good performance and high oil washing efficiency, so it is an ideal main agent for well washing fluid.

2.2.3 Formulation Optimization of Flushing Fluid

In order to improve the solubility of well-washing fluid and reduce the foaming rate of formula system, and prevent the damage of reservoir caused by water lock when well-washing fluid enters the reservoir, functional additives with the ability of reducing interfacial tension and improving seepage flow are introduced into formula system, which can improve the oil-washing efficiency of well-washing fluid and reduce the comprehensive performance of reservoir damage at the same time.

In the experiment, two kinds of auxiliaries were selected, the concentration was 0.01, 0.02, 0.03, 0.04, 0.05%, and the main agent concentration was 0.6%. With the increase of the auxiliary concentration, the foam gradually decreased. When the auxiliary concentration exceeded 0.05%, the solution was layered after being placed in static solution, and the concentration of the auxiliary agent was 0.02% when combined with the actual conditions.

2.3 Basic Formula of Well Flushing Fluid

Water-based flushing fluids are obtained with low volatility and non-flammability. The recommended formula is 2% octylphenyl polyoxyethylene ether +8% sodium dodecyl sulfate +0.02% additive +88.98% water.

3 Performance Evaluation of High Efficiency Well Flushing Fluid System

3.1 Interfacial Tension Measurement

Interfacial tension is an important parameter for evaluating the performance of flushing fluid [6]. Referring to the Standard SY/T5370-1999 of China Petroleum and Natural Gas Industry, the dynamic interfacial tension between surfactant and oil and water was measured by the rotary drop interfacial tensiometer, thus the optimum dosing concentration could be recommended.

In order to ensure the applicability of the basic formula of well washing fluid, three kinds of typical crude oils were selected in the experiment, namely, high wax content, high asphaltene content and ordinary crude oil.

The interfacial tension of oil and water decreases obviously after adding well washing fluid. When the concentration of reagent is more than 0.6%, the interfacial tension of three kinds of crude oil changes little and tends to be stable, which indicates that 0.6% should be the critical concentration to change the interfacial tension. Considering the cost of dosing, the dosing concentration was set at 0.6%.

3.2 Stability

If the well flushing fluid has better demulsification effect on crude oil, it will not affect the demulsification of crude oil, and has no negative impact on the later stage of crude oil into the production process. Laboratory experiments refer to the enterprise standard Q/SY HB 0913-2015 "Technical Specification for Crude Oil Demulsifier", which adds 0.6% crude oil and 0.6% flushing fluid to the plugged colorimetric tube according to the ratio of 1:9, shakes hands for 5 min, makes crude oil and flushing fluid mix evenly, and then puts them on the desktop for 5 h, observes the experimental phenomena, and makes blank experiments with clean water.

After fluid mixing, the well-washing fluid and crude oil can be emulsified together, which shows that the well-washing fluid has emulsifying effect on crude oil. After 5 h, obvious stratification occurs in the colorimetric tube, which indicates that the well-washing fluid has good demulsification performance and will not have a negative impact on crude oil.

4 Field Application

4.1 Field Experiment Design

The function of high efficiency flushing fluid is to ensure that the tubing rod is cleaned before starting the tubing and reduce pollution. Therefore, the oil wells that cannot be cleaned by the existing flushing fluids and have more oil on the ground are selected as experimental wells and field tests are carried out. According to the performance of well flushing fluid and the requirements of construction site, the dosage of well flushing fluid is designed to be 40–50 m^3. The reverse circulation well washing process was adopted and 25 MPa pressure test was conducted before washing.

4.2 Experiments

The developed high-efficiency well-washing fluid has excellent performance and clean and smooth surface of oil pipe after cleaning operation. It can be proved that the well-washing fluid has a good cleaning effect on the impurities on the surface of pipe and rod in the well, and the well site is clean and there is no crude oil on the ground, which reduces the follow-up work of steam cleaning after the rod and tube start-up.

5 Conclusion

(1) Through laboratory experiments, a high efficiency well flushing fluid system has been developed. The system has high oil washing efficiency, good stripping, carrying and cleaning ability for crude oil on tubing rod, and the active water film formed by surfactant in the well washing fluid system can ensure the wettability of tubing rod and weaken the oil scale adsorption in later operation.

(2) Through field experiments, the cleaning effect of high-efficiency well-washing fluid at 60 °C is good, the surface of rod and tube is free of oil pollution, which

reduces the follow-up surface work, and there is no floor oil in the operation, thus realizing the requirements of green environmental protection and meeting the production needs.

References

1. Wenjun, Cai, Dayong, Wang: laboratory development of a W/O diesel-based well washing fluid. Guangdong Chem. Ind. **17**(39), 25–26 (2012)
2. Guo, T., Jiang, W.: Application of surfactant in chemical well washing of coal mines. Riyue Chem. Ind. **6**(3), 53–55 (1999)
3. Zheng, Y., Wang, L., Zhang, R., Li, T., Cao, X., Jiang, X., Sun, J., Xing, S.: Laboratory study of WF-II low-damage flushing fluid. Drill. Fluid Compl. Fluid **3**(23), 51–54 (2006)
4. Changzhen, Guo, H., Li, X., Liu, W., Wang, G.: Research and application of compound high efficiency washing fluid in Bohai oilfield. Inner Mongolia Petrochem. Ind. **3**, 8–10 (2016)
5. He, F., Li, X., Zhang, K., Zhang, J.: Development and application of HNCY-2 flushing fluid. Oilfield Chem. **10**(2), 116–119 (1993)
6. Yarong, Fu: Understanding of hot washing and wax removal in oil wells. Jianghan Oil Prod. Technol. **1**, 13–14 (1997)

Research and Application of Degradable Impervious Materials in Workover Field

Shuang-yan Zhang[1(✉)], Zhong-hai Qin[1], Hong-yu Gong[2], Li-min Liu[1], Yi-chao Li[1], Zhen-yong Zhou[3], Mei-mei Du[1,2,3], Meng Liu[1], and Tao Sun[1]

[1] Engineering Technology Research Institute of Huabei Oilfield Company, Renqiu Hebei, People's Republic of China
{cyy_zhangsy, cyy_qzh}@petrochina.com.cn

[2] The Third Exploit Factory of Huabei Oilfield Company, Hejian Hebei, People's Republic of China

[3] Exploration and Development Research Institute of Huabei Oilfield Company, Renqiu Hebei, People's Republic of China

Abstract. It is difficult to do innocuous treatment of the woven polyvinyl chloride geofilm impervious cloth which is used in the workover field. In order to solve those problems, researchers study the environmental protection degradable impervious cloth to prevent oil and water from leaking during the oil well operation, while secondary pollution will not be formed. So that it can achieve the purpose of green oil well operation at last. Based on the characteristics of degradability and impermeability, researchers discover the main functions and shortcomings of impervious cloth and research on the screening of degradable materials, the design of indoor formula, the preparation process and the post-treatment technology. Developing products are suitable for the use of workover and harmless treatment in later stage. Among many degradable materials, PLA and PBAT have excellent degradability, suitable mechanical strength, large scale of production and close price to traditional materials, so they are optimal for the production of composite impervious cloth. Combined with the characteristics of two kinds of materials, PLA non-woven fabric and PBAT/PLA film were selected as raw materials to produce composite impervious cloth. Through the experiment and optimization combination, the composite impervious cloth with different area, thickness and strength could be processed and produced. After well workover treatment site laying and trial in

---

J. Lin (ed.), *Proceedings of the International Petroleum and Petrochemical Technology Conference 2019*, pp. 299–312, 2020.
https://doi.org/10.1007/978-981-15-0860-8_25

oil field, result shows that the practical application effect is enough good, that can effectively prevent crude oil leakage and protect workover site environment. Based on the biodegradable impermeable cloth developed by environmental protection materials, the problem that the ordinary impervious cloth not being degraded cause the secondary pollution is solved. So that we can build the last barrier to prevent and control the pollution during workover treatment.

Keywords: Green workover · Degradable · Anti-seepage · Environmental protection

According to the regulation of well control rules, workover operations in oil field need to lay impervious cloth under wellhead, tubing bridge seat and sucker rod bridge seat to prevent crude oil from falling to the ground. A large amount of impervious cloth is consumed every year, and the impervious cloth must be recovered and centralized after operation. At present, the anti-seepage cloth used in the field is mainly made of composite geomembrane which is made up of polyvinyl chloride as raw material [1], which has the advantages of low price, easy processing, good mechanical properties, anti-aging, etc. It can be used to prevent seepage from waste oil and wastewater or collect them during workover operation in oil field, and to reduce or even prevent waste oil and wastewater from polluting the farmland, river and groundwater. The disadvantage is that it is difficult to degrade under the condition of landfill. The polyvinyl chloride impervious cloth can reduce the pollution of waste liquid but becoming a new pollution itself. The harmless treatment will greatly increase the production cost of oil field. Therefore, it is urgent to develop a new natural degradable impervious material to replace the polyvinyl chloride impervious cloth, and to provide the environment-friendly impervious material for the green workover operation.

1 Selection of Degradable Impervious Materials

The degradable impervious cloth used in the workover operation needs to be laid on a well site with an area of several hundred square meters. Therefore, within selecting materials, we mainly consider mechanical properties, impervious function and degradation performance as the technical properties. At the same time, economy is considered synthetically, it requires raw materials cheap and readily available, being produced by mature process. Based on the above screening principle, the common biodegradable materials [2, 3] are compared and selected.

Poly-β-Hydroxyalkanoate (PHAs) is a kind of biological polyester produced by microbial fermentation engineering from natural plant starch. It has the advantage of completely biodegradability, hydrophobicity, good permeability, and thermoplastic process ability of petrochemical resins. The disadvantage is that the price is too high to be used in large quantities in the market. The molecular formula of PHAs is as Fig. 1.

Fig. 1 Molecular structure formula of Poly-β-Hydroxyalkanoate (PHAs)

Poly-ε-caprolactone (PCL) is a low melting point polymer obtained by ring-opening polymerization of ε-caprolactone. It has the advantage of it can be completely decomposed by microorganisms in anaerobic and aerobic environment, good hydrophobicity. The disadvantage is that the degradation rate is slow. The molecular formula of PLC is as Fig. 2.

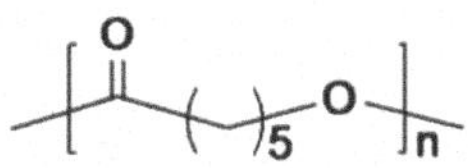

Fig. 2 Molecular structure formula of PLC

Polyadipic acid/butylenes terephthalate (PBAT) is a copolymer of butylenes adipate and butylenes terephthalate. It has the advantage of good ductility and elongation at break, excellent processing performance, excellent biodegradability, and very stable performance during normal storage and use. It is one of the best biodegradable materials in biodegradable plastics research and market application. The disadvantage is that degradation occurs only under certain microbial conditions such as composting. The molecular formula of PBAT is as Fig. 3.

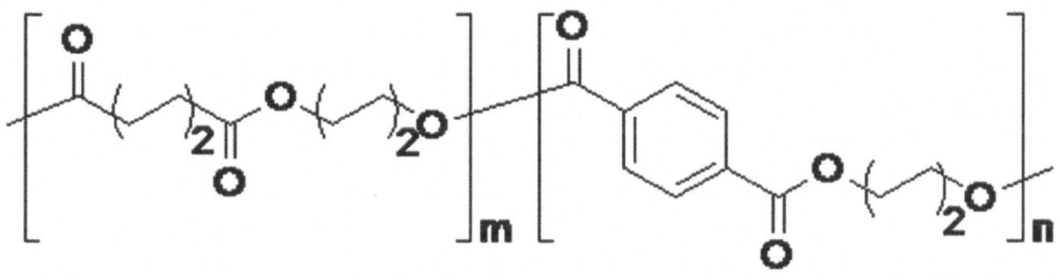

Fig. 3 Molecular structure formula of PBAT

Polylactic acid (PLA) is a kind of polyester synthesized from microbial fermentation product-lactic acid as monomer. It has the advantages of good moisture-proof, grease resistance and sealing, stable performance at room temperature, it can be completely degraded by microorganisms in nature after used, and it will eventually produce carbon dioxide and water, which is very beneficial to the protection of the environment. The disadvantage is that when the temperature is higher than 55 °C, the compost will degrade automatically. The molecular formula of PLA is as Fig. 4, and the physical properties of these degradable impermeable materials are shown in Table 1.

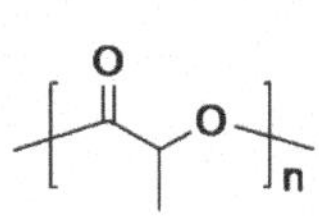

Fig. 4 Molecular structure formula of PLA

Table 1 Comparison of physical properties of several degradable impervious materials

Physical property	PHAs	PCL	PBAT	PLA
Tensile strength (MPa)	24	59	50	50
Elongation (%)	53	730	400	3
Glass-transition temperature (°C)	5	−60	65	58
Fusion temperature (°C)	160	60	130	150

These four kinds of biodegradable materials have been mass produced. Among them, Zhejiang Haizheng Group, in cooperation with Changchun Applied Chemistry Institute of the Chinese Science Academy, has developed a polylactic acid production line with an annual production capacity of 20,000 tons, which means that the production process of PLA in China is mature and the product price is close to those of traditional materials. PLA can be used in a large number of production practices. Based on the advantages and disadvantages of those four degradable materials, polylactic acid (PLA) and polyadipic acid/butylenes terephthalate (PBAT) are selected as raw materials for the preparation of degradable composite impervious cloth.

2 Formulation Design of Degradable Impervious Cloth

The biodegradable impervious materials using in workover field should fulfill three main indexes, which are mechanical property, impervious function and degradation performance. In view of the poor mechanical properties of single plastic film as a impervious cloth, which cannot meet the requirements for field workover treatment, researcher considers making impermeable cloth with composite material. Researcher designs PLA non-woven fabric and PBAT/PLA film in three-layer or five-layer composite. Between them, PLA adopts the form of non-woven cloth [4], which can guarantee tensile strength and increase tensile fracture strain and provide mechanical strength in composite material. PBAT adopts composite film form to provide impermeability in composite material [5]. Compared with the raw materials, the mechanical strength of the composite impervious materials is improved, and the superior properties of the two materials are synthesized.

The tensile strength of PLA is large, but the elongation at break is only 3%. Therefore, researchers use the non-woven PLA fabric of electrospinning technique. The elongation at break can be increased to more than 20% by changing the shape of the material. The strength and toughness of the material will not be easy to break by changing the shape of the material under the condition of retaining the original characteristics of PLA. The PLA non-woven fabric is used as the base material of the composite to ensure its mechanical strength.

The toughness of PBAT is very strong. The elongation at break is as high as 500 and it is not easy to be damaged. It can be used as the impervious layer of composite material, but the pure PBAT film is easily to deform and it is difficult to compound processing. The degradable supermarket shopping bags used in the market are produced by PBAT and PLA blending. On the one hand, it is solved the problem that

PBAT film is easily deformation, on the other hand, the price of PLA is lower than that of PBAT, so the cost of use is reduced. Therefore, researchers choose PBAT/PLA film instead of pure PBAT film as the impervious layer of composite material.

The traditional compound processing technology mostly uses the adhesive, which binds the multilayer materials together. The commonly using adhesives are petrochemical products, which are non-renewable and cannot be degraded. Biodegradable adhesives are generally expensive and costly to use, and they are not conducive to large-scale application. The melting point of PBAT/PLA films is 130 °C, while the heat resistance of PLA nonwoven fabrics is as high as 170 °C. By controlling the compound heating temperature, the PBAT/PLA film with low melting point can be melted as the adhesive, so that the two materials can be combined. The PLA nonwoven fabric and the PBAT/PLA film are completely biodegradable materials, which are formed by thermal composite processing. The degradation properties of the composites are ensured without the introduction of other materials.

3 Study on Preparation Process of Compound Impervious Cloth

3.1 Experimental Parameter Design

Researchers select three kinds of non-woven fabrics with different grams [6] (40 g/60 g/80 g) and two kinds of PBAT/PLA films of different thickness (30 um, 60 um), which are selected from the manufacturers. The non-woven fabric and the film are thermally recombined by "cloth-film-cloth", and the preparation parameters of degradable impermeable cloth are shown in Table 2.

Table 2 Design of laboratory preparation parameters for degradable impervious cloth

Raw and processed material	PLA non-woven fabric (40 g, 60 g, 80 g); PBAT/PLA film (30 um, 60 um)
Composite process	Three layers
Complex method	PLA non-woven fabric + PBAT/PLA film + PLA non-woven fabric
Heating temperature, °C	130, 140, 150, 160
Compound pressure, KPa	5 ~ 50
Recombination time, s	60 ~ 480

3.2 Experimental Results and Analysis

3.2.1 Tensile Property Test

Tensile property of composite impervious materials under different experimental conditions has been tested according to the GB/T 1040.1-2006, including tensile strength (σ_m), tensile fracture strain (ε_b) and tensile maximum load (F_m) [7]. The stretching rate is 5 mm/min, and the sample's sizes is 10 mm × 100 mm. Researchers make five test samples as a group, and take the average test results.

1. The effect of heating temperature to tensile properties of composite impervious cloth

The results of tensile test at 130, 140, 150 and 160 °C are shown in Table 3. "—" represents that the material can't carry out the normal tensile test.

The "60 g + 30 um + 60 g" three-layer composite impervious cloth is heated to 130 °C and the thermal recombination rate is relatively slow. After 120 s, the PLA non-woven fabric and the PBAT/PLA film could still be torn apart. At the heating temperature of 140 °C, the PLA nonwoven fabric could not be completely separated from the PBAT/PLA film after 120 s, and the PLA nonwoven layer is torn apart after being heated at 140 °C. Under the heating condition of 150 °C, the composite deformation and the smoothness of the sample are poor. After 120 s, the PLA non-woven fabric adheres completely to the PBAT/PLA film and some edges tend to become transparent. Under the heating conditions of 150 and 160 °C, serious deformation occurs in the composite. After 60 s of composite, the PLA nonwoven fabric and the PBAT/PLA film are completely bonded together, the specimen becomes hard and some edges tends to become transparent. This is because the processing temperature of PLA nonwoven fabric is 150 °C, and the compound temperature is higher than its processing temperature, which leads to the melting of PLA nonwoven fabric. Therefore, the composite temperature should be lower than 150 °C, otherwise there will be composite defect. The composite strength at 140 °C is better than that at 130 °C.

2. The effect of PLA nonwoven fabric weight on tensile properties of composite impervious cloth

The effect of PLA non-woven fabric weight on the properties of the composite impervious cloth is tested. The tensile test results from 40 g/60 g/80 g PLA are shown in Table 4. With the increase of PLA non-woven fabric weight, the tensile strength doesn't increase, and even decreases slightly, indicating that the increase of PLA non-woven fabric weight has no effect on the tensile strength.

As the three kinds of composite films are recombined, the dimensions are regular and it hasn't shrinkage deformation, which meets the test requirements. The maximum tensile load of "80 g + 30 um + 80 g" three-layer composite membrane is obviously increase comparing with which of "60 g + 30 um + 60 g" three-layer composite film, but the tensile strength is basically the same. This is because the tensile strength is the quotient of the maximum tensile load and cross-sectional area, while the 80 g PLA non-woven fabric is thicker than the 60 g PLA non-woven fabric, and the composite thickness is larger.

3. The effect of PBAT/PLA film thickness on tensile properties of composite impervious cloth

The influence of the thickness of PBAT/PLA film on the properties of the composite impervious cloth is tested. The PLA non-woven fabric is used at the weight of 40 g/60 g/80 g and the thickness of the PBAT/PLA film is 60 um. The tensile test results are shown in Table 5.

With the prolongation of heating time, the PBAT/PLA film of 60 um melts and different thickness of PLA nonwoven fabric mix together to form the core composite

Table 3 Test results of tensile properties of composite impervious cloth at different temperatures

Sample	σ_m, MPa	F_m, N	ε_b, %	Compound temperature, °C	Recombination time, s
60 g + 30 um + 60 g	9.6 ± 0.84	36.5 ± 3.17	26.6 ± 3.78	130	120
60 g + 30 um + 60 g	10.9 ± 0.34	37.0 ± 1.16	28.4 ± 2.88	140	120
60 g + 30 um + 60 g	–	–	–	150	120
60 g + 30 um + 60 g	–	–	–	160	120

Table 4 Test results of tensile properties of different PLA nonwoven heavy composite films

Sample	σ_m, MPa	F_m, N	ε_b, %	Compound temperature, °C	Recombination time, s
40 g + 30 um + 40 g	12.5 ± 0.59	30.0 ± 1.47	26.2 ± 2.95	140	120
60 g + 30 um + 60 g	10.9 ± 0.34	37.0 ± 1.16	28.4 ± 2.88	140	120
80 g + 30 um + 80 g	11.3 ± 0.58	50.7 ± 2.53	26.8 ± 7.26	140	120

Table 5 Test results of mechanical properties of composite impervious cloth with different thickness of PBAT/PLA films

Sample	σ_m, MPa	F_m, N	ε_b, %	Compound temperature, °C	Recombination time, s
40 g + 60 um + 40 g	16.0 ± 0.88	32.0 ± 1.75	22.2 ± 2.86	140	480
60 g + 60 um + 60 g	17.2 ± 0.60	49.9 ± 1.74	18.2 ± 5.36	140	480
80 g + 60 um + 80 g	12.0 ± 1.19	41.9 ± 4.18	23.6 ± 2.51	140	480

layer at 140 °C, which can provide good mechanical strength. The results show that "60 g + 60 um + 60 g" can provide higher mechanical strength than "40 g + 60 um + 40 g". However, "80 g + 60 um + 80 g" could not improve the mechanical strength of the composite impervious cloth, which is related to the match between PLA nonwoven fabric and PBAT/PLA film, at the same time the raw material cost increases a lot. Compared with "40 g + 30 um + 40 g", the maximum tensile load of "40 g + 60 um + 40 g" three-layer composite impervious cloth increases slightly, indicating that the extension of the composite time has little effect on the tensile maximum load. The tensile strength increases by 4 MPa up to 16 MPa, which is due to the decrease of the film thickness with the increase of the composite time.

Compared with the mechanical properties of braided PVC composite geomembrane (Table 6) in the field, "60 g + 60 um + 60 g" composite impervious cloth is recommended as the best formulation considering cost performance.

Table 6 Test results of tensile properties of woven polyvinyl chloride composite geomembrane

Sample	σ_m, MPa	F_m, N	ε_b, %
Woven polyvinyl chloride cloth	15.6 ± 2.8	51.1 ± 3.7	19.6 ± 1.4

3.2.2 Impervious Performance Test [8]

The impervious performance test of "60 g + 60 um + 60 g" composite impervious cloth is carried out to test whether the material's impermeability can achieve the requirements.

1. Performance test of impervious water

Test methods: The composite impervious cloth is placed flat on the table, and adds 0.5 ml distilled water to its surface, observing the state of the water which droplet on the composite impervious cloth. After 10 min of rest, the composite impervious cloth is flipped over, and the surface of the composite impervious cloth is wiped by filter paper. Researchers observe whether the filter paper absorbs water or not, then taking pictures. It can be seen from Fig. 5 that the water droplets are aggregated on the composite

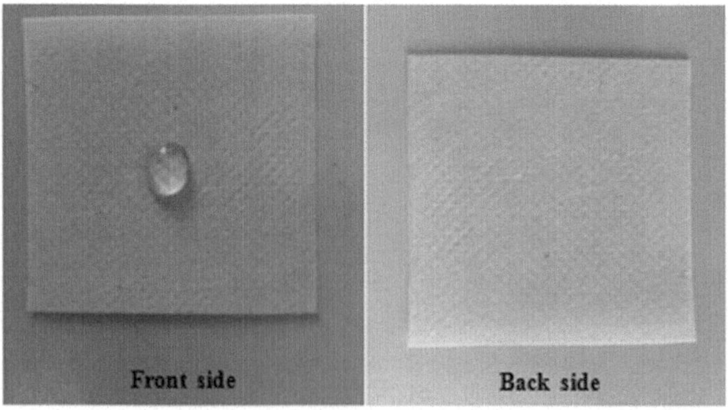

Fig. 5 Water anti-seepage performance test

impervious cloth, but there is no water seepage on the back, which is related to the surface free energy of the water. With the nature of the water and the structural characteristics of the material itself, the composite membrane can completely block water.

2. Performance test of impervious crude oil

Test methods: the composite impervious cloth is placed flat on the heating table, and the 1 ml crude oil is picked up with glass rod, dropped onto the composite film. The heating table is heated to 50 °C, so as to imitate that the composite impervious cloth is used on the high temperature days in summer. The crude oil is heated and flowing, and the fluidity of the crude oil is greatly increased. It also increases the possibility of permeating the composite impervious cloth. Researchers observing the state of crude oil on the composite impervious cloth, removing the composite membrane after 30 min, wiping the other side of the composite membrane with filter paper, seeing if the filter paper is contaminated with crude oil, verifying whether the composite impervious cloth can prevent oil leakage, taking pictures and recording the result. The test results are shown in Fig. 6. At 50 °C, the crude oil wouldn't leak, and the composite impervious cloth can play a good role in blocking crude oil.

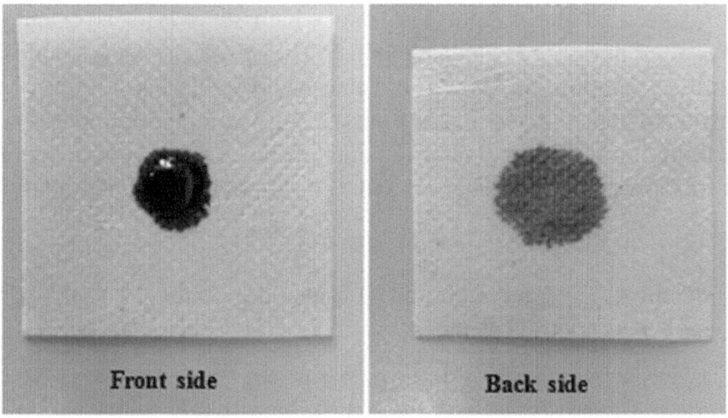

Fig. 6 Oil anti-seepage performance test

4 Study on the Technology of Post-degradation Treatment [9, 10]

4.1 Alkali Degradation Test of Composite Impervious Cloth

The sample of "60 g + 60 um + 60 g" three-layer composite impervious cloth is cut into 2 cm × 2 cm and researchers make up 0.25 mol/L, 1 mol/L, 2 mol/L concentration sodium hydroxide solution. The sample is soaked in sodium hydroxide solution and the properties of the sample are observed at different times, and the results of alkali degradation in composite impervious cloth are shown in Table 7.

Table 7 Experimental results of alkali degradation in composite impervious cloth

Time/concentration	0.25 mol/L	1 mol/L	2 mol/L
1 day	Partial dissolution	Partial dissolution	Partial dissolution
2 day	Partial dissolution	Partial dissolution	Complete dissolution
3 day	Partial dissolution	Complete dissolution	–
7 day	Partial dissolution	–	–
12 day	Complete dissolution	–	–

4.2 Natural Degradation Experiment

30 g culture soil (soil in flower bed) and 100 ml distilled water are used to prepare mixed solution. After full mixing, they are equally divided into 8 beaker, and 8 pieces of "60 g +60 um +60 g" composite impervious cloth sample with weight about 1 g are obtained. After cleaning with deionized water, which the initial weight records as W_o, is added to each beaker, so that it is completely covered with culture soil. Under natural environment condition, the composite impervious cloth sample from beaker is taken out every 10 days, and the weight loss rate which records ω is calculated by dry weighing W_t, after deionized water cleaning. The results of degradable impervious cloth's natural degradation are shown in Table 8.

It is considered that the "60 g + 60 um + 60 g" composite impervious cloth has good degradability and meets the requirements of the experimental design through the experiments of rapid alkali degradation and natural composting degradation.

Table 8 Experimental results of natural degradation of degradable impervious cloth

Order number	Time, d	W_o, g	W_t, g	ω, %
1	10	0.1277	0.1277	0
2	20	0.115	0.0982	14.6
3	30	0.1183	0.0682	42.3
4	40	0.1152	0.0581	49.6
5	50	0.1334	0.0561	57.9
6	60	0.1489	0.0502	66.3
7	70	0.1204	0.0119	90.1
8	80	0.1055		100

5 Field Application Effect

Based on the above research results, two samples of composite impervious cloth have been produced and tested in 4 working wells during the workover operation. All of them are not damaged, which shows that they meet the requirements of workover. The using effect of the degradable and anti-seepage cloth is shown in Table 9.

Table 9 Application effect of biodegradable and impervious cloth in workover operation

Sample cloth	Well number	Well parting	Time, d	Condition of service
1#	W20-10	Water well	14	Without damage
	A421-1	Water well	5	Without damage
2#	Q42-32x	Oil well	8	Surface adhesion oil and wax, without damage
	Q28-10x	Oil well	24	Without damage

6 Conclusion

1. The "60 g + 60 um + 60 g" composite impervious cloth can meet the requirements of impermeability, which can not only guarantee that the crude oil can't be permeated, but also can completely prevent water leakage.
2 The "60 g + 60 um + 60 g" three-layer composite impervious cloth sample cloth is tested by degradation experiment and field application to meet the requirements of green workover.
3 PLA and PBAT, the raw materials for making degradable impervious cloth, have been mass produced. Through the laboratory sample preparation process research, two sample cloths have been effectively applied in the field. It shows the possibility of further enlargement.

References

1. Liu, L., Chu, C.: Properties and application of geotextiles. Nonwovens **9**(4), 11–14 (2001)
2. Fang, Z., Wang, W., Xia, B.: Research progress and application of biodegradable polymer materials. Res. Dev. Degradable Plast. **8**, 33–38 (2006)
3. Ping, Y.: Advances in polylactic acid biodegradable composites. Guangxi Light Ind. **122**(1), 22–23 (2009)
4. Lin, L.: Biodegradable polylactic acid nonwoven fabric. New Mater. Chem. Eng. **27**(8), 24–26 (1999)
5. Si, P.: Preparation and degradation of PLA/PBAT thin films. Plast. Technol. **43**(10), 68–72 (2015)
6. Niu, X.: Truzler will showcase a number of new non-woven fabric technologies. China Textile News **3**, 1 (2014)

7. Tao, J., Hu, D., Liu, L., et al.: Thermal, mechanical and biodegradable properties of PLA, PPC and PHBV blends. Ion Exch. Adsorpt. **26**(1), 59–67 (2010)
8. Tian, H.: Preparation and Waterproof and Moisture Permeability of Polyvinylidene Fluoride/Polyurethane Fiber membrane. Tung Wah University (2015)
9. Dong, Q.: Progress in degradation of polylactic acid. Plastics **3**, 2–4 (2011)
10. Zhang, Y., Liu, X., Ma, Q.: Research progress of polylactic acid/adipic acid-terephthalic acid-butylene ester copolymer whole biodegradable blend. Chin. Plast. **28**(3), 7–15 (2014)

Effect of Viscosity Index Improver and Base Oil on Soot Dispersion Performance of Modern Diesel Engine Oil

Qing-hong Xia(✉)

Research Institute of Petroleum Processing, SINOPEC, Beijing, China
xiaqh.ripp@sinopec.com

Abstract. Incompletely burned diesel fuel tends to generate soot into the crankcase, resulting in a sharp increase in oil viscosity. Modern diesel engine oils are required to have a good soot dispersion performance, that means the viscosity of the oil will not increase significantly after being mixed with the soot. Carbon black was used as a simulant. The dispersion performance of the oil was evaluated by the viscosity growth rate. Investigate the change in viscosity before and after the oil is added to carbon black. Investigate the effects of different base oils and viscosity index improvers on the viscosity growth rate of the test oils after adding carbon black. Ethylene-propylene copolymer type viscosity index improver (OCP) has the worst soot dispersion performance. Polymethacrylate type viscosity index improver (PMA) has the best soot dispersion performance. Hydrogenated styrene/diene copolymer viscosity index improver (HSD) soot dispersion performance centered. Under the premise of adding the same viscosity index improver, the API Group II and the API Group III base oils disperse better than the API Group I base oils. Systematic study on the effect of viscosity index improvers and base oils on the soot dispersion performance of diesel engine oil, explore the mechanism. Provide strong technical support for the development of modern diesel engine oil.

Keywords: Diesel engine oil · Base oil · Viscosity index improver · Dispersibility performance

J. Lin (ed.), *Proceedings of the International Petroleum and Petrochemical Technology Conference 2019*, pp. 313–318, 2020.
https://doi.org/10.1007/978-981-15-0860-8_26

1 Introduction

High-grade diesel engine oil requires high soot dispersibility. The high-power heavy-duty diesel engine adopts direct injection technology, and the fuel injection amount is large. The diesel fuel that is not completely burned easily generates soot into the crankcase [1, 2]. After the soot enters the diesel engine oil, if the soot cannot be effectively dispersed in the oil, the soot will aggregate to form a network structure, and a small amount of soot can cause the oil viscosity multiply [3, 4]. The degree of viscosity growth is related to the soot dispersion performance of diesel engine oil [5]. The Mack T-8 bench test of diesel engine oil is mainly used to investigate the dispersing performance of oil soot [6]. From CF-4 (Mack T8-A), CG-4 (Mack T8) to CH-4 (Mack T8-E), the mass fraction of soot contained in the oil increased from 2.0 and 3.8% to 4.8%. Modern diesel engine oils place higher demands on the soot dispersibility of oil products. The influence of base oil and viscosity index improver on the dispersion performance of oil soot has great practical significance for the development and application of modern diesel engine oil.

2 Test Method

Test method for determining physical and chemical indicators of oil products:

Kinematic viscosity: GB/T 265.

Viscosity index: GB/T 2541.

The simulation test method is as follows.

Soot dispersibility: The test of soot dispersibility uses carbon black as a stimulant. After adding a certain mass fraction of carbon black to the test oil, test the sample's viscosity at 100 °C, then calculate the viscosity growth rate, use the viscosity growth rate to distinguish the dispersion of the sample.

2.1 Soot Formation Process

The formation of soot is related to air temperature, oxygen concentration, type of fuel, and ratio of air to combustion. In the engine, the higher the preheating temperature of the air, the lower the oxygen content in the air, the more uneven mixed of the fuel and air, the more the fuel crack, then more soot will be produced.

The formation of soot can be roughly divided into two stages. The first stage: producing a soot core from low molecular weight unsaturated hydrocarbons, polymerized by chemical reaction; the second stage: the generated soot core is grown into soot particles through liquefaction polymerization and agglomeration. Soot particles generally have a structure similar to graphite [7]. In the following experiments, carbon black was used as a simulant for soot.

2.2 Investigation on Dispersion Performance of Viscosity Index Improver

Viscosity index improver is an oil-soluble polymer compound, Adding a viscosity index improver to a base oil with low kinematic viscosity can improve the kinematic viscosity and improve the viscosity of the base oil. The oil viscosity is not too high at low temperatures, ensuring that the engine can start smoothly; the oil has a sufficient viscosity at high temperatures and lubricate the engine well. The viscosity index improver commonly used in diesel engine oil mainly has ethylene-propylene copolymer (OCP), Polymethacrylate (PMA), Hydrogenated styrene/diene copolymer (HSD), etc. [8].

Ethylene-propylene copolymer type viscosity index improver can be directly synthesized from ethylene and propylene, it is also possible to apply a high shear stress to a high molecular weight ethylene propylene rubber at a certain temperature and degrade to a predetermined molecular mass to obtain. The relative molecular mass of the ethylene-propylene copolymer type viscosity index improver is 30,000 to 100,000.

The polymethacrylate type viscosity index improver is formed through solution polymerization which contains a methacrylate monomer, the initiator and a relative molecular mass regulator. The relative molecular mass of polymethacrylate type viscosity index improver used to modulate multi-stage internal combustion engine oil is about 150,000.

Hydrogenated styrene/diene copolymer can be obtained by anionic polymerization of styrene with butadiene or isoprene and catalytic hydrogenation to saturate the remaining double bonds in the molecular chain. The hydrogenated styrene/diene copolymer type viscosity index improver has a relative molecular mass of 50,000 to 100,000.

Different viscosity index improvers have different effects on the dispersibility of diesel engine oil. Adding different viscosity index improvers to the same base oil to adjust the viscosity of the oil to near level, then add carbon black, Investigate the effect of different viscosity index improvers on the soot dispersion performance of oil. Test results are shown in Table 1.

Table 1 Dispersion properties of different types of viscosity index improvers

Viscosity index improver	Viscosity growth rate /%
OCP	3964
PMA	2423
HSD	3505

To the polarity, the base oil is relatively strong, the viscosity index improver is relatively weak, and the carbon black is the weakest. Therefore, carbon black easily adsorbs viscosity index improver in base oil. That is, the functional group of the viscosity index improver easily interacts with the surface of the carbon black, affecting the aggregation state of the carbon black. The test results show that the viscosity index improver type has an effect on the dispersion properties of the oil.

It can be seen from Table 1 that the OCP type viscosity index improver has the worst dispersion performance; the PMA type viscosity index improver has the best dispersion performance. The dispersion property of HSD type viscosity index improver is middle.

2.3 Investigation on the Dispersion Performance of Basic Oil

The base oil is the carrier of the additives, which occupies a large proportion in the lubricating oil and has a great influence on the performance of the oil.

In the 1990s, the American Petroleum Institute (API) and the Association Technique de l"Industrie Européenne des Lubrifiants (ATIEL) proposed a lubricant base oil classification, which is still in use today. The API base oil classification standard classifies the main chemical composition (saturated hydrocarbon, sulfur content) and viscosity index of the base oil as the main basis for classification, and classifies the base oil into five categories. The base oil classification is shown in Table 2.

Table 2 API-1509 base oil classification standard

Classification	Saturated hydrocarbon /%	Viscosity index	Sulfur content /%
I	<90	80–120	>0.3
II	≥90	80–120	<0.3
III	≥90	>120	<0.3
IV	Polyalphaolefin (PAO)		
V	All non-I, II, III, IV base oils		

Class I base oil is produced by the traditional "old three sets" process, and the production process is basically based on physical processes, without changing the hydrocarbon structure. Group II base oils are generally produced by solvent refining and hydrogenation processes, with less impurities than Class I base oils and high saturated hydrocarbon content. The Group III base oil is produced by a full hydrogenation process and is a hydrogenated base oil with a high viscosity index. Class III base oil has the characteristics of high viscosity index and low volatility. Group IV base oil is a polyalphaolefin (PAO) synthetic oil. Paraffin decomposition and ethylene polymerization are the common production methods. These base oils do not contain S, P and metals. The pour point is extremely low, usually below −40 °C. The viscosity index generally exceeds 140. Poor ability to dissolve polar additives due to its low polarity. Other synthetic oils (synthetic hydrocarbons, esters, silicone oils, etc.), vegetable oils, and reclaimed base oils other than Group I to Group IV oils are collectively referred to as Group V base oils.

The following tests were carried out using APII, II, III base oils. The performance of the base oil used is shown in Table 3.

Table 3 Base oil viscosity and viscosity index

Classification	Kinematic viscosity (40 °C)/ (mm^2/s)	Kinematic viscosity (100 °C)/ (mm^2/s)	Viscosity index
I	32.05	5.24	101
II	34.46	6.01	118
III	31.93	5.73	121

Using carbon black dispersion test to simulate Mack T-8 engine test. Evaluating the dispersibility of the sample after adding carbon black with a mass fraction of 3% to the test oil, Differentiating the dispersion of the sample by using the viscosity growth rate of the test oil relative to the new oil at 100 °C.

There are four main processes for carbon black dispersion in oil: carbon black agglomerates are destroyed into smaller particles; carbon black particles are wetted; redistribution of carbon black particles; Self-polymerization of carbon black particles. Wetting and redistribution of carbon black particles are affected by the interaction of carbon black particles with base oils. After the uniform dispersion system formed between carbon black and oil, when the surface tension and surface energy of the oil are lowered, the wetting of the carbon black particles and the oil is promoted, the carbon black and the oil are sufficiently contacted, and the carbon black particles are improved in the oil, improve the redistribution rate and uniformity of carbon black particles in oil, thereby improving the dispersion properties of oil. Add a certain mass fraction of carbon black to the API base oil, API II base oil and API III base oil respectively. Investigate the effect of base oil on dispersion performance. The results are shown in Table 4.

Table 4 Effect of base oil on dispersibility

Classification	Viscosity growth rate /%
I	37
II	30
III	29

It can be seen from Table 4 that the different types of base oils have different soot dispersing ability. Under the premise of adding the same viscosity index improver, the viscosity growth rate from API class I base oil to API class III base oil gradually decreases. It shows that API II and API III base oils disperse better than API I base oils. Mainly because the polarity of aromatic hydrocarbons are more polar than alkanes and are easily polarized. Carbon black in base oils with high aromatic content is more likely to aggregate. As a result, the API Group I base oil viscosity growth rate is higher than the other two types of base oil.

3 Conclusion

Viscosity index improver has an effect on the dispersion properties of oils. the OCP type viscosity index improver has the worst dispersion performance; the PMA type viscosity index improver has the best dispersion performance. The dispersion property of HSD type viscosity index improver is middle.

Under the premise of adding the same viscosity index improver, The viscosity growth rate from API Group I base oil to API III base oil is gradually decreasing. The soot dispersion performance of API II and API III base oils is better than that of API I base oils.

Acknowledgements. Thanks to my colleagues for their help in the experiment.

References

1. Gautam, M., Chitoor, K., Durbha, M., Summers, J.C.: Effect of diesel soot contaminated oil on engine wear—investigation of novel oil formulations. Tribol. Inter. **32**(12), 687–699 (1999)
2. George, S., Balla, S., Gautam, V., Gautam, M.: Effect of diesel soot on lubricant oil viscosity. Tribol. Inter. **40**(5), 809–818 (2007)
3. Shank, G., Goshorn, K., Cooper, M., Van Dam, W., Richards, S.: A history of mack engine lubricant tests from 1985–2005: Mack T-7 through mack T-12. SAE Trans. **114**(4), 1289–1307 (2005)
4. Bardasz, E.A., Cowling, S.V., Ebeling, V.L., et al.: Understanding soot mediated oil thickening through designed experimentation-part 1: Mack EM6-287, GM 6.2L. SAE Paper 952527 (1995)
5. George, S., Balla, S., Gautam, V., et al.: Effect of diesel soot on lubricant oil viscosity. Tribol. Int. **40**, 809–818 (2007)
6. Mc Geehan, J.A., Franklin, T.M., Bondarowicz, F., et al. New diesel engine oil category for 1998: API CH-4. SAE Paper 981371 (1998)
7. Dobbins, R., Fletcher, R.A., Chang, H.C.: The evolution of soot precursor particles in a diffusion flame. Combust. Flame **115**(3), 285–298 (1998)
8. Goldblatt, I., Mchenry, M., Henderson, K., et al.: Lubricant for use in diesel engines. US Patent 6,187,721 (2001)

Development of Combined Washing and Distillation Technology

Ji-peng Dong[1], Wei-wen Wang[2], Xiu-ling Guo[1], Jian-long Li[1], and Guang-hui Chen[1](✉)

[1] College of Chemical Engineering, Qingdao University of Science and Technology, Shangdong, China
{dongjipeng0302, XLGuo0606, longtengchen01}@163.com, ljlong@qust.edu.cn

[2] Graduate School, Qingdao University of Science and Technology, Shangdong, China
wwang@qust.edu.cn

Abstract. The washing and distillation tower (WDT) is a general equipment which can combine wet washing with distillation. In order to solve the problems of incomplete removal of solid particles, low efficiency of gas-liquid separation, small flexibility and easy blockage of internals of conventional WDT, a combined washing and distillation technology (CWDT) was developed by coupling inertial dust removal with flow-guided sieve tray, dual-flow tray and micro-bubble mass transfer technology (MBT). Experimental research, by laying a MBT mesh up the tray, was carried out based on the new MBT to examine the change of the mean bubble diameter, pressure drop and total gas hold-up under different operating conditions. The experimental results show that the mean bubble diameter can be reduced from 20 mm–57 mm to 1 mm–3.5 mm, greatly increases the washing effect and gas-liquid mass transfer efficiency. The technology has been successfully applied to the synthesis of organosilicon monomer and the heat recovery process of crude gas, and achieved good results, which proves the feasibility and advancement of the technology.

Keywords: Combined washing and distillation technology · Micro-bubble mass transfer technology · Organosilicon monomer · Crude gas

This paper was prepared for presentation at the 2019 International Petroleum and Petrochemical Technology Conference in Beijing, China, 27–29, March, 2019.

J. Lin (ed.), *Proceedings of the International Petroleum and Petrochemical Technology Conference 2019*, pp. 319–330, 2020.
https://doi.org/10.1007/978-981-15-0860-8_27

1 Introduction

Washing and distillation tower (WDT) as a common equipment which couples wet washing with distillation operation has been widely used in petroleum refining, petrochemical, coal chemical and other processes [1–3]. It is usually locates behind the cyclone separator and has two functions. Firstly, to further remove the solid particles (such as solid materials, catalyst particles, etc.) contained in the gas by liquid washing. Secondly, under some special conditions, the tower needs to have the distillation function to roughly separate the gas components as to reduce the subsequent separation load. The plate tower is mostly adopted. When the gas mixture contains dust enters the WDT, solid particles will be blocked by the liquid during the gas passes through the liquid layer and come to the bottom of the tower with the heavy components. Simultaneously, heat and mass exchange between the gas components on the tray is realized to roughly separate the heavy and light components.

Actually, the separated solid particles will cause tray blockage and lead to unstable and discontinuous production [4, 5]. The washing and separation effect did not meet the design requirements, which was mainly due to large bubbles formed when the gas passing through the liquid layer on the tray, making the gas-liquid exchange insufficient and solid particles difficult to be completely removed. As a result, the gas at the top of the tower contains solid particles and the light component at the bottom of the tower exceeds the requirement.

In order to overcome the shortcomings mentioned above, a combined washing and distillation technology (CWDT) was developed. It combines cheveron baffle with flow-guided sieve tray, dual-flow tray and microbubble mass transfer technology (MBT) to solve the disadvantages such as incomplete removal of solid particles, low gas-liquid separation efficiency, small operating flexibility and easy blockage of internals.

2 Combined Washing and Distillation Technology (CWDT)

Figure 1 shows the structure of a CWDT tower. For strengthen the washing and distillation process, cheveron baffle [6], dual-flow tray [7, 8], flow-guided sieve tray [9, 10] are locate from bottom to top in the tower. Meanwhile, micro-bubble mass transfer device is added on the flow-guided sieve tray or dual-flow tray according to the technical requirements.

When it working, the gas mixture enters the tower from the air inlet and passes through the cheveron baffle firstly. As the cheveron baffle has a certain opening ratio, when the liquid phase flowing from the upper tray and the dusty gas simultaneously pass through the holes on the baffle, the gas is wet washed and most solid particles are washed down. As the cheveron baffle has blocking effect, and its surface is relatively wet, solids can be collected and flow down with the liquid. Finally, the liquid together with the solid particles flows out the tower.

The dual-flow tray is set above the cheveron baffle. The gas passes through the sieve holes, and the liquid also leaks through the holes to strengthen gas-liquid contact, which is beneficial to gas-solid separation. As a result of maintaining a certain level of liquid on the tray, when the gas passes through, the gas and liquid are in full contact to

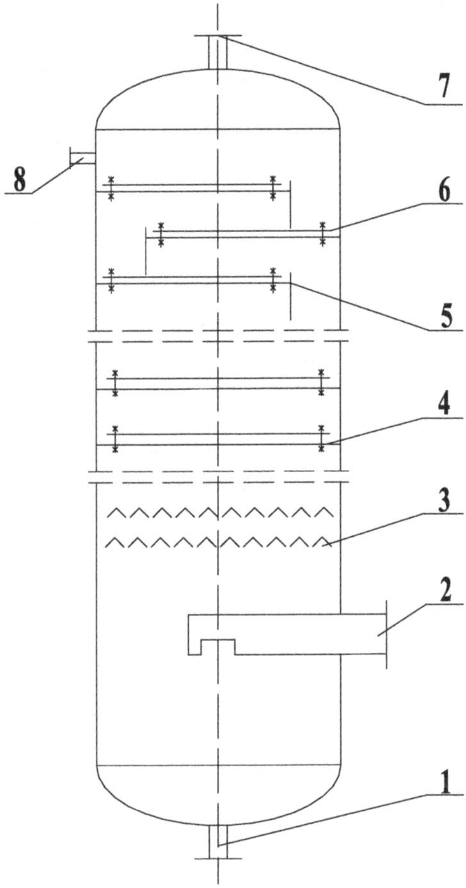

Fig. 1 The structure of CWDT tower. (1) liquid outlet; (2) gas inlet; (3) cheveron baffle; (4) dual-flow tray; (5) flow-guided sieve tray; (6) MBT; (7) gas outlet; (8) liquid inlet

remove finer particles. The key point is that the diameter of sieve holes and opening ratio are large, which can effectively avoid the deposition and blockage of solid particles.

The flow-guided sieve tray is set above the dual-flow tray. The gas pushes the liquid through the guided hole, which can reduce the height drop of the liquid layer and adjust the height of the foam, so as to restrain entrainment and avoid deposition of ultra-fine particles.

In order to eliminate the ultra-fine particles enveloped by large bubbles in the foam of the tray, a MBT is set up on the tray. The device can reduce the diameter of the large bubbles generated when the gas passes through the liquid layer to 1/20 of the original ones. The contact probability between the ultra-fine particles and the liquid is increased, which is conducive to their separation. In addition, the specific surface area of liquid layer is increased, which enhances the gas-liquid mass transfer and improves the separation efficiency of the tray.

3 Experimental Research of MBT

This paper puts forward a simple and feasible structural improvement to breaking large bubbles, laying a MBT mesh in the foam layer, forcing the large bubbles passing through the mesh to break into small ones close to the mesh size, so as to enhance the mass transfer process on the plate. In order to investigate its performance, the hydrodynamic experiments of MBT were carried out.

3.1 Experimental Procedure

The geometric dimensions of column and the tray are given in Table 1 and the experimental setup used in this paper is shown in Fig. 2. Tap water as liquid phase with Rhodamine B (a tracer) is pumped into the top of the tower from the tank through the flow meter, meanwhile air as the gas phase is flowed into the bottom of the tower by the draught fan. Tap water and air cross-flow in the tower. Under the constant liquid flow rate, the changes of hydrodynamics on the tray under different gas velocity (from 2.31 to 3.45 m s^{-1}) rate were measured and compared with or without MBT. The parameters such as bubble distribution and bubble diameter were taken by CCD and analyzed by PIV system during the experiment.

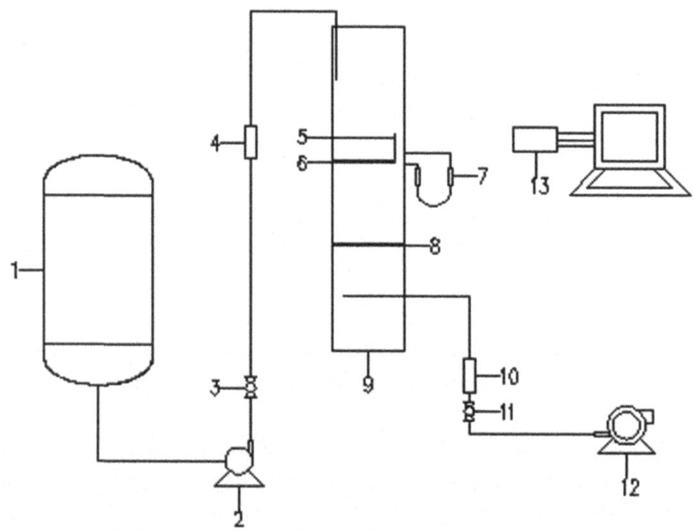

Fig. 2 Experimental flow chart: (1) tank; (2) liquid pump; (3) 11-valve; (4) 10-flow meter; (5) MBT; (6) sieve tray; (7) U pressure meter; (8) gas distributor; (9) bottom outlet; (12) draught fan; (13) CCD camera & computer

The structure of the tower and the geometric dimensions of sieve tray are shown in Table 1 and Fig. 3 respectively. The gas distributor has 55 holes of 5 mm diameter, arranged in a triangular pitch. The MBT is a square mesh made of stainless steel with a side length of 0.47 mm and a wire diameter of 0.14 mm which is placed up to the sieve tray 0.03 m. The structure is shown in Fig. 4.

Table 1 Experiment structure parameter

Program	Parameter
Material	Organic glass
Column height	600 mm
Column diameter	φ 130 mm × 5 mm
Weir height	50 mm
Weir length	74 mm
Hole diameter	φ 5 mm
Opening ratio	9.5%

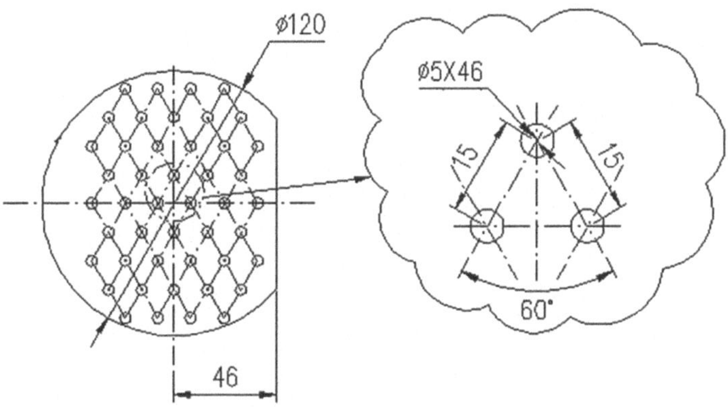

Fig. 3 The geometric dimensions of sieve tray

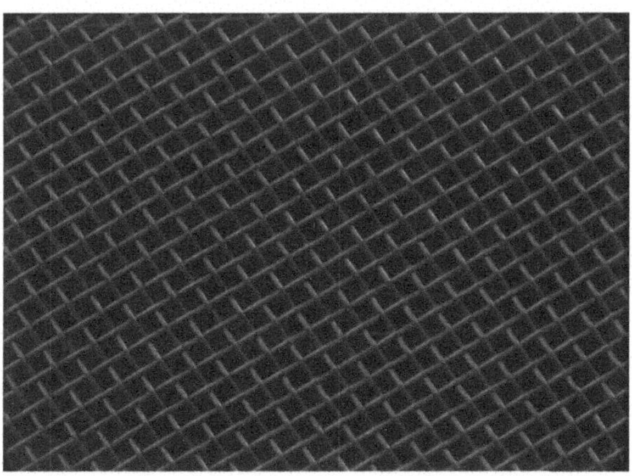

Fig. 4 The photo of MBT mesh

3.2 Results and Discussion

Mean Bubble Diameter. The mean diameter distribution of bubbles determines the specific surface area of gas-liquid mixing, which has an important influence on the gas-liquid mass transfer effect. It is related to gas holdup, interfacial concentration and surface tension etc. When the diameter of bubbles is larger, the volume is bigger and the rising speed is faster, which are not conducive to gas-liquid mass transfer.

With a superficial gas velocity of 3.03 m s^{-1}, Fig. 5 shows the radial distribution of bubbles at three heights of 0.02 m ($z_1 = 0.02$ m), 0.05 m ($z_2 = 0.05$ m) and 0.08 m ($z_2 = 0.05$ m) above the tray respectively. It can be seen that the large bubbles are concentrated in the center of the tray, while the small ones are dispersed in the low gas holdup area near the tower wall. With the increase of the axial height, the central bubbles coalesce and become larger. The mean diameter of bubbles on sieve tray is close to 50 mm (between 20 and 60 mm), while it dropped down to 1 to 3 mm on the MBT sieve tray. The smaller bubble diameter means that the specific surface area of gas-liquid contact increases significantly, which is beneficial to the intensification of gas-liquid mass transfer process.

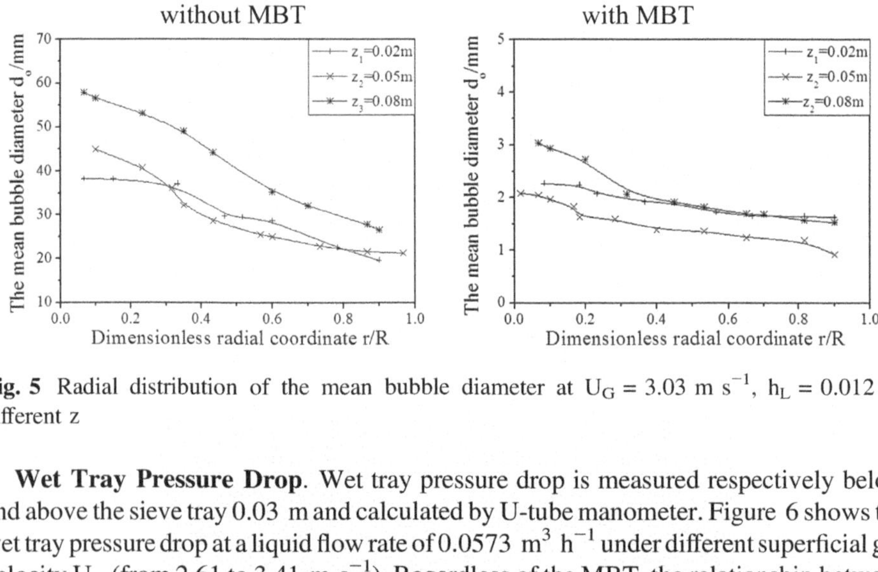

Fig. 5 Radial distribution of the mean bubble diameter at $U_G = 3.03$ m s^{-1}, $h_L = 0.012$ m, different z

Wet Tray Pressure Drop. Wet tray pressure drop is measured respectively below and above the sieve tray 0.03 m and calculated by U-tube manometer. Figure 6 shows the wet tray pressure drop at a liquid flow rate of 0.0573 m^3 h^{-1} under different superficial gas velocity U_G (from 2.61 to 3.41 m s^{-1}). Regardless of the MBT, the relationship between pressure drop and U_G is similar. With the increase of U_G, the pressure drop also increases gradually. When the gas volume increases to a certain extent, the pressure drop of the plate will increase significantly, which indicates the gas-liquid contact becomes to bubbling state. The pressure drop of the MBT sieve tray to an average of 499 Pa, 56% higher than that of the sieve tray. This is due to the forced contraction of bubbles when they through the MBT, which is equivalent to the secondary expansion of air flow. In this case, the drag coefficient is larger, and the dissipation energy of gas increases, leading to the increase of pressure drop. Higher pressure drop causes more energy cost, so it is only suitable for the special location of the column.

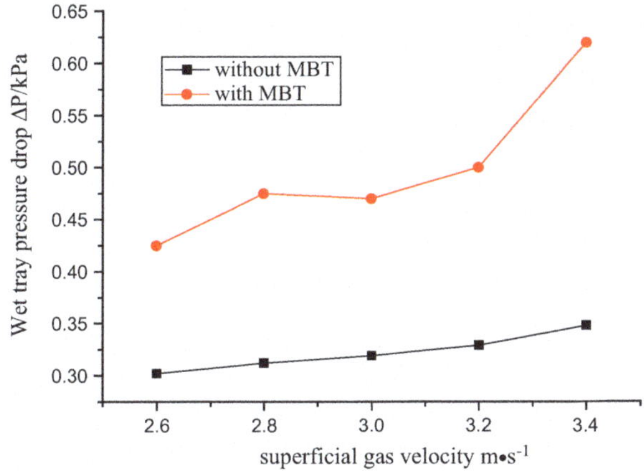

Fig. 6 Effect of U_G on wet tray pressure drop at liquid flow rate of 0.0573 m^3 h^{-1}, ■ without MBT, ● with MBT

Total Gas Hold-Up. The total gas hold-up (ε_g) is calculated according to the following formula,

$$\varepsilon_g = (H - H_L)/H$$

where, H is the overall foam layer height and H_L is the clear liquid height.

For sieve stray, when H_L is 0.012 m and the superficial gas velocity U_G from 2.36 to 3.03 m s^{-1}, the gas holdup varies from 46 to 52%. Meanwhile, with the increase of U_G, both of the overall foam layer height and gas holdup are increase. However, under the same U_G, the gas holdup of the MBT sieve tray is 4.29% higher than that of the sieve

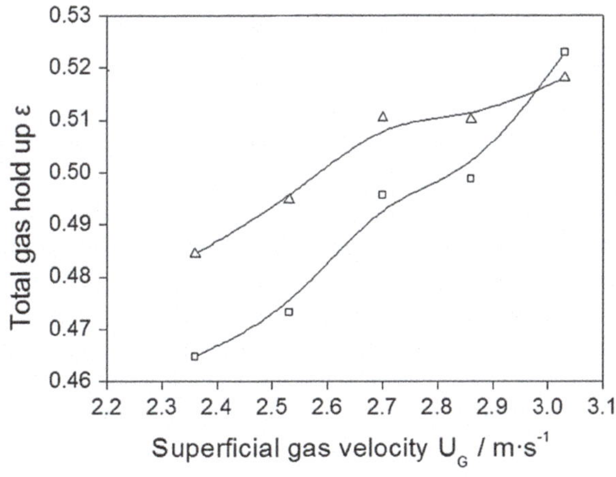

Fig. 7 Effect of superficial gas velocity on gas hold-up at liquid flow rate of 0.0573 m^3 h^{-1}, □ without MBT, △ with MBT

tray, ranging from 48 to 51%. This is due to the existence of MBT, which makes the bubbles break into small ones, the buoyancy is smaller, the residence time in liquid phase is prolonged, and the volume fraction of gas is increased. The detail of the total gas holdup is shown in Fig. 7.

4 Industrial Application

4.1 Purification Tower of Organosilicon Monomer

The purification tower of organosilicon monomer synthesis process is used to separate the ultra-fine particles carried in the crude monomer synthesis gas by wet washing, and to separate most of the high boiling components from the products [11–14]. As mentioned above, there are many disadvantages in the operation process such as tray blockage and low gas-liquid separation efficiency. So the author designed a new purification tower based on CWDT. The structure of the new tower shows in Fig. 8.

The new tower is similar to that shown in Fig. 1, except that a liquid distributor is added in the bottom of the tower. This is mainly due to the fact that when synthesis gas passes through the liquid layer at the bottom, the large diameter of the inlet pipe is easy to produce large bubbles and the gas distribution effect is poor, causing liquid level and airflow fluctuation and even local strong airflow will impact the tray internals. Increasing the gas distributor can enhance the gas distribution effect and improve gas-solid separation efficiency.

Figure 9 shows the change of capacity in the purification tower of Shandong Jinling chemical co., ltd. in October 2017 within 24 days. The theoretical design capacity of the equipment is 330 t/d (13.75 t/h). According to the data, the minimum capacity of the equipment when start-up is about 50% of the theoretical design value, but the maximum output (October 18) is close to 200%, that is, the operational flexibility is between 50 and 200% of the theoretical design.

4.2 The Crude Gas Purification and Heat Recovery Process Based on the CWDT

Coal coking is an important technology and widely used in the field of coal chemical industry. Coal is isolated from air for high-temperature dry distillation, which generates coke, crude gas, coke powder, etc. [15]. The temperature of crude gas is about 650–750 °C. It is composed of many components such as tar, hydrogen, methane, carbon monoxide, benzene, etc. The coking temperature of heavy tar in the crude gas is about 400–500 °C, and the tar mixed of coke powder is easy to block pipes and equipments.

The traditional cooling process of crude gas mostly uses ammonia water directly spraying, which causes a lot of heat waste and waste-water. Recent years, many surplus heat recovery technologies have been developed, such as heat pipe, conduction oil jacket technology, etc., but due to coking, scaling and other reasons, long-term stable operation can not be achieved [16–18]. The author has developed a new heat recovery

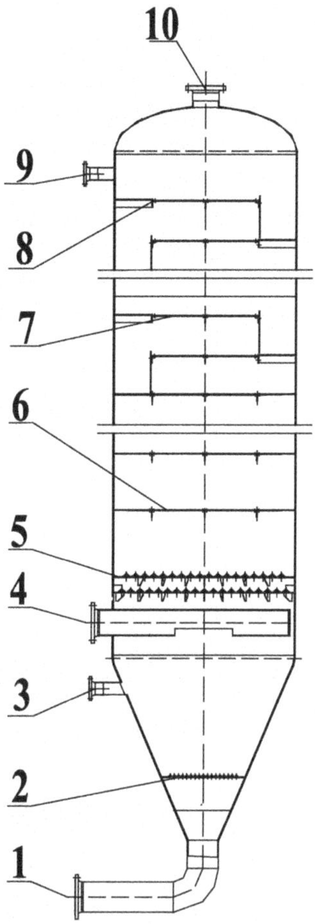

Fig. 8 The structure of new purification tower (1) gas inlet; (2) distributor; (3) liquid outlet; (4) gas inlet; (5) cheveron baffle; (6) dual-flow tray; (7) flow-guided sieve tray; (8) MBT; (9) liquid inlet; (10) gas outlet

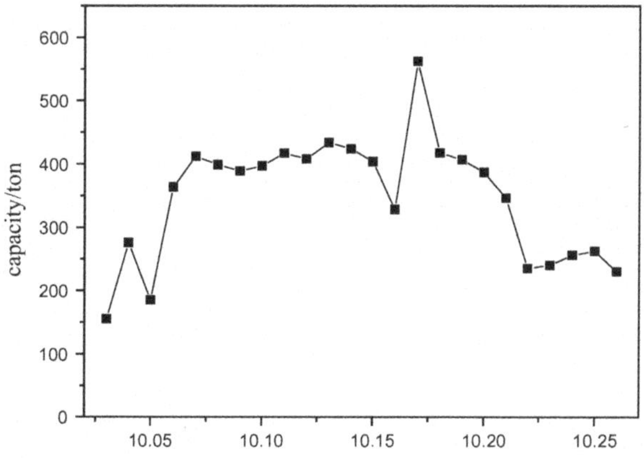

Fig. 9 The capacity of purification tower based on CWDT

process based on CWDT which can remove the coke powder completely, separate gas and tar clearly and produce steam by recover the surplus heat contained in the crude gas.

The main innovation of the new technology is to combine heavy oil washing, tar removal and heat transfer directly in a distillation tower. The specific process is shown in Fig. 10. The crude gas from the coke oven enters the bottom of the distillation tower and passes through the cheveron baffle to remove the majority of dust, then continues to rise through the tray and exchange heat with the high boiling point wash oil from the top, removes the remaining dust simultaneously. The heated wash oil is extracted from the tower through the side-draw, and cooled by the heat exchange system outside the tower, then it circulates back to the top of the rectification tower and continues to wash the crude gas.

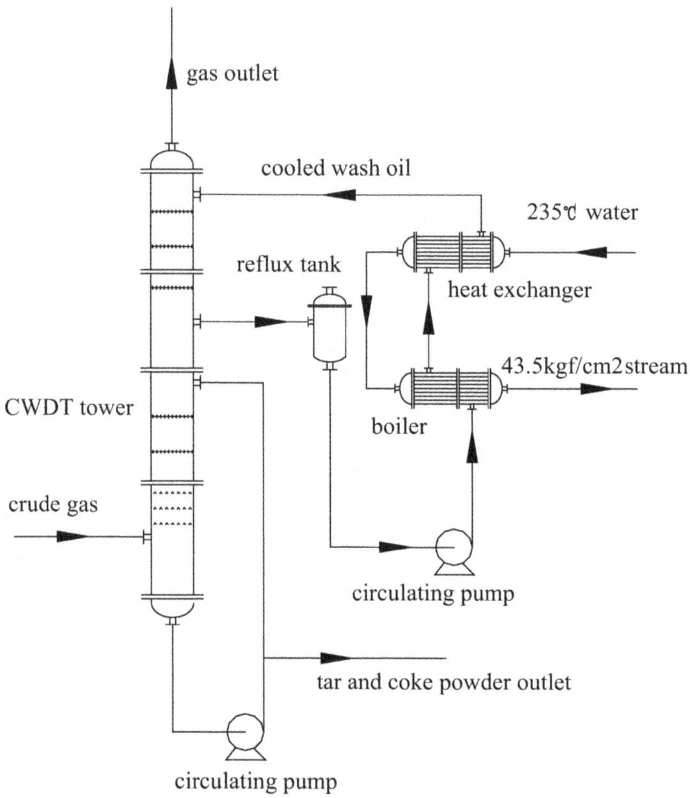

Fig. 10 The crude gas purification and heat recovery process based on the CWDT

Compared with the traditional process, the new process greatly simplifies the original ammonia water spraying process, saves water and eliminates the production and subsequent treatment of huge waste water. Meanwhile, through the side-draw in the middle section of the tower, the high boiling point wash oil can produce 43.5 kgf/cm^2 steam after heat transfer by the exchanger. By calculation, 10 kN m^3/h crude gas can generate 6.3 t/h steam of 43.5 kgf/cm^2.

5 Conclusions

(1) In this paper, a CWDT was developed by coupling various tower internals with MBT.

(2) Experiment results show that the average diameter of bubbles decreases to 1–3 mm, pressure drop increases by 56% in some areas and gas holdup increases by 4.29% when MBT sieve tray is installed. It illustrates that MBT tray can significantly improve the gas-liquid mass transfer effect, but will lead to increased energy consumption, so it is only suitable for the special location of the tower.

(3) The combined technology has been applied to the synthesis of organosilicon monomer with operational flexibility between 50 and 200%,and both the washing and distillation effect meets the design requirements.

(4) The application of the CWDT in the heat recovery process of crude gas can greatly simplify the ammonia spraying process, significantly reduce the water consumption and eliminate the generation of waste water and subsequent treatment, and simultaneously produce high-pressure steam via side-draw to recover huge energy.

Acknowledgements. The project is supported by the transformation project of scientific and technological achievements of Qingdao (no. 16-6-2-50-nsh).

References

1. Cheng, H., Yu, G.C.: The present status of distillation technology and its trend to development. Chem. Eng. **29**(1), 52–55 (2001)
2. Cheng, H., Zhou, M., Yu, G.C.: Recent development of enhancement of gas-liquid mass transfer. Prog. Chem. **13**(4), 315–322 (2001)
3. Qin, L., Jin, H.B., Yang, S.H.: Effect of sieve plates structure on hydrodynamic parameters in multistage bubble column reactor. Chem. Eng. **38**(9), 15–18 (2010)
4. Liu, Y., Liu, J.P., Miao, Y.L., et al.: Cause analysis and optimization of PVC dry tail gas scrubbing tower. China Chlor-Alkali **7**, 21–22 (2017)
5. Xu, M.: Modification on scrubber stray in coal gasification plant. Large Scale Nitrogenous Fertilizer Ind. **41**(2), 132–136 (2018)
6. Wang, X.B., Lu, Y.G., Liu, Z.Z.: Analysis on blockage problem of recycle oil heat exchanger in the delayed coking unit and improvement measures. Sci. Technol. Chem. Ind. **22**(3), 52–55 (2014)
7. Zhang, J.J., Wang, Y.F., Mao, X.J., et al.: Performance characteristics of two-phase direct contact condensation on dual-flow tray. Chem. Eng. **43**(8), 38–43 (2015)

8. Gao, Y.R., Liu, Y.G., Huang, J.: New drop-in cross-flow shower tray application. Nitrogenous Fertilizer Technol. **32**(2), 52–54 (2013)
9. Xu, J.N., Pi, Y., Li, Q.S., et al.: Study and application of highly efficient flow guided sieve tray and BHSII type packing. Chem. Ind. Eng. Prog. **24**(7), 800–804 (2005)
10. Li, Q.S., Guo, F., Zhang, W.L., et al.: Research and industrial application of the high efficiency flow-guided sieve tray. Chem. Ind. Eng. **32**(5), 1–7 (2015)
11. Tian, L.L., Wang, J.J., Gu, X.P., et al.: Advances in synthetic technology of organosilicon monomer. Mod. Chem. Ind. **24**(12), 23–26 (2004)
12. Duan, J.H., Wang, W.J., Fan, J.L., et al.: Simulation and optimization of organosilicon monomer separation processes. Chem. Eng. **40**(8), 69–73 (2012)
13. Li, J.L., Zhong, W.: Research progress of separation and purification of dimethyldichlorosilane. Chem. Ind. Eng. Prog., 89–94 (2011)
14. Wang, W.L., Wang, T.: Advances of synthetic methylchlorosilane technology in China. Silicone Mater. **22**(1), 1–5 (2008)
15. Tang, Z.G., Wen, Y.M.: Progress of distillation technology in coking. Chem. Ind. Eng. Prog. **28**(11), 1883–1889 (2009)
16. Sun, B.D., Xian, K.J., Li, M.Z.: Technology research of waste heat recovery and utilization. Energy Metall. Ind. **34**(5), 44–47 (2015)
17. Ma, T.Z., Jiang Z.Y.: Heat pipe research and development in China. In: Proceedings of the International Heat Pipe Conference, Tsukuba (2002)
18. Zhang, Z., Yu, H.L., Yang, D.W.: Analysis on problems of deposited graphite in ascension pipe of coke oven during coke-oven gas heat recovery. Clean Coal Technol. **18**(1), 79–81 (2012)

Validity of Using Fatigue-Based Span Limits in Wave Dominated Environments

Funke Elizabeth Dacosta-Salu and Oludare Jeremiah(✉)

Feddo Pty. Ltd., Perth, Australia
{funke.dacosta-salu, dare.Jeremiah}@feddogroup.com

Abstract. A subsea pipeline has a unique role to produce oil and gas from an offshore petroleum field, connecting a petroleum facility at the open sea and a near shore terminal at the coast. Pipelines are usually installed on the seabed which have uneven surface. Free spans or suspended spans normally occur in subsea pipelines due to the irregularity of seabed and by scouring phenomena around the installed non-buried pipeline. The span could be as a result of scour, obstruction on the seabed, undulating terrain and topography, sand waves and natural profile. The maximum allowable span is usually considered when design is being carried out. The allowable free span length based on input data and constants given by DNVGL F105 were obtained and compared with the maximum allowable free span length for the response model. The result obtained for the in line and cross flow was 20 m while the maximum allowable span length was 78.2 m. A Mathcad spreadsheet was developed for this calculation and this can be used as an analytical tool for calculating free span. The author would like to compare the results obtained using this spreadsheet with other standard free span softwares in the future.

Keywords: Span · FLS · ULS · VIV · MAFSL

1 Introduction

The successful extraction of fluid from wells cannot be completed without the fluid passing through pipelines from seabed to production facility for purification purpose and then transported to end users. These pipes are laid on seabed which may have

---

J. Lin (ed.), *Proceedings of the International Petroleum and Petrochemical Technology Conference 2019*, pp. 331–342, 2020.
https://doi.org/10.1007/978-981-15-0860-8_28

undulating slope leading to free span. In many deep-water pipeline projects, and also in intermediate water in harsh environments, free spans are one of the major challenges. The complex combination of uneven/erodible seabed and near-seabed hydrodynamics potentially leads to very expensive span intervention works. Free span is referred to as the unsupported region of the pipeline. It can also be expressed as a situation where there is a lost in contact between the pipeline and the seabed over a significant distance

Fig. 1 Free spanning pipeline on an uneven seabed [2]

[1]. Figure 1 shows a free span session in a pipeline.

Span can lead to pipeline deformation or crack [1] due to repeated vibration around the welded area.

Over the free spans the sea current can cause Vortex Induced Vibrations (VIV) as a result of the hydrodynamic sea current loads. VIV forces have an oscillatory nature and they can cause vertical and horizontal oscillations of free span portions of the pipeline [3]. This VIV will ultimately result in fatigue failure [1].

The action of wave and current on the unsupported weight of the pipeline can be catastrophic if not properly designed.

Hence, free span pipeline design is done to ensure the pipeline's integrity is maintained against permanent load as a result of the seabed functional load generated by internal pressure and temperature and dynamic loads from direct wave and current [4]. Due to the presence of current and wave action, free span pipelines are generally subjected to dynamic loading which may lead to fatigue. Bending and sagging of the pipeline will also be experienced at the span region [4].

Knowledge about free span and its mitigation is essentially important when it is absolutely difficult to prevent span as seabed have different topography. The nature of the slope of the seabed can be natural; it can be as a result of scour or landslide due to erosion or other similar climate disaster during operation. Seabed without span at installation can develop span during pipeline operation. An example of this is when a pipeline rest on a sandy seabed in shallow to intermediate water [4]. This scenario can pose a problem to the free span integrity management as a result of the change in free span configuration during the operating life of the pipeline [4].

The cost of making seabed corrections and free span intervention is quite substantial [5]. A way of reducing the cost is generating a reliable method of analyzing fatigue is essential [5]. Free span can also constitute a great danger to human health when the pipeline is pulled by the fishing trawl during fishing.

Typically, the maximum allowable free span length (MAFSL) is usually calculated based on the information obtained during the geotechnical and geophysical survey. In the event that the allowable free span length is exceeded, a correction means of reducing the span is employed. This corrective means include the use of trenching, rock dumping [6]. This is a very expensive means and hence an accurate span evaluation is important. It is desired that the free span should be maintained at allowable length at installation and during operation or throughout the design life. A number of tools have been used to calculate MAFSL. This includes the DNV FAT FREE, SAGE Profile, VIVANA, SHEAR 7, ANAPIPE-VIV, ORCAFLEX and ABACUS.

These softwares are used to determine what portion of the pipeline will exceed the maximum span and may likely fail due to this. The determination of the technique to correct span is therefore proffered. The methods of correcting span are the use of concrete mattress, grout bags, rock dumping and the use of sand bags.

Vortex is usually formed in regions where span is experienced which are as a result of the action of wave and current. Vortices sheet are generated around a pipe or cylindrical shape as a result of the flow of wave and current around the object [6]. This can lead to resonance or vibration which is normally called Vortex induced Vibration. Fatigue is a common phenomenon experienced here. Figure 1 shows pipeline vortex shedding. The welded part constitutes an area where stress is concentrated. Hence this could lead to failure. The fatigue limit state is the approach used when considering the effect of vortex shedding-induced vibration and direct wave action in pipeline span. The vibration can be inline or cross flow based on the direction of vibration with respect to the axis of the pipe. Figure 2 shows the inline and cross flow directionality of vibration.

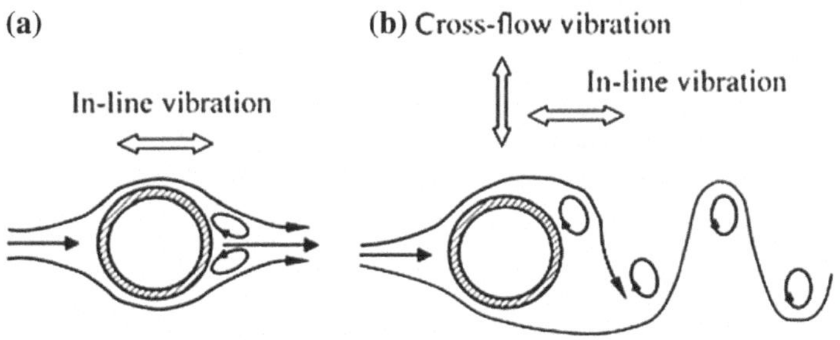

Fig. 2 Vortices and vibrations of free span section: **a** symmetrical vortices and in-line vibration; **b** asymmetrical vortices and in-line and cross-flow vibrations [6, 7]

The significance of this study is to show the reasons for using FLS for the estimation of MAFSL in a wave dominated environment as against the use of ULS or other methods. It will also show how MAFSL can be done using hand calculation. A MathCAD Spreadsheet will be developed to enable users who do not have access to the expensive softwares be able to do the calculations.

According to DNVGL [8], the fatigue Limit State determines the fatigue life of a pipeline while the design using Ultimate Limit State determines extreme stress in the pipeline in order to prevent buckling.

The scope of the report is the confirmation of the FLS in a wave dominated environment and this will be done with the aid of MathCAD software. DNVGL standard was used for the spreadsheet development.

Some works have been done in the past on the numerical analysis of free span in pipe but none has evaluated free span in a wave dominated environment. Some of these literatures are reviewed.

The work by Project Consulting Services, Inc. [9] developed a CAM for the calculation of MAFSL in pipeline in the static and dynamic state. The result of this was compared to the result obtained from CAESAR II. It was concluded that the results obtained from CAESAR II falls within the range of the values of CAM given the same parameters. This calculation did not take the effect of FSL in a wave dominated environment into consideration.

The presentation given by Taufik [1] on Pipeline Free Span Analysis made calculations in-order to determine the critical span length. It is expected that for pipeline design, the span length should not exceed the critical length. Although this gives a simplified formula for the calculation of critical span length, the calculation was focused on the effect of current on the pipeline.

The work by Jishang et al. [10] on the calculation of maximum allowable free span length and safety assessment of the DF1-1 submarine pipeline. An investigation of the DF1-1 was done. The free span length was determined for a current dominated environment.

Sollund et al. [5], made an investigation on wave induced damage in multi span pipeline and a way of improving the analysis was given. A time domain algorithm was developed in the paper which helped to study the dynamic response of direct wave loading on multi span pipelines. The report was able to give an improved frequency domain that helps to predicts multi-mode response. The report shows that the frequency domain can give an accurate fatigue life estimate of multi span pipelines.

This present research work aims to discuss the behavior of free span pipelines using fatigue-based span limits in wave dominated environments.

2 Methodology

Calculation was carried out for a pipeline in shallow water. A subsea field is considered shallow if the water depth is about 200 m [6]. DNVGL is the guideline that was used to develop this spread sheet. The water depth used for this calculation is 95 m.

The parameters used for this analysis are given in the tables below as obtained from [10] but modified for the purpose of the analysis (Tables 1, 2).

Table 1 Flowline parameters

S. No.	Parameters description	Value
1	Pipe outer diameter	0.5588 m
2	Pipe inner diameter	0.527 m
3	Overall diameter including concrete coating	0.70 m
4	Pipe wall thickness	0.0159 m
5	Concrete coating thickness	0.060 m
6	Marine growth thickness	0.0106 m
7	Density of concrete coating	3040 kg/m^3
8	Density of steel	11391 kg/m^3
9	Density of pipeline content	57 kg/m3
10	Density of seawater	1020 kg/m3
11	Kinematic viscosity of seawater	1.05×10^{-6} m^2s^{-1}
12	Young modulus	2.07×10^{11} Pa
13	Young modulus of concrete	25×10^9 Pa
14	Specified minimum yield strength of the pipe	360 MPa
15	Maximum allowable operating pressure	10 MPa
16	Gap between pipeline and seafloor	0.1932 m
17	Acceleration due to gravity	9.81 m/s^2
18	Linear thermal expansion coefficient	1.16×10^{-5} K^{-1}
19	Minimum water depth	73 m
20	Seawater density	1.02×10^3 kg/m^3
21	Significant wave height	6 m
22	Maximum wave height	11.3 m
23	Peak period	12.4 s
24	Current velocity	0.8 m/s
25	At reference point	1 m
26	Mean current velocity at pipe height	0.8 m/s
27	Given wave velocity at pipe height	1.0 m/s
28	Relative angle between the pipeline direction and the current flow direction	90°
29	Relative angle between the pipeline direction and the wave flow direction	90°
30	Wave spreading parameter	8
31	Spectral peakedness	3.3
32	Pipe roughness	1/20 m
33	Trench depth	0 m
34	Design temperature	60 K
35	Ambient seawater temperature	20 K
36	Design pressure	13.79 MPa

Table 2 Constants [8]

1	Seabed roughness parameter	5×10^{-6} m
2	Soil type (soft)	5
3	Safety factor for inline	1.4
4	Safety factor for cross flow	1.4
5	Boundary condition (for single span) tip	3
6	Pipe ends. Ends	1
7	Safety zone	1
8	Load condition for stress check (operation)	3

2.1 Input Design Parameters

The calculation involves the screening of the span length(s) which is done first to determine the maximum span along the pipeline. This is followed by determining if the fatigue life is satisfied or not. This is thereafter followed by knowing if the stresses are susceptible to buckling or not. Figure 3 shows the main aspects of free span, the main parameters required and the major outcomes.

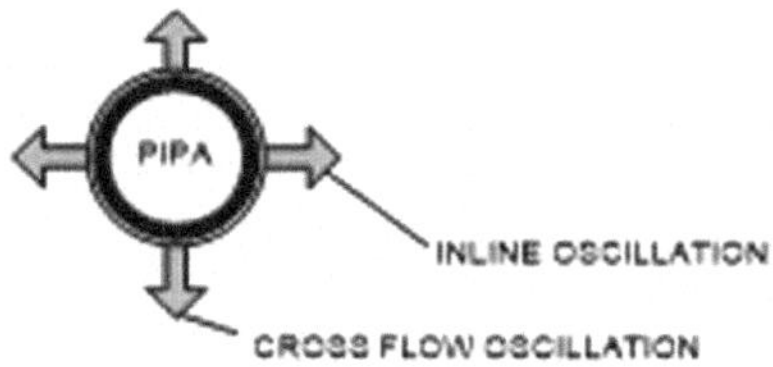

Fig. 3 Inline and cross flow vibration [1]

The steps followed in this report is to enable ease of understanding and ability to achieve the required result which is the allowable free span length. The first step is to know the flow regime using the current flow velocity ratio and identify the flow regime based on the classification.

The flow regime is determined by calculating the current flow velocity ratio given by Eq. (1)

$$\alpha = Uc/(Uc + Uw) \tag{1}$$

where U_c is the current velocity normal to the pipe, U_W is the significant wave induced velocity amplitude.

$0.5 < \alpha < 0.8$ wave dominant and current is superimposed by wave [8] (Fig. 4).

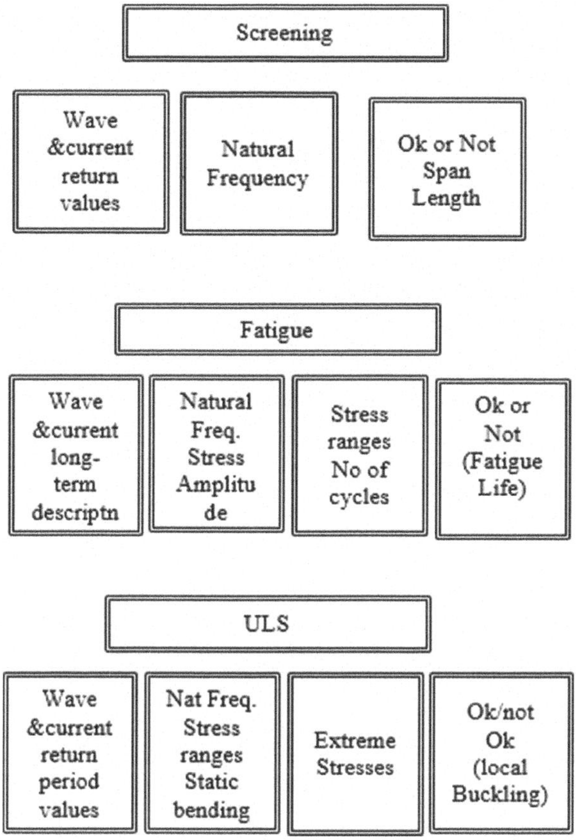

Fig. 4 Main aspect of free span, key parameters and main results [8]

The submerged weight was calculated using Eq. (2)

$$\mathrm{Wd} + \mathrm{Wg} + \mathrm{Wmg} - \mathrm{Ww} \tag{2}$$

where Wd is the dry pipe weight, Wg is the content weight, Wmg is the marine growth weight and Ww is the weight displaced by water.

The value was gotten to be 2261 kg/s^2.

The specific gravity was calculated using the total dry pipe weight divided by the weight displaced by the water volume. This was obtained to be 1.587.

The maximum wave period was calculated based on the input using the formula given below. This is important if the peak period has not been given in the parameters. The calculation may be omitted if the necessary information is made available in the input data.

The generalized Philip's constant, Frequency, Spectral width parameter, density function, frequency transfer function, and wave induced velocity spectrum values were gotten based in the Eqs. (3), (4), (5), (6), (7), and (8) respectively. These can be found in Paragraph 3.3.3, 3.3.5 and 3.36 [8]. The values obtained were used to calculate the spectral moment and near bed significant wave velocity and zero up crossing period with Eqs. (9), (10), and (11).

$$\alpha := \frac{5}{16} \cdot \frac{H_{.S}^2 \cdot \omega_{.p}^4}{g1^2} \cdot (1 - 0.287 \cdot ln(\gamma)) \tag{3}$$

where α is the Philip's constant, ω_p is the angular spectral peak frequency with the formula given by Eq. (4), $H_{.s}$ is the significant wave height, g1 is the gravitational acceleration, γ is the peak enhancement factor

$$\omega := \frac{2 \cdot \pi}{T_{.p}} \tag{4}$$

$T_{.p}$ is denoted as peak period [8].

$$\sigma(\omega) := \begin{vmatrix} 0.07 & \text{if } \omega \leq \omega_{.p} \\ 0.09 & \text{otherwise} \end{vmatrix} \tag{5}$$

where $\sigma(\omega)$ is the spectral width parameter [8].

$$S_{.\eta\eta}(\omega) := \alpha \cdot g1^2 \cdot \omega^{-5} \cdot e^{\frac{-5}{4} \cdot \left(\frac{\omega}{\omega_{.p}}\right)^{-4}} \cdot \gamma^{e^{-0.5 \cdot \left(\frac{\omega - \omega_{.p}}{\sigma(\omega) \cdot \omega_{.p}}\right)^2}} \tag{6}$$

where $S_{.\eta\eta}(\omega)$ is the density function [8].

$$G(\omega) := \begin{vmatrix} \frac{\omega \cdot \cosh kk(\omega) \cdot (FOD + g_{.p})}{\sinh(kk(\omega) \cdot d)} & \text{if } \omega < \omega_{.max} \\ 0 & \text{otherwise} \end{vmatrix} \tag{7}$$

where $G(\omega)$ is the frequency transfer function from sea surface elevation to wave-induced flow velocities at pipe [8], $g_{.p}$ is the gap between the pipeline and the seafloor, FOD is the final outer diameter of the pipe and $kk(\omega)$ is the wave number.

$$S_{.UU}(\omega) := G(\omega)^2 \cdot S_{.\eta\eta}(\omega) \tag{8}$$

where $S_{UU}(\omega)$ is the wave-induced spectrum at the pipe level [5].

$$Mn(n) := \int_0^{\omega_{.max}} \omega^n \cdot S_{.UU}(\omega) d\omega \tag{9}$$

where Mn(n) is the spectral moment of order n [8].

$$U_{.S} := 2 \cdot \sqrt{Mn(0)} \cdot \frac{m}{sec} \quad (10)$$

where $U_{.s}$ is the significant flow velocity at pipe level [8].

$$T_{.u} := 2 \cdot \pi \cdot \sqrt{\frac{Mn(0)}{Mn(2)}} \cdot sec \quad (11)$$

where $T_{.u}$ is the mean zero up crossing period of oscillating flow at pipe level [8].

Reduction factor was calculated based on the formula in Eq. (12) and can be seen in paragraph 3.4.3 [8]. The result from the reduction factor was used to obtain the wave velocity.

$$R_{.D} := \sqrt{\int_{\frac{-\pi}{2}}^{\frac{\pi}{2}} w(\beta) \cdot \sin(\theta_{.w} - \beta)^2 d\beta} \quad (12)$$

where w(β) is the directional energy spreading function and θ_w is the main wave direction.

The current velocity was calculated based on the formula in Eq. (13) and can be seen in paragraph 3.4.6 [8].

$$U_{.W} := U_{.S} \cdot R_{.D} \quad (13)$$

U_s is the near bed significant wave velocity [8].

The Euler buckling length (using the traditional method) was calculated based on Eq. (14). This was done after the effective axial load, actual load in pipe wall, actual stress in pipe wall, hoop stress and Euler constant have all been determined.

$$L_{.Eulert} := \begin{vmatrix} \left(\sqrt{\frac{EI_{.R} \cdot b}{N_{.eff}}} \cdot \pi\right) & \text{if } N_{.eff} < 0N \\ \infty \cdot m & \text{otherwise} \end{vmatrix} \quad (14)$$

where N_{eff} is the effective axial load, EI_R is the moment of inertial (resistance), and b is the Euler constant.

The Euler length for inline and cross flow were obtained from the formula given in Eqs. (15) and (16).

$$Leulerin(Tip) := root\left[N_{.eff} - \left(\frac{Pein(L)}{C_{.2}(Tip)}\right), L\right] \quad (15)$$

$$\mathrm{Leulercf(Tip)} := \mathrm{root}\left[\mathrm{N}_{.\mathrm{eff}} - \left(\frac{\mathrm{Pecf(L)}}{\mathrm{C}_{.2}(\mathrm{Tip})}\right), \mathrm{L}\right] \tag{16}$$

where Pein(L) is the Euler buckler load for inline direction, Pecf(L) is the Euler buckler load for cross flow direction, $C_{.2}$(Tip) is the boundary condition coefficient and L-Span Length.

Reduced Velocity for in line and cross flow were determined using Eqs. (17) and (18) which were obtained from paragraphs 4.3.5 in [8].

$$\mathrm{V}_{.\mathrm{Rdin}}(\mathrm{U}_{.\mathrm{c}}, \mathrm{U}_{.\mathrm{W}}, \mathrm{L}) := \frac{\mathrm{U}_{.\mathrm{W}} + \mathrm{U}_{.\mathrm{c}}}{\mathrm{f}_{.\mathrm{o\_in}}(\mathrm{L}) \cdot \mathrm{FOD}} \cdot \gamma_{.\mathrm{f}} \tag{17}$$

where U_w is the wave velocity, U_c is the current velocity, fo_in(L) is the natural frequency for in line and γ_f is the frequency factor

$$\mathrm{V}_{.\mathrm{Rdcf}}(\mathrm{U}_{.\mathrm{c}}, \mathrm{U}_{.\mathrm{W}}, \mathrm{T}_{.\mathrm{u}}, \mathrm{L}) := \frac{\mathrm{U}_{.\mathrm{W}} + \mathrm{U}_{.\mathrm{c}}}{(\mathrm{f}_{.\mathrm{o\_cfr}}(\mathrm{L})) \cdot \mathrm{FOD}} \cdot \gamma_{.\mathrm{f}} \tag{18}$$

where fo_cfr(L) is the natural frequency for cross low.

The result obtained from the above was used to determine the maximum allowable span length for the inline and cross flow directions.

The maximum free span length in the inline direction was thereafter obtained using Eq. (19) and the result was compared with the threshold current flow velocity value in the flow regime α.tr = 0.5

$$\mathrm{L}_{.\mathrm{inF105}}(\mathrm{U}_{.\mathrm{c}}, \mathrm{U}_{.\mathrm{w}}) := \left| \begin{array}{ll} \mathrm{L}_{.\mathrm{max}}(\mathrm{U}_{.\mathrm{c}}, \mathrm{U}_{.\mathrm{w}}) & \text{if } \alpha(\mathrm{U}_{.\mathrm{c}}, \mathrm{U}_{.\mathrm{w}}) \geq \alpha_{.\mathrm{tr}} \\ \left[\left(\frac{\infty}{10}\right)\mathrm{m}\right] & \text{otherwise} \end{array} \right. \tag{19}$$

where L_{max} is the maximum span and $\alpha(U_c,U_w)$ is the current flow velocity.

An important calculation that helped to determine if the analysis falls within the wave criterion of the response model is given in Eq. (20) This can be seen in paragraph 2.3.5 [8].

$$\mathrm{Wave\_criterion} := \left[\frac{\mathrm{U}_{.\mathrm{c\_100year}}}{(\mathrm{U}_{.\mathrm{w\_1year}} + \mathrm{U}_{.\mathrm{c\_100year}})}\right] \tag{20}$$

where Uc_100 year is the current velocity at 100 years return period and Uw_1 year is the wave velocity at 1 year return period.

After all calculations have been made, the results were obtained and comparison is done to know if the span is within the allowable limit.

3 Results

The current flow ratio, α, gave a value of 0.522 which means the pipe is in a wave dominated environment and the current is superimposed by wave [8]. Hence the effect of wave is more significant in the pipe than current. This is usually predominant in shallow waters.

The result of the maximum allowable free span was obtained using Euler equations and equations obtained from DNVGL F105 [8].

The results for the Euler buckling equations are given in Table 3. The maximum free span length was 38.02 m in the cross-flow direction. The inline and traditional method of calculating Euler buckling length gave the values of 37.65 m and 33.07 m respectively.

Table 3 Calculated span lengths results using Euler buckling

	Euler buckling lengths
Method	Length (m)
Traditional	33.07
Inline	37.65
Cross flow	38.02

The results of the Maximum calculated free span for the input parameters given above in the inline and cross flow directions according to DNVGL F105 [8] was 20 m. This was compared with the maximum allowable free span length for the model which gave a value of 78.2 m as shown in Table 4.

Table 4 Validity and comparison of maximum span lengths

	Span length	
	Max calculated span length (m)	Maximum allowable span length for model (m)
Inline	20	78.2
Cross flow	20	78.2

4 Conclusion

The reason for using fatigue limit span in a wave dominated environment has been established. An analytical procedure for determining the maximum allowable span length was determined based on the input data and the constants from DNVGL F105. The result showed that the maximum allowable span length is not exceeded in comparison with the response model.

A MathCad spreadsheet was developed for this analysis and can be used for subsequent analysis in conditions where the commercial softwares are too expensive for the design engineers.

In the nearest future, the author would like to verify the results obtained by the spread sheet with the standard softwares to know if there is any variation.

Acknowledgements. My profound appreciation goes to the CEO of FEDDO Integrated Service, Engr Dare Jeremiah for his words of support, assistance and encouragement towards the completion of this research paper. I would also like to appreciate my colleagues for their cooperation, support and understanding towards the successful completion of this research paper.

References

1. Taufik, A.: Pipeline Free Span Analysis. Lecture Note. https://www.scribd.com/presentation/317113760/Docfoc-com-249591587-Class-10-Free-Span-Analysis-pptpp. Last accessed 12 Nov 2018
2. Abeele, A.V., Boel, F., Hill, M.: Fatigue analysis of free spanning pipelines subjected to vortex induced vibrations. In: Proceedings of the 32nd International Conference on Ocean, Offshore and Arctic Engineering, OMAE2013-10625, France (2013)
3. Marcus, et al: A comparative study of a free span pipeline through numerical simulations. In: Proceedings of the ASME 2014 33rd International Conference on Ocean, Offshore and Arctic Engineering OMAE2014-24069, USA (2014)
4. Drago, M., Mattioli, M., Bruschi, R., Vitali, L.: Insights on the design of free-spanning pipelines. In: Philosophical Transactions of the Royal Society A 373: 201401111. Royal society, Publishing, Italy. http://dx.doi.org/10.1098/rsta.2014.0111 (2015)
5. Sollund, H., Fyrileiv, O., Vedeld, K., Hellesland, J.: Improved assessments of wave-induced fatigue for free spanning pipelines. App. Ocean Res. Res. Gate (2016)
6. Bai, Y., Bai, Q.: Subsea Pipelines and Risers, 1st edn. Elsevier, UK (2005)
7. Xiao, Z.G., Zhao, X.L.: Prediction of natural frequency of free spanning subsea pipelines. Int. J. Steel Struct. 81–89 (2010)
8. DNVGL RP F105. Free Spanning Pipeline. June (2017)
9. Project Consulting Services, Inc.: Analysis and Assessment of Unsupported Subsea Pipeline Spans. https://www.scribd.com/doc/52047688/analysis-and-assessment-of-unsupported-subsea-pipeline-spans. Last accessed 30 Dec 2019
10. Jishang X., et al.: Calculation of maximum allowable free span length and safety assessment of the DF1-1 submarine pipeline. J. Ocean Univ. China, Res. Gate (2010)

Sealing Capability Evaluation and Study of Operation Optimization of One Natural Gas Storage in North China

Xue-jiao Zhang(✉) and Xu-qing Ma

Beijing Gas Group Research Institute, Beijing, China
1978156377@qq.com

Abstract. A gas storage is usually an artificial gas field or gas reservoir formed by re-injecting natural gas which is from the long-distance pipeline into the underground space. A gas storage is generally built in the vicinity of a city. A certain gas storage of this paper which is situated in North China is an important basis and guarantee for relieving tension of using natural gas for citizens and industries of cities in winter in North China. In this paper, the effectiveness of the gas storage is evaluated by studying the geological conditions of the gas storage, the period of fault activity, the relative relationship between the two faults, the sealing coefficient of the fault and the closure of the cap. Combined with the two conditions on the demand and the transportation of natural gas, the gas injection time and production time are determined together with injection and production balance time of the gas storage. After that, four different gas storage operation schemes are achieved whose peak regulation intensity is 1.2:1, 1.4:1, 1.6:1, and 1.8:1 respectively. Through the software simulation calculation comprehensively, many indicators are obtained such as the design operation gas volume time, liquid production volume, the number of required wells, the average gas production volume in the stage of gas production peak and so on. Also, the optimal operation plan of the gas storage is established. The above research results show that the gas storage has good sealing properties and is a good choice for building the gas storage. The optimized operation scheme can ensure the stable and orderly operation of the gas storage and meet the design requirements of peaking regulation. The construction of the gas storage is of great significance for maintaining national energy security and improving the imbalance between natural gas supply and consumption.

This paper was prepared for presentation at the 2019 International Petroleum and Petrochemical Technology Conference in Beijing, China, 27–29, March, 2019.

J. Lin (ed.), *Proceedings of the International Petroleum and Petrochemical Technology Conference 2019*, pp. 343–350, 2020.
https://doi.org/10.1007/978-981-15-0860-8_29

Keywords: Gas storage · Sealing capability · Operation scheme

1 Introduction

Underground gas storage can better solve the problem of urban gas non-uniformity and play a role in seasonal peaking regulation. When an accident such as a sudden major natural disaster or a pipeline leak occurs on the gas mains, the gas supply will be interrupted. At this time, the gas storage can also be used as an emergency reserved gas source. This can greatly improve the reliability of the gas supply. Underground gas storage has large gas storage capacity, safe and reliable, and has been widely adopted by countries all over the world. At present, there are more than 600 natural gas underground gas storages in use around the world, and the working gas capacity exceeds 300 billion square meters. This ratio is equivalent to 10% of the global annual gas consumption. The use of depleted gas reservoirs as gas storage is the most widely used gas storage method.

The history of underground gas storage can be traced back to the beginning of the last century. In 1915, the United States conducted gas storage experiments in Ontario. At present, the number of natural gas underground gas storages owned by the United States ranks first in the world. In the United States, most natural gas underground gas storage is a type of depleted gas (oil). The gas storage reservoirs constructed with water layers are mainly distributed in the central and western regions, while the gas storage reservoirs constructed with salt caves are mainly distributed along the Gulf Coast. Some salt caverns have also been developed in the northeast, midwest and south-western regions of the United States. The idea of using abandoned mines as natural gas storage has also been tried, but industrial applications are still being tested. The construction of underground gas storage in Russia started late, but its development is very fast. Underground gas storage in Western Europe, including France, is also relatively developed. In particular, France, Germany, Italy and the United Kingdom have more than 25% of the available gas.

2 Sealing Capability Evaluation of Gas Storage

2.1 Geological Conditions of Gas Storage

The main target layer of the gas storage reservoir has a thickness of more than 150 m, mainly composed of coarse sandstone containing gravel, and the reservoir contains a small amount of fine-silt sandstone. The deposits in the target layer are dominated by Tertiary sediments. The target layer forms a rift basin under the stress of the extension zone. The structure is slower, and the dip angle of the formation is 3°–7°. The target layer has a depth of about 1010 m, a structural width of about 200 m, and a trap area of about 1.5 km^2. The plane is superimposed and contiguous, and there is a uniform oil-water interface between the fault blocks, which is the bottom-water reservoir of the gas-top oil ring.

2.2 Analysis of the Fault Sealing of Gas Storage

Fault sealing refers to the nature of the fracture zone and the fault upper and lower disk rocks that prevent the fluid from continuing to flow due to the difference in displacement pressure. The mechanism of fault closure is capillary closure. Fault closure has the characteristics of lateral closure and vertical closure. Fault sealing is an important basis for the site selection of underground gas storage.

The main target layer structure of the gas storage is relatively intact and is surrounded by faults. The fault of this structure ceases to act, which acts as a sealing effect on the oil and gas water system. Through the seismic section, it can be found that the fault does not extend into the Neogene strata. During the drilling process, there was no such phenomenon as drilling during the drilling. When the fault was drilled, the drilling fluid did not leak. Based on the above reasons, it can be considered that the fault judges that the fault is closed in the vertical direction.

The gas storage is surrounded by 4 faults. The two discs of the four faults are sandstone and mudstone, which can prevent fluid flow. The drilling test results show that the faults have different oil-water interface and pressure system. This indicates that the two sides of the fault belong to different oil and gas water systems and pressure systems. This is also the good sealing of the fault.

A plurality of faults is developed inside the target layer, and the faults divide the target layer into several small regions. The faults of these faults are small. The lithology on both sides of the fault is sandstone with good connectivity. According to the test results, these faults all have a uniform oil-water interface, and the original formation pressure is the same. During gas well production, the pressure changes in these areas are only close. Therefore, it can be considered that the fault inside the target layer does not have a seal.

The closure of the section can be evaluated by calculating the ratio of the thickness of each mudstone layer in the section to the distance of the fracture. This method is simple, efficient, accurate and practical. Using this method to quantitatively calculate the fault, the results show that the value of the boundary fault is greater than 85%. The fault in the target layer is less than 30%. This also proves that the outer fault of the target layer belongs to the closed fault, while the fault seal inside the target layer is weak, and the flow resistance of the fluid in the inner fault is small.

In the early stage of the operation, gas storage pressure monitoring was carried out to evaluate the sealing of the fault. The pressure of the well A1 outside the gas storage tank was kept near the initial pressure value, and the bottom hole pressure did not change during the gas injection process. The pressure monitoring results inside the gas storage tank show that the pressure fluctuation of the gas storage in the gas storage and storage balance period is small. The injected gas is slowly gas-filled, and the fault inside the target layer is not closed, and the boundary fault is well sealed during the injection and production process.

2.3 Cap Formation Sealing Capability

The sealing mechanism of the cap layer is physical closure, which means that the capillary force of the mud cap layer is used to prevent the migration of oil and gas.

Capillary forces are related to pore throat radius, hydrocarbon properties, and media temperature conditions. A large set of mudstone is developed above the gas reservoir of the gas storage. Mudstones are thicker than 50 m and are distributed throughout the area. The mudstone has a porosity of 6.2–7.6% and a permeability of 0.45–1.2 mD. The mudstone has a fine particle size and a median particle size of no more than 15 μ. Mudstone has the characteristics of stable distribution, large thickness and pure lithology. This mudstone is highly impermeable and separates the gas layer from its overlying water layer. The thickness of the cover layer is much larger than the breakage of the outer fault, which ensures the capping ability of the cover layer.

3 Gas Storage Operation Plan Optimization

3.1 Gas Storage and Gas Production

At the beginning of gas injection, the pressure of the gas storage is low. As the amount of gas injection increases, the pressure in the gas storage reservoir continues to rise. The gas injection process is shown in Fig. 1. During the gas injection in the gas storage, the natural gas transported by the natural gas pipeline is transferred to the injection and production station of the gas storage tank after being distributed by the distribution station. The injection and production station measure, separates, filters and pressurizes the natural gas. After that, natural gas is injected into the gas well through the gas tree.

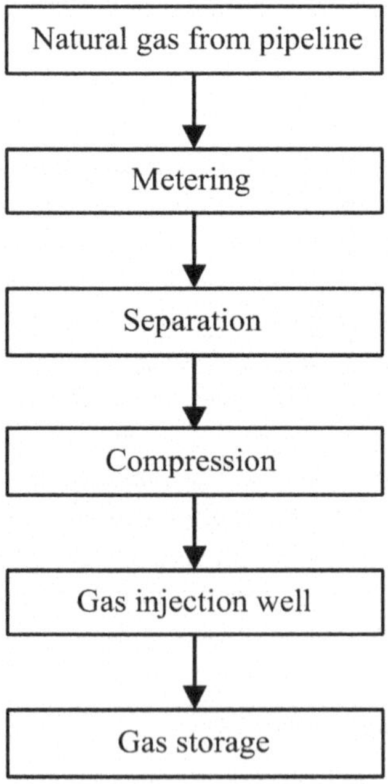

Fig. 1 Gas injection process

During the gas storage in the gas storage, the natural gas is produced by the gas well. After separation, dehydration, metering, etc., it is transported through the gas pipeline and then enters the distribution station. The gas production process is shown in Fig. 2.

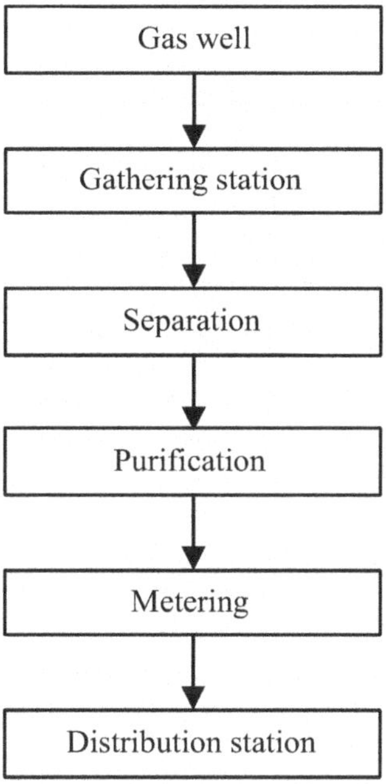

Fig. 2 Gas production process

3.2 Determination of Operating Parameters of Gas Storage

The gas storage is rebuilt from abandoned oil and gas reservoirs. The formation of the gas storage is seriously deficient and the formation energy is low. The gas storage has been unable to meet the pressure of the pipeline. In order to convert the reservoir into an underground gas storage, it is necessary to inject gas first, and then gas, in order to achieve the purpose of peaking gas.

According to the characteristics of gas consumption in North China, combined with the change in gas demand, the operation cycle of gas storage is determined. The gas storage is used for gas production in the winter and gas injection in the summer. At the end of the gas production period and the end of the gas injection period, a balance period should be reserved. Therefore, the design gas production period is 100 days, the gas injection period is 240 days, and the equilibrium period is 25 days. The specific arrangements are shown in Table 1.

Table 1 Gas storage runtime

S. No.	Status	Time	Days/d
1	Gas injection	February 22nd–October 19th	240
2	Gas production	November 1st–February 8th	100
3	Balance	February 9th–February 21st October 20th–October 30th	25

3.3 Gas Storage Operation Plan Optimization

According to the design principle of the gas storage plan, under the same operating environment, the working gas volume is 3.6×10^8 m^3 as the constraint condition. According to the peak intensity of the gas storage, three sets of schemes were designed. The peak-to-valley ratios of the peaks were 1.2:1, 1.4:1, 1.6:1, and 1.8:1, respectively. The peak period of the peak design of the scheme was December of the annual gas production period. From the 20th to the 10th of January, a total of 21 days, the rest of the time is a relatively low production and stable production period, a total of 79 days.

The production results of the 4 operating schemes can be obtained by software simulation calculation, as shown in Table 2.

Table 2 Operation of the four schemes

Plan number	1	2	3	4
Peaking ratio	1.2:1	1.4:1	1.6:1	1.8:1
Low-yield gas production/$m^3 \times d^{-1}$	260	240	236	210
High yield gas production/$m^3 \times d^{-1}$	312	336	377	378
Maximum daily gas production/$m^3 \times d^{-1}$	340	362	384	397
Number of injection wells	11	10	8	12
Number of cycles	8	4	5	6
Water production	more	less	less	more

The peaking regulation ratio of Option 2 is 1.4:1. When operating according to this scheme, the gas storage will reach the design working capacity after 4 cycles. Ten injection wells are needed. The highest daily gas production is 362 $m^3 \times d^{-1}$. The average daily gas production during the peak period is 336 $m^3 \times d^{-1}$. The average daily gas production during the low peak period is 240 $m^3 \times d^{-1}$. Gas wells produce less water. Considering the above factors, it is recommended that the gas storage be operated according to the plan with a peak ratio of 1.4:1.

4 Conclusions

(1) By analyzing the active period of the fault, the relationship between the two plates of the fault, the sealing coefficient of the fault, and other parameters, it is confirmed that the sealing of the gas reservoir boundary fault is good.

(2) Many faults inside the gas storage are not closed, which indicates that the inside of the gas storage is a complete gas storage system.
(3) The upper layer of the gas storage tank is covered with a huge mudstone layer. The mudstone caprock has a large thickness and a stable plane distribution, and has good sealing performance.
(4) By comparison, when the peaking ratio is 1.4:1, the gas storage is in good condition.

Acknowledgements. I would like to express my gratitude to all people who have contributed to this article, especially to the leaders and colleagues of Beijing Gas Group Research Institute.

References

1. Lv, K.: Feasibility study on underground gas storage reservoir in Jilin oilfield. China University of Petroleum (East China) (2015)
2. Guo, Z.: Study on trap effectiveness evaluation of reconstructing underground gas storage in block Lei61. Petrochem. Ind. Technol. **25**(12), 86 (2018)
3. S, Yao: Technologies of gas injection and production in Wen96 underground gas storage. Corros. Prot. Petrochem. Ind. **35**(04), 26–29 (2018)
4. Berlendis, S., Lascourreges, J.F., Schraauwers, B., et al.: Anaerobic biodegradation of BTEX by original bacterial communities from an underground gas storage aquifer. Environ. Sci. Technol. **44**(9), 3621 (2010)
5. Ma, S., Zhou, H.C., Zhou, H.C.: Gas storage in porous metal-organic frameworks for clean energy applications. Chem. Commun. **46**(1), 44–53 (2010)
6. Chen, Q., Luo, M., Hammershøj, P., et al.: Microporous polycarbazole with high specific surface area for gas storage and separation. J. Am. Chem. Soc. **134**(14), 6084–6087 (2012)
7. Lu, W., Yuan, D., Zhao, D., et al.: Porous polymer networks: synthesis, porosity, and applications in gas storage/separation. Chem. Mater. **22**(21), 5964–5972 (2010)
8. Ross, D.J.K., Bustin, R.M.: The importance of shale composition and pore structure upon gas storage potential of shale gas reservoirs. Mar. Pet. Geol. **26**(6), 0–927 (2009)
9. Farha, O.K., Yazaydın, A.Ö., Eryazici, I., et al.: De novo synthesis of a metal–organic framework material featuring ultrahigh surface area and gas storage capacities. Nat. Chem. **2**(11), 944–948 (2010)
10. Slatt, R.M., O'Brien, N.R.: Pore types in the barnett and woodford gas shales: contribution to understanding gas storage and migration pathways in fine-grained rocks. AAPG Bull. **95**(12), 2017–2030 (2011)
11. Biloé, S., Goetz, V., Guillot, A.: Optimal design of an activated carbon for an adsorbed natural gas storage system. Carbon **40**(8), 1295–1308 (2002)
12. Rabbani, M.G., Sekizkardes, A.K., Kahveci, Z., et al.: A 2D mesoporous imine-linked covalent organic framework for high pressure gas storage applications. Chem. Eur. J. **19**(10), 3324–3328 (2013)
13. Lai, G., Margot, F., Secomandi, N.: An approximate dynamic programming approach to benchmark practice-based heuristics for natural gas storage valuation. Oper. Res. **58**(3), 564–582 (2010)
14. Rabbani, M.G., El-Kaderi, H.M.: Synthesis and characterization of porous benzimidazole-linked polymers and their performance in small gas storage and selective uptake. Chem. Mater. **24**(8), 1511–1517 (2012)

15. Enright, G.D., Udachin, K.A., Moudrakovski, I.L., et al.: Thermally programmable gas storage and release in single crystals of an organic van der Waals host. J. Am. Chem. Soc. **125**(33), 9896–9897 (2003)
16. Chang, K.J., Talu, O.: Behavior and performance of adsorptive natural gas storage cylinders during discharge. Appl. Therm. Eng. **16**(5), 359–374 (1996)
17. Yang, W., Greenaway, A., Lin, X., et al.: Exceptional thermal stability in a supramolecular organic framework: porosity and gas storage. J. Am. Chem. Soc. **132**(41), 14457–14469 (2010)
18. Xiang, Z., Cao, D., Lan, J., et al.: Multiscale simulation and modelling of adsorptive processes for energy gas storage and carbon dioxide capture in porous coordination frameworks. Energy Environ. Sci. **3**(10), 1469–1487 (2010)
19. Etminan, S.R., Javadpour, F., Maini, B.B., et al.: Measurement of gas storage processes in shale and of the molecular diffusion coefficient in kerogen. Int. J. Coal Geol. **123**(1148), 10–19 (2014)
20. Oku, T., Narita, I.: Calculation of H_2 gas storage for boron nitride and carbon nanotubes studied from the cluster calculation. Phys. B Condens. Matter **323**(1), 216–218 (2002)
21. Gallo, M., Glossmanmitnik, D.: Fuel gas storage and separations by metal–organic frameworks: simulated adsorption isotherms for H_2 and CH_4 and their equimolar mixture. J. Phys. Chem. C **113**(16), 6634–6642 (2009)
22. Qiu, Y., Deng, H., Yang, S., et al.: Syntheses, crystal structures, and gas storage studies in new three-dimensional 5-aminoisophthalate praseodymium polymeric complexes. Inorg. Chem. **48**(9), 3976 (2009)
23. Comotti, A., Bracco, S., Valsesia, P., et al.: 2D multinuclear NMR, hyperpolarized xenon and gas storage in organosilica nanochannels with crystalline order in the walls. J. Am. Chem. Soc. **129**(27), 8566–8576 (2007)
24. Najibi, H., Chapoy, A., Tohidi, B.: Methane/natural gas storage and delivered capacity for activated carbons in dry and wet conditions. Fuel **87**(1), 7–13 (2008)
25. Teatini, P., Castelletto, N., Ferronato, M., et al.: Geomechanical response to seasonal gas storage in depleted reservoirs: a case study in the Po River basin, Italy. J. Geophys. Res. **116** (F2), 490–500 (2011)
26. Arab, P., Rabbani, M.G., Sekizkardes, A.K., et al.: Copper(I)-catalyzed synthesis of nanoporous azo-linked polymers: impact of textural properties on gas storage and selective carbon dioxide capture. Chem. Mater. **26**(3), 1385–1392 (2014)

Analysis and Countermeasures for the Integrity Damage Factors of Underground Gas Pipeline in a Certain City of North China

Xu-qing Ma and Xue-jiao Zhang(✉)

Beijing Gas Group Research Institute, Beijing, China
1978156377@qq.com

Abstract. Underground gas pipeline integrity means that the pipeline is always in a fully reliable service state. This paper analyzes the damage of the integrity of underground gas pipelines in a city in North China in recent years, and summarizes the main causes of the integrity damage of gas pipelines. Pipeline corrosion, private modification of pipelines, and third-party damage are the main causes of the integrity damage of the gas pipeline network in a city in North China. Among them, illegal construction of some units and construction of municipal infrastructure are the main sources of third-party damage. In order to effectively protect the integrity of urban underground gas pipelines, this paper has formulated countermeasures based on the actual situation of the city. Combined with the actual situation of a city in North China, the DC potential gradient method and the dense interval potential method were used for corrosion detection. Once the pipeline is found to be corrosive, repairment is carried out immediately. In response to third-party damage issues, the urban management and other law enforcement agencies will be joint to timely supervise and guide the high-risk construction of the third-party site such as drilling in an orderly and safe way. Through supervision, hidden trouble on the pipelines occupation caused by the third-party construction units will be eliminated. Pipeline corrosion inspection should be conducted regularly to check whether the pipeline is aging. By strengthening daily supervision, it can effectively reduce pipeline

J. Lin (ed.), *Proceedings of the International Petroleum and Petrochemical Technology Conference 2019*, pp. 351–357, 2020.
https://doi.org/10.1007/978-981-15-0860-8_30

damage caused by pipeline aging and other reasons. The research in this paper provides reference and guidance for the management and maintenance of underground gas pipeline integrity.

Keywords: Gas pipeline integrity · Integrity damage · Countermeasures

1 Introduction

The underground gas pipeline of a city in North China consists of 3 trunk lines and 126 branch lines with a total length of 1624.55 km[1–5]. The pipeline has an annual gas transmission capacity of 0.8 billion cubic meters per year.

Underground gas pipeline integrity means that underground gas pipelines are always in a fully reliable service state[6–8]. As shown in Fig. 1, the integrity of underground gas pipelines includes three levels of meaning. (1) The connectivity of the pipeline is complete throughout the life of the pipeline. (2) During the whole life cycle of pipeline operation, the functions such as opening and closing of pipeline valves are always in a controllable state. (3) The ability to handle various accidents throughout the life cycle of pipeline operation[9–13].

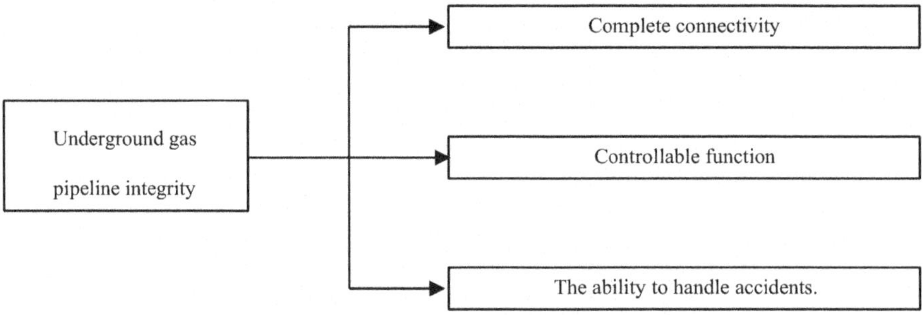

Fig. 1 The meaning of underground pipeline integrity

Conducting pipeline integrity studies can identify and evaluate the risks faced in pipeline operations and develop appropriate risk control strategies. During the pipeline operation, the safety level of pipeline operation can be continuously improved, and the risk level of pipeline operation is always controlled within a reasonable range [14–18].

2 Analysis of Integrity Damage of Underground Pipelines in a Certain Area of North China

2.1 Pipeline Damage Statistics

As shown in Table 1, the various forms of damage to the gas pipeline network in a city in North China have been counted in recent years. It can be seen from the table that pipeline corrosion, privately modified pipeline, and third-party damage are the main causes of the integrity damage of the gas pipeline network in a city in North China. In addition, the design is not scientific and reasonable, the construction quality is not up to standard, etc., which will lead to the destruction of the integrity of the gas pipeline network. However, the above reasons include many situations and a low proportion.

Table 1 Causes of pipe network integrity damage

No.	Kind	Proportion/%
1	Pipeline corrosion	32
2	Privately modified pipeline	27
3	Third party destruction	29
4	Other	12

2.2 Privately Modified Pipeline

Some residents have a weak sense of safety in using gas. In the process of home decoration, gas pipelines are sometimes arbitrarily changed according to individual wishes. In order to save money, some residents employ units that do not have relevant qualification certificates and capabilities to carry out gas facility changes. After the completion of construction, neither the pressure acceptance nor the relevant departments are required to conduct the test. This kind of privately altered pipeline often does not meet the standards set by the state, and there are high risks. A gas leak accident caused by private demolition has occurred in a city in North China. Fortunately, it was disposed of in time without causing serious consequences.

2.3 Third Party Damage

Due to the lack of legal and regulatory and related knowledge, some third-party construction units lack awareness of protecting urban gas pipeline networks. These units did not even apply for construction permits, nor did they coordinate with the gas pipeline network protection department for blind construction. According to the statistics on the damage of gas pipeline network in a certain district of Beijing, drilling, excavation and other operations are the main construction methods for third-party construction units to destroy urban gas pipeline networks. This type of construction has the characteristics of short construction period and high liquidity, which is difficult to prevent in time. In order to create a beautiful and clean living environment for the public, water supply, heat and other pipelines are mostly arranged underground. In recent years, the city has gradually increased the intensity of building subway facilities. A large number of subway lines are built every year. These municipal infrastructure construction and construction processes inevitably increase the risk of gas pipeline damage. According to the statistics of gas pipeline network damage, the main reasons for the destruction of gas pipeline network in municipal infrastructure construction include two aspects. First, the actual position of the gas pipeline does not match the drawing. The second is the pipeline pressure. When constructing near the cultural protection unit, some gas pipelines adjusted their orientation and location according to the site conditions, but did not update the construction drawings in time. In some construction sites, in order to facilitate on-site construction, sand and gravel are illegally piled up above the pipeline, causing ground subsidence and causing pipeline damage.

2.4 Pipeline Corrosion

Metal corrosion refers to changes between the metal material and the surrounding environment, and causes damage to the metal itself and the surrounding environment. Corrosion of metallic materials is a spontaneous, inevitable gradual process. Corrosion of the steel pipe body will seriously reduce the strength, pressure bearing capacity and reliability of the pipe. This will also shorten the service life of the steel pipe, increase the safety risk of pipeline operation and increase maintenance costs. Pipeline corrosion also shortens maintenance and pipe replacement cycles, threatening the safety of the entire gas distribution system.

The use of gas in a city in North China is earlier, some pipe networks have been built for a long time, and the anti-corrosion insulation layer outside the metal pipelines is damaged. Some greenbelt trees are closer to the gas pipeline. As the trees grow, the roots of some plants penetrate the pipeline anti-corrosion layer, which increases the corrosion of the pipeline. Natural precipitation and green watering make these pipelines in a humid environment all year round, and the corrosion rate is faster than other pipelines.

3 Countermeasures to Strengthen the Integrity Management of Underground Pipelines in a Certain Urban Area of North China

3.1 Corrosion Detection and Treatment

There are several methods for detecting the anti-corrosion layer of buried steel pipe. There are multi-frequency tube current method, Pearson method, DC potential gradient method and dense interval potential method. Because different types of detection methods have different application ranges, it is especially important to choose the correct detection method. Combined with the actual situation of a city in North China, the DC potential gradient method and the dense interval potential method are generally used for corrosion detection in the application process. The dense interval potential method can evaluate the effectiveness of the protection method and determine the influence range of the stray current to detect the damage of the anti-corrosion layer. This method is not suitable for the detection of pipe corrosion in frozen soil layers. The detection effect in the rock area or rock backfill is general. This method requires an impressed current to provide a cathodic protection system. The DC potential gradient method provides a more accurate positioning. This method can effectively determine the location and damage of the damaged layer. This method also analyzes the corrosive activity of the metal at the break. Similar to the dense interval potential method, the DC potential gradient method is also not suitable for use in frozen soil layers.

3.2 Third Party Damage Control

Analysis of the gas pipeline network damage in a city in North China, in many cases, the responsible subject lacks the knowledge and safety awareness of safe gas, and does not recognize the potential danger of their behavior. It is recommended that the joint community neighborhood committee and the Municipal Emergency Management Bureau and other units strengthen the promotion of gas safety knowledge and popularize relevant laws and regulations. By enhancing the safety awareness of residents and construction units, the probability of gas pipeline damage accidents is reduced from the source.

Before the construction of various projects, actively coordinate with the departments of water affairs and heat, optimize the construction plan, and reduce the possibility of damage to the gas pipeline network. During the construction process, joint law enforcement departments such as urban management will supervise and guide the site, eliminate potential safety hazards such as pipeline pressure, and guide high-risk construction such as drilling. Do a good job in emergency disposal plan. Once there is danger, you should find it early and deal with it early. After the completion of the construction, timely check the gas pipelines that may be affected nearby to ensure that no danger occurs.

3.3 Pipe Network Maintenance

Ensure the safe operation of the city gas pipeline network, with emphasis on protection. Regularly carry out pipeline corrosion inspection, check whether the household gas equipment is aging, privately modify, verify the actual position of the gas pipeline and the corresponding situation of the drawings, which can effectively reduce the pipeline damage caused by corrosion, aging, illegal construction and other reasons. In the maintenance process of gas pipelines, it is necessary to set and improve the safety warning signs in time so that individuals or construction units can contact the gas pipeline transportation management department to deal with emergencies in a timely manner.

4 Conclusions

(1) Through statistical analysis of the damage of the integrity of underground gas pipelines in a city in North China in recent years, it is found that pipeline corrosion, private modification of pipelines, and third-party damage are the main reasons for the integrity of the pipeline network.
(2) Combined with the actual situation of a city in North China, the DC potential gradient method and the dense interval potential method are often used for corrosion detection.
(3) Through the joint law enforcement department, supervise and guide the site on site, eliminate potential safety hazards such as pipeline pressure, and guide high-risk construction such as drilling.
(4) By regularly conducting pipeline corrosion inspection, it can effectively reduce pipeline integrity damage caused by corrosion, aging and other reasons.

Acknowledgements. I would like to express my gratitude to all people who have contributed to this article, especially to the leaders and colleagues of Beijing Gas Group Research Institute.

References

1. Zhi, T.L., Zhi, L.W., Dong, J.L.: Study on low-strain integrity testing of pipe-pile using the elastodynamic finite integration technique. Int. J. Numer. Anal. Meth. Geomech. **37**(5), 536–550 (2013)
2. Song X.: Evaluation and integrity management corrosion protection for buried steel pipe. Harbin Institute of Technology (2016)
3. Zhou T.: Integrity management of city gas pipeline. Chongqing University (2009)
4. Jin T.: Research on the Integrity Management of the Underground Gas Pipeline. South China University of Technology (2013)
5. Wols, B.A., Daal, K.V., Thienen, P.V.: Effects of climate change on drinking water distribution network integrity: predicting pipe failure resulting from differential soil settlement. Procedia Eng. **70**, 1726–1734 (2014)
6. Lee, J.H., Kim, D.H.: Integrity evaluation of pipe welding zones using wavelet transforms, and specific sensitivities based on SH-EMAT pulse-echo method. Int. J. Precis. Eng. Manuf. **15**(10), 2051–2057 (2014)
7. Ray, P., Srinivasan, B., Balasubramaniam, K., et al.: Monitoring pipe wall integrity using fiber Bragg grating-based sensing of low-frequency guided ultrasonic waves. Ultrasonics **90**, 120 (2018)
8. Brown, M.J.P., Moore, I.D., Fam, A.: Performance of a cured-in-place pressure pipe liner passing through a pipe section without structural integrity. Tunn. Undergr. Space Technol. Incorporating Trenchless Technol. Res. **42**(5), 87–95 (2014)
9. Yeom, K.J., Kim, W.S., Oh, K.H.: Integrity assessment of API X70 pipe with corroded girth and seam welds via numerical simulation and burst test experiments. Eng. Fail. Anal. **70**, 375–386 (2016)
10. Guidara, M.A., Bouaziz, M.A., Schmitt, C., et al.: Structural integrity assessment of defected high density poly-ethylene pipe: burst test and finite element analysis based on J-integral criterion. Eng. Fail. Anal. **57**, 282–295 (2015)

11. Baker, J.: Application of the threat matrix to improving the efficiency of risk assessments for the integrity management of subsea pipeline systems. Underw. Technol. **34**(3), 129–134 (2017)
12. Černý, I., Mikulová, D., Sís, J.: Examples of actual defects in high pressure pipelines and probabilistic assessment of residual life for different types of pipeline steels. Procedia Struct. Integrity **7**, 431–437 (2017)
13. Meriem-Benziane, M., Abdul-Wahab, S.A., Zahloul, H., et al.: Finite element analysis of the integrity of an API X65 pipeline with a longitudinal crack repaired with single- and double-bonded composites. Compos. B Eng. **77**, 431–439 (2015)
14. Rathod, D.W., Pandey, S., Singh, P.K., et al.: Microstructure-dependent fracture toughness (J IC) variations in dissimilar pipe welds for pressure vessel system of nuclear plants. J. Nucl. Mater. **493**, 412–425 (2017)
15. Pearson, M.J., Pipe, A.G., Mitchinson, B., et al.: Implementing spiking neural networks for real-time signal-processing and control applications: a model-validated FPGA approach. IEEE Trans. Neural Netw **18**(5), 1472 (2007)
16. Rakin, M., Medjo, B., Arsić, M., et al.: API J55 steel casing pipe with an initial surface crack under internal pressure—determination of fracture parameters. Key Eng. Mater. **488–489**, 577–580 (2012)
17. Norintan Farina, H., Hafiz, A., Safwan, A. et al.: An innovative approach for tubing integrity monitoring with the application of electromagnetic tool and caliper tool in Tandem (2015)
18. Tada, N., Nohara, M.: Evaluation of semi-ellipsoidal wall thinning on back surface of plate by direct-current potential difference method. Int. J. Str. Integrity **6**(6), 714–724 (2015)

Study on Collection of Weak Signals About Corrosion of Floor of Large Tank

Wen-cai Liu[1,2,3(✉)], Xi Chen[1], Zhi-ming Yuan[4], and Xu Sun[5]

[1] CNPC Research Institute of Safety & Environment Technology, Beijing 102206, China
wencailiu@cnpc.com.cn
[2] China University of Petroleum, Beijing 102249, China
[3] Xisha Tun West China Petroleum Technology Innovation Base, Room A-609, Shahe Town, Changping District, China
[4] China National Oil and Gas Exploration and Development Corporation, Beijing 100034, China
[5] Petrochina Xinjiang Marketing Company Akesu Branch, Xinjiang 843000, China

Abstract. Storage tank is one of the important oil and gas storage facilities, and tank floor leakage is one of the main causes of storage tank leakage, this issue has become the focus of attention of enterprises. Based on the results of long-term monitoring of sound emission of different oil storage tanks, the characteristics and corrosion laws of static and dynamic corrosion of oil storage tank bottom plate are obtained. This paper studies the secondary amplification technology and the detection method of the inner sensor of the tank to improve the reliability of weak signal recognition of corrosion on the bottom plate of large tank.

Keywords: Acoustic emission · Corrosion · Weak signals · Tank

1 Introduction

Traditionally, in acoustic emission (AE) technique for detecting corrosion of tank floor, sensors are installed on outer tank wall near the floor to collect acoustic signals. This technique is not good due to less and weak signals[1]. Therefore, two methods are

J. Lin (ed.), *Proceedings of the International Petroleum and Petrochemical Technology Conference 2019*, pp. 358–373, 2020.
https://doi.org/10.1007/978-981-15-0860-8_31

proposed to enhance intensity of AE signals about corrosion of tank floor and improve measurability. The first method is to develop secondary amplification technique for AE detection[2]; the second method is to develop and apply immersed system for collecting weak acoustic signals about corrosion of tank floor.

2 Experimental Study on Development and Application of Secondary Amplification Technique

Since signals about corrosion of tank floor are weak and nowadays large tanks are increasingly popular in China, it is difficult to detect central zone of tank floor [3]. When signals arrive at outer wall of tank, sensors receive signals with low amplitude. To enhance intensity of signals, secondary amplifier is developed. To investigate performance of secondary amplifier in collection of weak acoustic signals, a set of comparative tests are carried out on simulative tank and steel plate. Common amplifier related information is shown in Table 1. DP3I and R3α sensors are both common sensors for AE detection. They are of resonance type with best response at 30 kHz. For the reason of supply voltage, DP3I sensor with built-in front amplifier can connect secondary amplifier only and cannot connect 2/4/6 front amplifier; while R3α sensor can connect 2/4/6 front amplifier.

Table 1 Models and data of amplifiers

Model	Amplification factor	Band width (kHz)	Power supply
2/4/6 front amplifier	Adjustable at 3 levels: 20, 40, 60 dB	0–1200 plug-connected filter	From AE card
Secondary amplifier	20 dB	Pass band	From AE card

2.1 Experiment on Secondary Amplification Technique for Tank Detection

2.1.1 System and Procedure

An R3α sensor (No. 1 sensor) and a DP3I sensor (No. 2 sensor) are installed on simulative tank. No. 1 sensor connects 2/4/6 front amplifier, secondary amplifier and AE collection card in sequence; No. 2 sensor connects secondary amplifier and AE collection card in sequence. 40 dB gain is selected for 2/4/6 front amplifier; 20 dB gain is selected for secondary amplifier. The threshold is 35 dB as shown in Figs. 1 and 2. Lead is broken at perpendicular bisector of connecting line between No. 1 and No. 2 sensors. Amplitudes of signals received by No. 1 and No. 2 sensors are observed and recorded.

2.1.2 Results and Analysis

Amplitudes of signals received by No. 1 and No. 2 sensors are shown in Table 2. Two channels receive signals with almost same amplitudes. DP3I sensor is an integrated sensor with 40 dB gain in built-in amplifier and 20 dB gain in secondary amplifier, totally 60 dB gain. R3α sensor without built-in amplifier connects front amplifier with

Fig. 1 Experiment on secondary amplification of corrosion of simulative tank floor

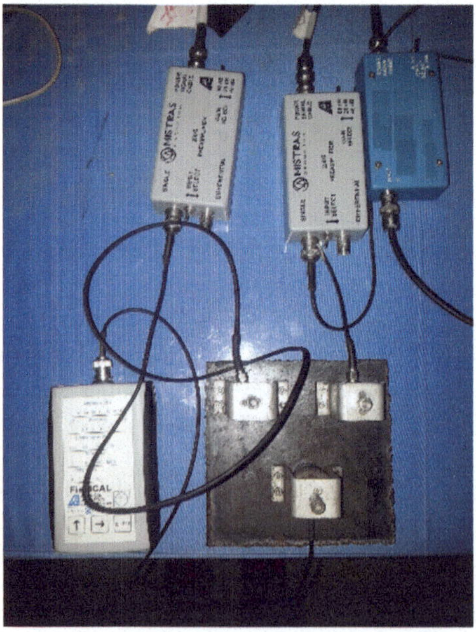

Fig. 2 Experiment on performance of secondary amplification

Table 2 Amplitudes of signals received in different channels

Channel	Amplitude of breaking lead signal/dB		
	1	2	3
1	97	96	94
2	97	95	95

40 dB gain and secondary amplifier with 20 dB gain, totally 60 dB gain too. Therefore, amplitudes of gains in Channel 1 and Channel 2 are same, and amplitudes of signals are almost same too.

2.2 Experiment on Performance of Secondary Amplification Technique

2.2.1 System and Procedure

In order to further define performance of secondary amplification technique, additional experiment is carried out on steel plate, i.e. common material of tank. Three R3α sensors are coupled on steel plate and are numbered as 1, 2, 3 respectively. No. 1 sensor connects 2/4/6 front sensor (set as 20 dB, 40 dB, 60 dB gain respectively) and AE collection card in sequence. No. 2 sensor connects 2/4/6 front amplifier with 40 dB gain, secondary amplifier with 20 dB gain and AE collection card in sequence. No. 3 sensor is placed at perpendicular bisector of connecting line between No. 1 and No. 2 sensors. It connects FieldCAL signal generator which emits simulative AE signals with different amplitudes. The threshold is 35 dB. Amplitudes of signals received by No. 1 and No. 2 sensors are observed and recorded.

2.2.2 Results and Analysis

Amplitudes of simulative AE signals and amplitudes of signals received by No. 1 and No. 2 sensors are shown in Table 3. It shows that for R3α sensor without built-in front amplifier, if it connects front amplifier with 40 dB gain which is generally used for tank detection and then connects a secondary amplifier, the performance is same as direct use of 2/4/6 front amplifier with 60 dB gain. Therefore, secondary amplifier is mainly designed for DP3I sensor with built-in sensor which is generally used for AE detection.

Table 3 Amplitudes of signals in different channels

Channel	Amplitude of signal/dB		
3 (simulative signal source)	40	60	80
1 (+20 dB)	\	39	59
1 (+40 dB)	40	59	79
1 (+60 dB)	59	79	99
2 (+40 +20 dB)	59	79	99

Although with secondary amplification technique, intensity of AE signal can be enhanced, the precondition is that AE sensors installed on outer wall of tank can receive the signal, i.e. when the signal about corrosion of tank floor arrives at the sensor, the amplitude of signal should be higher than the preset threshold. If when AE signal, especially signal from central zone of tank floor, arrives at sensor, its amplitude is attenuated below the threshold, the signal may still be lost. Therefore, this method has limitation. So study on immersed collection of weak signals about corrosion of tank floor is performed. Firstly, experimental feasibility study is carried out.

As an auxiliary analytic measure of AE detection technique, the technique for locating AE source of corrosion of tank floor can define activity of AE source so that it can somewhat predict detection results. However, due to complexity of sound speed, variety of signal types and limitation of threshold values etc., the accuracy of location may be affected, resulting in big error between detected defect position and actual defect position. Additionally, since large tank design becomes popular, if sound source of defect is present on tank floor, when it is transmitted through medium and detected by the sensor, due to large volume of tank, this process may cause increased attenuation of signal amplitude and loss of location events. Therefore, it cannot properly define activity of defects, affecting confidence of safety assessment on corrosion of tank floor. A new location method is urgently demanded to solve above-mentioned problem. This method can reduce location error as well as properly define activity of defects; it can combine time difference location and zone location for use in detecting small tank as well as meet zone location requirements for large tank. This chapter proposes a new method for location on tank floor to solve above-mentioned problem.

3 Location and Identification of AE Sources on Tank Floor

AE signals are mainly classified into two modes, namely continuous mode and burst mode. Continuous-mode signal location is mainly applied to locate leakage source. This section focuses on burst-mode signal source location. Burst-mode signal location includes two types, i.e. time difference location and zone location. Time difference location includes 1D (linear) location, 2D (planar) location and 3D location. Traditional tank floor sound source location is mainly planar arbitrary triangle location which is a 2D location method. Zone location mainly includes two types, i.e. separate channel control and location in order of signal arrival.

3.1 Planar Arbitrary Triangle Location

Three sensors are placed in same plane, namely 1# (x_1, y_1), 2# (x_2, y_2), 3# (x_3, y_3). As required, 1# sensor is located at origin of reference coordinate system. There is an emission source at $S(x, y)$ with distance r_1, r_2, r_3 from 1#, 2#, 3# sensors respectively. Location layout is shown in Fig. 3.

Additionally, t_1, t_2, t_3 are respectively the time to arrive at 1#, 2#, 3# sensor. So the difference between the distance from sound source S to 1# and the distance from S to 2#, 3# is:

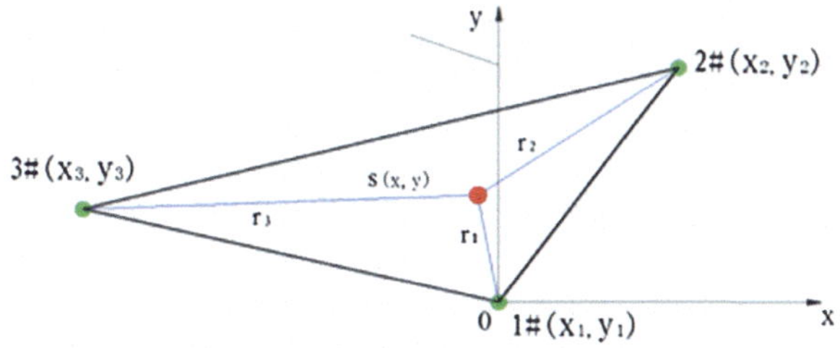

Fig. 3 Planar arbitrary triangle location

$$\delta_1 = \Delta t_{21} \cdot v = (t_2 - t_1) \cdot v \tag{1}$$

$$\delta_2 = \Delta t_{31} \cdot v = (t_3 - t_1) \cdot v \tag{2}$$

then

$$r_2 = r_1 + \delta_1 \tag{3}$$

$$r_3 = r_1 + \delta_2 \tag{4}$$

where v is sound speed, which depends on transmission medium; Δt_{21}, Δt_{31} are respectively the difference between the time for signal to arrive at 1#, 2# and at 1#, 3#. Then circles are drawn respectively with 1#, 2#, 3# as the center and r_1, r_2, r_3 as radius, the intersection point being S(x, y), forming the system of equation as below:

$$(x - x_1)^2 + (y - y_1)^2 = r_1^2 \tag{5}$$

$$(x - x_2)^2 + (y - y_2)^2 = r_2^2 \tag{6}$$

$$(x - x_3)^2 + (y - y_3)^2 = r_3^2 \tag{7}$$

Substitute Eq. (5) into Eqs. (6), (7) to get:

$$2(xx_2 + yy_2) = \left(x_2^2 + y_2^2 - \delta_1^2\right) - 2r_1\delta_1 \tag{8}$$

$$2(xx_3 + yy_3) = \left(x_3^2 + y_3^2 - \delta_2^2\right) - 2r_1\delta_2 \tag{9}$$

Make polar coordinate conversion for Eqs. (8) and (9); allow $x = r_1 \cos\theta, y = r_1 \sin\theta$ and $L_1 = x_2^2 + y_2^2 - \delta_1^2, L_2 = x_3^2 + y_3^2 - \delta_2^2$ to get:

$$2r_1(x_2 \cos\theta + y_2 \sin\theta + \delta_1) = L_1 \tag{10}$$

$$2r_1(x_3 \cos\theta + y_3 \sin\theta + \delta_2) = L_2 \tag{11}$$

When $x_2 \cos\theta + y_2 \sin\theta + \delta_1 \neq 0$, $x_3 \cos\theta + y_3 \sin\theta + \delta_2 \neq 0$, the following can be deduced from polar coordinate Eqs. (10), (11):

$$\begin{aligned} r_1 &= \frac{L_1}{2(x_2 \cos\theta + y_2 \sin\theta + \delta_1)} \\ &= \frac{L_2}{2(x_3 \cos\theta + y_3 \sin\theta + \delta_2)} \end{aligned} \tag{12}$$

Multiply two sides of the equation to get:

$$\begin{aligned} &(L_1x_3 - L_2x_2)\cos\theta + (L_1y_3 - L_2y_2)\sin\theta \\ &= L_2\delta_1 - L_1\delta_2 \end{aligned} \tag{13}$$

To facilitate cosine transform, allow

$$M = \left[(L_1x_3 - L_2x_2)^2 + (L_1y_3 - L_2y_2)^2\right]^{1/2} \tag{14}$$

Divide two sides of Eq. (13) by Eq. (14) to get:

$$\begin{aligned} &\frac{(L_1x_3 - L_2x_2)\cos\theta}{M} + \frac{(L_1y_3 - L_2y_2)\sin\theta}{M} \\ &= \frac{L_2\delta_1 - L_1\delta_2}{M} \end{aligned} \tag{15}$$

Make cosine transform for Eqs. (4)–(15); allow $\cos\varphi = \frac{L_1x_3 - L_2x_2}{M}$, $\sin\varphi = \frac{L_1y_3 - L_2y_2}{M}$; with previous assumptions of $x_2 \cos\theta + y_2 \sin\theta + \delta_1 \neq 0$, $x_3 \cos\theta + y_3 \sin\theta + \delta_2 \neq 0$, it is obvious that coefficients of $\cos\theta$ and $\sin\theta$ are both smaller than 1; based on cosine law, the form $\cos(\theta - \phi) = k$ can be used, where

$$k = \frac{L_2\delta_1 - L_1\delta_2}{M} \tag{16}$$

and

$$\tan\varphi = \frac{L_1y_3 - L_2y_2}{L_1x_3 - L_2x_2} \tag{17}$$

Because $\tan\phi$ can be calculated with the time difference Δt for AE signal to arrive and the transmission speed v measured at coordinates (x_2, y_2) and (x_3, y_3), i.e. positions of 2# and 3# sensors, angle φ can be determined within the range $[-\pi, \pi]$. With assumption that M is positive, there are two solutions for θ within $[-\pi, \pi]$. To obtain valid solutions, θ value must allow r to be positive. Then the position of sound source S can be determined.

3.2 Zone Location with Separate Channel Control

When amplitude of AE signal is attenuated excessively or number of AE channels is limited so that it is difficult to satisfy conditions for time difference location, zone location is selected to locate AE source. Since tank is characterized with large volume and long diameter of floor, when performing AE detection, amplitude of signal is attenuated excessively. Therefore, zone location should be used for AE detection on floor of large tank, while zone location in combination with time difference location is used for small tank.

Separate channel control means that depending of attenuation of signal, test piece is divided into several zones; a sensor is installed near the center of each zone; each sensor mainly receives AE wave within combined attenuation zone. It is not necessary to install sensors in certain array. Spacing between sensors can be large. But within detected zone, signal from sound source at any position can be received by at least one sensor. Figure 4 shows the principle of location. R is radius of detection zone of sensor. Radius length depends on the attribute of transmission of sound wave through different materials and the frequency range of sensor etc. Red zone is detected zone of tank floor.

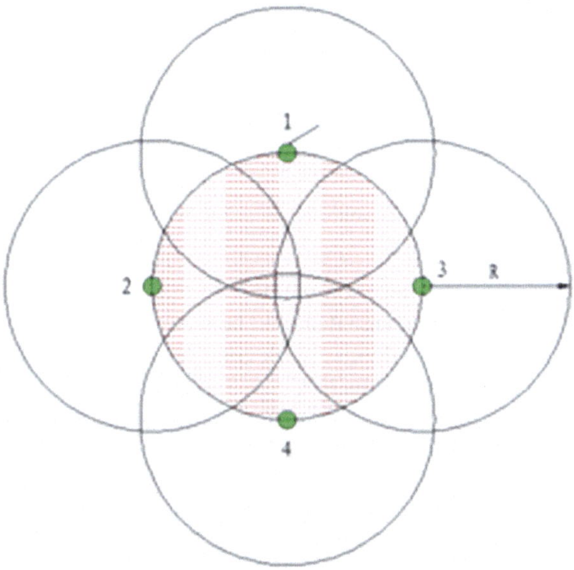

Fig. 4 Zone location with separate channel control

4 Immersed AE Monitoring on Corrosion of Tank Floor

In traditional AE detection on corrosion of tank floor, sensors are arranged as below: depending on size and type of tank, some sensors are arrayed evenly around tank wall near the floor; triangle time difference location is used. Figure 5 shows planar arbitrary triangle location on tank floor. Six sensors are evenly coupled around the tank. Any three sensors enable triangle location, forming multiple sets of triangle for location.

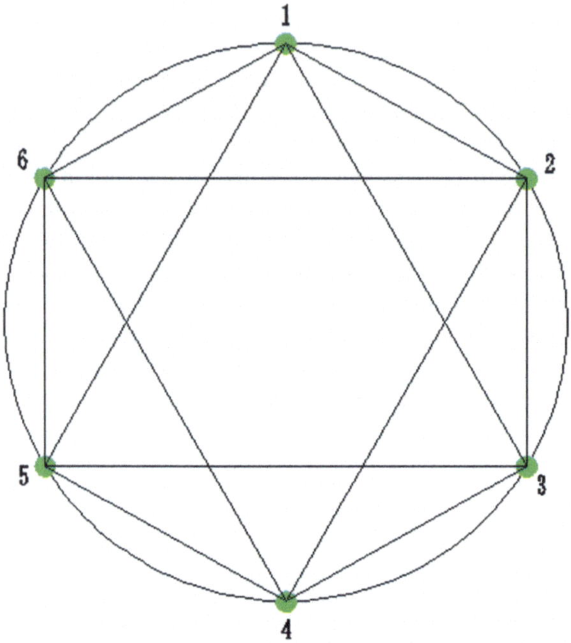

Fig. 5 Triangle location on tank floor

Although with conventional location method, multiple sets of triangle location can be formed, to locate an AE event, signal should be received by at least three sensors. As can be seen, central zone of tank floor is weak zone for monitoring. For tank with large volume, based on experimental results in Chap. 3, signals about corrosion of tank floor are weak with small amplitude. After attenuation in transmission process, signals about corrosion defect may be omitted. Moreover, study shows that AE signals from tank floor transmit in tank through three paths: first path: signals transmit directly through liquid in tank in form of spherical wave to AE sensors on tank wall, as shown in Fig. 6a; second path: signals transmit through tank floor in form of cylindrical wave for some distance; after they get to welding joint and the pattern changes, some signal energy transmits into liquid in tank, and finally transmits in liquid in form of spherical wave to AE sensors on tank wall, as shown in Fig. 6b; third path: signals transmit through tank floor in form of cylindrical wave to welding joint with tank wall and the pattern changes; some signal energy transmits to the wall, and finally transmits to AE sensors through the wall, as shown in Fig. 6c. Due to such multi-path effect and

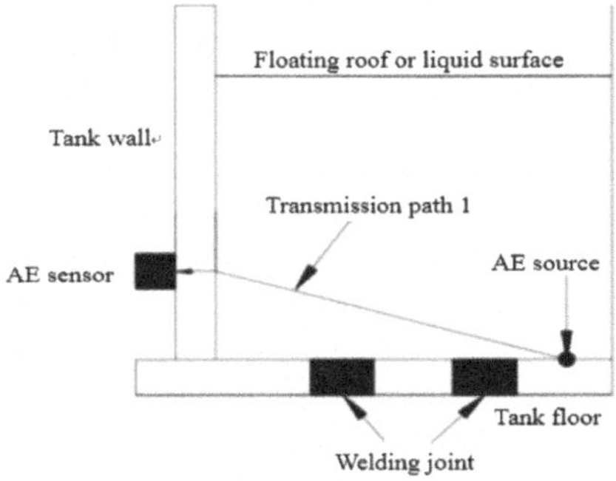

(a) Transmission path 1

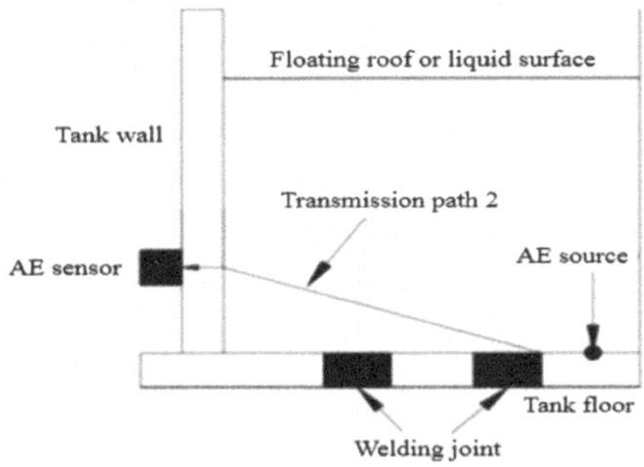

(b) Transmission path 2

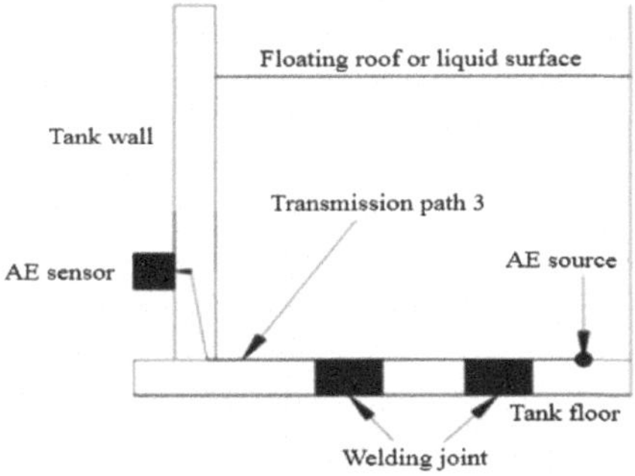

(c) Transmission path 3

Fig. 6 Transmission path for AE source on tank floor

attenuation and dispersion of sound wave in transmission process, when time difference location is used to reversely deduce position of sound source, it may cause big error in tank floor location. In addition, location is intended to identify AE sources of corrosion of tank floor, find zones with high activity and determine severity of corrosion, but not to determine exact position of each AE event.

Therefore, immersed AE monitoring on tank floor is proposed. In this technique, sensors are arrayed on central zone of tank; these sensors and those installed on outer tank wall jointly monitor corrosion of tank floor. This technique improves sensitivity of AE detection, accuracy of sound source identification and reliability of assessment. As shown in Figs. 7 and 8, when a sensor is placed on central zone of tank, attenuation of sound source S is halved. Based on actual operation of tank, two AE techniques for full monitoring on tank floor, i.e. immersed and embedded techniques, are proposed.

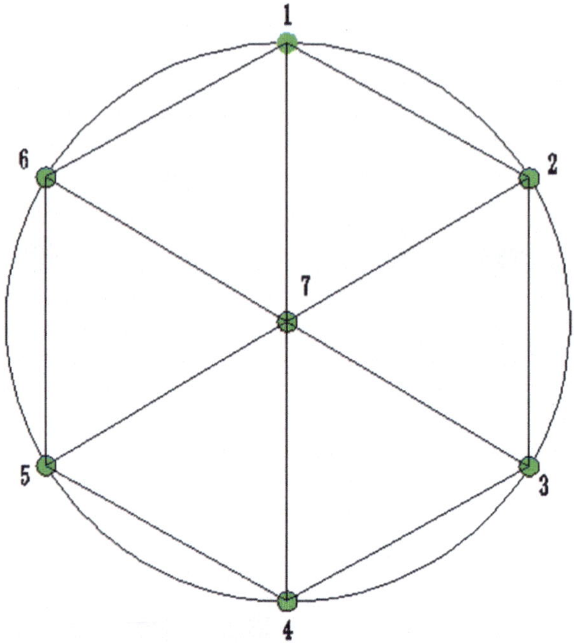

Fig. 7 Triangle location with one sensor on center of tank floor

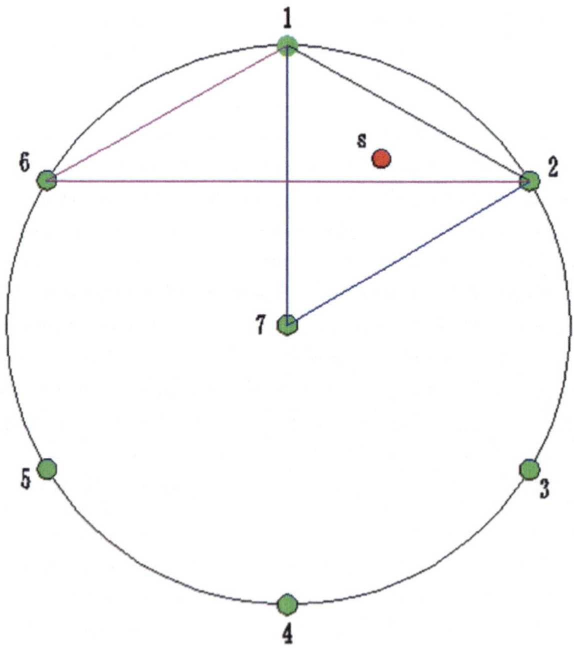

Fig. 8 Comparison of sound source location at position S on tank floor

5 Immersed AE Full Monitoring on Corrosion of Tank Floor

For an existing tank, to place AE sensors at central zone of tank floor, they should be put into the tank. So, immersed AE full monitoring technique is used.

5.1 Transmission Attribute

5.1.1 Attenuation Experiment Protocol

In this experiment, attenuation of breaking lead AE wave through water in tank and through metal tank wall is measured to figure out attenuation regularity and make analysis and comparison. In the experiment, signal of breading lead with hardness 2H (0.5 mm) is used. Lead is broken near 1# sensor. 12 heights are calibrated with scale of 100 mm. Each height is calibrated for three times to reduce error of calibrated value. Amplitude changes are recorded. Experimental devices are connected as shown in Fig. 9.

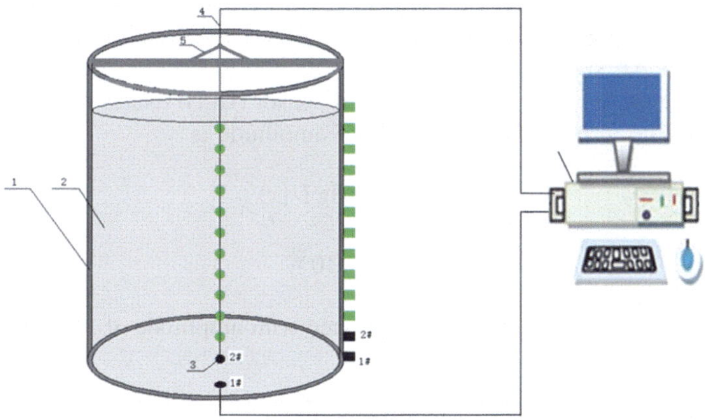

Fig. 9 Devices for sound wave attenuation experiment (① is experiment on sound attenuation through water; ② is experiment on attenuation through metal tank wall; 1: small vertical tank; 2: clean water; 3: underwater sensor; 4: signal wire; 5: metal wire clamp; 6: AE detection analyzer)

Two types of sensor are used. First type is DP3I of PAC with working frequency range 20–220 kHz, peak frequency 31.74 kHz, 40 dB gain, temperature range −65 to 175 °C, coupled with vacuum grease. Second type is new R.45IUC underwater sensor of PAC with working frequency range 20–220 kHz, peak frequency 22.461 kHz, 40 dB gain, temperature range −35 to 65 °C, 5 m-long coaxial cable, coupled with water.

Since the experiment is performed indoors, there are no obvious external interference signals. So threshold value is set as 30 dB. With multiple tests and sizes of tank, PDT, HDT and HLT values are finally determined, as shown in Table 4.

Table 4 Data setting for AE detection system

Threshold/dB	Sampling rate SPS	PDT/us	HDT/us	HLT/us
30	2048	300	600	600

5.1.2 Theoretical Calculation of Attenuation Coefficient

When AE wave transmits in medium, energy is attenuated mainly in three ways, namely attenuation by diffusion, attenuation by scattering and attenuation by absorption. Attenuation of AE wave in water is mainly in way of diffusion, i.e. as transmission distance increases, sound pressure in sound field is distributed differently; sound wave is diffused so that energy and pressure of sound wave on unit area decrease gradually. Dissimilarly, attenuation though metal tank wall is combination of three ways of attenuation. But due to limitation of experimental conditions, this paper focuses on comparison of attenuation of sound wave by diffusion in two media. So, with the following formula for attenuation of AE wave by diffusion, coefficients of attenuation of AE signals through water and metal are calculated:

$$\alpha = \frac{1}{x_1 - x_0} 20 \ln \left[\frac{U(x_0)}{U(x_1)} \right] \tag{18}$$

where: α is attenuation coefficient, dB/mm; x is distance from measuring point, mm; U (x_0) is peak voltage of AE signal at output end of sensor at sound source, μv; $U(x_1)$ is peak voltage of AE signal at output end of sensor at measuring point, μv.

Peak voltage of AE signal at output end of sensor (U(x)) relates to amplitude of AE signal (A(x)). The formula for calculation of amplitude is

$$A(x) = 20 \lg U(x) \tag{19}$$

$$U(x) = 10^{\frac{A(x)}{20}} \tag{20}$$

So attenuation coefficient α can be expressed with amplitude of AE signal (A(x)) as below

$$\alpha = \frac{1}{x_1 - x_0} \ln 10[A(x_0) - A(x_1)] \tag{21}$$

5.1.3 Results of Experiment and Analysis on Regularity

Experiment on attenuation through water is carried out as per experiment protocol. Table 5 shows attribute of attenuation of breaking lead AE wave through water.

Table 5 Attenuation of breaking lead AE wave through water

Distance D (mm)	100	200	300	400	500	600	700	800	900
n = 1	96	95	94	92	91	90	88	87	85
n = 2	96	95	93	92	91	89	88	87	85
n = 3	96	95	94	92	90	88	87	86	85
Mean	96	95	93.7	92	90.7	89	87.7	86.7	85
Mean amplitude of source	98	98.3	99	98.3	98.3	97	98	98	97.3
α dB/mm	0.036	0.034	0.037	0.036	0.035	0.033	0.034	0.033	0.031

Experiment on breaking lead attenuation through tank wall is carried out as per experiment protocol. Table 6 shows attribute of attenuation of breaking lead AE wave through metal tank wall.

Table 6 Attenuation of breaking lead AE signal through tank wall

Distance D (mm)	100	200	300	400	500	600	700	800	900
n = 1	95	93	92	89	86	82	79	76	73
n = 2	95	94	91	89	86	83	79	76	73
n = 3	95	93	92	88	85	82	80	76	73
Mean	95	93.3	91.7	88.7	85.7	82.3	79.3	76	73
Mean amplitude of source	98.3	98	98.3	97.3	97	98	98.3	98	97
α B/m	0.048	0.05	0.051	0.049	0.052	0.05	0.054	0.053	0.056

Based on data in Tables 5 and 6, attenuation graph and attenuation coefficient graph for breaking lead AE wave through water and metal tank wall are drawn. Abscissa indicates distance between measuring point and sound source. Ordinate respectively indicates the amplitude of sound wave measured at measuring points and the attenuation coefficient calculated based on formula 21, as shown in Figs. 10 and 11.

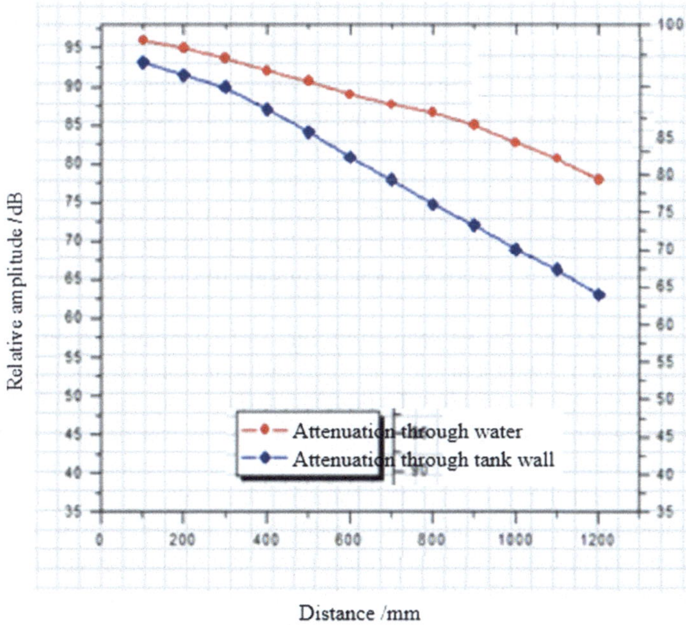

Fig. 10 Attenuation through different media

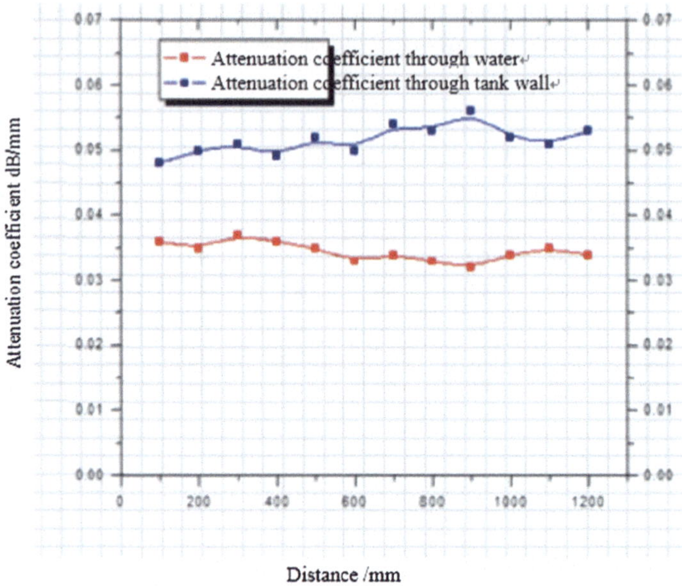

Fig. 11 Attenuation coefficient through different media

(1) Based on Fig. 10, as can be seen from attenuation graph for breaking lead AE wave through water, whole curve trends downwards linearly and is smooth. Within height range of 1200 mm, amplitude of signal is attenuated from averagely 98 to 78 dB with loss nearly 20 dB, mean attenuation being 1.67 dB per 100 mm.

(2) As can be seen from attenuation graph for breaking lead AE wave through metal tank wall, similarly, whole curve trends downwards linearly. Within height range of 1200 mm, amplitude of signal is attenuated from averagely 98 to 64 dB with loss nearly 34 dB, mean attenuation being 2.83 dB per 100 mm. Based on comparison between two attenuation curves, within same range of distance, amplitude of AE wave through water is attenuated less than that through metal tank. Based on experimental data, mean difference of attenuation is 1.16 dB per 100 mm. Therefore, it is a measurable technique to put sensor into water for AE detection.

(3) Based on Fig. 11, as can be seen from curve of attenuation coefficient through water, attenuation coefficient for AE wave through water is not big, and attenuation through water is stable (fluctuation of data may be caused by error in coupling between collector and sensor). Whereas, as can be seen from curve of attenuation coefficient through tank wall, attenuation coefficient of AE wave through tank wall is bigger than that through water. Based on experimental data, mean attenuation coefficient through tank wall per 100 mm is 0.017 dB/mm bigger than that through water. It further proves that it is a measurable and feasible technique to perform AE detection in medium in tank.

5.2 Experiment on Immersed AE Monitoring on Corrosion of Simulative Tank Floor

Purified water solution is contained in simulative tank. A corrosion source is placed on floor. R.45IUC underwater sensor 1 is placed in tank. Three DP3I sensors, numbered as 2, 3, 4, are evenly installed on outer tank wall near floor. 4 sensors jointly monitor corrosion source.

Monitoring results show that sensors #1, #2, #3, #4 all can receive AE signal from corrosion source. Due to small diameter of simulative tank, amplitudes of signals received by 4 channels are not highly different, as shown in Fig. 12.

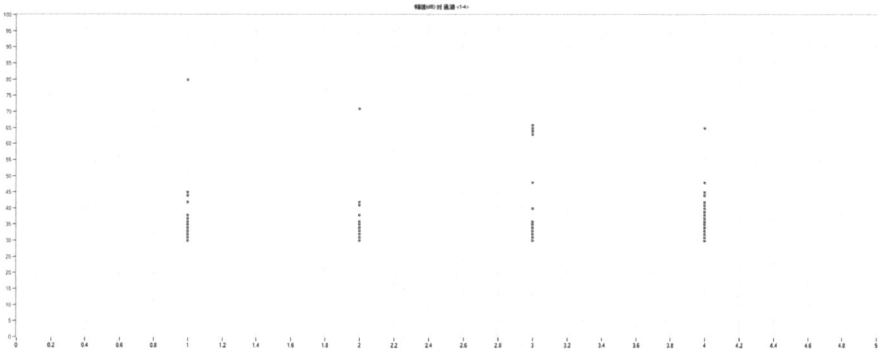

Fig. 12 Amplitudes of signals in different channels

Select an AE event. Take sensors #1 and #2 as examples. Signals from corrosion source received by AE sensors installed in tank and on outer wall are compared. As can be seen from Fig. 9, amplitude of signal received by sensor #1 is slightly higher than that by sensor #2, and signal profile of the former is clearer than that of the latter. It indicates that signal received by the sensor in tank mixes with less noise signals and has higher signal-to-noise radio.

Experimental results show that signal received by underwater sensor has better signal-to-noise radio, interference resistance and sensitivity. Based on analysis on waveform (frequency domain, time domain), it contains more information.

Acknowledgements. This work is supported by the national key research and development program of China "Typical storage of dangerous chemicals safety warning and protection integration key technology research and application demonstration (2016YFC0801200)".

References

1. Qiu, F., Dai, G., Zhang, Y.: Study on acoustic source identification and positioning method of tank bottom panel. Appl. Acoust. **34**(4), 364–372
2. Dai, G., Zhang, Y., Zhao, Y.: Study on the identification method of corrosion acoustic emission source of tank bottom plate. Nondestruct. Test. **37**(2), 14–18 (2015)
3. Zhang, Y., Zhao, Y., Liu, Y.: Study on prediction method of corrosion residual life of tank bottom plate based on data analysis of risk and acoustic emission. Press. Vessel **31**(1), 52–56

Study on Corrosion Behavior of 20# Seamless Steel Pipeline in High Sulfur Environment

Wen-tao Qu[1(✉)], Kai-lun Guo[1], Song Qing[2], and Xin Liu[3]

[1] School of Mechanical Engineering, Xi'an Shiyou University, Xi'an 710065, Shaanxi, China
wtqu@xsyu.edu.cn, gklellen1992@sina.com

[2] The Fifth Gas Production Plant, Changqing Oilfield Company, PetroChina, Xi'an 710021, Shaanxi, China
qsongl_cq@petrochina.com.cn

[3] South Sulige Operating Division of Changqing Oilfield Company, PetroChina, Xi'an 710021, Shaanxi, China
liuxin001_cq@petrochina.com.cn

Abstract. The corrosion behavior of 20# seamless steel pipe in the high-sulfur gas well of the fifth gas production plant with Petro China Changqing was investigated. The 20# seamless pipe with 159 × 7 mm was buried underground with 1 m in November 2013. The methods of scanning electron microscope (SEM), microhardness and X-ray diffraction (XRD) were used to evaluated the corrosion behavior of 20# seamless pipe. The microhardness of the pipe is 228 HV, yield strength is 419 MPa, tensile strength is 573 MPa and the elongation is 18.3%. The results showed that the FeS_2 is the main corrosion product on the inner and outer surfaces of the pipe based on the composition of the corrosion products. The hydrogen blistering and hydrogen induced cracking were also observed on the corrosion surface, which could be caused corrosion damage to the pipeline. These research results could provide theoretical basis and technical guidance for the corrosion protection and the hydrogen embrittlement prevention of 20# seamless steel in the high-sulfur gas well of the fifth gas production plant.

Keywords: 20# seamless steel · Gas pipeline · High H_2S content · Corrosion

J. Lin (ed.), *Proceedings of the International Petroleum and Petrochemical Technology Conference 2019*, pp. 374–386, 2020.
https://doi.org/10.1007/978-981-15-0860-8_32

1 Introduction

In the mining of high-sulfur gas wells, the corrosion behavior of equipment by H_2S is significantly greater than that of ordinary sulfur-containing gas wells. At the same time, some researchers found that the risks existing in mining and production management are also greater than those of ordinary sulfur-containing gas wells [1]. Corrosion not only causes huge economic losses to the development and production of oil and gas fields, and causes environmental pollution and threatens personal security. The corrosion problem of pipelines caused by H_2S is widespread in the process of natural gas production and transportation, which leads to serious corrosion problems of the pipelines body, even leads to piercing or breaking of pipelines, and destroying the gas transmission [2–5]. The wet H_2S is an acidic substance, the corrosion process of metal is accompanied by the formation of the sulfide film on the metal surface. The corrosion forms of metal mainly include electrochemical corrosion, hydrogen embrittlement and sulfide stress corrosion cracking [6]. H_2S and corrosive media react with the metal pipelines should pass through the FeS_x films and the residual oxide [7, 8]. The damage of hydrogen embrittlement would lead to serious failures of downhole pipeline, surface equipment, and wellhead device. What is more serious is blowouts or fire accidents [9].

Taking the pipeline of the fifth gas production plant with Petro China Changing as the research object, the fifth gas production plant is affiliated to the lower stratum of the Sulige gas field, across the Inner Mongolia Autonomous Region and the two provinces of Shaanxi. The average H_2S content of the natural gas collected by the fifth gas production plant is 207.2 mg/m^3 and the maximum content is 1196.4 mg/m^3 [10]. In order to reduce the occurrence of corrosion accidents in the fifth gas recovery plant, increase natural gas production and economic benefits, it is necessary to strengthen the research on H_2S corrosion in the fifth gas recovery plant, so as to reduce the corrosion of H_2S in the process of natural gas mining, transportation and processing. The research on H_2S corrosion to further avoid accidents, casualties and economic losses.

2 Materials and Methods

2.1 Material

The 20# seamless pipe is one of the common used pipeline in the oil and gas wells. In November 2013, the 20# seamless pipe with the specification model number 159×7 mm was placed in the high-sulfur gas well of the fifth gas production plant for a period of time, and taken out in November 2017. The sampling site was the outlet line of the separator of Sudong 4X station, and the pipeline was buried 1 m deep. In order to facilitate sampling during the test, the seamless steel pipe was cut by wire cutting as shown in Fig. 1.

2.2 Matrix Hardness, Composition and Mechanical Properties

The hardness of the matrix was tested by a FM800 micro Vickers hardness tester. After removing the corrosion products on the inner and outer surfaces of the specimens of

(a) outside (b) internal

Fig. 1 The 20# seamless pipe in environments of high H_2S content after placement for a certain time

Fig. 1 by mechanical method, the specimens with the size of 10 mm × 10 mm × 1 mm were cut by the method of electric discharge wire cutting. The surfaces of the specimens were polished with silicon carbide water sandpaper with 400, 600, 800, 1000, 1200, and 1500#, then the polished specimens were immersed in absolute ethanol for 5 min using ultrasonic equipment and finally dried with cold air. When the hardness was measured, the force applied of 4900 N was used, the holding time was 20 s, and 8 points were tested for each specimen. The matrix composition was analysed by a D/Max 2500 multi-function of X-ray diffraction (XRD) before matrix hardness testing.

The mechanical tensile properties of the matrix are sampled on the specimen shown in Fig. 1 according to the requirements of the standard GB/T 2281-2010. The size of standard tensile specimen is shown in Fig. 2, and the specimen is shown in Fig. 3 which was machined by a wire cutting. The tensile strength, yield strength and elongation of the specimen were obtained.

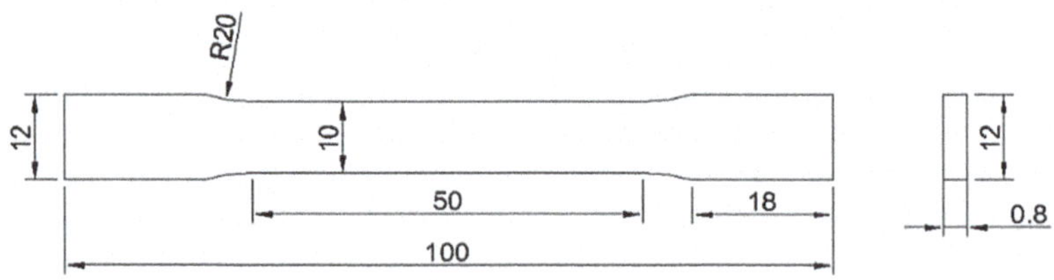

Fig. 2 Schematic diagram of tensile specimens

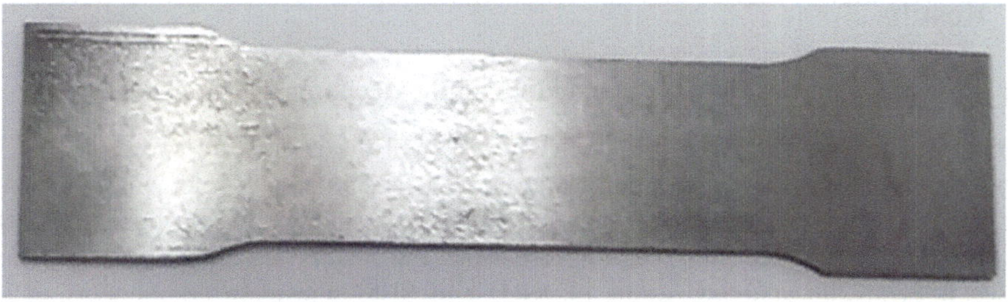

Fig. 3 Picture of tensile specimen

2.3 Corrosion Products

Sampling the corrosive material on the inner and outer surfaces of the specimen according to the position shown in Fig. 4. The corrosion morphology of the specimen was taken by field emission scanning electron microscopy (including the energy spectrum test function) and the composition of the corrosion product was analysed.

In order to confirm the phase composition and composition of the corrosion products, the test was carried out using a D/Max 2500 multi-function of XRD.

(a) outside (b) internal

Fig. 4 Schematic diagram of sampling

3 Results

3.1 Matrix Composition, Microhardness and Mechanical Properties

The specimen was first subjected to XRD test, and the results are shown in Fig. 5. From the figure, the main phase composition of the specimen is Fe element. In order to confirm the matrix composition, the specimen was studied by spectroscopy under SEM. The energy spectrum scanning area and the results are shown in Fig. 6 and Table 1. From the analysis results of XRD and SEM, the main components of the specimen are composed of the main element Fe and the secondary elements C, Si and Mn, etc., which conform to the chemical composition of 20# seamless steel [11].

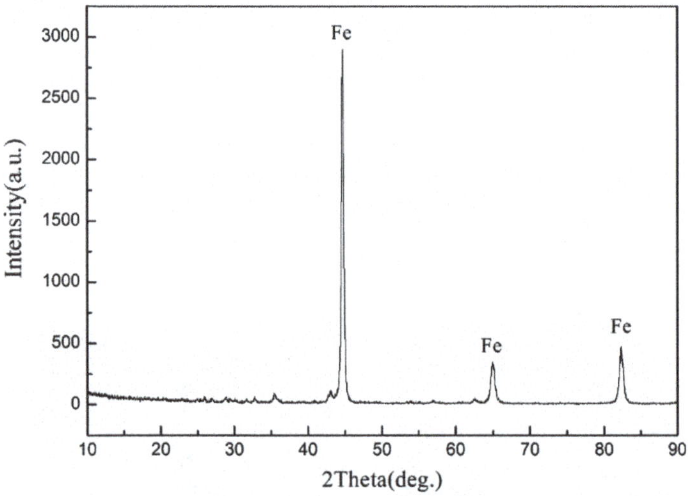

Fig. 5 X-ray diffraction spectra of specimen matrix

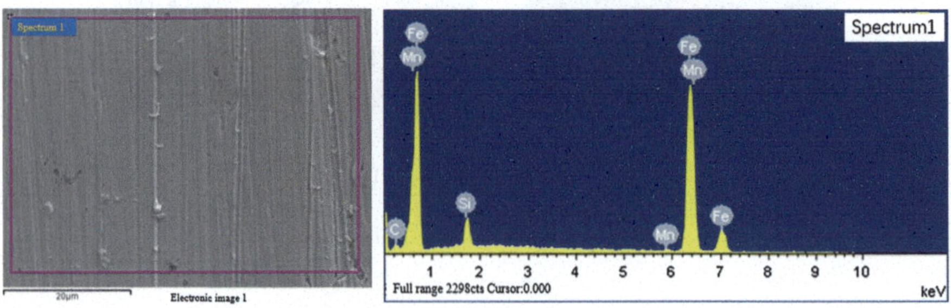

Fig. 6 EDS spectra of specimen matrix

Table 1 EDS results of specimen matrix

Element	C K	Si K	Mn K	Fe K
wt%	2.56	0.58	0.71	93.99
at.%	10.63	1.15	0.64	83.87

The microhardness of the matrix is shown in Table 2. The average microhardness of the matrix is 228 HV. The microhardness of the specimen indicates that the composition and microstructure of the matrix are uniform.

Table 2 Microhardness of 20# seamless pipe (HV)

	1	2	3	4	5	6	7	8	Average
1#	234.03	215.98	225.79	212.28	218.55	231.15	219.90	213.17	221.36
2#	234.20	234.20	226.80	231.23	243.45	223.97	221.04	223.97	229.86
3#	226.2	224.71	230.67	217.46	250.59	230.76	232.31	233.87	230.82

The tensile curves of the matrix specimen are shown in Fig. 7. The mechanical properties of the different specimens are consistent, the average yield strength is 419 MPa, the average tensile strength is 573 MPa, and the average elongation is 18.3%. The results show that the 20# seamless steel pipe used in the paper meets the requirements of GB/T 9711-2017 Steel Pipe for Oil and Gas Industrial Pipeline Transportation [11].

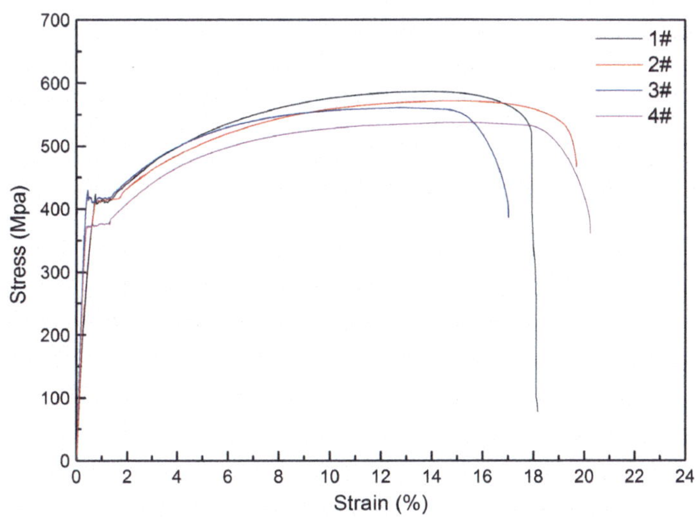

Fig. 7 The strain-stress curves of the 20# seamless pipe

3.2 Corrosion Morphology and Composition

Outer surface topography and composition

Figure 4a shows the corrosion of the macroscopic corrosion of the specimen, and can be divided into two major regions, it is different from top to bottom. The two ends of the specimen being corrosion products of rust colour, and the middle region containing black corrosion products. According to Fig. 4a, region 1 and region 4 of specimen were selected for SEM corrosion morphology analysis, and the scanning results are shown in Fig. 8. Observation with the naked eyes showed that the outer surface of the specimen was covered with corrosion products, and there are spherical particle products or bulges in all regions, which is due to hydrogen gas generated during H_2S corrosion, resulting in hydrogen bubbling [12]. At the same time, significant cracking can be observed on the corrosion product film in region 1 (Fig. 8b), which is caused by hydrogen induced cracking [13].

In order to further determine the composition of the above corrosion product film, the above region was subjected to an energy spectrum test, and the test results are shown in Fig. 9 and Table 3. The energy spectrum showed that the main components of the corrosion product film are Fe, Fe_3C, scale, iron oxide and iron sulfide, while Fe and Fe_3C are mainly the components of the matrix. During the corrosion process, Fe_3C remains with the dissolution of ferrite and becomes a corrosion product deposited on the surface of the specimen. Therefore, it can be seen from the energy spectrum analysis that the corrosion product film on the outer surface of the specimen is mainly iron oxide and iron sulfide. The iron oxide is caused by the oxidation of the steel pipe, and the iron sulfide is caused by the corrosion of H_2S.

When the SEM is used to perform the energy spectrum test on the specimen, unknown corrosive substances may be generated due to the influence of the test environment and the specimen itself. Therefore, in order to more accurately describe the composition of the corrosion product film, the composition and phase composition of the corrosion product film were further analysed by XRD, as shown in Fig. 10. The figure shows that the main components of the corrosion product film on the outer surface of the specimen are iron oxide (Fe_2O_3 and Fe_3O_4), FeS_2 and $CaCO_3$, and $CaCO_3$ is mainly caused by scale caused by long-term contact with water outside the pipe. Consequently, considering the energy spectrum and XRD results, the main corrosion product on the outer surface of the specimen in the high sulfur environment is FeS_2.

Inner surface topography and composition

The corrosion morphology of the inner surface of the specimen was sampled in the five regions shown in Fig. 4b. From the macroscopic morphology (Fig. 4b), the corrosion products on the inner surface of the specimen were basically the same, therefore, only the specimen of the region 3 was subjected to shape observation and its composition was analysed. Figure 11 is a SEM corrosion topography of the specimen of region 3. From the figure, corrosion products of different morphologies can be observed, and the thickness of the corrosion product film is also inconsistent. Due to hydrogen induced cracking, in Fig. 11a, not only the acicular corrosion products and the fracture can be observed.

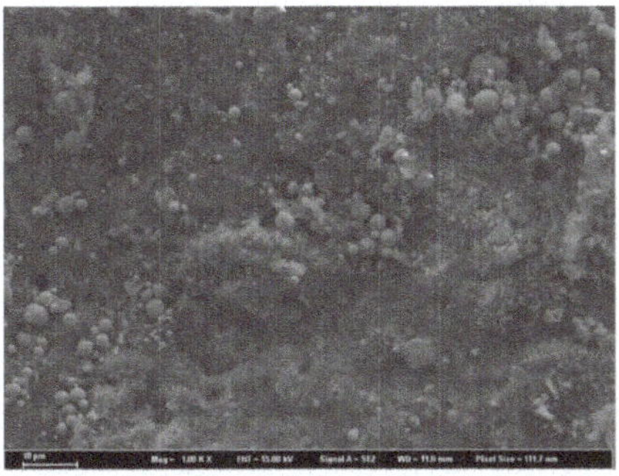

(a) area 1

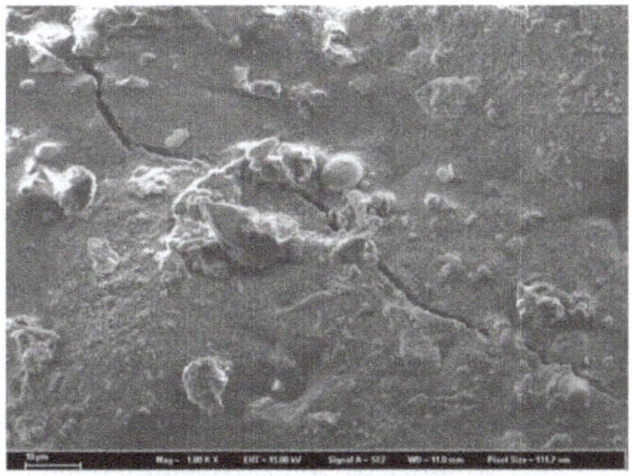

(b) amplification of (a)

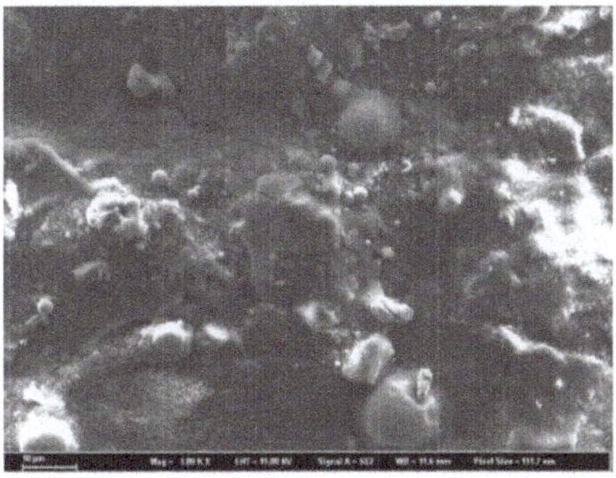

(c) area 4

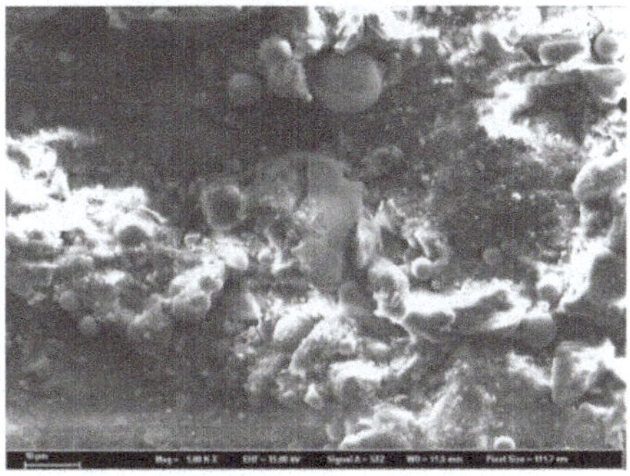

(b) amplification of (c)

Fig. 8 SEM morphology of the corrosion products of 20# seamless pipe for outside surface

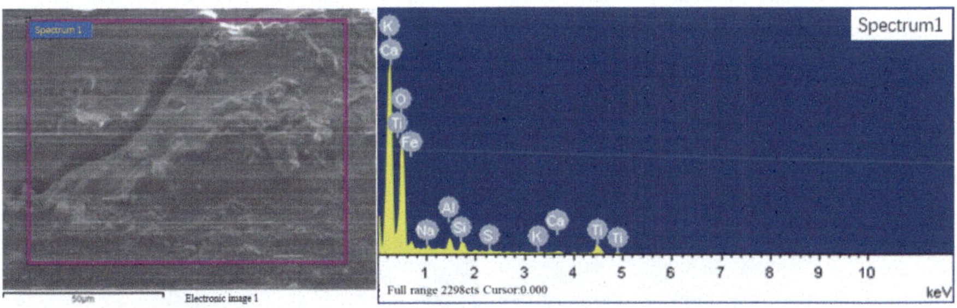

(a) area 1

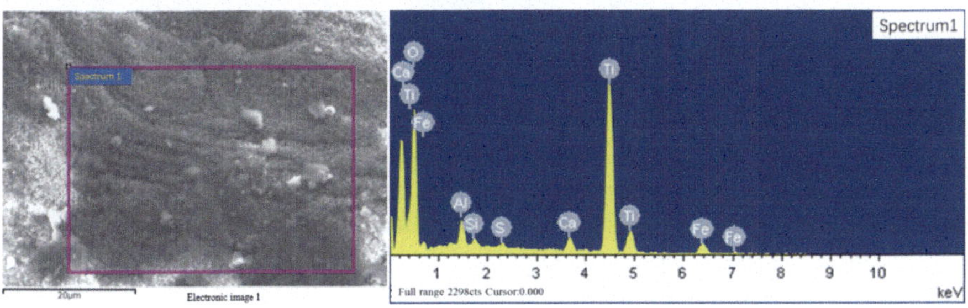

(b) area 4

Fig. 9 The EDS results of the 20# seamless pipe for outside surface

Table 3 The EDS results of the 20# seamless pipe for outside surface

Region	Element	O K	Na K	Al K	Si K	S K	K K	Ca K	Fe K
1	wt%	86.21	1.79	4.01	3.74	1.01	1.01	1.48	0.75
	at.%	92.02	1.34	2.53	2.28	0.54	0.44	0.63	0.23
4	wt%	72.96	–	5.07	1.42	1.09	–	5.02	14.45
	at.%	87.43	–	3.6	0.97	0.65	–	2.4	4.96

Similarly, in order to determine the composition of the surface corrosion product film in the specimen, the corrosion product film was subjected to an energy spectrum test using SEM, and the results are shown in Fig. 12 and Table 4. The results of the energy spectrum test show that the main chemical composition of the corrosion product film on the inner surface is composed of Fe, O and S elements, and the composition is corresponding to iron oxide and iron sulfide.

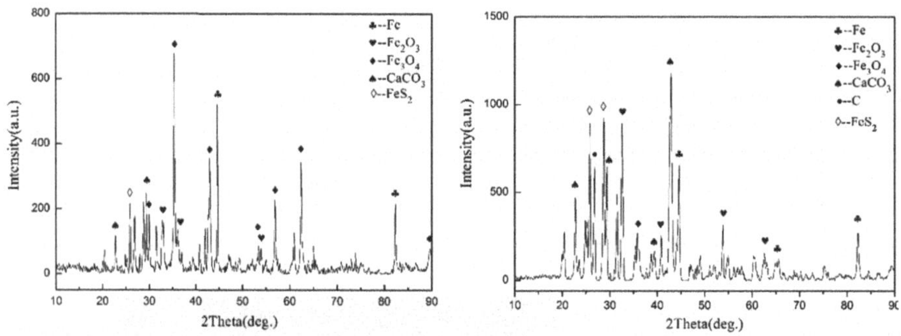

(a) area 1 (b) area 4

Fig. 10 XRD spectrum of the 20# seamless pipe for outside surface

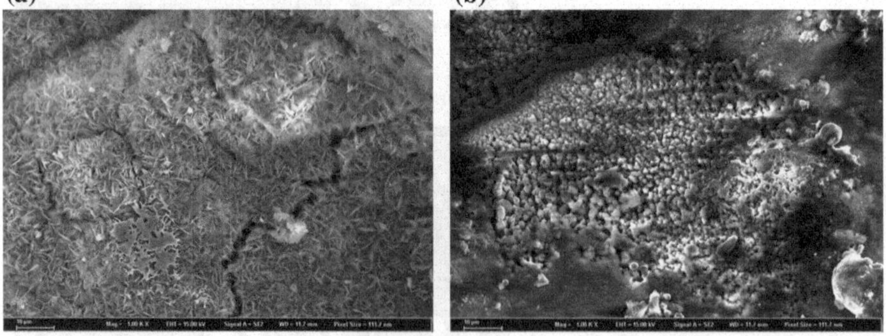

(a) (b)

Fig. 11 SEM morphology of the corrosion products of 20# seamless pipe for internal surface

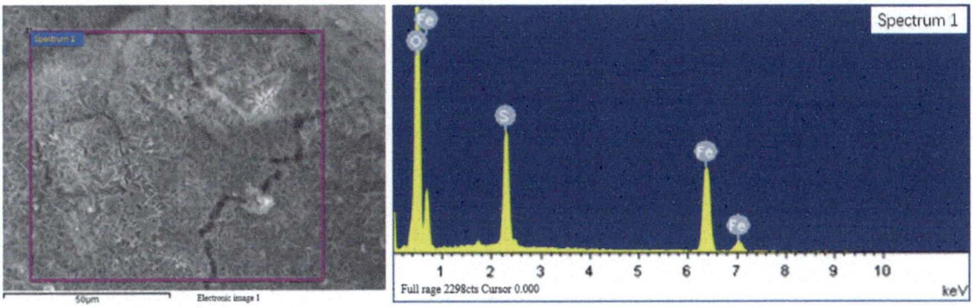

Fig. 12 The EDS results of the 20# seamless pipe for internal surface

Table 4 EDS results of the 20# seamless pipe for internal surface

Element	O K	S K	Fe K
wt%	42.26	11.43	46.31
at.%	69.01	9.32	21.67

The phase composition of the inner surface corrosion product film was further analysed by XRD, and the results are shown in Fig. 13. The XRD diffraction results indicate that the main phase composition of the surface corrosion product film of the specimen are FeS_2, Fe_2O_3 and $CaCO_3$. By further verifying the red product in the macroscopic corrosion morphology of the inner surface of the specimen shown in Fig. 4b, mainly from the colour of Fe_2O_3.

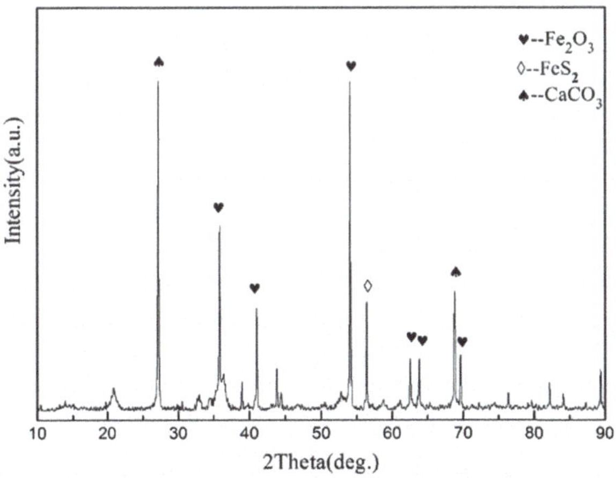

Fig. 13 XRD spectrum of the 20# seamless pipe for internal surface

4 Discussion

Many studies have shown that dry H_2S does not have corrosive damage to metal materials, and only wet H_2S is corrosive. When H_2S is dissolved in water, the dissociation reaction as shown below occurs immediately [6, 7].

$$H_2S \rightarrow HS^- + H^+ \tag{1}$$

$$HS^- \rightarrow {S_2}^- + H^+ \tag{2}$$

For steel pipelines, wet H_2S will react with Fe in the pipeline, resulting in corrosion of steel materials [14].

Anode reaction: $Fe - 2e \rightarrow {Fe_2}^+ \quad {Fe_2}^+ + {S_2}^- \rightarrow FeS \downarrow$

Cathodic reaction: $2H + 2e \longrightarrow H2 \uparrow$

The above chemical reaction formula show that the ionized products of wet H_2S are HS^- and ${S_2}^-$, and the ionized products are adsorbed on the surface of the oil and gas pipeline. After that, an adsorbed complex ion $Fe(HS)^-$ is formed. At the same time, the H atom is a strong depolarizer, and it is easy to obtain electrons at the cathode, which greatly reduces the bonding strength of the metal bonds between the Fe atoms, thereby further accelerating the above mentioned anode reaction, resulting in accelerated corrosion of steel. On the basis of the above cathode reaction, a part of H atoms form H_2 after electrons are obtained, and the other H atoms are adsorbed on the surface of the pipeline and diffused into the matrix due to sufficient energy. In the diffusion infiltration process of H atoms, H atoms are trapped by these defects when encountering defects such as defects at the grain boundary or phase boundary, crystal dislocations, and tensile stress concentration regions. With the increase of H_2 pressure at the defect, hydrogen bubbling is gradually formed on the surface of the steel pipe (Fig. 8); when the pressure is increased to a certain extent, stress concentration is formed at the edge of the defect, causing cracks between the interfaces to cause cracks (Figs. 8b and 11a) [14–17]. Both the hydrogen bubbling and hydrogen induced cracking mentioned above can cause the gas pipeline to be destroyed or even fail, which may bring the risk of hydrogen embrittlement to the pipeline.

The corrosion products of steel after being corroded by H_2S are mainly Fe_9S_8, Fe_3S_4, FeS_2 and FeS [3]. The final corrosion products depend on the PH of the environment and the concentration of H_2S, etc. [18–20], and the corrosion product in this paper is FeS2. Some studies suggest that when the concentration of H_2S is low, the surface of the steel can form a dense FeS film, and can prevent the passage of Fe ions, which reduces the corrosion rate of the steel to some extent; when the concentration of H_2S is high, a black loose layered or powdery film is formed, such as the Fe_2S film in this case, and Fe ions can easily pass through the film and form a macroscopic primary battery with Fe, thereby further accelerating the corrosion rate of the steel [1, 13, 15]. This indicates that the H_2S mass concentration in the high-sulfur gas well of the fifth gas production plant is relatively high.

5 Conclusions

According to the above test results and analysis, the collecting gas pipeline of 20# seamless steel pipe used in the high-sulfur gas well of the fifth gas production plant is easy to fail under the corrosion of H_2S.

(1) The performance of the gas pipeline of 20# seamless steel pipe used in the high-sulfur gas well of the fifth gas production plant meets the relevant national standards.
(2) The corrosion of the gas pipeline in the high-sulfur gas well of the fifth gas production plant is typical H_2S corrosion. The main corrosion product of the inner and outer surfaces of the pipeline after removing scale is FeS_2.
(3) By observing the corrosion morphology of the inner and outer surfaces of the specimen, it is known that H_2 is generated during the H_2S corrosion process, causing hydrogen bubbling and hydrogen induced cracking on the surface of the gas pipeline.

References

1. Gan, L.J., Huang, F., Zhao, X.Y., et al.: Hydrogen trapping and hydrogen induced cracking of welded X100 pipeline steel in H_2S environments. Int. J. Hydrogen Energy **43**(4), 2293–2306 (2018)
2. Lei, X.W., Wang, H.Y., Mao, F.X., et al.: Electrochemical behaviour of martensitic stainless steel after immersion in a H_2S-saturated solution. Corros. Sci. **131**, 164–173 (2018)
3. Huang, F., Cheng, P., Zhao, X.Y., et al.: Effect of sulfide films formed on X65 steel surface on hydrogen permeation in H_2S environments. Int. J. Hydrogen Energy **42**(7), 4561–4570 (2017)
4. Han, P., Chen, C.F., Yu, H.B., et al.: Study of pitting corrosion of L245 steel in H_2S environments induced by imidazoline quaternary ammonium salts. Corros. Sci. **112**, 128–137 (2016)
5. Zhao, W., Zou, Y., Kenji, M., et al.: Corrosion behavior of reheated CGHAZ of X80 pipeline steel in H_2S-containing environments. Mater. Des. **99**(5), 44–56 (2016)
6. Ai, Z.J., Fan, Y.W., Zhao, Q.K.: Review on H_2S corrosion of oil gas tubing and its protection. Surf. Technol. **44**(9), 108–115 (2015)
7. Wang, F., Wang, L.X., Liu, Z.Y., et al.: Study on stress corrosion behavior of TP110TS oil pipeline steel in the sour gas field environment. Surf. Technol. **44**(3), 57–62 (2015)
8. Zhou, C.H., Chen, X.Y., Wang, Z., et al.: Effects of environmental conditions on hydrogen permeation of X52 pipeline steel exposed to high H_2S-containing solutions. Corros. Sci. **89**, 30–37 (2014)
9. Qu, W.T., Liu, S.Y., Xu, Z.S.: Research on corrosion rate of sucker rod steel 20NIi2Mo in wet H_2S environment. Mater. Sci. Forum **814**, 319–324 (2015)
10. Chen, X.K., Lu, K.L., Xiao, Y., et al.: Research on diffusion regularity of continuous leakage of sulfur natural gas pipeline. J. Xi'an Univ. Sci. Technol. **35**(4), 403–408 (2015)
11. GB/T 9711-2017: Petroleum and natural gas industries—steel pipe for pipeline transportation systems (2017)

12. Wang, P.Y., Lv, Z.G., Zheng, S.Q., et al.: Tensile and impact properties of X70 pipeline steel exposed to wet H_2S environments. Int. J. Hydrogen Energy **40**(35), 11514–11521 (2015)
13. Bai, P.P., Zhao, H., Zheng, S.Q., et al.: Initiation and developmental stages of steel corrosion in wet H_2S environments. Corros. Sci. **93**, 109–119 (2015)
14. Li, X., Zhu, Q.J., Zhou, N., et al.: Oil-gas pipe corrosion and protection. Surf. Technol. **46** (12), 206–217 (2017)
15. Huang, H., Ma, H.L., He, R.Y., et al.: Analysis of the impact of natural gas pipeline internal corrosion caused by H_2S. Pipeline Tech. Equip. **3**, 34–36 (2014)
16. Chen, Y.X., Zheng, S.Q., Zhou, J., et al.: Influence of H_2S interaction with prestrain on the mechanical properties of high-strength X80 steel. Int. J. Hydrogen Energy **41**(24), 10412–10420 (2016)
17. Wang, D., Yuan, S.J., Wu, X.W., et al.: Research progress of CO_2/H_2S corrosion in oil and gas pipelines and the protection techniques. Surf. Technol. **45**(3), 31–37 (2016)
18. Zhou, C.S., Zheng, S.Q., Chen, C.F., et al.: The effect of the partial pressure of H_2S on the permeation of hydrogen in low carbon pipeline steel. Corros. Sci. **67**, 184–192 (2013)
19. Fragiel, A., Serna, S., Pérez, R.: Electrochemical study of two microalloyed pipeline steels in H_2S environments. Int. J. Hydrogen Energy **30**(12), 1303–1309 (2015)
20. Qu, W.T., Sun, X.G., Guo, J.S., et al.: Analysis of N80 tubing failure for CBM well dtainage. Mater. Sci. Forum **850**, 971–976 (2016)

Research and Application of Shell Changbei Gas Well Corrosion Monitoring Technology

Bao-yu Wu[1,2(✉)], Xin-fa Zhang[1,2], and Lei Liu[3]

[1] Drilling & Production Engineering Technology Research Institute, Chuanqing Drilling Engineering Co. Ltd., CNPC, Xi'an 710018, China
wubaoyu@cnpc.com.cn

[2] National Engineering Laboratory for Exploration and Development of Low-Permeability Oil & Gas Fields, 710018 Xi'an, China

[3] Changbei Branch of PetroChina Changqing Oilfield, 710018 Xi'an, China

Abstract. Effective corrosion monitoring technology is an important basis for accurately evaluating the corrosion resistance of gas wellbore string materials, exploring the corrosion mechanism and formulating anti-corrosion measures. In this study, two kinds of corrosion specimens, N80 and P110, carried by downhole electronic fishing device, were put into the depth of 955 and 1498 m target wells for 8760 h dynamic corrosion weightlessness test, and the fluid characteristics in gas wells were analyzed. At the end of the experiment, the specimens were taken back and the corrosion rate was calculated. The corrosion morphology was observed by SEM, EDS, XRD and other materials characterization techniques. The element composition and phase composition of the corrosion products were determined. The results show that the experimental wells are mainly hydrogen depolarization electrochemical corrosion caused by the dissolution of CO_2 corrosive gas associated with natural gas in a small amount of condensate water and uniformly adsorbed on the surface of carbon steel in the form of water film. The main corrosion products are typical cubic crystal $FeCO_3$. The corrosion products of N80 and P110 specimens at 1498 m are uniformly and compactly adhered to the surface of carbon steel, and the corrosion rate is less than 0.01 mm/a, which can effectively prevent the penetration of corrosion media. However, the corrosion products of 995 m/P110 specimens are loose and easy to fall off, and the grain is sparse. The corrosion tendency of 995 m/N80 specimens is local, and the corrosion of the tubular

J. Lin (ed.), *Proceedings of the International Petroleum and Petrochemical Technology Conference 2019*, pp. 387–400, 2020.
https://doi.org/10.1007/978-981-15-0860-8_33

column will be accelerated with the increase of water production in the later mining process. This technology provides technical support and theoretical basis for the implementation of anti-corrosion measures for gas wellbore string.

Keywords: Corrosion monitoring · Corrosion product film · Material characterization · Local corrosion

1 Introduction

Well S141 in Changbei Block of Yulin Gas Field is located in the northeast of Yishan Slope of Ordos Basin. The main producing stratum is Shan 2 Member of Shanxi Formation of Upper Paleozoic Permian. Its burial depth is between 2800 and 3000 m. Shan 2 Member is mainly composed of river, delta sandstone and mudstone and delta coal measures strata. Its lithology is composed of medium-coarse quartz sand, quartz sandstone, gravel sandstone and coal seam with a thickness of 55–70 m. The well has been normally developed for 20 years. With the increase of water produced by gas wells, the corrosion gases such as CO_2, H_2S and corrosive bacteria such as SRB in high salinity formation water will cause electrochemical corrosion of downhole string. Therefore, it is the basic guarantee for safety production in the later stage to know the corrosion situation of pipe string material, to master its corrosion mechanism and to formulate effective anti-corrosion measures [1, 2].

The well was drilled in 1995 and has been in operation for 20 years since 1999. In order to accurately grasp the service conditions of oil and casing materials in gas wells in this area and provide pertinent anti-corrosion measures for prolonging the maintenance period in the later period, a multi-arm caliper imaging logging tool (MIT) was used to inspect the tubing in the field. It was found that the remaining wall thickness of the tubing in some sections was about 95–97%, and the uniform corrosion was not serious, but there was local corrosion. The data showed that the local pitting corrosion depth was more than 3 mm. See Fig. 1. In order to accurately evaluate the corrosion resistance of tubing, a variety of wellbore corrosion monitoring techniques were used to evaluate and study.

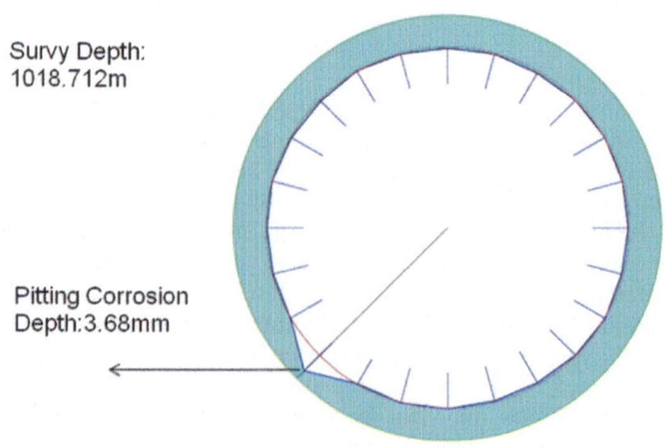

Fig. 1 Pipe MIT test chart

2 Experiment Procedure

2.1 Experimental Instruments and Materials

The main instruments used in the experiment are shown in Table 1.

Table 1 Main experiment instruments

Serial number	Instrument name	Instrument type	Manufacturer
1	Gas chromatograph	7890A	Agilent
2	ICP-AES	iCAP-7000	Thermo fisher
3	Potentiometric titrator	T70	METTLER TOLEDO
4	Multiparameter tester	SG23	METTLER TOLEDO
5	Electronic fishing device	ANT-B	ANT
6	Analytical balance	PRACTUM224-1CN	SARTORIUS
7	SEM	600FE	FEI Quanta
8	EDS	INCAENERGY 350	OXFORD
9	XRD	D8 ADVANCE	Bruker

Experimental materials: zinc acetate, iodine, potassium iodide, Tianjin Komeo Chemical Reagent Co., Ltd.; sodium thiosulfate titration solution standard solution (0.1000 mol/L), Beijing Century Orco Biotechnology Co., Ltd.

2.2 Wellbore Fluid Characteristics Analysis

(1) The gas components in laboratory wells were analyzed by chromatograph, and the cylinders collected to be measured were connected with the quantitative tube of the chromatograph. The percentage content of each component was separated and detected after nitrogen was brought into the column; (2) The content of hydrogen sulfide was determined by iodometry. After 30 L gas sample was absorbed by zinc acetate on site, the content of hydrogen sulfide was determined by iodine solution absorption and titration with sodium thiosulfate standard solution in the room; (3) Potentiometric titration and inductively coupled plasma emission spectrometry (ICP spectral lines K-766.491 nm, Na-588.995 nm, Ca-317.933 nm, Mg-285.213 nm, Ba-455.403 nm, Sr-421.552 nm, Fe-259.940 nm) were used to determine the main scaling ions in the produced water; (4) The parameters of pH, conductivity and TDS of water samples are measured by multi-parameter tester in turn, and the fluid characteristics of wellbore are analyzed comprehensively.

2.3 Off-Line Corrosion Monitoring and Analysis of Gas Wells

Based on the analysis of wellbore fluid characteristics and the determination of corrosion influencing factors, field experiments were carried out. The off-line corrosion monitoring method for gas wells was selected in this study. Two kinds of carbon steel specimens, N80 and P110, with the same material as the experimental wellbore string, were selected (size: $5.0 \times 2.5 \times 0.2$ cm). In order to ensure the stability of experimental data, five specimens of the same material constituted a group of parallel experiments with four groups of corrosion tests. The tablets were (a): 995 m/N80, (b): 995 m/P110, (c): 1498 m/N80, (d): 1498 m/P110, respectively. On the spot, the electronic salvage device with corrosion test pieces is put into the target well depth of the wellbore with a steel wire working vehicle, and the continuous corrosion weight loss experiment with a period of 8760 h is carried out. After the experiment is completed, the test pieces are taken out from the wellbore in turn with the steel wire working vehicle, sealed and packed with oxygen insulation, and then brought back to the laboratory for treatment: (1) Rinse with clean water and wipe off with filter paper, and remove the surface of the sample with acetone. The oil stains were soaked in acid cleaning solution for 5 min, and the surface corrosion products were removed by degreasing cotton. Finally, the oil stains were dehydrated in anhydrous ethanol and dried to observe the corrosion macromorphology and calculate the corrosion weight loss. (2) Observing the surface micro-corrosion morphology of corroded specimens by scanning electron microscopy (SEM); (3) Analyzing the micro-element composition of corrosion products by X-ray energy spectrum analysis (EDS); (4) Determining the phase composition of corrosion products by X-ray diffraction (XRD); Calculating the corrosion rate of specimens by corrosion weightlessness method; and analyzing the corrosion products by material characterization technology to determine the corrosion mechanism. The corrosion resistance of pipe string materials under field working conditions was comprehensively evaluated [3–5].

3 Results and Discussion

3.1 Wellbore Fluid Effect

From Table 2, it can be seen that the gas samples in Well S141 are mainly low molecular alkanes with CH_4 as the main component, but with 1.7% CO_2. As a common corrosive gas, CO_2 is easily dissolved in formation water to form CO_3^{2-}, HCO_3^-, which causes electrochemical corrosion on the carbon steel matrix of oil casing. The influencing factors of CO_2 corrosion are mainly divided into medium effect and environmental impact. The medium effect includes the pH of corrosion medium, chloride content, solubility of $FeCO_3$ corrosion products and so on [6, 7]. The environmental impact includes wellbore pressure, temperature, flow rate and experimental period. At the same time, H_2S was not detected in the gas sample components, which could eliminate the corrosion behavior caused by H_2S.

From Table 3, it can be seen that the water produced by the experimental well is $CaCl_2$ water type according to the calculation of the main ion concentration. Although the water does not contain Ba^{2+} and Sr^{2+}, the sulfate content is higher. According to the analysis of scaling mechanism, when the oil field water contains sulfate, carbonate, chlorine and other scaling ions, the temperature and pressure changes in the wellbore will destroy the original water quality balance, and the scaling crystals will gradually precipitate. It adsorbs and grows on the surface of metal, especially on the uneven surface of pipe string, which causes corrosion under scale. The average pH of water is 6–7, which belongs to neutral medium. The corrosion product film on metal surface and scale salts such as calcium sulfate and calcium carbonate are not easy to be dissolved and destroyed in neutral medium. In addition, the chloride content in the

Table 2 Gas Analysis Results of Well S141

Well number	CH_4 (%)	C_2H_6 (%)	C_3H_8 (%)	iC_4H_{10} (%)	nC_4H_{10} (%)	iC_5H_{12} (%)	nC_5H_{12} (%)	C_6 + (%)	Total hydrocarbon
S141	92.75	4.02	0.52	0.08	0.08	0.03	0.02	0.52	98.02
CO_2 (%)	He (%)	H_2 (%)	N_2 (%)	Specific gravity	ρ (g L^{-1})	Critical pressure (MPa)	Critical temperature (K)	H_2S (mg m^{-3})	H_2O (ppm)
1.70	0.11	0.03	0.24	0.61	0.7378	4.65	198.42	0	0

Table 3 Water quality analysis results of produced water samples from well S141

Conductivity ($uS\ cm^{-2}$)	PH	Ion concentration($mg\ L^{-1}$)								Salinity ($mg\ L^{-1}$)	Water type
		Na^+, K^+	Ca^{2+}	Mg^{2+}	Ba^{2+}, Sr^{2+}	Cl^-	HCO_3^-	SO_4^{2-}	ΣFe		
11,050	6.3	3547.8	1184.4	47.9	0.0	4887.2	314.1	3974.0	68.0	13,955.2	$CaCl_2$

water sample is 4887.2 mg/L, and chloride ion can accelerate the destruction of corrosion product film and expose metal matrix to further form pitting corrosion in high temperature environment. However, the comprehensive daily average water production of the experimental well is low, and mainly consists of a small amount of condensate water. With the lower water content, the CO_2 content in the gas phase component is higher, which will increase its solubility in condensate water and produce HCO_3^-. Water-film electrolyte evenly covers the material surface of the pipe string, and forms electrochemical corrosion. According to the change of HCO_3^- concentration, it shows cathodic accelerated hydrogen evolution corrosion or accumulation of corrosion products to inhibit corrosion.

3.2 Corrosion Test Results

From Fig. 2, it can be seen that (a), (b), (c), (d) of N80 and P110, which were salvaged from two locations of 995 and 1498 m, corroded to different degrees on the surface of four groups of metal corrosion specimens, and black corrosion products were formed. The macroscopic characteristics of corrosion products were observed, pittless corrosion and other local corrosion phenomena were found, and the main corrosion type was uniform corrosion. However, the apparent characteristics of the specimens at different depth monitoring locations are different. Some corrosion products of N80 and P110 specimens at 995 m position easily fall off after being invaded by acid cleaning solution and expose the metal matrix, but a few corrosion products still adhere to the surface of the matrix firmly without falling off, as shown in Fig. 2a and 1b; N80 and P110 specimens at 1498 m position have no obvious change after being invaded by acid cleaning solution. The surface of the specimens is basically covered by black acid insoluble corrosion products, and the metal matrix can not be observed. See Fig. 2c, d. Comparing and observing the appearance characteristics of two kinds of material specimens installed at the same location are also different, and the corrosion product coverage area of P110 material surface is larger. After comprehensive comparison, the order of corrosion product coverage area was $S_{1498\mathrm{m/P110}} > S_{1498\mathrm{m/N80}} > S_{995\mathrm{m/P110}} > S_{995\mathrm{m/N80}}$. The dense corrosion product film can isolate the corrosion medium and protect the matrix of the pipe string metal, while the loose corrosion product will accelerate the material corrosion.

The test pieces were weighed after drying and constant weight. The uniform corrosion rate is calculated according to the formula ($V_{corr} = \frac{8.76 \times 10^4 \times \Delta m}{S \times t \times \rho}$), the results are shown in Table 4.

From Table 3, it can be seen that the corrosion rates of (a), (b), (c), (d) correspond to the depths of 995 and 1498 m are less than 0.02 mm/a and slightly corroded, which indicates that the corrosion resistance of the two materials is better under the conditions of two gas wellbore conditions with different temperature and pressure. The corrosion rates of N80 and P110 specimens at 1498 m are 0.0071 mm/a and 0.0074 mm/a, respectively, which are lower than those at 995 m. Combining with the actual production situation, no anticorrosive and scale-proof chemicals were injected into the wellbore during the corrosion monitoring experiment [8]. It shows that under the influence of formation temperature and pressure, the corrosion product film on the surface of 1498 m specimens is more compact. It can isolate the electrochemical corrosion of condensate water on metal matrix more effectively. The experimental data agree with the experimental phenomena of macroscopic corrosion morphology.

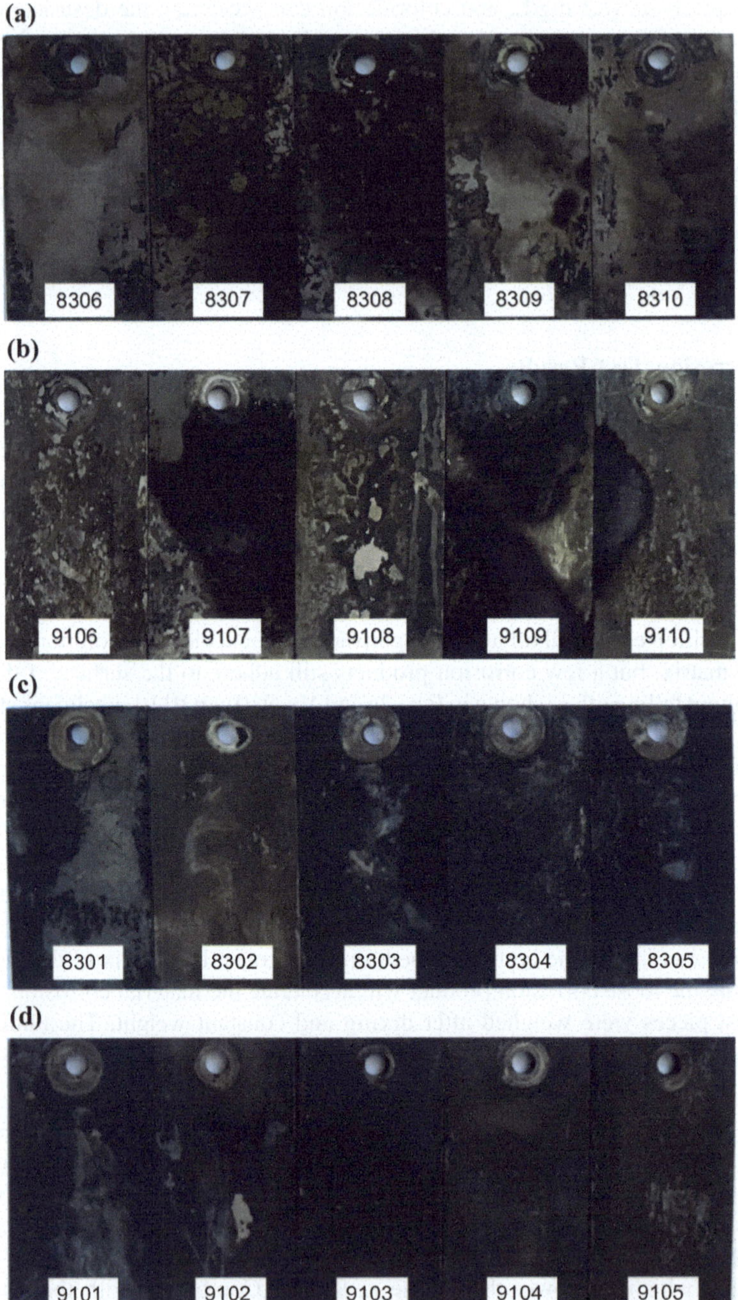

Fig. 2 Macroscopic corrosion morphology of specimens: **a** 995 m/N80; **b** 995 m/P110; **c** 1498 m/N80; **d** 1498 m/P110

Table 4 Corrosion monitoring data by weightlessness method

Strata depth (m)	Number	Δm (g)	S (cm^2)	t (h)	V (mm a^{-1})	V_{avg} (mm a^{-1})
1498	8301	0.1399	28.2766	8760	0.0063	0.0071
	8302	0.1456	28.3430	8760	0.0066	
	8303	0.1549	28.2384	8760	0.0070	
	8304	0.1652	28.1622	8760	0.0075	
	8305	0.1834	28.7465	8760	0.0082	
	9101	0.1550	28.5554	8760	0.0070	0.0074
	9102	0.1588	28.5016	8760	0.0071	
	9103	0.1624	28.6691	8760	0.0073	
	9104	0.1750	28.7250	8760	0.0078	
	9105	0.1738	28.5985	8760	0.0078	
995	8306	0.2987	28.1537	8760	0.0136	0.0145
	8307	0.3214	27.8224	8760	0.0148	
	8308	0.3011	28.2167	8760	0.0137	
	8309	0.3085	28.2910	8760	0.0140	
	8310	0.3657	28.3408	8760	0.0165	
	9106	0.3867	28.6008	8760	0.0173	0.0151
	9107	0.2577	28.6266	8760	0.0115	
	9108	0.3631	28.5308	8760	0.0163	
	9109	0.4053	28.5843	8760	0.0182	
	9110	0.2666	28.5358	8760	0.0120	

3.3 Corrosion Mechanism Analysis

In order to analyze the corrosion behavior of condensate water on two kinds of materials and the corrosion resistance mechanism of metal materials under different temperature and pressure conditions, the corrosion products were characterized by scanning electron microscopy (SEM) in the laboratory, as shown in Fig. 3. Observation of Fig. 3a–c shows that the surface of 995 m/P110 specimens is covered by corrosion product films with different grain sizes. The outer grain size is smaller, and the exposed bottom grain size is larger. Most of them are spherical grains with a size of 10 μm and without edges and corners. In Fig. 3d–f, the dense corrosion product film on the surface of 995 m/N80 specimens is composed of a large number of smaller grains (about 2 μm), but there are no accumulation between grains, and there are cracks. It is observed that a large number of small holes with 4–6 μm are dispersed on the surface of the corrosion product. The reason is the local corrosion formed by Cl^- penetrating corrosion product film in condensate water. When the water content of gas wells increases, the small holes will continue to expand and cause local pitting corrosion. From Fig. 3g–i, it can be seen that there are no holes on the surface of 1498 m/N80 specimens. Under 5000× scanning electron microscopy, it can be clearly observed that the corrosion product film is intersected by cubic crystals with distinct edges and angles and intersected on the surface of the matrix. The surface of 1498 m/P110 specimens in

Fig. 3j–l is identical with 3(h) and uniform without holes. The 5000× observation surface is compactly deposited with small grains of about 1–3 μm in size. The surface of different carbon steels at 1498 m has formed a denser corrosion product film, which can effectively isolate the corrosive medium, especially Cl^- penetration, avoid pitting

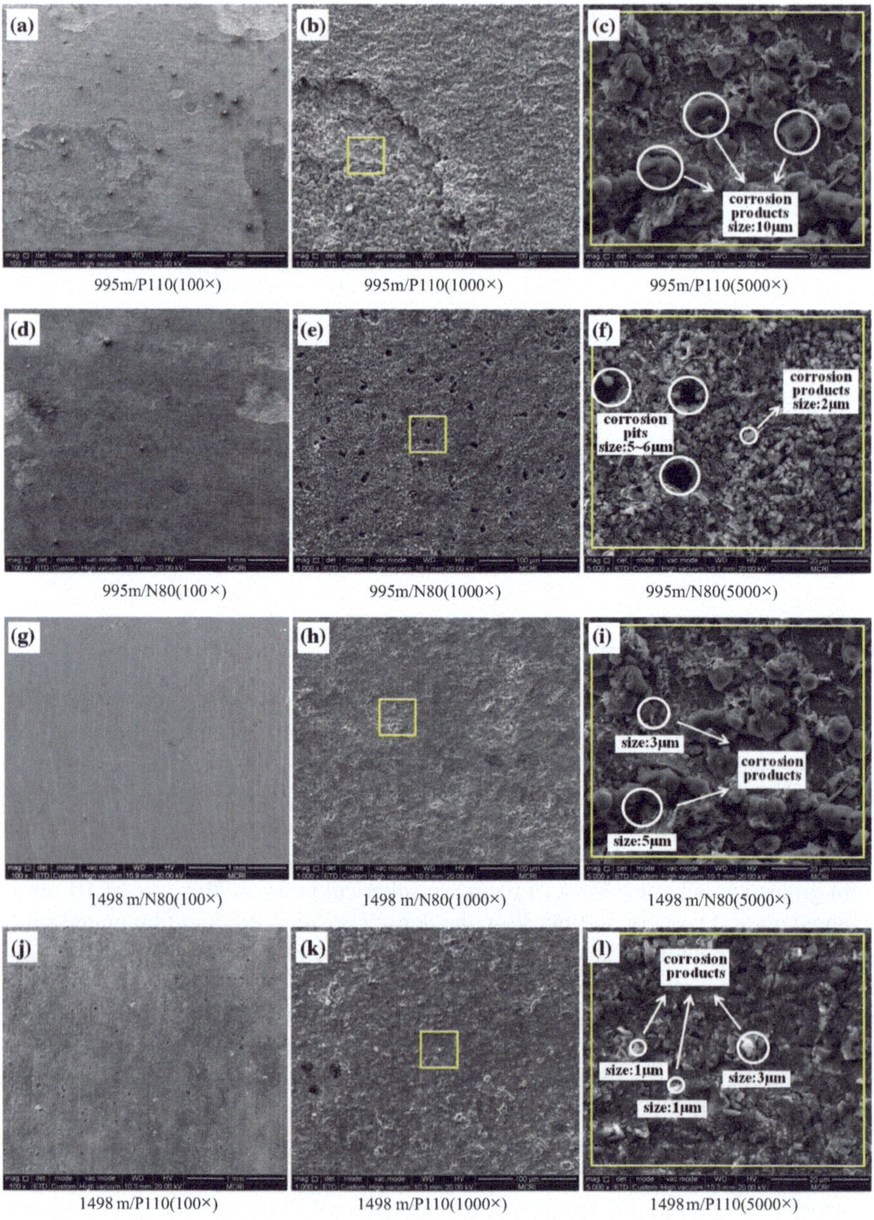

Fig. 3 Microscopic corrosion morphology of corrosion specimens (SEM)

corrosion, effectively reduce corrosion, and macro. The observed phenomena of corrosion morphology are consistent.

From Figs. 4 and 5, EDS elemental analysis and semi-quantitative analysis of elements, it can be seen that Fe, C and O elements were detected in the surface micro-areas of four groups of specimens with corresponding N80 and P110 materials at 995 and 1498 m locations respectively, and the weight percentages were similar, indicating that ferrous carbonate and various forms of iron oxides were formed after corrosion of the metal matrix. Among them, the sample of Fig. 4a contained 0.5% Mn element. Figure 4b the sample contains 0.14% Mn and 0.32% Si. In API standard, the weight ratio of Mn in N80 (36MnV) and P110 (20CrMo) ranges from 0.3 to 1.7%. The weight percentage of Si element is 0.17–0.35%. The Mn and Si detected in Fig. 4a, b micro-zones are from carbon steel matrix, indicating that the surface of the pipe at 995 m is not uniformly covered by corrosion products, while the surface of the pipe at 1498 m is not covered by metal elements other than Fe, indicating that the corrosion products cover uniformly and do not fall off [9, 10]. The composition of corrosion product films of four groups of specimens was analyzed by X-ray diffraction, as shown in Fig. 5. The peak energy of Fig. 5a–d XRD spectra shows that all the peaks except a small amount of Fe are $FeCO_3$, which indicates that under the conditions of temperature and pressure of 995 and 1498 m, the associated CO_2 of natural gas dissolves in condensate water and forms a water film on the surface of N80 and P110 tubular columns, forming a corrosion electrolytic cell. The corrosion products are relatively pure $FeCO_3$ phase. However, because the temperature and pressure at 1498 m are relatively high, the corrosion product film deposited on the metal surface is more uniform and compact, and the $FeCO_3$ crystal is not dissolved, so it can effectively isolate the pitting attack of chloride in condensate water without local corrosion, so the uniform corrosion rate of

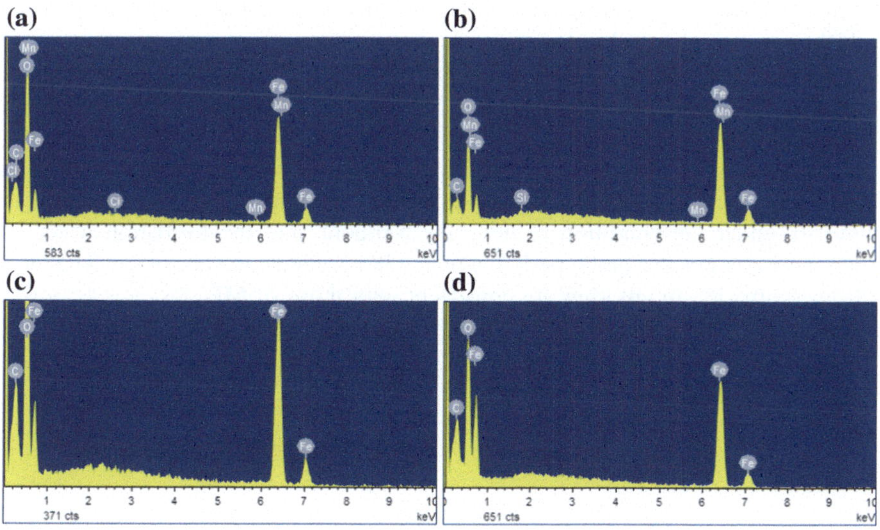

Fig. 4 Element analysis map of microzone eds for corrosion specimens: **a** 995 m/P110; **b** 995 m/N80; **c** 1498 m/N80; **d** 1498 m/P110

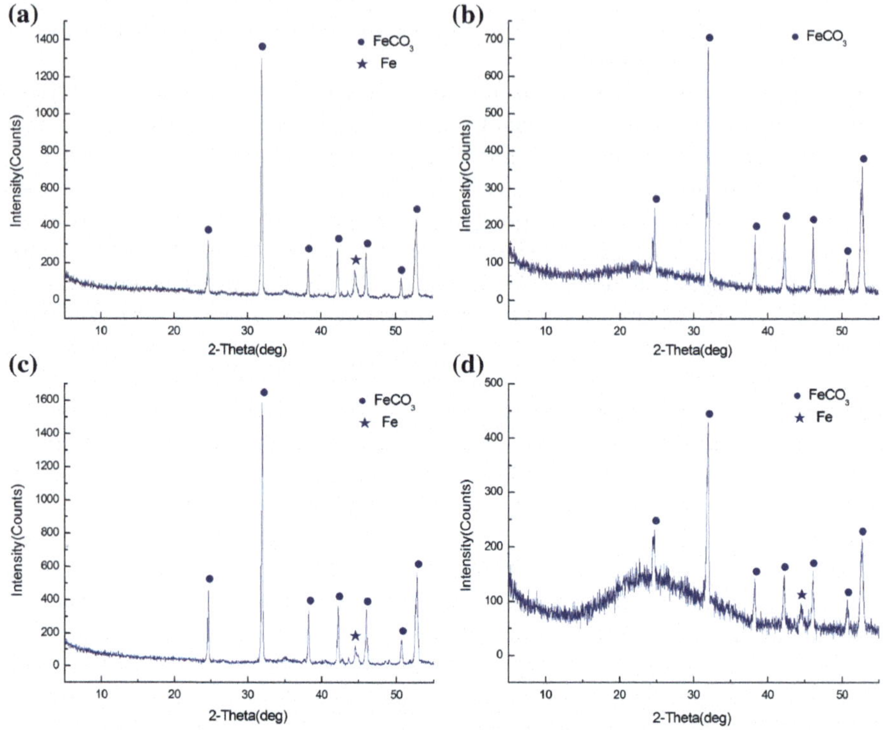

Fig. 5 XRD analysis of corrosion product composition of specimen: **a** 995 m/P110; **b** 995 m/N80; **c** 1498 m/P110; **d** 1498 m/N80

1498 m specimen is lower than that at 995 m, which is completely consistent with the weight loss test results (Table 5).

By measuring the depth of local corrosion pits in 3D, it is found that the maximum pitting depth is 70 μm, as shown in Fig. 6. According to the deepest pitting corrosion data, the calculated local corrosion rate is 0.07 mm/a, but the local average pitting depth of MIT test chart is more than 3 mm. Combined with the "deep excavation" effect after the local pitting formation, it shows that the local corrosion depth increases nonlinearly under the condition of field gas wellbore, and the formation water production increases in subsequent period, which will accelerate the corrosion. The results of off-line corrosion monitoring are consistent with those of MIT test.

Table 5 Weight percentage of EDS element semi-quantitative analysis in microarea of corroded specimens

Element	Weigh(%)			
	(a)	(b)	(c)	(d)
C	9.04	4.29	10.57	12.42
O	28.68	17.53	35.61	25.47
Cl	0.16	/	/	/
Mn	0.43	0.01	/	/
Si	/	0.32	/	/
Fe	61.70	77.86	53.82	62.12
Total	100.00			

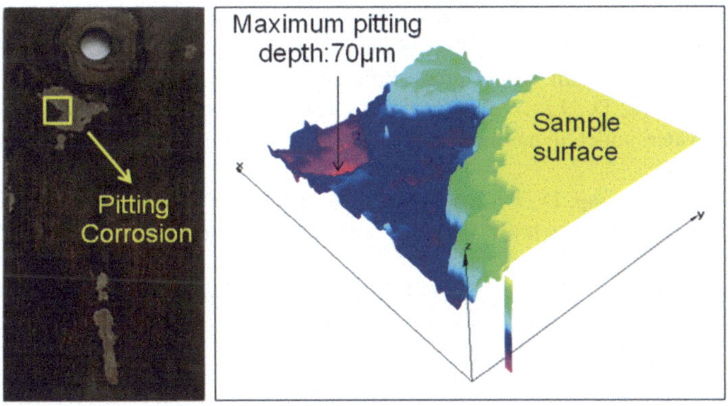

Fig. 6 Local corrosion 3D scan

4 Conclusion

(1) The water production of Well S141 in Changbei Gas Field is low, mainly consisting of low-content condensate water, low salinity of water quality, less corrosive components such as chloride. The corrosion product film at 1498 m is denser than that at 995 m, and the corrosion product $FeCO_3$ phase is dominant, which indicates that high temperature and high pressure conditions are more conducive to the dense deposition of corrosion product grains.

(2) The uniform corrosion loss of specimens with different well depths and different materials is small, but local corrosion appears on the surface, and the deepest corrosion depth is 70 μm. It can be predicted that the depth of corrosion pit will continue to increase. The experimental phenomena are consistent with the results of MIT test. The next step is to take targeted anti-corrosion measures to restrain local corrosion.

References

1. Olorunniwo, O.E., Imasogie, B.I., Afonja, A.A.: Evaluation of pipeline corrosion in sour-gas environment. Anti-Corros. Methods Mater. **54**(6), 346–353 (2007)
2. Li, W., Pots, B.F.M., Zhong, X., Nesic, S.: Inhibition of CO_2 corrosion of mild steel—study of mechanical effects of highly turbulent disturbed flow. Corros. Sci. **126**, 208–226 (2017)
3. Li, J.L., Ma, H.X., Zhu, S.D., Qu, C.T., Yin, Z.F.: Erosion resistance of CO_2 corrosion scales formed on API P110 carbon steel. Corros. Sci. **86**, 101–107 (2014)
4. Sun, J.B., Sun, C., Zhang, G.A., Zhao, W.B., Wang, Y.: Effect of water cut on the localized corrosion behavior of P110 tube steel in supercritical CO_2/oil/water environment. Corrosion **72**, 1470–1482 (2016)
5. Li, D.G., Feng, Y.R., Bai, Z.Q., et al.: Characteristics of CO_2 corrosion scale formed on N80 steel in stratum water with saturated CO_2. Appl. Surf. Sci. **253**(2), 8371–8376 (2007)
6. Yin, Z., Zhao, W., Bai, Z., et al.: Characteristics of CO_2 corrosion scale formed on P110 steel in simulant solution with saturated CO_2. Surf. Interface Anal. **40**(9), 1231–1236 (2008)
7. Li, D.G., Feng, Y.R., Bai, Z.Q.: Effect of temperature on the CO_2 corrosion product films of N80 carbon steel. J. Chin. Soc. Corros. Prot. **28**(6), 369–373 (2008)
8. Zhang, G.A., Liu, D., Li, Y.Z., Guo, X.P.: Corrosion behavior of N80 carbon steel in formation water under dynamic supercritical CO_2 condition. Corros. Sci. **120**, 107–120 (2017)
9. Esmaeely, S.N., Young, D., Brown, B.N., Nesic, S.: Effect of incorporation of calcium into Iron carbonate protective layers in CO_2 corrosion of mild steel. Corrosion **73**, 238–246 (2017)
10. Yin, Z.F., Feng, Y.R., Zhao, W.Z., et al.: Effect of temperature on CO_2 corrosion of carbon steel. Surf. Interface Anal. **41**(6), 517–523 (2010)

Experimental Research on Anti-corrosion Process of BG110-3Cr Casing Zinc Alloy

Bao-yu Wu[1,2(✉)], Zai-jun Chen[1,2], and Xiang-zhe Jin[1,2]

[1] Drilling & Production Engineering Technology Research Institute, Chuanqing Drilling Engineering Co. Ltd., CNPC, Xi'an 710018, China
wubaoyu@cnpc.com.cn

[2] National Engineering Laboratory for Exploration and Development of Low-Permeability Oil & Gas Fields, Xi'an 710018, China

Abstract. In order to study the effect of processing technology of Radium zinc anode for direct casting corrosion protection on the outer surface of BG110-3Cr casing on the properties of casing matrix and anode material. The mechanical properties and SEM fracture morphology of two sets of tensile specimens of casing material obtained by heating and casting heat conduction of resistance furnace were tested by metal material tensile test method, and then metallographic specimens were prepared for observation of microstructure. According to the properties of cast anode material, the content of melting elements in alloy was detected by ICP method and direct reading spectroscopy method respectively, and the content of melting elements in alloy was determined by electrochemical method. The polarization curves of the alloys were measured at 20 °C, 40 °C, 60 °C and 80 °C respectively, and the electrochemical properties were evaluated by comparing the polarization curves with those of the national standard anodes. The experimental results show that: (1) Rm and Rp0.2 of BG110-3Cr steel decrease to 913.1 MPa and 799.4 MPa respectively after heating up to 600 °C, the equiaxed dimples with uniform fracture surface become pits, the yield strength decreases, the carbide particles in tempered sorbite aggregate, the strength decreases, but the toughness increases, which is higher than API standard; (2) Under the influence of heat transfer and heating up of casting, the white distribution in the metallographic structure of the casing is 1–3 μm. Carbide particles, measured Rm and Rp0.2 increased to 1015.1 MPa and 890.5 MPa, respectively. (3) No macrosegregation and microscopic segregation were observed in the metallographic structure of the cast anode material. The

This paper was prepared for presentation at the 2019 International Petroleum and Petrochemical Technology Conference in Beijing, China, 27–29 March 2019.

J. Lin (ed.), *Proceedings of the International Petroleum and Petrochemical Technology Conference 2019*, pp. 401–413, 2020.
https://doi.org/10.1007/978-981-15-0860-8_34

polarization curve showed that the corrosion potential decreased with the increase of temperature. The corresponding corrosion potential at 80 °C was −1.09 V. No passivation was observed and the electrochemical performance was stable. The results provide technical support and theoretical basis for the processing of temperature-resistant anodes for cast anticorrosive casing and their application in the field of cathodic protection of deep well casing.

Keywords: Casting · Radium zinc anode · Heat conduction · Tempering heat treatment · Mechanical properties

1 Introduction

In Hade block of Tarim oilfield, oil and gas are deeply buried, formation pressure and temperature are on the high side, and high salinity formation water is rich in CO_2, H_2S, SRB and other soluble corrosive gases and corrosive substances. Under this bad condition, when cement sheath is not completely sealed or cementing quality is not up to standard, BG110-3Cr casing matrix will be exposed to corrosive media, accelerating corrosion, resulting in increasing number of casing damage wells year by year, affecting normal production and causing huge economic losses [1, 2]. Practice has proved that, in view of the casing damage caused by corrosion, adopting sacrificial anode cathodic protection measures and reasonably arranging anode materials on the outer surface of casing can effectively slow down corrosion [3]. However, there are two problems in the actual implementation of the current casing cathode technology: (1) When using the national standard anode material or the improved anode alloy, the material performance failure caused by the high temperature in the wellbore will occur, which limits its application. Taking the national standard Zn–Al–Cd alloy as an example, when immersed in formation water above 60 °C, the potential reversal of the anode material will occur due to intergranular corrosion, and the cathodic protection of the casing material can not be carried out. (2) The sacrificial anode is usually connected to the casing by clamp locking, which leads to poor conductivity.

Based on the preliminary test and demonstration, a new type of zinc-based alloy anode material was developed by adding alloying elements and changing the alloy composition ratio on the basis of the national standard zinc anode, which can meet the requirements of high temperature environment in deep wells. It was also innovatively proposed that the anode material should be cast directly on the surface of casing by casting method to form an anode ring to make the connection between the anode and casing closer. Enhance the conductivity and facilitate current transmission.

During the casting process of casing, infrared thermal imager was used to track the temperature changes of melting furnace, ladle, casting liquid and inner surface of casing in turn. It was found that the temperature of casing would rise rapidly and keep warm until it cooled to room temperature in air after mould opening due to the influence of heat conduction of casting liquid. Its process is essentially a non-standardized secondary tempering heat treatment of carbon steel, which will affect its mechanical properties [4]. In order to reduce the influence of heat transfer on the properties of casing matrix, the process parameters such as melting temperature, casting time, casting temperature and holding time must be precisely controlled. At the same time, the non-melting and segregation of anode alloy elements caused by low temperature should be prevented, and the electrochemical properties should be reduced. The relevant contents are seldom mentioned in the research reports. Therefore, the applicability of the anode casting process is comprehensively demonstrated and evaluated by means of reasonable design experiments, auxiliary mechanical properties testing, fracture morphology analysis, microstructure observation and electrochemical performance testing. It provides technical support and theoretical basis for the popularization and application of this technology in the field of cathodic protection technology for deep well casing.

2 Experiment Procedure

2.1 Experimental Instruments and Materials

The main instruments used in the experiment are shown in Table 1.

Table 1 Main experiment instruments

Serial number	Instrument name	Instrument type	Manufacturer
1	Electronic universal testing machine	AG-IS 250kN	SHIMADZU
2	Metallurgical microscope	DSX510	OLYMPUS
3	SEM	600FE	FEI Quanta
4	Direct reading spectrometer	MAXx	SPECTRO
5	ICP-OES	iCAP-7000	Thermo Fisher
6	Electrochemical workstation	Interface 1010E	GAMRY
7	Box type resistance furnace	SX2-2.5-10NP	YIHENG
8	Analytical balance	PRACTUM224-1CN	SARTORIUS
9	Infrared thermal imager	GTC400C	BOSCH

Main materials: oil casing (steel grade: BG110-3Cr, specification: 139.70 m 10.36 mm, length: 11.36 m); purity of more than 99% of aluminum ingots, zinc ingots, cadmium grains, indium grains, magnesium powder; Preparing anticorrosive sleeve: Remove the coating on the surface of sleeve, set the temperature of smelting furnace at 500 °C, heat zinc ingot to melt, add weighted Al–Mg binary alloy, raise the furnace temperature to 650 °C, stir and dissolve evenly, then cool down to 500 °C, add indium, cadmium and refining agent NH_4Cl, When all the elements are mixed evenly, the anode ring is formed by casting on the outer surface of the casing through the die.

2.2 Experimental Design of Casing Performance Testing

In the casting process, the temperature fluctuation of the casing matrix will occur rapidly due to the influence of heat conduction and air cooling. The temperature of the

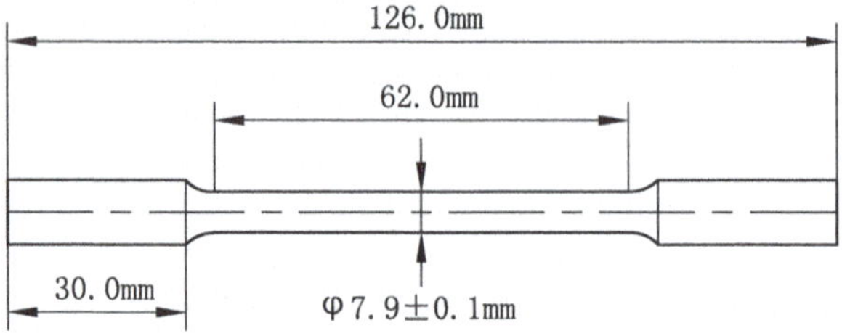

Fig. 1 Cylindrical tensile specimens

inner surface of the casing can be recorded by infrared thermal imager in the casting area, but the temperature change of the casing surface in the casting fluid-covered area can not be measured accurately. However, the inner surface is affected by gas convection heat transfer after contacting with air, and the temperature is lower than that of the outer surface. By measuring the temperature fluctuation of casting liquid, the upper and lower limits are determined to be 600 °C and 550 °C respectively. According to the heat transfer effect and heat loss of the contact between casting liquid and casing matrix, the upper limit of temperature rise on the outer surface of casing in casting area is 600 °C, so 550 and 600 °C are chosen as the temperature rise points of simulated heat treatment.

In the first step, three parallel samples were cut on the ordinary BG110-3Cr casing and processed into standard circular tensile specimens according to GB/T 228.1-2010 standard requirements. The dimensions (diameter: 7.9 ± 0.1 mm, standard distance L: 62.00 mm) are shown in Fig. 1. Casting heat transfer of casing was simulated by box resistance furnace. Two of the specimens were heated to 550 °C and 600 °C respectively, then kept for 30 min. Mechanical properties, tensile fracture SEM analysis and metallographic analysis of the three specimens were measured after air cooling to room

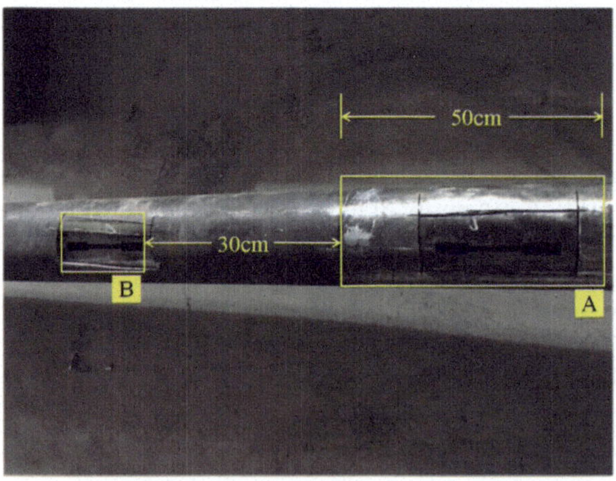

Fig. 2 Sampling position of casing surface: Area A is 50 cm long, which is the casing matrix exposed after stripping and coating the anode after the anode casting; Area B is a common casing 30 cm away from area A to ensure that it is not affected by heat transfer

temperature. In the second step, on the basis of the first experimental demonstration, the experimental areas A and B are randomly selected on the surface of the casing, and the mechanical properties of the stretching parts are tested after cutting out a sample, as shown in Fig. 2.

2.3 Anodic Chemical Composition Testing

Inductively Coupled Plasma Emission Spectrometry (ICP): Matrix matching method in the national standard method was used to detect the content of molten metal elements in the metal powder of zinc alloy anode material after casting by iCAP7000. The experimental parameters were set as follows: 40 MHz frequency; 1.2 kW power; 99.99% argon as cooling gas; gas flow rate (plasma 12 L/min, atomizer 1 L/min, auxiliary gas 0.7 L/min); radial observation height 8 mm; spectral lines (nm) Al: 308.21, Cd: 214.44, Mg: 279.55, In: 230.61 with less interference and high sensitivity; zinc alloy powder weighed accurately by analytical balance 0.1009 g The alloy was dissolved in a mixture of 10 mLHCl + 1 mLHF + 1 mLHNO_3 at low temperature, then cooled to room temperature, and the volume was set to 100 mL with deionized water. The content of metal elements in the dilution solution was measured, and the molten element ratio of the alloy was calculated. Direct-reading spectroscopy: Side-cut zinc alloy and on its surface, 2 kW high spark power excitation sample was selected for detection.

2.4 Anode Electrochemical Performance Test

According to ASTM G5*, using Interface 1010E electrochemical workstation and double-deck water bath plate corrosion cell, the electrochemical polarization curves of zinc alloy materials were tested by simulating formation temperature changes (20, 40, 60, 80 °C). The influence of ambient temperature on the thermal sensitivity of anode materials was characterized by the change of corrosion potential of the polarization curves. Parameters: SCE (reference electrode), platinum electrode (auxiliary electrode) and working electrode (5 cm × 2.5 cm × 0.2 cm size specimens processed with cast zinc alloy anode, and 1 cm^2 as working area) constitute a three-electrode system together; corrosion electrolyte solution is collocated as simulated formation water. The scanning range of potentiodynamic scanning method was $E_{corr} \pm 150$ mV, and the scanning speed was 0.166 mV/s.

3 Results and Discussion

3.1 Test Results of Casing Performance

Table 2 is the experimental data of mechanical properties of three parallel samples made of BG110-3Cr casing heated by box resistance furnace. The tensile strength and yield strength of three groups of specimens under 45 kN tensile strength stress and 40 kN yield strength stress are given respectively. The tensile strength and yield strength of 2# specimens increased to 935.3 MPa and 822.5 MPa, respectively, but the

Table 2 Testing of mechanical properties of stretching parts heated by resistance furnace

Sample number	Experimental condition	Tensile strength stress (kN)	Tensile strength (MPa)	Yield strength stress (kN)	Yield strength (MPa)	Break elongation (%)	Section shrinkage (%)
1#	25 °C	45.9	933.9	40.3	820.6	17.9	72.9
2#	550 °C	45.9	935.3	40.3	822.5	19.5	73.6
3#	600 °C	44.9	913.1	39.3	799.4	18.5	73.0
/	API standards	45.0	862.0	40.0	758–965	/	/

tensile strength and yield strength of 3# specimens began to decrease significantly, but 913.1 and 799.4 MPa still met API standards. The necking phenomenon and obvious plastic deformation occur in the fracture area of the three tensile parts. See Fig. 3b. The macro-morphology of the specimen is cup-conical, and there is obvious fiber area. See Fig. 4. The fracture morphology is all equiaxed dimples, which belongs to typical ductile fracture. The fracture morphology of 1# sample and 2# sample is observed by 10,000× and 40,000× scanning electron microscopy. Shallow dimples with sizes less than 1 micron are evenly distributed, and the number of dimples in 2# samples is less, as shown in Fig. 4c, f. Observation of 3# samples shows that dimple size increases significantly and depth increases. With the increase of temperature, local dimples aggregate and grow to form oval pits of 10 μm, as shown in Fig. 4i. The increase of dimple size indicates that more severe plastic deformation occurs during dimple formation, more energy is absorbed during fracture, more fully deformed and better plasticity, and the corresponding yield strength value is low, which is consistent with the results of mechanical properties test (Fig. 5).

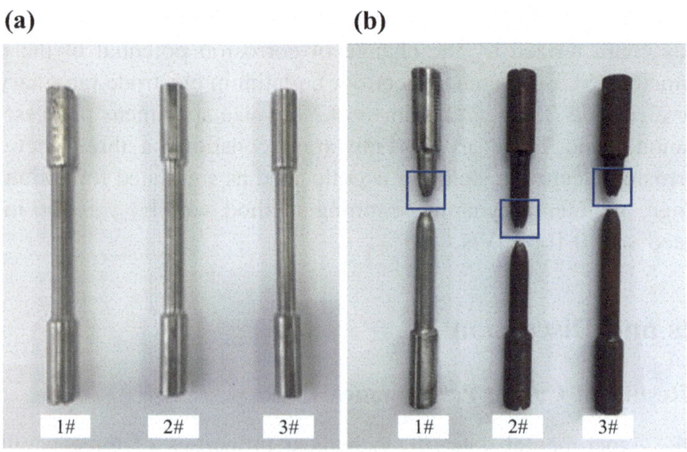

Fig. 3 Experimental stretching piece: **a** before test; **b** after test

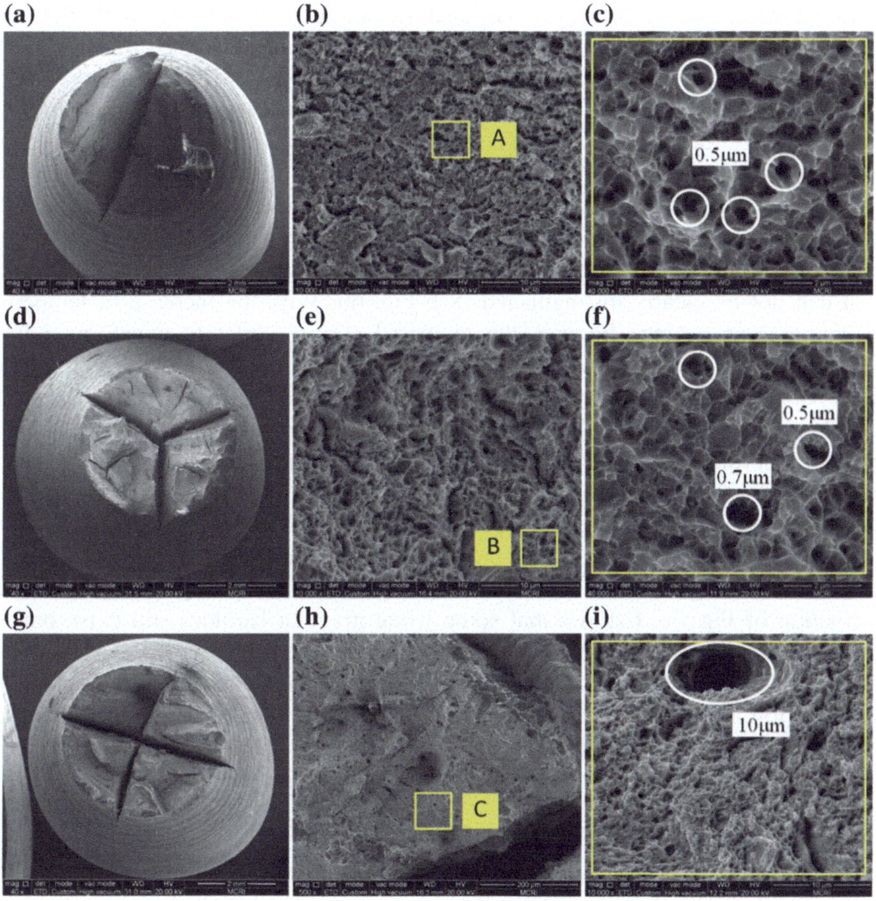

Fig. 4 Fracture morphology of tensile parts after heating (SEM): **a** 1#25 °C (40×); **b** 1#25 °C (10,000×); **c** 1#25 °C (40,000×); **d** 2#550 °C (40×); **e** 2#550 °C (10,000×); **f** 2#550 °C (40,000×); **g** 3#550 °C (40×); **h** 3#550 °C (500×); **i** 3#550 °C (10,000×)

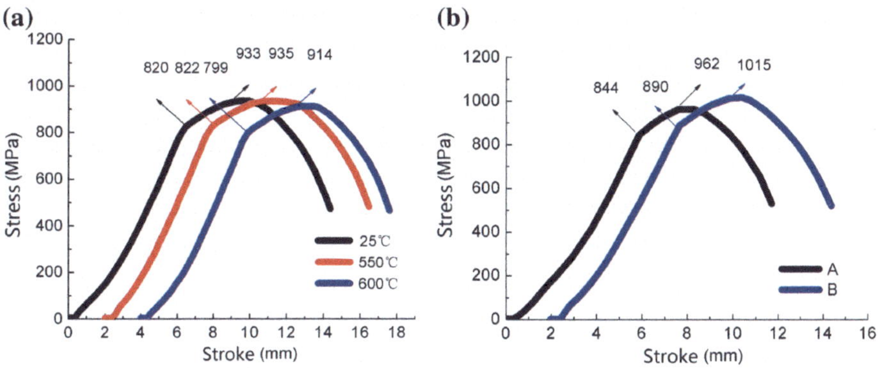

Fig. 5 Stress-strain curves of specimens corresponding to different heating modes: **a** resistance furnace heating; **b** heat conduction heating in casting

BG110-3Cr seamless steel pipe is heat treated by normalizing at 930 °C and quenching at 920 °C and oil cooling at 200 °C. The structure obtained by this process is mainly lath tempered martensite and a small amount of ferrite [5–11], which satisfies the strength and enhances the toughness at the same time. Figure 6a is the martensite metallographic structure of blank BG110-3Cr casing steel without resistance furnace heating. In essence, the rapid heating, insulation and cooling process of casing matrix in casting process is a secondary tempering heat treatment process for BG110-3Cr. After tempering temperature exceeds 500 °C, the structure is mainly tempered sorbite. At this time, the carbide particles are small and distributed evenly in the structure, which can not be clearly distinguished. See Fig. 6b. With the increase of tempering temperature in resistance furnace, the tempered sorbite structure transformed from tempered martensite and troostite continues to grow when tempered temperature rises to 550 °C. See Fig. 6c, d. Fine white round carbides with uniform distribution of 1–3 um in tempered sorbite can be clearly observed. The structure has good strength and toughness coordination. The metallographic structure has good tensile strength and yield strength with 2# experimental results are consistent. The yield strength and tensile strength of 3# specimens decreased rapidly by 21 and 23 MPa after heating up to 600 °C. The reason is that the finer the carbide particles in tempered sorbite, the higher the hardness and strength of the alloy, but the toughness of the alloy is slightly worse. Observation of Fig. 6e, f shows that some white granular carbides still exist, but the white carbon precipitated from the other part of tempered sorbite remains. At this time, quenching stress is basically eliminated, martensite decomposition and austenite transformation are basically completed, and the structure tends to be stable under this temperature condition, the fracture mode of the structure changes from brittle fracture to ductile fracture. Therefore, 3# elongation and section shrinkage increase after fracture.

Table 3 shows that the tensile strength and yield strength of 2# specimens increase slightly, reaching 1015.1 MPa and 890.5 MPa respectively, which meets the standard requirement. It shows that the matrix temperature rise caused by casing surface casting does not affect the mechanical properties of 2# specimens.

3.2 Test Results of Anodic Chemical Composition

As shown in Table 4, the contents of elements in zinc alloys were determined by inductively coupled plasma emission spectrometry (ICP-OES) and direct reading spectrometry (DRS), respectively. The relative error between the average value and the amount of added material is less than 10%, the element loss is less, the yield is stable and meets the design requirements. The results show that the alloying elements added by this smelting process are uniformly distributed in the melting liquid at high temperature without macrosegregation, and the inner and outer surfaces of the casting samples are observed by metallographic microscope with fine grains and uniform distribution without micro segregation (see Fig. 7).

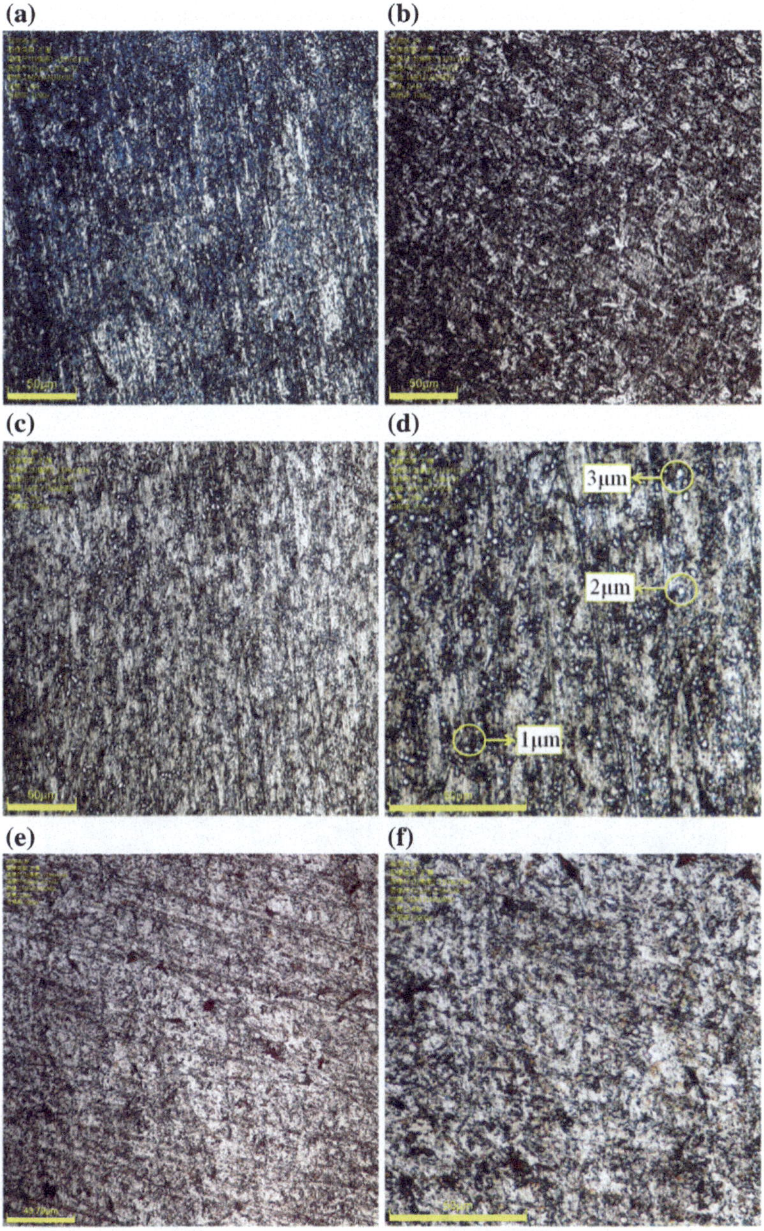

Fig. 6 Metallographic structure corresponding to different heat treatment processes: **a** tempered martensite; **b** tempered troostite; **c** tempered sorbite; **d** tempered sorbite (2000×); **e** carbide grain growth; **f** carbide grain growth (2000×)

Table 3 Testing of mechanical properties of casting heat transfer drawing pieces

Sample number	Sampling area	Tensile strength stress (kN)	Tensile strength (MPa)	Yield strength stress (kN)	Yield strength (MPa)	Break elongation (%)	Section shrinkage (%)
1#	A	45.9	962.7	40.5	844.2	16.5	71.6
2#	B	46.1	1015.1	40.9	890.5	17.5	72.0
API standards		45.0	862.0	40.0	758–965	/	/

Table 4 Elemental analysis of zinc alloy

Element	Material addition (wt %)	ICP			DRS		
		Value (wt %)	RSD (%)	δ (%)	Value (wt %)	RSD (%)	δ (%)
Al	0.221	0.217	0.5	1.8	0.226	2.9	2.3
Cd	0.092	0.090	2.1	2.2	0.088	3.7	4.4
Mg	0.062	0.059	1.1	4.8	0.056	0.9	9.7
In	0.063	0.067	2.6	6.4	0.068	1.5	9.5

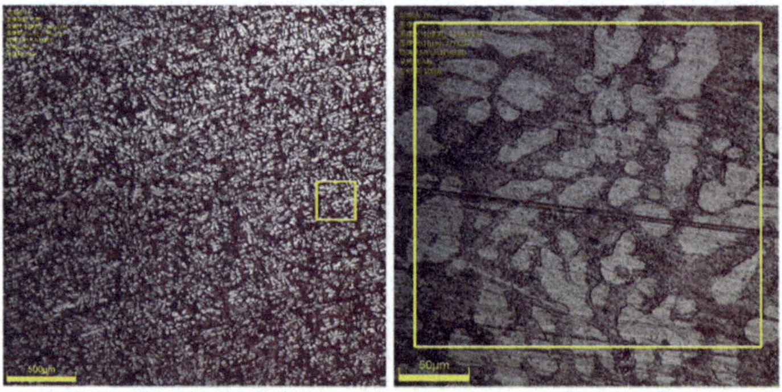

Fig. 7 Metallographic structure of zinc alloy

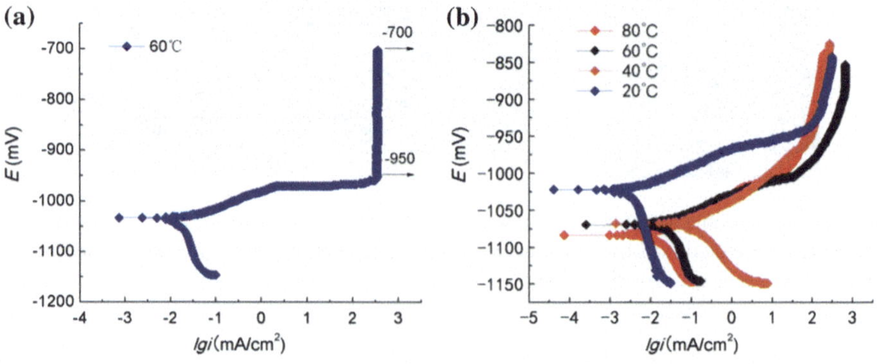

Fig. 8 Electrochemical polarization curve of zinc alloy: **a** national Standard Zn–Al–Cd alloy; **b** new zinc alloy

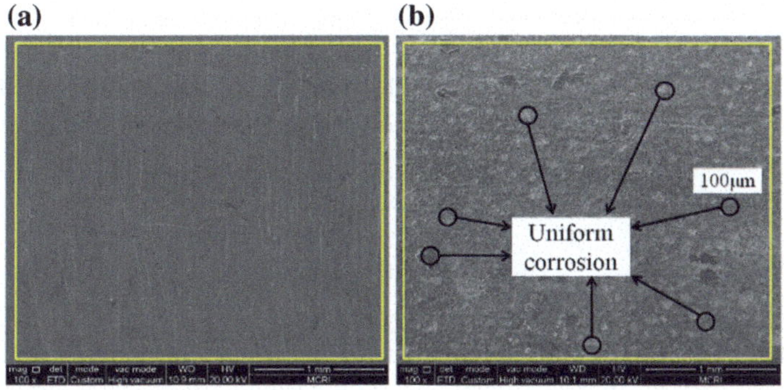

Fig. 9 Microscopic corrosion morphology of zinc alloy (SEM): **a** before corrosion; **b** after corrosion

3.3 Test Results of Electrochemical Performance of Anode

Figure 8a is the polarization curve of standard Zn–Al–Cd alloy in simulated formation water with high salinity at 60 °C When the corrosion potential is in the range of −0.95 to −0.70 V, the current density changes very little and enters the anode passivation area. The anode material can not continue to output current at this temperature, and can not protect the casing matrix under this condition. Because the corrosion electrochemical reaction of zinc alloy is affected by temperature, the main product of zinc anode corrosion reaction at low temperature is $Zn(OH)_2$. When the temperature of corrosive medium rises to 55 °C, the corrosion product $Zn(OH)_2$ covered on the surface of the anode will be transformed into zinc oxide, while zinc oxide has electronic conductivity, which makes the zinc anode reversed into cathode protection with the increase of temperature, and the bushing into anode will be corroded. Figure 8b is a fitting of polarization curve of zinc anode tested in simulated casing water at 20–80 °C. With the increase of temperature, the corrosion potential of zinc anode decreases gradually, and the corrosion potential of zinc anode at 80 °C reaches −1.09 V. The anodic polarization curve is smooth, the polarizability of the electrode is small, the resistance of the electrode reaction is small, and the passivation phenomenon occurs unaffected by temperature. The reason is that adding appropriate proportion of In into the national standard zinc anode alloy can greatly improve the discharge performance of zinc and increase its activation potential area, which is conducive to making full use of the active substance zinc during discharge. Moreover, the passivation potential of zinc-in alloy electrode is more positive than that of pure zinc electrode, and the passivation current in the passivation zone is less than that of pure zinc. Therefore, zinc alloy is not easy to passivate after adding in element, but the passivation film formed after passivation is more stable and corrosion resistance is enhanced [12]. At the same time, low content of Cd can form solid solution with impurities in zinc anode, and further weaken the self-corrosion of new alloy. The measured current efficiency is more than 90%. The micro-morphology of the alloy surface before and after spot A corrosion was observed by scanning electron microscopy. It was found that a large number of shallow circular corrosion pits of 100 μm were evenly distributed in 10 mm^2, and no local pitting corrosion was observed. The structure of the anode surface coating did not change significantly after corrosion. This indicates that the anode material can be uniformly dissolved in the environment with formation temperature lower than 80 °C, thus providing continuous protection for the casing. Protect current and restrain electrochemical corrosion of casing (Fig. 9).

4 Conclusion

(1) Through mechanical properties testing, fracture morphology analysis and metallographic observation, it is found that the secondary heat treatment effect produced by direct casting sacrificial anode on casing surface has a direct impact on mechanical properties because of its large fluctuation in temperature in a short time. Resistance furnace will air-cooled the specimen to room temperature at 600 °C, Rm and Rp0.2 values are significantly reduced, affecting the mechanical properties; through strict control of casting process, the performance of the measured casing affected by heat conduction is slightly improved, which meets the standard requirements.

(2) The effects of casting process on the properties of zinc alloy sacrificial anode materials were studied by electrochemical polarization curve test and chemical composition test. It was found that the electrochemical properties of the alloys tested by polarization curves at 20, 40, 60 and 80 °C were not affected by temperature, and the corrosion potential reached −1.09 V without passivation. ICP and direct reading spectroscopy showed less loss of metal elements in the alloys, and the yields were stable, which met the design requirements, and were not observed by metallographic microscopy. The macrosegregation and microsegregation phenomena show that the process can meet the quality requirements of anode castings.

(3) The anticorrosive technology of direct casting zinc anode on the outer surface of BG110-3Cr casing ensures the electrochemical performance of the anode material and improves the mechanical properties of the casing matrix slightly.

References

1. Qiu, Z.X., Chang, Z.L., et al.: Major corrosion factors in the CO_2 and H_2S coexistent environment and the relative anti-corrosion method: taking Tazhong I gas field, Tarim Basin, as an example. Pet. Explor. Dev. **39**(2), 238–242 (2012)
2. Wang, T., Yang, S.L., Zhu, W.H., et al.: Law and countermeasures for the casing damage of oil production wells and water injection wells in Tarim Oilfield. Pet. Explor. Dev. **38**(3), 31–32 (2011)
3. Edavan, R.P., Kopinski, R.: Corrosion resistance of painted zinc alloy coated steels. Corros. Sci. **51**(10), 2429–2442 (2009)
4. Wang, X.D., Wang, B.F., Cao, J.G., Liu, H.J.: Research on quenching process for P110 oil casing. Hot Work. Technol. **40**(6), 185–186 (2011)
5. Chen, J.D., Mo, W.L., Wang, P., Lu, S.H.: Effects of tempering temperature on the impact toughness of steel 42CrMo. Acta Metall. Sin. **48**(10), 1186–1193 (2012)

6. Birol, Y.: Impact of grain size on mechanical properties of $AlSi_7Mg_{0.3}$ alloy. Mater. Sci. Eng.: A (Struct. Mater.: Prop. Microstruct. Process.) **559**, 394–400 (2013)
7. Elvins, J., Spittle, J.A., Sullivan, J.H., Worsley, D.A.: The effect of magnesium additions on the microstructure and cut edge corrosion resistance of zinc aluminium alloy galvanised steel. Corros. Sci. **50**(6), 1650–1658 (2008)
8. Hui, W.J., Dong, H., Wang, M.Q., et al.: Influence of quenching temperature on mechanical properties of low alloy high strength Cr-Mo-V steel. Heat Treat. Met. **27**(3), 14–16 (2002)
9. Jian, L.I., Yang, J., Zhou, H.M., Kuang, X.G., Yang, L., Guo, X., et al.: Corrosion and mechanical properties of zat10 deformed zinc alloy in different corrosive environments. J. Cent. South Univ. **43**(9), 3380–3385 (2012)
10. Brown, S.G.R., Barnard, N.C.: 3d computer simulation of the influence of microstructure on the cut edge corrosion behaviour of a zinc aluminium alloy galvanized steel. Corros. Sci. **48** (8), 2291–2303 (2006)
11. El-Sayed, A.R.: Inhibitive action of ferricyanide complex anion on both corrosion and passivation of zinc and zinc-nickel alloy in the alkaline solution. J. Power Sources **196**(15), 6573–6582 (2011)
12. Sullivan, J., Weirman, C., Kennedy, J., Penney, D.: Influence of steel gauge on the microstructure and corrosion performance of zinc alloy coated steels. Corros. Sci. **52**(5), 1853–1862 (2010)

Development and Application of Integrated Corrosion Monitoring System in Refining Unit

Liang-chao Chen[1], Jian-feng Yang[1(✉)], and Xin-yuan Lu[2]

[1] Beijing University of Chemical Technology, North Third Ring Road 15, Chaoyang District, Beijing 100029, China
2004500017@mail.buct.edu.cn

[2] China Special Equipment Inspection and Research Institute, Building 2, Hepingjie Xiyuan, Chaoyang District, Beijing 100029, China

Abstract. For a large part of accidents in refining enterprise are caused by equipment and pipeline corrosion. By enhancing corrosion monitoring management on process and equipment, enterprise can improve the effect of corrosion protection work, and guarantee the safe and stable operation of unit. In order to manage the on-line monitoring, on-line and manual fixed-point thickness detecting and chemical analysis of corrosion monitoring methods commonly used in refining enterprise, this paper develops the integrated corrosion monitoring system, which achieves data acquisition, data management and data analysis etc. in various types of corrosion monitoring. Based on the method of Integrity Operating Windows, and taking the circuit as the evaluation object, the integrated corrosion monitoring system can do correlation analysis to all kinds of monitoring results, and judge the root cause of the corrosion occurred from multi aspects by combining with the basic data of equipment and process, which provides guidance and basis for the enterprise decision-making on corrosion management. After more than one year of operation, the system has played an important role in finding abnormal data analysis in many times, judging corrosion position quickly, and guiding enterprises to take corresponding anticorrosion measures rapidly, which effectively prevent the safety accidents.

Keywords: Corrosion monitoring · System development · Corrosion evaluation

J. Lin (ed.), *Proceedings of the International Petroleum and Petrochemical Technology Conference 2019*, pp. 414–426, 2020.
https://doi.org/10.1007/978-981-15-0860-8_35

1 Introduction

In order to effectively control the corrosion problem of refining plant, many corrosion control measures have been adopted by refinery and chemical enterprises, including process anticorrosion, material upgrading, surface treatment and so on (Javaherdashti 2000). Corrosion monitoring technology, which is used to evaluate the anticorrosion effect of the process, control the operation state of the equipment and make early warning for equipment corrosion failure, has been widely used in refining and chemical enterprises (Chen et al. 2017). On-line corrosion monitoring, on-line thickness detecting and other real-time monitoring systems based on the Internet of things and chemical analysis, fixed-point thickness detecting and other in-service monitoring methods are most widely used, which play an active role in providing data support for plant technicians to carry out corrosion protection work (Li et al. 2016). However, there are still some problems in the management and application of corrosion monitoring data in China, such as non-standard data management, poor data correlation of various systems and low application level.

In this paper, an integrated corrosion monitoring system is established for the on-line corrosion monitoring, on-line and manual fixed point thickness detecting and chemical analysis monitoring methods. The system is based on the SOA architecture and the whole business target to realize the monitoring data collection, the related system integration and the data analysis utilization, and then the main function modules including the corrosion monitoring, the thickness detecting, the chemical analysis and the alarm and analysis decision are established. Based on the idea of circuit in risk assessment and the method of Integrity Operation Window (IOW), the system establishes data analysis and intelligent correlation model from two aspects of process and equipment. After a certain time of application, the system realizes the function of monitoring, warning and evaluation and decision-making effectively.

2 Corrosion Monitoring Management in Process Industry

In order to realize the requirement of integrated corrosion monitoring management, it is necessary to collect, process, analyze, compare and integrate all kinds of monitoring data. Then by making comprehensive use of the data, correlating and analyzing the data intelligently, all kinds of abnormal data and emergency warning can be realized to guide managers to find corrosion problems or hidden dangers quickly and to take effective measures in time which would reduce the risk of accidents caused by corrosion.

From the point of view of data management and application, the corrosion monitoring system needs to meet the basic functions of collecting and storing monitoring data on the one hand, and to correlate and analyze the data effectively on the other hand.

2.1 Basic Functional Requirements of Corrosion Monitoring System

According to the process and technical characteristics of daily corrosion monitoring work, three data collection and management modules, namely "on-line corrosion monitoring", "chemical analysis" and "fixed-point thickness detecting", should be set up. A large amount of monitoring data is uploaded to the monitoring data center server from the unit server. At the same time, the application client needs to establish a three-level application layer (decision layer, management layer, executive layer) to input and manage the data in the data server through privilege management.

(1) On-line corrosion monitoring

The process corrosion data are obtained by the probe inserted into the pipeline. It mainly includes on-line monitoring of pH value of process condensate and on-line monitoring of corrosion rate of medium corrosion status.

(2) Chemical analysis

Chemical analysis is an effective means to monitor corrosion status and predict corrosion trend of the unit. The content of pH, Fe^+, Cl^-, H_2S, NH_3–N in condensate water, and acid value, sulfur content, total chlorine content, nitrogen content and chlorine content in medium can be detected by chemical analysis.

(3) Fixed-point thickness detecting

Because of the convenience and economy of the fixed-point thickness detecting technology, it has been the most widely used corrosion monitoring technology in refinery and chemical enterprises at present. The detecting principle is ultrasonic, which can be divided into on-line thickness detecting and manual thickness detecting. The fixed-point thickness detecting reflects the corrosion state of the equipment and pipeline by measuring the thickness reduction of the wall.

2.2 Data Application and Analysis Requirements

The corrosion monitoring system needs to apply the monitored data to realize the state evaluation and provide the decision information of the application for the process and equipment.

The process evaluation mainly uses the data from on-line monitoring and chemical analysis. By monitoring the effect of electric desalination and anticorrosion, the evaluation of corrosion control of system medium can be realized.

The state evaluation of the equipment and pipeline is based on the data of on-line monitoring and fixed-point thickness detecting methods. Combining with the corrosion rate of the on-line monitoring probe under the same circuit, the most reasonable corrosion state of the circuit is calculated, and the residual life of the equipment and pipeline is evaluated.

3 Development of the Integrated Corrosion Monitoring System

3.1 System Architecture

Corrosion and anticorrosion management data come from the basic information of the equipment itself on the one hand, and from the state monitoring data on the other hand, which need to be integrated in many aspects (Black et al. 2000). Therefore, the system design is based on SOA technology architecture to realize data interaction and integration with ERP, MES and other systems, to realize the collection and storage of corrosion monitoring data, as well as to realize corrosion monitoring, fixed-point thickness detecting, chemical analysis, alarm and analysis decision-making, system management and other main functions of data interaction application. Finally, the application layer for different levels of users is formed, as shown in Fig. 1.

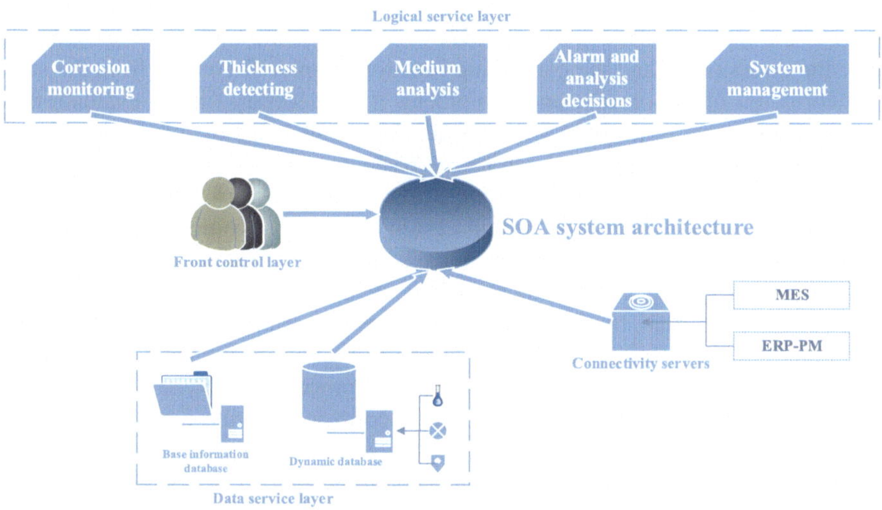

Fig. 1 Architecture of integrated corrosion monitoring system

Each monitoring function module can store, browse and calculate the collected data, and provide various trends analysis and comparative results at the same time.

The corrosion monitoring function module can analyze the collected monitoring data, get corrosion loss, various corrosion rate and trend curve display. It can also realize the comparison of corrosion trend changes of multiple monitoring points. And the overall corrosion situation of a certain section of corrosion circuit can be comprehensively analyzed.

The function module of chemical analysis realizes the comprehensive information management of chemical analysis data, including the functions of browsing, analyzing, contrasting and alarming for all kinds of chemical data. By setting the standard value and the critical value of every index, the data are analyzed, summarized and alerted based on the integrality operation window technology.

The fixed-point thickness detecting module collects the on-line thickness detecting data, and imports the manual thickness detecting data. By analyzing and calculating, the module can realize the function of data query and data report output, and calculate the corrosion rate of the measuring point as well as display it with the trend graph.

The alarm and evaluation decision-making module is the core function of the system. The idea of circuit from risk assessment and the concept of IOW are used to manage corrosion in both equipment and process. The anticorrosion effect monitoring and warning of the process, the comprehensive analysis of the corrosion rate of the equipment circuit and the calculation and prediction of the residual life are realized, at the same time the corresponding decision suggestions are given.

3.2 Evaluation and Decision-Making Model

The biggest challenge for the China petrochemical enterprise to apply the monitoring system is the lack of effective use of data (Guo 2015), the traditional systems only realize the data storage and the computing function simply. Therefore, from the stance of anticorrosion of the process and equipment, it is necessary for the evaluation and decision-making mode not only to make full use of all kinds of monitoring data for correlation analysis, but also to combine the basic data of equipment and operation.

Figure 2 shows the data flow direction of the integrated corrosion monitoring system analysis decision model. The system design of the process model is mainly based on on-line corrosion monitoring and chemical analysis data. Combined with basic data and operation data, the content of corrosive elements in medium and operational condition are controlled by IOW principle. The corrosion prevention and control suggestions and monitoring optimization are given by monitoring the corrosion rate in real time to master the anticorrosion effect of the process. The system design of the equipment model is mainly based on on-line corrosion monitoring and fixed-point thickness detecting data. Combined with basic data and operation data, the comprehensive corrosion rate calculation and the accurate evaluation of residual life of each component in the circuit are carried out to guide the upgrading and maintenance of the equipment and pipeline.

Detailed process evaluation and decision-making model is shown in Fig. 3. The monitoring data and chemical analysis data of the process are used to judge whether the critical value is within a reasonable range or beyond the critical value. After the system judgment, an alarm is made and some pertinent adjustment suggestions are put forward according to the exceeding conditions. If the adjustment can not be alleviated or the value is in excess of the limit for a long time, the system puts forward some suggestions in the aspects of anticorrosion adjustment and so on. At the same time, according to the recent fixed-point thickness detecting reading results of equipment and pipeline in the same process, if the over-limit state has caused the corrosion of the equipment, then some comprehensive decision suggestions for anticorrosion and material optimization would be provided.

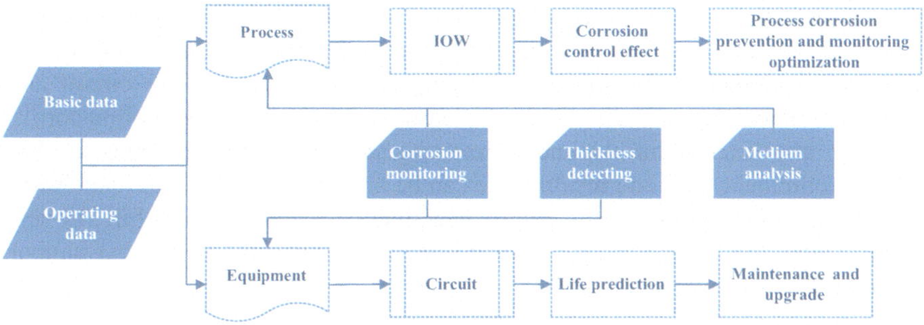

Fig. 2 Data flow of evaluation and decision-making model

The equipment and pipeline corrosion evaluation and decision-making model is shown in Fig. 4. The health state of the equipment is the most important in the process of production and operation (Wang and Chen 2014). Therefore, the acquisition of state information of the equipment and pipeline in time and the identification of the accurate evaluation and prediction can guarantee the safe and stable operation of the unit.

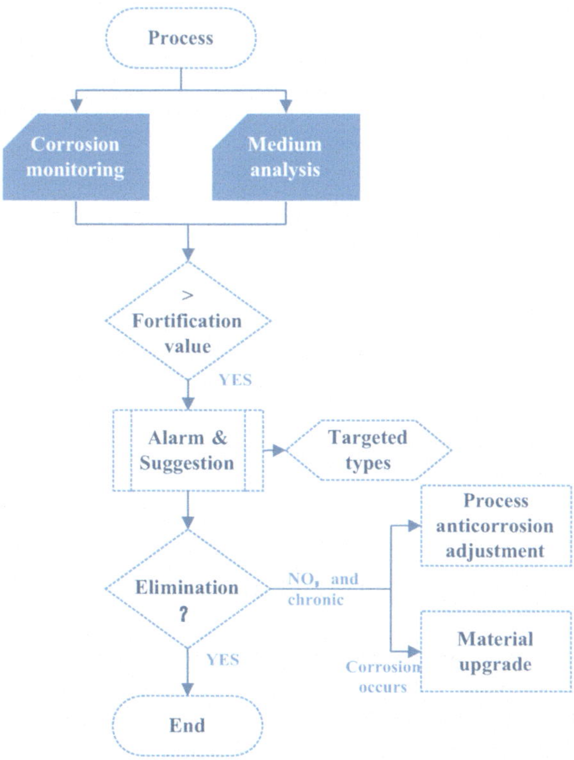

Fig. 3 Process corrosion evaluation and decision-making model

Corrosion is the main form of equipment and pipeline damage. The corrosion rate and residual life obtained through effective monitoring and calculating are the main contents of equipment corrosion management. The system takes the circuit as the management object, by analyzing and calculating synthetically of the on-line corrosion monitoring and fixed-point thickness detecting data the corrosion rate, evaluated thickness, retirement limit and residual life can be obtained, which could make sure the corrosion state of the whole circuit can be effectively evaluated. Meanwhile, by setting up alarm value, the system can put forward decision suggestion of process anticorrosion, inspection and maintenance plan and so on.

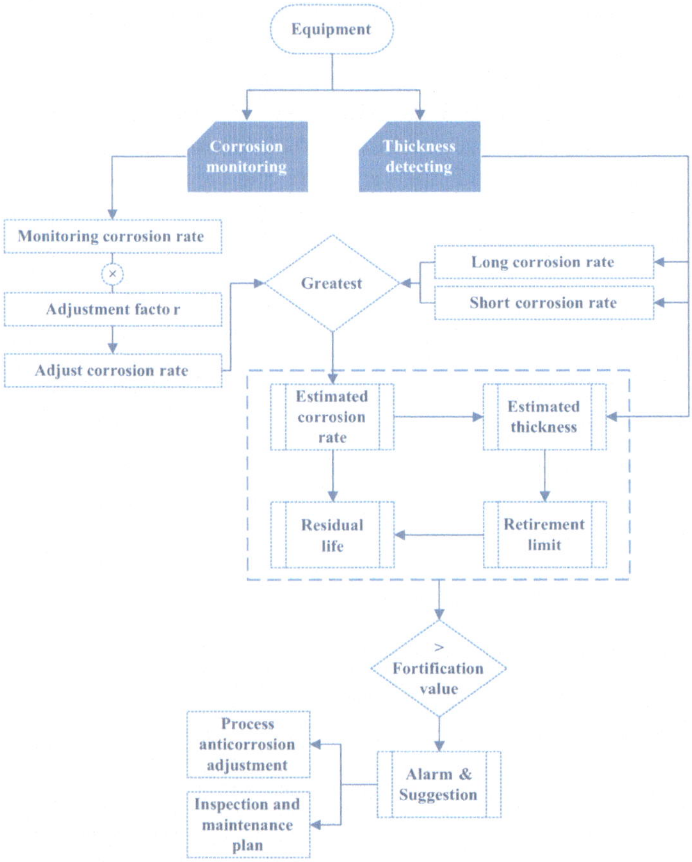

Fig. 4 Equipment and pipeline evaluation and decision-making model

The long corrosion rate and short corrosion rate of the equipment and pipeline in the state assessment model are calculated as shown in Fig. 5. The long corrosion rate reflects the average corrosion rate of the measuring point since its operation, and the short corrosion rate can clearly reflect the fluctuation of corrosion rate in a short period of time.

$$\boxed{\text{Long corrosion rate}} = \frac{\text{Initial wall thickness} - \text{Latest detecting thickness}}{\text{Latest detecting date} - \text{Commissioning date}}$$

$$\boxed{\text{Short corrosion rate}} = \frac{\text{Difference of latest two detecting thickness}}{\text{Difference of latest two detecting date}}$$

Fig. 5 Calculation method of long and short corrosion rate

Adjusting corrosion rate is calculated from the on-line monitoring corrosion rate. Although on-line monitoring directly reflects the corrosion rate of material in the medium, it is different from the corrosion rate of the equipment itself. Through the statistical analysis of the relationship between the corrosion rate of more than 500 on-line monitoring probes and the actual pipeline corrosion rate of the monitoring circuit, the adjustment factor is obtained, as shown in Fig. 6. The adjustment ratio is defined as the ratio of the maximum long corrosion rate to the average corrosion rate of the circuit. The ratio reflects the fluctuation of corrosion in the whole circuit. The larger the ratio is, the more uneven the corrosion situation of the circuit is. The ratio can not only be used to adjust the on-line monitoring corrosion rate, but also can be used to increase or decrease the corrosion monitoring points by defining the range of the ratio.

$$\boxed{\text{Adjustment factor}} = 0.8 \times \begin{cases} 1.25 & (\text{Adjustment ratio} > 0.5) \\ 1 & (\text{Adjustment ratio} \leq 0.5) \end{cases}$$

$$\boxed{\text{Adjustment ratio}} = \frac{\text{Max point long corrosion rate}}{\text{Circuit average corrosion rate}} = \frac{\text{Max point long corrosion rate}}{\text{Total circuit loss} \div \text{Total circuit exposure time}}$$

Fig. 6 Calculation method of adjustment factor and adjustment ratio

The estimated thickness is used to estimate the thickness of points that are not measured effectively or measured in time, as shown in Fig. 7.

$$\boxed{\text{Estimated thickness}} = \text{Latest detecting thickness} - \text{Estimated corrosion rate} \times \left(\text{Calculate date} - \text{Latest detecting date}\right)$$

Fig. 7 Calculating method of estimated thickness

The method of calculating the residual life of equipment and pipeline is shown in Fig. 8. Before calculating, it is necessary to determine the retirement limit wall thickness of the components. The system provides a variety of calculation methods, if the corrosion allowance of the components has been obtained in the system, then the difference between wall thickness and corrosion allowance is taken as the retirement limit thickness, otherwise the system calculates the retirement limit thickness according to the relevant standards, including ASME B31.3, API 653 and ASME VIII.

$$\boxed{\textbf{Residual life}} = \textbf{Calculate date} + \frac{\textbf{Estimated thickness} - \textbf{Retirement limit thickness}}{\textbf{Estimated corrosion rate}}$$

Fig. 8 Calculation method of residual life

4 System Application

The integrated corrosion monitoring system has played a better role since its operation. Through the data analysis of on-line corrosion monitoring, fixed-point thickness detecting and chemical analysis, the corrosion situation of the process and equipment can be fully understood in time and residual life of equipment and pipeline can be predicted. The system can also find and analyze the abnormal data in time, and find the exact location of corrosion quickly, at the same time give out the corresponding anticorrosion measures, which effectively prevent the occurrence of accidents. Typical application cases are as follows:

(1) The corrosion rate of the probe at the outlet of an atmospheric tower top heat exchanger suddenly increased from 0.002 to 0.766 mm/a, and the Fe^{2+} value from chemical analysis exceed the standard obviously (standard value is 3.0, the measured value is 5.6). The system alerted and prompted to check the corrosion inhibitor injection condition or adjust the amount of the inhibitor. It was found that the failure of the injection pump resulted in the insufficient injection of the inhibitor in the atmospheric top system and the sharp increase of the corrosion rate in a short period of time. After being repaired, the injection recovered and the probe corrosion rate decreased to 0.028 mm/a, as shown in Fig. 9.

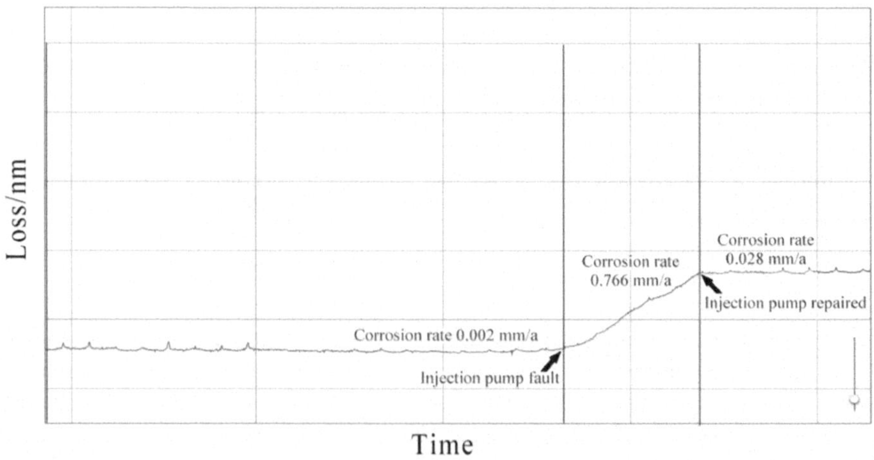

Fig. 9 Corrosion loss and corrosion rate trend at atmospheric tower top

(2) The corrosion rate of the corrosion monitoring probe at the entrance of the top heat exchanger of the fractionator of a FCC unit increased month by month this year, from an average corrosion rate of 0.1437 mm/a at the beginning of the year to an average corrosion rate of 0.4771 mm/a in the middle of the year. However, there was no abnormality in the on-line and fixed-point thickness detecting, so the system alerted overrun rate and prompted to increase the number of fixed point thickness detecting of the system pipeline, and analyze the corrosive medium in the oil and water outlet of the heat exchanger to determine the corrosion cause. Through the comprehensive detection of the pipeline wall thickness, it was found that the two pipe elbows in the system had a certain thinning by corrosion.

(3) An on-line corrosion rate monitoring probe was installed on the wall of a fractionator in a FCC unit. The probe was made of carbon steel. The corrosion rate of the probe has been kept at about 0.3 mm/a since the probe was officially operated, and after 4 months the metal loss of the probe was completed and the probe then failed, as shown in Fig. 10. The initial warning by the monitoring system was to take anticorrosive measures and increase detecting points. With the increase of corrosion and probe failure, the system prompted to upgrade the material.

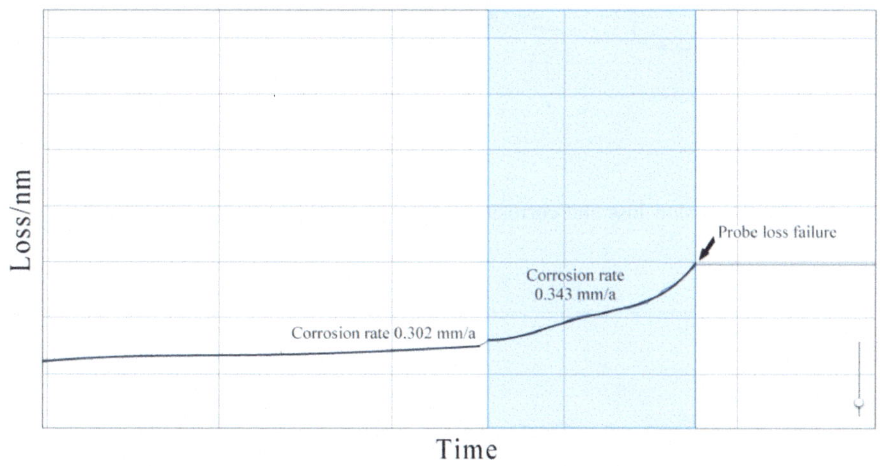

Fig. 10 Corrosion loss and corrosion rate trend at the fractionator

(4) The inlet and outlet pipeline of the atmospheric tower top air cooler has always been prone to corrosion (Wu et al. 2018). The integrated corrosion monitoring system has been used in an atmospheric and vacuum unit for a few months, and all the on-line corrosion rate monitoring, chemical analysis and the fixed point thickness detecting at air cooler outlet pipeline exceed the defensive value, the corrosion rate of the medium and pipeline was on the high side (Figs. 11 and 12),

the iron ion content was also in excess of the standard (Fig. 13). Then, some suggestions were put forward from the aspects of technology, material and monitoring by the system. Considering the complexity of the corrosion problem, the management set up a special technical group, and improved the form and location of the injection agent. The corrosion rate and iron ion decreased obviously, then the system alarm disappeared and the corrosion was effectively controlled.

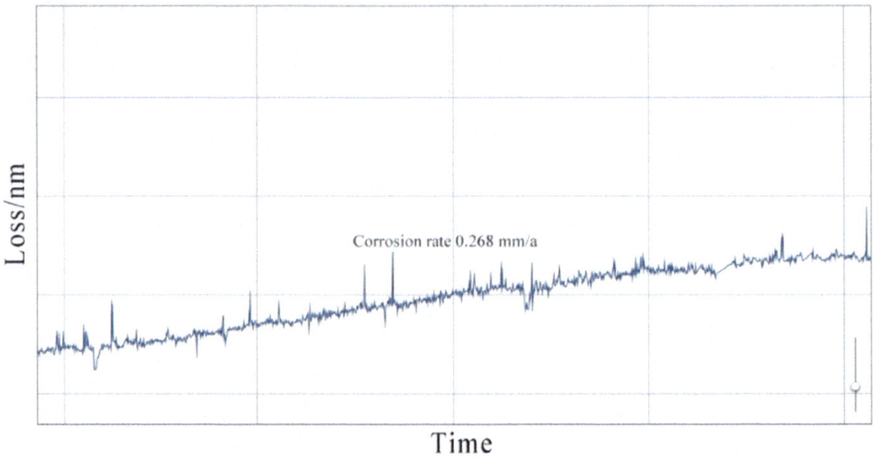

Fig. 11 Corrosion loss and corrosion rate trend at the air cooler outlet pipeline

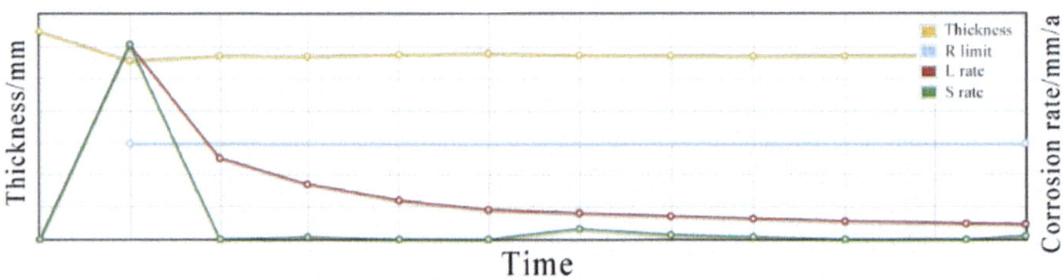

Fig. 12 Trend of thinning point thickness and corrosion rate

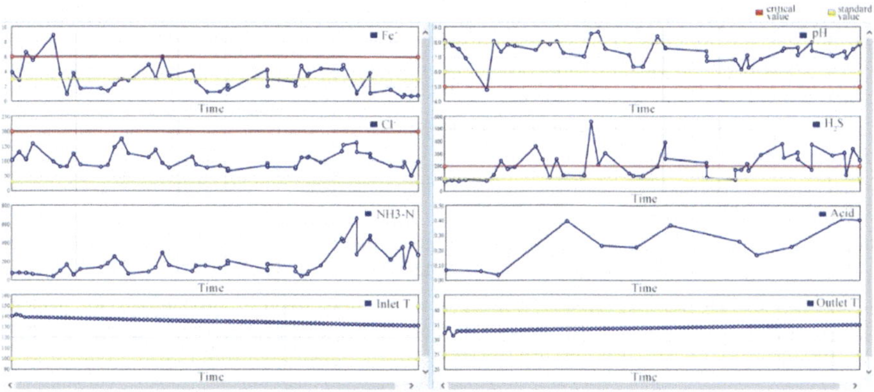

Fig. 13 Trend of chemical analysis in different types

5 Conclusion

This paper innovatively proposes the corrosion evaluation and decision-making model, and develops integrated corrosion monitoring system. The main contents are as follows:

(1) By analyzing the requirements of corrosion monitoring management, three basic functional are determined, including on-line corrosion rate monitoring, fixed-point thickness detecting and chemical analysis, then the data application and analysis requirements are also determined.
(2) The system data analysis, integration and management architecture are based on SOA. Then based on IWO and risk management method the evaluation and decision-making models of process and equipment are established. The functions of process anticorrosion effect, equipment and pipeline state evaluation and life prediction are realized.
(3) Through the application, the corrosion rate of medium and equipment is monitored, the anticorrosion effect of the process is evaluated, the place where the corrosion problem occurs is quickly forewarned, then the corresponding suggestions are provided through the comprehensive analysis.

Acknowledgements. This research was supported by The National Key Research and Development Program of China (2018YFC0809004).

References

Black, I., Eden, D.C., Franklin, P.: Advanced technology equipment reliability solutions incorporating corrosion management and condition monitoring expertise & technology. Online Conference: Corrosion Control for the Future (2000)

Chen, F., Fu, D., Zhou, K., Wu, J., Du, C., Liu, Y.: Development and application of resistance probe corrosion monitoring technology. Corros. Sci. Protect. Technol. **29**(6), 669–674 (2017)

Guo, L.: Establishment of corrosion protection management system in refining and chemical enterprises. Pet. Chem. Equip. **18**(04), 77–79 (2015)

Javaherdashti, R.: How corrosion affects industry and life. Anti-Corros. Methods Mater. **47**(1), 30–34 (2000)

Li, J., Liang, G., Zhou, S.: Corrosion monitoring and management system of refining unit. Corros. Prot. **37**(10), 847–851 (2016)

Wang, G., Chen, J.: Research on integrity management technology of static equipment in petrochemical enterprises. Mod. Chem. Ind. **34**(06), 10–12 (2014)

Wu, X., Ren, R., Cheng, W.: Corrosion analysis and management of atmospheric top system in atmospheric and vacuum distillation unit. Corros. Protect. Petrochem. Ind. **35**(02), 30–33 (2018)

Application and Improvement of Ensemble Kalman Filter Method in Production Data Analysis

Wen-ting Yue[1,2(✉)] and John Yilin Wang[1]

[1] Department of Energy and Mineral Engineering, The Pennsylvania State University, University Park, PA 16803, USA
yuewenting@petrochina.com.cn
[2] Department of Overseas Strategy and Development Planning, Research Institute of Petroleum Exploration and Development (RIPED), PetroChina, Beijing 100083, China

Abstract. Production data analysis (PDA) is a subject to determine reservoir properties and predict future performance of single wells. It is normally conducted with either analytical methods or numerical history-matching methods. However, the analytical models normally have limited accuracy due to simplifying assumptions, while the history-match based on numerical simulations sacrifices computation load to increased accuracy. Therefore, in this study, the authors incorporated Ensemble Kalman Filter (EnKF) to improve PDA considering its superior efficiency in predicting system state and uncertainties. In this work, we applied the EnKF algorithm to single-well reservoir models to estimate permeability, skin factor, and drainage area. First, we tested the model accuracy after comparing the EnKF estimates to known reservoir properties. Next, we evaluated the cases with large estimation error and then adjusted the initial uncertainties and covariance of the static parameters. With confirmed improvements of property estimation in synthetic cases, the model is finally applied in a field study for further verifications. The results from this study confirmed that EnKF method could be an efficient solution for the modern PDA. They also indicate that accuracy issues are sometimes present when estimating skin factor and reservoir permeability simultaneously: large error exists in the property estimates, their uncertainties are overly reduced, and thus analysis and predicts are affected. By increasing the initial uncertainty bounds and adding minimum threshold values for the covariance, the property estimates could be

This paper was prepared for presentation at the 2019 International Petroleum and Petrochemical Technology Conference in Beijing, China, 27–29, March, 2019.
This paper was selected for presentation by the IPPTC Committee following review of information contained in an abstract submitted by the author(s). Contents of the paper, as presented, have not been reviewed by the IPPTC Technical Committee and are subject to correction by the author(s). The material does not necessarily reflect any position of the IPPTC Technical Committee, its members. Papers presented at the Conference are subject to publication review by Professional Team of Petroleum Engineering of the IPPTC Technical Committee.
J. Lin (ed.), *Proceedings of the International Petroleum and Petrochemical Technology Conference 2019*, pp. 427–451, 2020.
https://doi.org/10.1007/978-981-15-0860-8_36

improved, and reasonable uncertainty bounds are preserved. The methodology from this study is applicable for full-field evaluations.

Keywords: Oil and gas · Production data analysis · Ensemble Kalman Filter · Numerical history matching · Characterization of oil/gas reservoirs · Field application

Nomenclature

Letters

A_x	Cross-sectional area normal to x direction, ft^2
A_y	Cross-sectional area normal to y direction, ft^2
B_g	Gas formation volume factor, RB/scf
B_o	Oil formation volume factor, RB/stb
B_w	Water formation volume factor, RB/stb
c	Component
B	Formation volume factor, reservoir volume/volume at standard conditions
$\boldsymbol{C}_D$	Covariance matrix of the measured data
$\boldsymbol{C}_Y^f$	The prior covariance matrix for the state variables
$\boldsymbol{d}_m^o$	Measurement data used for assimilation
$\boldsymbol{d}_m$	Simulation model output at each assimilation time step
k	Reservoir permeability, md (in the context of governing equation)
k_r	Relative permeability, fraction
k_{rl}	Relative permeability to phase l, dimensionless
k_{rg}	Relative permeability to gas, dimensionless
k_{ro}	Relative permeability to oil, dimensionless
k_{rw}	Relative permeability to water, dimensionless
k_s	Permeability of the damaged/stimulated area, md
k_x	Permeability in the direction of the x axis, md
k_y	Permeability in the direction of the y axis, md
$\boldsymbol{K}$	Kalman Gain
$\overrightarrow{\dot{m}}_c$	Mass flux for component c, lbm/(D-ft^2)
$\dot{m}_{cx}$	c component of mass flux in x direction, lbm/(D-ft^2)
$\dot{m}_{cy}$	c component of mass flux in y direction, lbm/(D-ft^2)
m_{V_c}	Mass of component c per unit volume of rock, lbm/ft^3
N_d	Number of observations
N_k	Ensemble size
N_x	The number of the state variables
N_y	$N_y = N_\theta + N_x + N_d$
N_θ	The number of poorly known parameters
$\boldsymbol{O}$	$N_d \times (N_\theta + N_x)$ null matrix
P_{cgo}	Gas/oil capillary pressure, psi
P_{cow}	Oil/water capillary pressure, psi
P_g	Gas pressure, psi
P_o	Oil pressure, psi
P_w	Water pressure, psi

q_{gsc} Gas production rate at standard conditions, scf/D
q_{m_c} Rate of mass depletion for component c through wells/lbm/D
q_{mt_c} Rate of mass depletion for component c between phases, lbm/D
q_{osc} Oil production rate at standard conditions, STB/D
q_{wsc} Water production rate at standard conditions, STB/D
r_s Radius of the damage/stimulated area around a wellbore, ft
$\mathrm{r_w}$ Well radius, ft
R_s Solution gas oil ratio, scf/STB
S_g Gas saturation, fraction
S_o Oil saturation, fraction
S_w Water saturation, fraction
t Simulation time, days
Δt Simulation time step, days
u Superficial velocity, RB/(D-ft^2)
$\overrightarrow{u_c}$ Component c superficial velocity vector, RB/(D-ft^2)
u_{cx} x component of component c superficial velocity, RB/(D-ft^2)
u_{cy} y component of component c superficial velocity, RB/(D-ft^2)
V_b Gridblock bulk volume, ft
$\boldsymbol{x}$ Reservoir state vector (including reservoir pressures and phase saturations), $x = x(t) \in \mathbb{R}^{N_x}$
$\boldsymbol{y}_m^k$ State vector of ensemble number k at time m
$\boldsymbol{Y}_m$ State vectors of the entire ensembles at time m
$\overline{\boldsymbol{Y}_m}$ Matrix containing ensemble mean in each column at time m
$\overline{y_m^1}$ Ensemble mean of each member at time m

Greek Symbols
ε_m^o Measurement error, which is a zero-mean Gaussian noise $\varepsilon_m^o \sim \mathcal{N}(0, R)$
α_c Volumetric conversion factor, 5.614583
β_c Transmissibility conversion factor, 1.127
γ_g Gas gravity, psi/ft
γ_l Gravity of phase l, psi/ft
γ_o Oil gravity, psi/ft
γ_w Water gravity, psi/ft
ε_m Model error
$\boldsymbol{\theta}$ Static reservoir parameters, $\theta \in \mathbb{R}^{N_\theta}$
Δ Difference, difference operator
μ_g Gas viscosity, cp
μ_o Oil viscosity, cp
μ_w Water viscosity, cp
ρ_c Density of component c, lbm/ft^3
ρ_g Gas phase density, lbm/ft^3
ρ_{gsc} Gas phase density at standard condition, lbm/ft^3
ρ_w Water phase density, lbm/ft^3
ϕ Porosity, fraction

Δx Difference along x direction, ft
Δy Difference along y direction, ft

Subscripts
b Grid bulk
c Component
fg Free gas
g Gas
k Ensemble member, $k = 1, 2, 3...N_k$
l Phase l or component l
m Assimilation time, $m = 1, 2, \ldots, N_m$
mt Mass transfer
o Oil
sc Standard condition
sg Solution gas
t Simulation time
w Water

Superscripts
f Forecasted
o Observed or measured
u Updated
T Transpose of
$\rightarrow$ Vector

1 Introduction

Oil and gas production data analysis refers to the process of matching model predictions with existing production data, and thus to identify possible issues with production, estimate reservoir properties, and forecast future production. It provides important information to understand fluid behavior. The commonly used PDA models include analytical models and numerical models. Analytical models, such as the material balance calculation and tank-type model, decline curve analysis, and type curve analysis, have been developed since 1945 (Arps 1945), and are widely used due to their low cost in computation (Cao et al. 2016; Fan and Ettehadtavakkol 2017; Wu et al. 2016; Zheng et al. 2016). These models, however, are built on multiple simplifying assumptions which are prone to error (Ertekin et al. 2001; Poston and Poe 2008). Any applications compromise the model assumptions might lead to erroneous conclusions.

Even if assumptions are well met, and production is carefully controlled to meet the requirements of the models, the curve-fitting process will face further challenges due to non-uniqueness of their results (Bhattacharya and Nikolaou 2013). Hence, these limitations can impact the efficiency of production data analysis.

Compared to analytical models, numerical history matching models present a more accurate and versatile solution for the modern PDA (Ampomah et al. 2016; Cai et al. 2017; Cokar et al. 2013; Mullins et al. 2017; Zhao and Liao 2012). In numerical models, the input data (e.g., geological description, fluid properties, and relative permeability/capillary pressure) are altered to match the observed data (e.g., fluid rates, pressures, tracers, and temperatures). However, the numerical simulation normally takes long time to run and the traditional matching methods suffer from run-time and computation issues: with either manual matching or Monte Carlo method, the entire history of production is simulated repeatedly, the results are compared with the recorded data, and then the models are improved by modifying the input data. This process may not be efficient since the majority of the calculations are wasted while only one single set of solution will be adopted. The gradient-based method avoids repeated calculations but requires complex derivation of the adjoint equations, which can largely increase the computational load of the matching process. Implementation of the gradient-based model requires decomposition of the physical model; therefore, it is not compatible with the available software. Neither analytical models nor numerical evaluations would consider the estimation uncertainty into account.

Ensemble Kalman filter (EnKF) initially proposed by Evensen (1994b, 2003, 2006) is a Monte-Carlo implementation of the Bayesian update problem in which an ensemble of models is used to update the parameters sequentially according to the chronological order in which the data are acquired and assimilated (Evensen 1994a, 2006; Naevdal et al. 2005). This method is free from complex derivation and the implementation of adjoint equations required by gradient-based matching methods. The sequential update method avoids repeated simulations of the entire production history. Its implementation thus turns out to be significantly efficient and is independent of any reservoir simulator (Liang 2007). Unlike other history matching methods, the EnKF-based method can also assess the simulation uncertainties by using an ensemble of models.

To tackle the issues faced in oil and gas PDA, Geir et al. (2002) used EnKF for permeability estimation in a simplified reservoir application. The measurement they used for assimilation are bottom-hole pressure data, oil and gas production data. Their estimations presented promising matching results. Naevdal et al. (2005) extended the application in a simplified 2D reservoir model for a North Sea field. The study also presents good matching results. Haugen et al. (2008) used EnKF to match the production history of a North Sea Field as a case study. Parameters evaluated in their work include reservoir permeability, porosity, and reservoir states, such as phase pressures and saturation. Measurements used in the assimilation included bottomhole pressure, oil production rate, water cut, and gas/oil ratio. The EnKF proves to improve the manually history matched model. Zhang et al. (2016) developed and applied a damped iterative EnKF algorithm to estimate relative permeability and capillary pressure for the tight formations by matching the lab record of a conventional measurement data from waterflooding experiments. They reported that the uncertainty ranges of parameters and production profiles were reduced.

In the single-well analysis, parameters of interest are normally different from that in the field scale. Aside from reservoir permeability, parameters such as skin factor and drainage area are important indications of production behavior; therefore, they are of primary concern in well testing and other production data analysis. Li et al. (2010) applied the EnKF method to integrate well-test pressure data into a 2-layer 5-spot heterogeneous reservoir model to estimate the skin factor and permeability/porosity distribution. Reasonable estimation is reported, and the doubly stochastic model that they used could improve the accuracy. However, the estimation of skin factor was not as successful. Han et al. (2014) utilized both microseismic and well-test data to estimate the porosity and log-permeability distribution of a synthetic reservoir model. They reported a good data match and large uncertainty reduction after applying EnKF method. However, well sand face pressure data and microseismic data used in their study are not always available in field applications. It will be more challenging to conduct production data analysis with limited data.

In this study, we seek for a versatile, accurate, and efficient model to analyze important reservoir/well parameters related to single-well production, especially when limited data are available. The EnKF algorithm was linked with a single-well reservoir model to estimate permeability, skin factor, and drainage area. We investigated the accuracy of the EnKF model through synthetic testing. Next, we evaluated the source of error and adjusted the initial uncertainty and covariance of static parameters. The field cases considered in this study were used to verify the model.

2 Governing Equations

2.1 Investigation Scope

Prior to the implementation of EnKF, an investigation scope is defined. This section describes the model of multiphase flow in the 2D drainage of a well where the main driving force for recovery is natural reservoir pressure. The model employed in this study reflects the response of a single producer to primary depletion in a four-spot drainage pattern with the configuration shown in Fig. 1.

2.2 Mass Balance of Flow in Porous Media

A cumulative material balance on a control volume V_b that contains oil, water, and gas phases in the time interval Δt was carried out. By assuming that oil and water are not soluble with each other, the three-phase, slightly compressible flow displacement in the matrix block then can be described as (Ertekin et al. 2001):

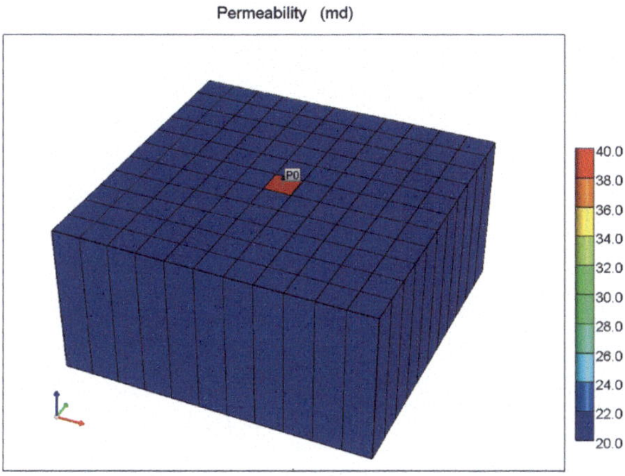

Fig. 1 A representation of the investigation scope

$$-\frac{\partial}{\partial x}(\dot{m}_{cx}A_x)\Delta \mathrm{x}-\frac{\partial}{\partial y}(\dot{m}_{cy}A_y)\Delta \mathrm{y}=V_b\frac{\partial}{\partial t}(m_{V_c})-q_{m_c}-q_{mt_c} \tag{1}$$

where $\dot{m}_c$ is the mobility of phase $c, c = o, w, fg$, and sg; x and y represent two directions; A_x and A_y are the two cross-sectional areas in x and y directions; q_m is the source/sink term of phase c (+ for source, − for sink); q_{m_c} is the rate of mass depletion for component c through wells, and q_{mt_c} is the rate of mass transfer of component c between phases. Other variables are defined in Nomenclature.

2.3 Development of Fluid Flow Model

By assuming oil and water are not soluble with each other, while gas may contain some amount of light oil, the reservoir is at a constant temperature, and the fluids are in thermodynamic equilibrium, we have the following definitions. For the oil, water, and free-gas components, it is possible to express mass flux as the product of density and Darcy's velocity for phase l ($l = o, w, g$); mass per unit volume of reservoir (or concentration) as the product of porosity, phase saturation, and phase density; and mass flow rate as the product of volumetric (phase)—flow rate and phase density. For the solution-gas component, mass flux, mass per unit volume, and mass-flow rate may be expressed in terms of the oil-phase properties R_s and B_o and the gas-phased density at standard conditions, $\rho_{gas\,c}$. Mass per unit volume, mass flux, and mass-flow rate are quantities that describe components ($c = o, w, g, fg$, or sg), whereas flux (volumetric velocity), volumetric-flow rate, density, formation volume factor (FVF), saturate, and solution-gas/oil ratio describe phases ($l = o, w, g$). The terms mentioned are explained in (Ertekin et al. 2001). When substituting these terms and Darcy's law for multiphase flow, the following equations for each phase can be obtained:

$$\frac{\partial}{\partial x}\left[\beta_c k_x A_x \frac{k_{ro}}{\mu_o B_o}\left(\frac{\partial P_o}{\partial x} - \gamma_o \frac{\partial Z}{\partial x}\right)\right]\Delta x + \frac{\partial}{\partial y}\left[\beta_c k_y A_y \frac{k_{ro}}{\mu_o B_o}\left(\frac{\partial P_o}{\partial y} - \gamma_o \frac{\partial Z}{\partial y}\right)\right]\Delta y$$
$$= \frac{V_b}{\alpha_c}\frac{\partial}{\partial t}\left(\frac{\phi S_o}{B_o}\right) - q_{osc} \tag{2}$$

$$\frac{\partial}{\partial x}\left[\beta_c k_x A_x \frac{k_{rw}}{\mu_w B_w}\left(\frac{\partial P_w}{\partial x} - \gamma_w \frac{\partial Z}{\partial x}\right)\right]\Delta x + \frac{\partial}{\partial y}\left[\beta_c k_y A_y \frac{k_{rw}}{\mu_w B_w}\left(\frac{\partial P_w}{\partial y} - \gamma_w \frac{\partial Z}{\partial y}\right)\right]\Delta y$$
$$= \frac{V_b}{\alpha_c}\frac{\partial}{\partial t}\left(\frac{\phi S_w}{B_w}\right) - q_{wsc} \tag{3}$$

$$\frac{\partial}{\partial x}\left[\beta_c k_x A_x \frac{k_{rg}}{\mu_g B_g}\left(\frac{\partial P_o}{\partial x} - \gamma_g \frac{\partial Z}{\partial x}\right) + \beta_c k_x A_x \frac{k_{ro} R_s}{\mu_o B_o}\left(\frac{\partial P_o}{\partial x} - \gamma_o \frac{\partial Z}{\partial x}\right)\right]\Delta x$$
$$+ \frac{\partial}{\partial y}\left[\beta_c k_y A_y \frac{k_{rg}}{\mu_g B_g}\left(\frac{\partial P_o}{\partial y} - \gamma_g \frac{\partial Z}{\partial y}\right) + \beta_c k_y A_y \frac{k_{ro} R_s}{\mu_o B_o}\left(\frac{\partial P_o}{\partial y} - \gamma_o \frac{\partial Z}{\partial y}\right)\right]\Delta y \tag{4}$$
$$= \frac{V_b}{\alpha_c}\frac{\partial}{\partial t}\left(\frac{\phi S_g}{B_g} + \frac{\phi R_s S_o}{B_o}\right) - q_{gsc}$$

where $S_o + S_g + S_w = 1$. All other variables are defined in Nomenclature. The other functional terms can be described as follows, in Eqs. (5)–(7):

Both the formation volume factors and viscosities are pressure-dependent. This means that:

$$\begin{aligned} B_l &= B_l(P_l) \\ \mu_l &= \mu_l(P_l), \quad l = o, w, g \end{aligned} \tag{5}$$

(a) The relative permeability curves are dependent on saturation:

$$k_{rl} = k_{rl}(S_w), \quad l = o, w, g \tag{6}$$

(b) The phase pressures can be defined by the capillary pressure:

$$\begin{aligned} P_{cow}(S_w) &= P_o - P_w \\ P_{cgo}(S_g) &= P_g - P_o \end{aligned} \tag{7}$$

The previous flow equations are strongly nonlinear partial differential equations. Therefore, numerical methods are required to obtain engineering solutions to the flow problem. A finite-difference approach is normally used to obtain the numerical solution to multiphase-flow systems. The flow equations are discretized using algebraic approximations of the second-order derivatives with respect to space and the first-order derivatives with respect to time. The detailed derivations can be found in the references (Ertekin et al. 2001).

3 Mathematical Formulation of the EnKF Method

3.1 EnKF Algorithm

Ensemble Kalman Filter (EnKF) is a Monte-Carlo implementation of the Bayesian update problem in which an ensemble of models is used to update the parameters sequentially according to the chronological order in which the data are acquired and assimilated. The EnKF methodology (Evensen 2003; Mandel 2009) is a two-step process: the first step is to forecast the reservoir behavior evolved forward in time using the reservoir simulation model, given by:

$$\text{Reservoir simulation model: } \boldsymbol{x}_t = f(\boldsymbol{x}_{t-1}, \boldsymbol{\theta}) + \varepsilon_m \in \mathbb{R}^{N_x}, \varepsilon_m \sim \mathcal{N}(0, Q) \tag{8}$$

$$\begin{aligned} \text{Initial state : } x(t_0) &= x_0, \\ \theta(t_0) &= \theta_0 \end{aligned} \tag{9}$$

where x_t represents the dynamic system state. $m = 1, 2, 3 \ldots t$. When simulating reservoir system, the dynamic state includes phase pressures and saturations. $\boldsymbol{\theta}$ is a scalar which includes static reservoir parameters, such as reservoir permeability, skin factor, and drainage area. They are called static because we assume these parameters do not change over time. ε_m is model error. When model error is neglected, the flow rate measurements $\boldsymbol{d}_m^o$ are related to the reservoir states through the following nonlinear relation:

$$\boldsymbol{d}_m^o = g(\boldsymbol{x}_m, \boldsymbol{\theta}) + \varepsilon_m^o \in \mathbb{R}^{N_d} \tag{10}$$

where superscript "o" stands for observed, meaning the measured oil rate data. From Eq. (10), measured oil rate data is the combination of model output $g(\boldsymbol{x}_m, \boldsymbol{\theta})$ based on estimated system state $(\boldsymbol{x}_m, \boldsymbol{\theta})$ and model error ε_m^o.

The second step is the data assimilation where the static and state variables which describe the system are updated to reduce the observation-data mismatch. When applying the EnKF method to nonlinear problems, we define an augmented state vector which includes static parameters $\boldsymbol{\theta}$ (e.g. local reservoir permeability, skin factor, and drainage area), system state $\boldsymbol{x}$ (e.g. reservoir pressure and phase saturations), and production data $\boldsymbol{d}$ (e.g. well production rates). The state variables for each simulation model form a state vector, and the ensemble of state variables forms an ensemble matrix. This gives us:

$$\boldsymbol{y}_{m,k} = \begin{Bmatrix} \boldsymbol{\theta} \\ \boldsymbol{x} \\ \boldsymbol{d} \end{Bmatrix}_{m,k} \in \mathbb{R}^{N_\theta + N_x + N_d} \tag{11}$$

where k is the ensemble member, $k = 1, 2, 3 \ldots N_k$.

Define the $N_d \times N_y$ matrix $\boldsymbol{H}$ by the following equation, where H is a matrix containing only 0's and 1's.

$$\boldsymbol{H} = \left[\boldsymbol{O}_{N_\theta + N_x} \quad \mathbf{I}_{N_d}\right] \tag{12}$$

where $\boldsymbol{O}_{N_\theta + N_x}$ is a zero matrix with dimension of $(N_\theta + N_x) \times N_d$, and $\mathbf{I}_{N_d}$ is a unit matrix with dimension of $N_d \times N_d$. N_y, in turn, equals to $N_x + N_d$. The relation between the state vector and the data will be given by the linear relation,

$$\boldsymbol{d}_m = \boldsymbol{H}\boldsymbol{y}_m + \varepsilon_m^o \tag{13}$$

It is convenient to introduce the matrix Ym, holding the ensemble members at time m,

$$\boldsymbol{Y}_m = \left\{\boldsymbol{y}_{m,1}, \boldsymbol{y}_{m,2}, \ldots, \boldsymbol{y}_{m,N_k}\right\} \in \mathbb{R}^{N_y \times N_k} \tag{14}$$

and $\overline{\boldsymbol{Y}_m}$ storing the ensemble mean in each column,

$$\overline{\boldsymbol{Y_m}} = \left\{\overline{y_{m,1}} \quad \overline{y_{m,2}} \quad \ldots \quad \overline{y_{m,N_k}}\right\} = Y_m 1_{N_k} \tag{15}$$

where $^{\mathbf{1}}\boldsymbol{N}_{\mathbf{k}}$ is the $N_k \times N_k$ matrix with all elements equal to $1/N_k$. Also, let us introduce the ensemble perturbation matrix, given by:

$$\boldsymbol{Y}_m' = \boldsymbol{Y}_m - \overline{\boldsymbol{Y}_m} = \boldsymbol{Y}_m(\boldsymbol{I} - 1_{N_k}) \tag{16}$$

where $\boldsymbol{Y}_m'$ will be used to calculate error covariance matrix in Eq. (19). At each data assimilation step, each state vector of the ensemble is updated using the ensemble approximation to the Kalman gain in the following manner, where the superscripts "*f*" and "*u*" stand for forecast and update (i.e. data assimilation) respectively:

$$\boldsymbol{Y}_{m,k}^u = \boldsymbol{Y}_{m,k}^f + \boldsymbol{K}\left(\boldsymbol{d}_{m,k}^o - \boldsymbol{H}\boldsymbol{Y}_{m,k}^f\right) \tag{17}$$

In this study, we used the perturbed-observation approach (Burgers et al. 1998; Houtekamer and Mitchell 1998) for which random observation perturbations with the same covariance as the true observation error are added to observations in each ensemble member. In Eq. (18), K is the ensemble approximation to the Kalman gain and is computed as

$$\boldsymbol{K} = \boldsymbol{C}_Y^f \boldsymbol{H}^T \left(\boldsymbol{H}\boldsymbol{C}_Y^f \boldsymbol{H}^T + \boldsymbol{C}_D\right)^{-1} \tag{18}$$

where $\boldsymbol{C}_D$ is the error covariance of the measured data. In hydrocarbon production, the error in measured data could come from inaccuracy of flow meter, etc. The collection of all ensemble state vectors is denoted by Y and is a matrix of the dimensions $N_y \times N_k$. In Eq. (19), $\boldsymbol{C}_Y^f$ denotes the prior covariance matrix for the state variables, which can be estimated from the ensemble using the following expression:

$$C_Y^f = \mathrm{Cov}\left(\mathbf{Y}_\mathrm{m}^\mathrm{f}\right) \approx \frac{1}{\mathrm{N_k} - 1} \mathbf{Y}_\mathrm{m}^{\mathrm{f}\prime} \left(\mathbf{Y}_\mathrm{m}^{\mathrm{f}\prime}\right)^\mathrm{T} \tag{19}$$

From Eq. (17), we can see that the updated state vector $Y_{m,k}^u$ is a combination of simulated state $Y_{m,k}^f$ and a correction term represented by $K\left(d_{m,k}^o - HY_{m,k}^f\right)$. The correction term is depended on a gaining factor K and the difference between the measured production data and the simulated production data at the same time step. From Eq. (17), the larger the difference is, the larger the correction term is, and vice versa. The correction term also depend on the Kalman Gain K. From Eq. (18), K depends on measurement error C_D (which represent the credibility of measured oil rate) and the covariance matrix for the state C_Y^f (which represent the credibility of simulated oil rates). System state, in turn, will be corrected with linear regression using Eq. (19).

3.2 EnKF Implementation

The procedure of implementing the EnKF algorithm in this study is briefly described as follows:

(1) Initialization.

The dynamic and static variables that describe the reservoir states are initialized by perturbing the initial state, and the resulting ensemble should resemble the true uncertainty (Haugen et al. 2006). We will discuss the detailed parameterizations and initializations of this study in the following subsection.

(2) Forecast:

To advance the ensemble of the updated reservoir simulation models from previous assimilation time step using the nonlinear propagator $f(\boldsymbol{x}, \boldsymbol{\theta})$ (i.e., reservoir simulator in this study) from time t_1 to time t_2 where new measurement data is available to be assimilated.

(3) Calculate Increments:

To assemble the static reservoir properties (i.e., local permeability, skin factor, and drainage area. $\boldsymbol{\theta}_{\boldsymbol{m}}$), reservoir state (i.e. phase pressures and saturations $\boldsymbol{x}_m = f(\boldsymbol{x}_{m-1}, \boldsymbol{\theta}_{\boldsymbol{m}-1})$), and simulated measurement data (i.e. $\boldsymbol{d}_m = g(\boldsymbol{x}_m, \boldsymbol{\theta}_{\boldsymbol{m}}))$ obtained from each model at time t_2 into the augmented state matrix in $\boldsymbol{Y}_m$ in Eq. (14). Calculate ensemble mean $\overline{\boldsymbol{Y}_m}$ using Eq. (15). Calculate ensemble perturbation matrix $\boldsymbol{Y}_m^{\prime}$ using Eq. (16). Calculate the prior covariance matrix C_Y^f using Eq. (19). Calculate Kalman gain K using Eq. (17). Finally, the increments are estimated by $K\left(d_{m,k}^o - HY_{m,k}^f\right)$ in Eq. (17).

(4) Update:

Update model realizations by combining forecasted system state with increments using Eq. (17).

(5) Proceed:

Proceed to the next time step, and follow the same procedures from (2) to (4).

3.3 Model Setups—Assumptions, Parameterization and Initialization

In this section, we define the parameters to be estimated, and the method to assign values for each ensemble member. We assume, the reservoir properties (static variables) unknown include local permeability, skin factor, and drainage area of a single-well. We assume the values of these parameters are initially obtained from core analysis, well logging, or drainage division. The initial ensembles are obtained by adding perturbations to the surveyed number. The perturbations are drawn from a normal distribution with a mean of zero and a specified covariance. We also assume the local reservoir permeability near a producer is homogeneous and isotropic, since the primary production studied in this research is not affected by inter-well connections, and hence cannot be used to evaluate the heterogeneity in the permeability distribution.

A drainage boundary is typically defined when pressure gradients caused by productions from adjacent wells meet. A drainage area can also be defined by natural barriers or traps. It is an important parameter to show the "potential" of a well. In our study, we simulate a commonly used square drainage area. We treat grid size as a parameter since it is directly written in the above governing equations. We can easily convert grid size to drainage area since grid numbers in both x and y directions are constant. The grid numbers we used in this study reflect an accurate response of a single-well production. This was done by a sensitivity study from several grid sizes.

Productions wells have a zone of altered permeability near the wellbore resulting from drilling or completion operations or stimulation. "Skin" is a coefficient used to quantify increased or decreased dimensionless pressure drop due to non-ideal conditions. The Hawkins' type skin in Fig. 2 assumes a simple 2-zone radial composite reservoir model. The steady-state flow exists in the "inner-zone" and any other relevant flow regimes can exist in the "outer zone". The stimulated or damaged permeability k_s only exists in a certain area adjacent of a wellbore, while additional pressure drops are accounted for in an outer area. To use EnKF method to estimate skin factor, Li and Han used a radial-type local grid refinement around the active well with 20 radial rings to simulate the skin (Li et al. 2010). In our study, we used the same skin type by adopting both local permeability (permeability in the outer zone) and near-wellbore permeability (permeability in the inner zone). We also assume the stimulated or damaged area only exist in the wellblock. Skin factor will be calculated using the following Peaceman's formula in Eq. (20)

$$S = \left(\frac{k}{k_s} - 1\right) \ln \frac{r_s}{r_w} \tag{20}$$

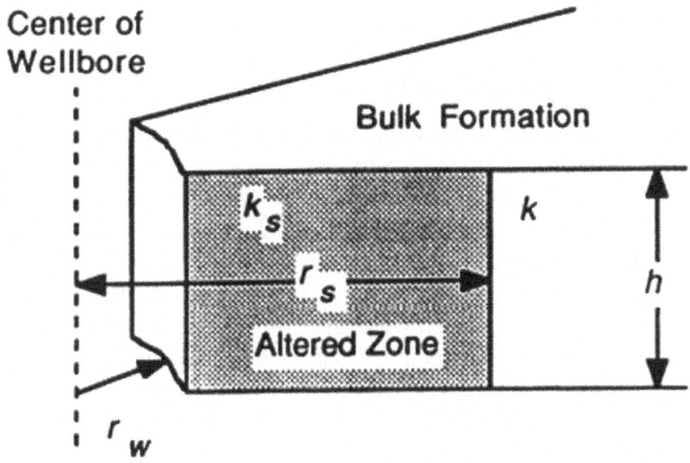

Fig. 2 The Hawkins' type skin factor (Blasingame 2007)

To initialize the realizations of reservoir pressures and phase saturations, we assume that these values are initially obtained from field measurements. Their initial ensembles are also obtained by adding perturbations to the surveyed number. The perturbations are drawn from a normal distribution with a mean of zero and a specified covariance. To initialize our model, the covariance of the pressure is 100 psi^2, and the covariance of the saturation is 0.0025. Other reservoir properties are not included in the EnKF estimates, and they are summarized in Table 1. The PVT data and the other rock properties are documented in the authors' previous study (Mandel 2009). The data apply for both synthetic case studies and field case studies in later sections.

Table 1 Information used into reservoir model-setup

Model description	Values
Layer thickness	15 ft
Number of grids in X-direction	11
Number of grids in Y-direction	11
Porosity	0.1
Reservoir depth	11400 ft
Initial reservoir pressure	5031 psi
Oil bubble point pressure	2640 psi
Initial oil saturation	0.8
Initial water saturation	0.2
Reservoir temperature	208 F

4 Results and Discussions

4.1 Accuracy Analysis of the Model

We used 30 randomly-generated synthetic cases to evaluate the accuracy of the EnKF-based PDA model. In each case, the reservoir permeability and the wellblock permeability are random numbers between 1 and 100 md, and the drainage area is a random number between 80 and 350 acres. The ensemble size we used is 50, which proves to be adequate based on a sensitivity test.

Prior to the test, we first prepared the production data as the measurements used to "correct" the model estimates. When generating the production profiles of each cases, the oil production is firstly maintained constant at 500 bbl/day. The well production is changed to a bottomhole flowing pressure (BHFP)-specified production as reservoir pressure releases. The specified BHFP is 500 psi. We used monthly production of ten years as the measurements, and the assimilation starts from the first measurement after the oil production starts to declines.

The following case is discussed in details. The true values, initial inputs, model estimations, absolute error, and error percentage of each estimated reservoir property are documented in Table 2. Figure 3 is the oil rate-data match of this case. 100 data points representing 100 assimilations are recorded and presented. The red dots are the "measured" oil rate data from the synthetic model. The blue solid curve is the ensemble mean at each assimilation step. The blue dashed lines enclose the estimate uncertainty within ± 3 standard deviations. We can see that due to adjustments of the reservoir properties, there are some fluctuations of the oil rate uncertainties during the first 8 data points. A good match is achieved after assimilating 10 data points, and the uncertainty quickly decreases within the first several assimilations.

Table 2 Summarization of the estimation example

	Reservoir permeability (md)	Well block permeability (md)	Drainage area (acres)
True value	16.99	22.27	316.65
Initial input	50.00	30.00	120.00
Model estimation	28.95	23.00	335.24
Absolute error	16.28	0.73	18.59
Error percentage (%)	95.82	3.3	5.8

Figure 4 is about the drainage area estimate. Drainage area is a reservoir constant, therefore the red solid line, which represent the true drainage area value, does not fluctuate with time. The blue solid line is the EnKF mean estimate. The blue dashed lines enclose its estimate uncertainty. In Fig. 4, the area estimate settled at around 335 acres and the uncertainty is only about ± 2 acres. The true drainage area value lies outside of the confidence interval. Figure 5 is about the permeability estimation of this case. Similarly, permeability is also considered a reservoir constant as is described in

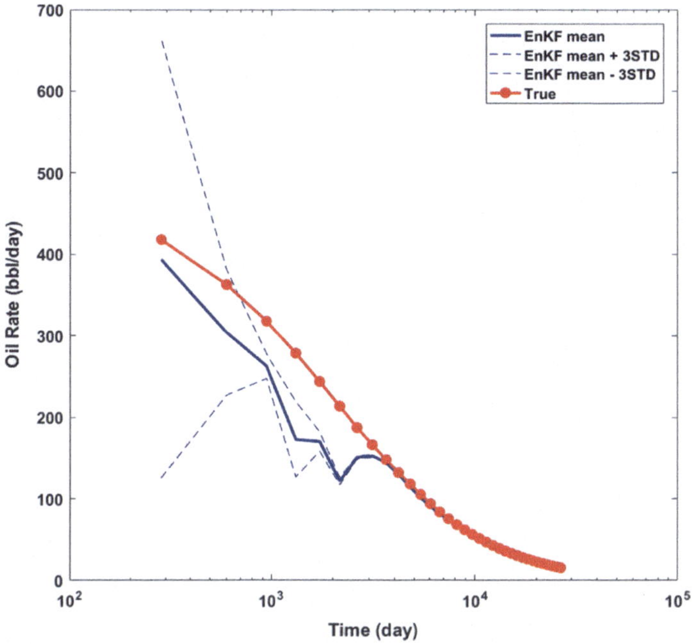

Fig. 3 Matching of the oil rate for the synthetic case. Blue solid line is the EnKF mean estimate. Blue dashed lines enclose estimate uncertainty. Red solid line with dots is the true production rate

previous sections. From Fig. 5, the uncertainties of permeability estimate quickly decrease after 4 or 5 assimilations. Although the mean estimate of the wellblock permeability is close to the true value, the mean estimate of the reservoir permeability is about 16 md larger than the true value, and the true value is not within the confidence interval.

Our observation from this test is, since EnKF is a naturally variance-decreasing method, the swift reduction of variances (of estimated reservoir properties) will impede the generation of enough observation (oil rate data) spread. The oil rate predictions, even though quickly fall into a good match with measured oil rates, but defined with an unwanted set of reservoir properties. Therefore, we can observe small uncertainties and large estimation (mean estimate) error.

we documented the test range and average error of the 30 synthetic cases in Table 3 to exemplify the accuracy and precision of the current model. The average error of drainage estimate is small compared to the range of the true value. The average error of permeability estimates, however, is more significant compared with that of drainage area. The histogram of the absolute error of the estimated local and wellblock permeability are documented in Figs. 6 and 7. From the histograms, we can find that the distribution of the absolute error is larger than that of the drainage area. For most of the cases, the error in the permeability estimations are within 8 md, but there are still cases with error as large as 35 md. The large error exit in cases with a large contrast of reservoir and near-wellbore permeability values. The estimation of the wellblock

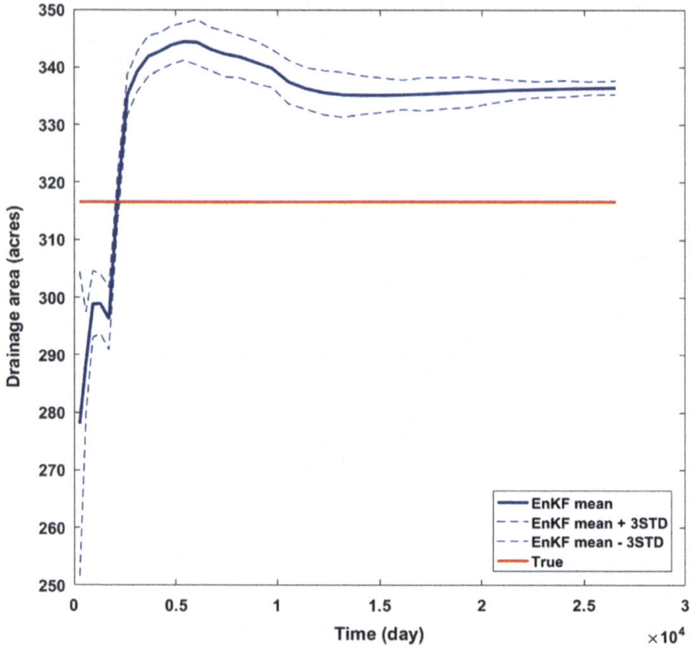

Fig. 4 Drainage area estimation of the synthetic case. Blue solid line is the EnKF mean estimate. Blue dashed lines enclose the estimate uncertainty. Red solid line with dots is the true production rate

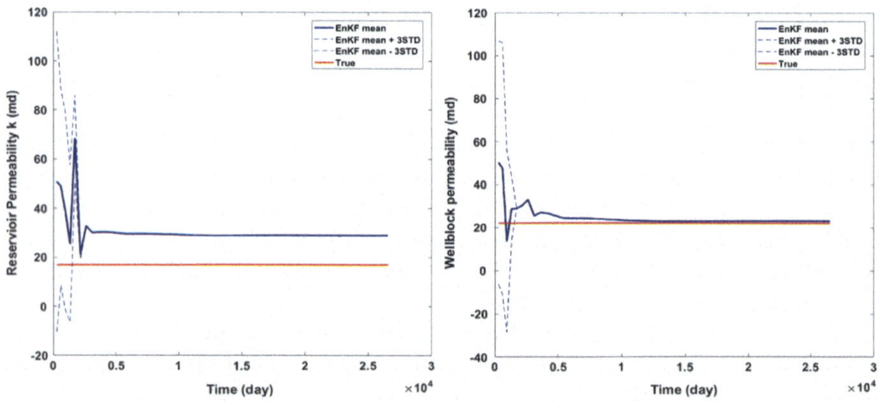

Fig. 5 Permeability estimation of the synthetic case. Blue solid line is EnKF the mean estimate. Blue dashed lines enclose the estimate uncertainty. Red solid line with dots is the true production rate

permeability has very similar accuracy with the estimation of reservoir permeability, since they are essentially the same reservoir property except that the locations of interest are different.

Table 3 Test range and average error in property estimation for 30 random cases

Properties	Range of the true values	Average error
Drainage area A (acres)	80.00–350.00	8.31
Reservoir permeability k (md)	5.00–120.00	11.94
Well block permeability ks (md)	5.00–120.00	11.74

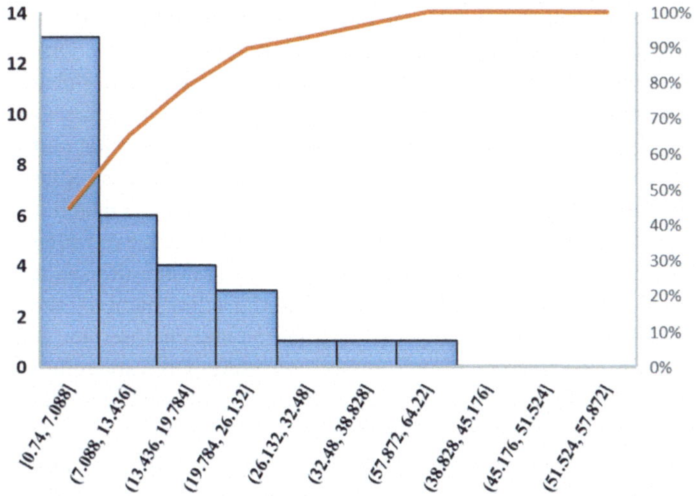

Fig. 6 Histogram of absolute error in estimated reservoir permeability

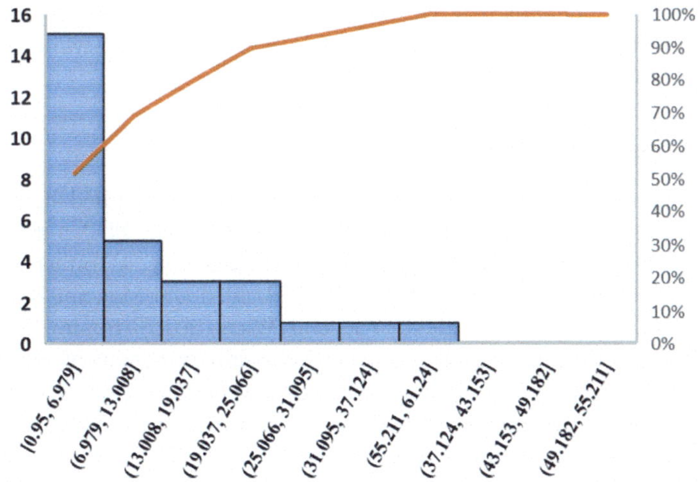

Fig. 7 Histogram of absolute error in estimated wellblock permeability

4.2 Model Improvements

Based on our previous observations, it is important to capture the early transient behavior to improve the property estimates. When estimating permeability field, Haugen et al. added stochastic noise to the rate data to larger spread and variance in the predicted ensemble and, thus, to increase weight on the measurements in the update step (Haugen et al. 2008). However, they concluded that the impact of this added noise on the assimilated results is marginal. We will present our results when an enlarged spread improved estimation of reservoir properties in our application.

In the following case, we manually adjusted the observation error. Observation error starts with a larger value (50% of first recorded oil rate) and linearly reduce to 0.01 bbl/day within 20 data points. The estimated covariance of the reservoir properties is hold constant when decrease to a threshold value. The value for permeability and drainage area are 4 md and 8 acres. The above adjustment ensured a larger spread in both predicted ensemble and forecasted ensemble. Figure 8 is the match of oil rate after the adjustment. From Fig. 8, a good match of oil rate is achieved after assimilating about 18 data points. Figure 9 is the drainage area estimation after adjustment. From Fig. 9, the mean estimate of drainage area after the adjustment is similar with the estimate before. The uncertainty, however, is enlarged and enclosed the true value. The enlarged uncertain (± 12 acres) is reasonable considering the magnitude of the true value (316.65 acres). Figure 10 is the permeability estimation after the adjustment. From Fig. 10, the mean estimates of permeability approaches to the true value as the assimilation proceeds. After assimilating about 550 data points, absolute error in ensemble mean is less than 1 md. After the improvements, we can see that several additional data points are needed to reach a good match of the oil rate. The mean estimates of the reservoir properties have significant improvements. The uncertainty of the permeability estimate is larger after adjustment, and the true value is covered by the confidence interval. The enlarged uncertainty is acceptable in field applications. The above adjustment improved the accuracy of EnKF mean estimate while conserved a reasonable uncertainty.

4.3 Field Cases

After testing and modifying the model with synthetic cases, we applied the model to the target field and assessed its ability in evaluating field cases. The PDA model is applied in a currently producing carbonate oil reservoir in Alabama. A former study documented the primary development of the field (Al Haddad and Mancini 2013; Heydari and Baria 2006; Mancini et al. 2008; Yue and Wang 2015). Figure 11 shows the location and structure of the light oil reservoir. We mainly present the evaluation of one well in this paper. Figure 12 shows the production and pressure data of well #8971514 in the above field. The data point marked with the arrow sign is treated as the anchor point, which is the first assimilating point. BHFP is 1632 psi at this point and is constant until the end of the record. Note that, only the oil rate data is assimilated while the other surveying data are missing. Measurement error is initiated from 250 bbl/month. Monthly oil production (1 data point per 30 days) is publicly available.

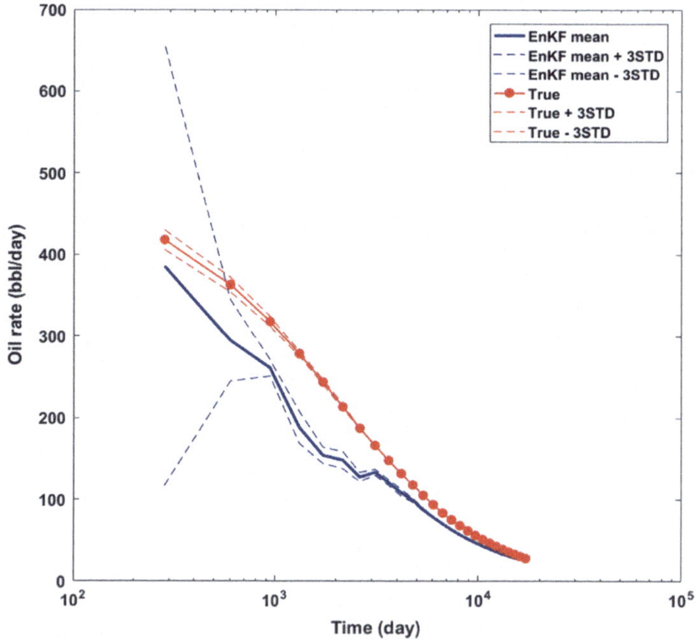

Fig. 8 Oil rate matching of the synthetic case after adjustment. Blue solid line is the EnKF mean estimate. Blue dashed lines enclose the estimate uncertainty. Red solid line with dots is the true production rate

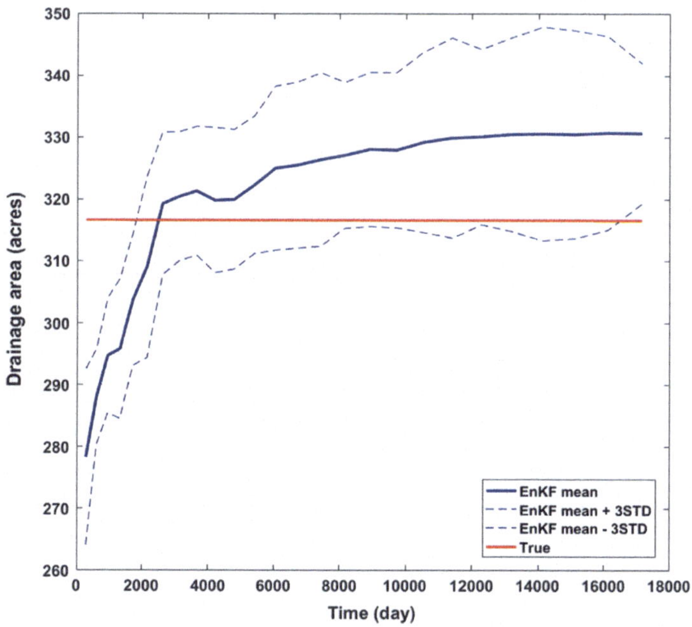

Fig. 9 Drainage area estimation of the synthetic case after adjustment. Blue solid line is the EnKF mean estimate. Blue dashed lines enclose the estimate uncertainty. Red solid line with dots is the true production

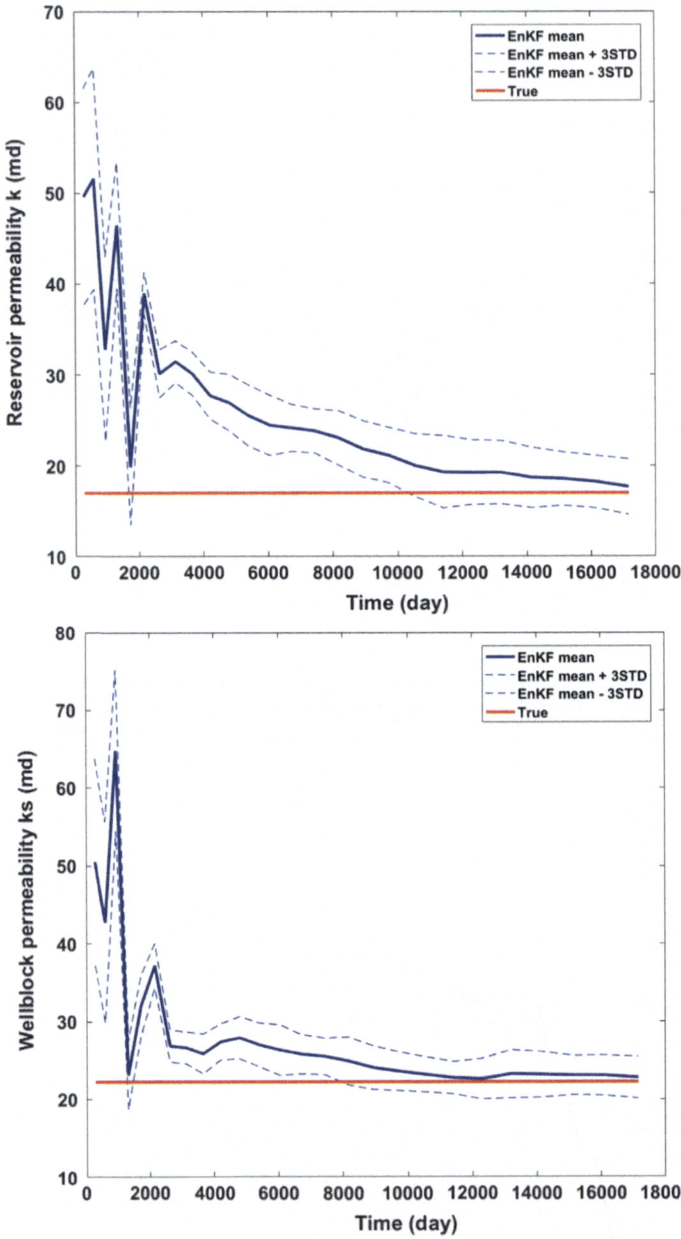

Fig. 10 Permeability estimation of the synthetic case after adjustment. Blue solid line is the EnKF mean estimate. Blue dashed lines enclose estimate uncertainty. Red solid line with dots is true production

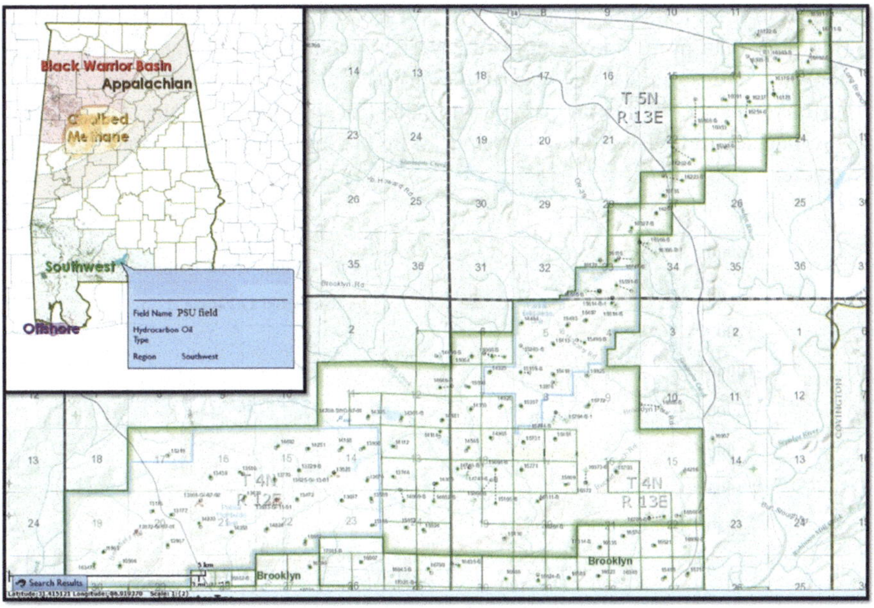

Fig. 11 Map showing the major structural features and approximate updip limit of the Smackover deposition and the location of the PSU fields (modified from SOGBA)

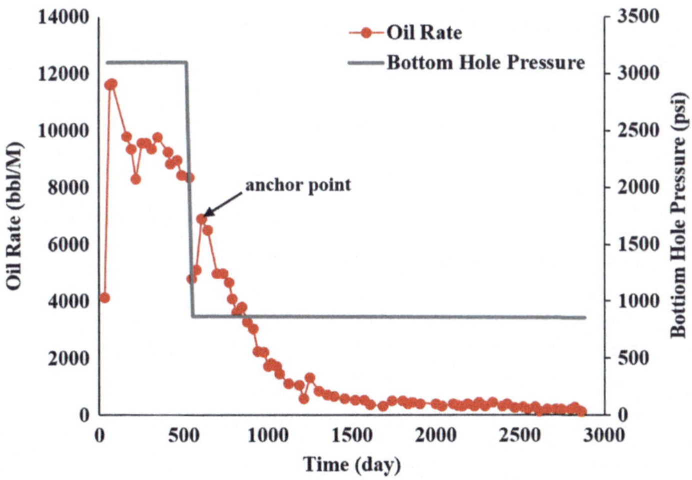

Fig. 12 Oil production rate and flowing bottom hole pressure of well 8971514 in the PSU field

We utilized 30 data points after the anchor point for assimilation purpose. The last 10 points are used to verify the estimation.

Figure 13 shows the matching of the oil rate for well 8971514. Notice the large estimation error after initiation of data assimilation quickly reduces as assimilation proceeds. After assimilating about 21 points, the oil rate obtains a very good match. From the 30th data point on, we hold reservoir permeability, skin factor, and drainage area constant and simulate the rest of the production (for about 300 days). Results indicate consistency of the prediction with the true production data. The estimated reservoir permeability is about 24.00 md, skin factor is 9.11, and drainage area is 161.82 acres.

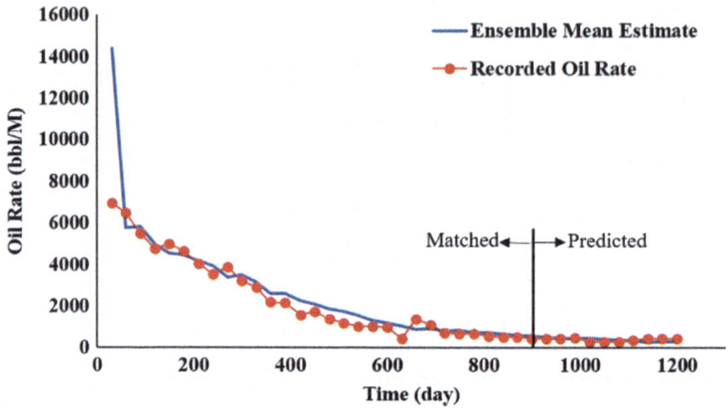

Fig. 13 Oil rate matching for well 8971514 in the PSU field

The model is used to estimate reservoir properties of additional wells in the target field which we are able to obtain production data and core permeability measurements (courteous of the operating company). We present the estimated reservoir properties of 4 wells in Table 4. The matched permeability values are close to core permeability, which validates the accuracy and applicability of the model.

Table 4 Evaluation of 4 wells using the EnKF-based PDA model

	Well name	Estimated permeability (md)	Skin	Drainage (acres)	Measured permeability (md)
1	8971514	24	9	162	27
2	9754004	15	−5	279	15
3	2784982	17	−5	378	18
4	9575068	20.4	−3.0	303.0	23

5 Conclusion

In this study, we focused our scope on a single-well and used the EnKF algorithm to estimate reservoir permeability, skin factor, and drainage area. We then modified the model to improve the accuracy of the method.

From the evaluations of the synthetic models, a good match of the oil rate is achieved in first several assimilations. The estimation of drainage area has high accuracy, but the estimation of skin (wellblock permeability) and reservoir permeability present some accuracy issues: large error exit in cases with the large contrast of reservoir and near-wellbore permeability values.

Based on our evaluation, the accuracy issue in reservoir permeability estimation and skin estimation is caused by reservoir model behavior and property definition: within a certain range, the variation of reservoir permeabilities will not generate an adequate spread of oil rates. Also, EnKF will naturally reduce covariance as assimilation proceeds, which further reduce spreads.

By manually enlarge model error and variance in the predicted ensemble, we put a larger weight on the measurements in the update step. The accuracy of the EnKF mean estimate is improved, and a reasonable uncertainty is preserved.

The above model and method prove to be applicable in real field evaluations. The accuracy is verified by core measurement. The estimated property fields capture a reasonable geological structure and spatial geological features.

By combining EnKF with single-well reservoir simulation, we proved that the model is capable of estimating reservoir properties of utmost interest in well testing without having to go through the complex preconditions of well testing. The EnKF-based in single-well PDA could improve accuracy and applicability issues in traditional analytical methods.

Acknowledgements. The authors would like to acknowledge Computer Modeling Group Ltd. for providing the CMG software for this study.

References

Al Haddad, S., Mancini, E.A.: Reservoir characterization, modeling, and evaluation of Upper Jurassic Smackover microbial carbonate and associated facies in Little Cedar Creek field, southwest Alabama, eastern Gulf coastal plain of the United States. AAPG Bull. **97**(11), 2059–2083 (2013). https://doi.org/10.1306/07081312187

Ampomah, W., Balch, R., Cather, M., Rose-Coss, D., Dai, Z., Heath, J., Mozley, P.: Evaluation of CO_2 storage mechanisms in CO_2 enhanced oil recovery sites: application to Morrow Sandstone reservoir. Energy Fuels **30**(10), 8545–8555 (2016). https://doi.org/10.1021/acs.energyfuels.6b01888

Arps, J.J.: Analysis of decline curves. Trans. AIME **160**(01), 228–247 (1945). https://doi.org/10.2118/945228-G

Bhattacharya, S., Nikolaou, M.: Analysis of production history for unconventional gas reservoirs with statistical methods. SPE J. **18**(5), 878–896 (2013). https://doi.org/10.2118/147658-PA

Blasingame, T.A.: "Skin Factor" Concept (2007)

Burgers, G., Jan van Leeuwen, P., Evensen, G.: Analysis scheme in the Ensemble Kalman Filter. Mon. Weather Rev. **126**(6), 1719–1724 (1998). https://doi.org/10.1175/1520-0493(1998)126%3c1719:ASITEK%3e2.0.CO;2

Cai, Y., Liu, D., Pan, Z.: Partial coal pyrolysis and its implication to enhance coalbed methane recovery: a simulation study. Energy Fuels **31**(5), 4895–4903 (2017). https://doi.org/10.1021/acs.energyfuels.7b00219

Cao, P., Liu, J., Leong, Y.-K.: A fully coupled multiscale shale deformation-gas transport model for the evaluation of shale gas extraction. Fuel **178**, 103–117 (2016). https://doi.org/10.1016/J.FUEL.2016.03.055

Cokar, M., Ford, B., Gieg, L.M., Kallos, M.S., Gates, I.D.: Reactive reservoir simulation of biogenic shallow shale gas systems enabled by experimentally determined methane generation rates. Energy Fuels **27**(5), 2413–2421 (2013). https://doi.org/10.1021/ef400616k

Ertekin, T., Abou-Kassem, J.H., King, G.R.: Basic applied reservoir simulation. In: Doherty, H. L. (eds.) Memorial Fund of AIME. Society of Petroleum Engineers (2001). Retrieved from http://books.google.com/books?id=M41dAQAACAAJ&pgis=1

Evensen, G.: Inverse methods and data assimilation in nonlinear ocean models. Physica D **77**(1–3), 108–129 (1994a). https://doi.org/10.1016/0167-2789(94)90130-9

Evensen, G.: Sequential data assimilation with a nonlinear quasi-geostrophic model using Monte Carlo methods to forecast error statistics. J. Geophys. Res. (1994b). https://doi.org/10.1029/94JC00572

Evensen, G.: The Ensemble Kalman Filter: theoretical formulation and practical implementation. Ocean Dyn. (2003). https://doi.org/10.1007/s10236-003-0036-9

Evensen, G.: Data Assimilation: The Ensemble Kalman Filter. Springer, New York, Inc. (2006). Retrieved from http://dl.acm.org/citation.cfm?id=1206873

Fan, D., Ettehadtavakkol, A.: Semi-analytical modeling of shale gas flow through fractal induced fracture networks with microseismic data. Fuel **193**, 444–459 (2017). https://doi.org/10.1016/J.FUEL.2016.12.059

Geir, Æ., Trond, M., Vefring, E.: Near-well reservoir monitoring through Ensemble Kalman Filter. In: Proceedings of SPE/DOE Improved Oil Recovery Symposium. Society of Petroleum Engineers (2002). https://doi.org/10.2118/75235-MS

Han, M., Li, G., Chen, J.: Assimilating microseismic and well-test data by use of EnKF for accurate reservoir characterization. SPE J. (2014). https://doi.org/10.2118/171554-PA

Haugen, V.E., Natvik, L.-J., Evensen, G., Berg, A., Flornes, K., Naevdal, G.: History matching using the Ensemble Kalman Filter on a North Sea field case. In: Proceedings of SPE Annual Technical Conference and Exhibition. Society of Petroleum Engineers. (2006). https://doi.org/10.2118/102430-MS

Haugen, V.E., Naevdal, G., Natvik, L.-J., Evensen, G., Berg, A., Flornes, K.: History matching using the Ensemble Kalman Filter on a North Sea field case. SPE J. **13**(4), 382–391 (2008). https://doi.org/10.2118/102430-PA

Heydari, E., Baria, L.R.: A microbial Smackover Formation and the dual reservoir-seal system at the Little Cedar Creek Field in Conecuh County of Alabama. GCAGS Trans. **55**, 294–320 (2006)

Houtekamer, P.L., Mitchell, H.L.: Data assimilation using an Ensemble Kalman Filter technique. Mon. Weather Rev. **126**(3), 796–811 (1998). https://doi.org/10.1175/1520-0493(1998)126%3c0796:DAUAEK%3e2.0.CO;2

Li, G., Han, M., Banerjee, R., Reynolds, A.C.: Integration of well-test pressure data into heterogeneous geological reservoir models. SPE Reservoir Eval. Eng. **13**(03), 496–508 (2010). https://doi.org/10.2118/124055-PA

Liang, B.: An Ensemble Kalman Filter Module for Automatic. The University of Texas at Austin (2007)

Mancini, E.A., Parcell, W.C., Ahr, W.M., Ramirez, V.O., Llinás, J.C., Cameron, M.: Upper Jurassic updip stratigraphic trap and associated Smackover microbial and nearshore carbonate facies, eastern Gulf coastal plain. AAPG Bull. **92**(4), 417–442 (2008). https://doi.org/10.1306/11140707076

Mandel, J.: A Brief Tutorial on the Ensemble Kalman Filter, 7. Atmospheric and Oceanic Physics (2009). Retrieved from http://arxiv.org/abs/0901.3725

Mullins, O.C., Zuo, J.Y., Pomerantz, A.E., Forsythe, J.C., Peters, K.: Reservoir fluid geodynamics: the chemistry and physics of oilfield reservoir fluids after trap filling. Energy Fuels **31**(12), 13088–13119 (2017). https://doi.org/10.1021/acs.energyfuels.7b02945

Naevdal, G., Johnsen, L., Aanonsen, S., Vefring, E.: Reservoir monitoring and continuous model updating using Ensemble Kalman Filter. SPE J. **10**(1), 66–74 (2005). https://doi.org/10.2118/84372-PA

Poston, S.W., Poe Jr., B.D.: Analysis of Production Decline Curves. Society of Petroleum Engineers (2008). Retrieved from http://books.google.com/books?id=rHxOPgAACAAJ&pgis=1

Wu, Y., Cheng, L., Huang, S., Jia, P., Zhang, J., Lan, X., Huang, H.: A practical method for production data analysis from multistage fractured horizontal wells in shale gas reservoirs. Fuel **186**, 821–829 (2016). https://doi.org/10.1016/J.FUEL.2016.09.029

Yue, W., Wang, J.Y.: Feasibility of waterflooding for a carbonate oil field through whole-field simulation studies. J. Energy Res. Technol. **137**(6), 064501 (2015). https://doi.org/10.1115/1.4030401

Zhang, Y., Song, C., Yang, D.: A damped iterative EnKF method to estimate relative permeability and capillary pressure for tight formations from displacement experiments. Fuel **167**, 306–315 (2016). https://doi.org/10.1016/J.FUEL.2015.11.040

Zhao, X., Liao, X.: Evaluation method of CO_2 sequestration and enhanced oil recovery in an oil reservoir, as applied to the Changqing Oilfields, China. Energy Fuels **26**(8), 5350–5354 (2012). https://doi.org/10.1021/ef300783c

Zheng, J., Ju, Y., Liu, H.-H., Zheng, L., Wang, M.: Numerical prediction of the decline of the shale gas production rate with considering the geomechanical effects based on the two-part Hooke's model. Fuel **185**, 362–369 (2016). https://doi.org/10.1016/J.FUEL.2016.07.112

Synchronized Switch Damping with Diode (SSDD) Based on a Network of Piezoelectric Elements for Improved Vibration Damping of Smart Structure

Dan Wu[1(✉)], Guo-wang Gao[1], Claude Richard[2], and Daniel Guyomar[2]

[1] Xi'an Shiyou University, Xi'an, China
wudan@xsyu.edu.cn
[2] LGEF EA 682, Université de Lyon, INSA-Lyon, Lyon, France

Abstract. The proposed work is based on the Synchronized Switch Damping technique (SSD) which has already been applied in reducing and suppressing structure vibration, a new approach called Synchronized Switch Damping with Diode (SSDD) for enhanced vibration damping of smart structure is developed, and it applies on energy redistribution with the piezoelectric elements network. The SSD technique reveals that the damping performance is strongly related to a build-up voltage which results from the continuous switching. It hints that the enhanced piezoelectric voltage results in the improved damping performance. In SSDD technique, diodes and switches are used to form a combinational network, and it is able to achieve a low-loss energy transfer from the source piezoelectric elements which extract the vibration energy of the structure to a target end piezoelectric element for increasing its operative energy aiming to improve a given mode damping. In this work, the simulation of a clamped plate with four piezoelectric elements modelled in the Matlab/SimulinkTM environment is shown. The damping performance can achieve an increase of 8 dB at least in some weak excitation case, for example, the coupled piezoelectric elements are excited in the pulse stimulus. The simulation results show the relationship between the damping performance on a given targeted mode and the established power flow. The importance of non-linear multi-actuator approach for improved vibration damping of extended smart structure is revealed by the experimental data.

Keywords: Structural vibration damping · Synchronized switch damping with diode (SSDD) · Energy transfer · Network of piezoelectric elements

1 Introduction

Structure vibration commonly exists in various mechanical systems, such as cars, planes, buildings and roads, etc. Large unwanted structure vibration could degrade the system stability and reliability. In recent years, the vibration control aiming at reducing the elastic vibration of mechanical structure energy has gained more and more attention. Among the many solution proposed in the switching technique relying on

J. Lin (ed.), *Proceedings of the International Petroleum and Petrochemical Technology Conference 2019*, pp. 452–470, 2020.
https://doi.org/10.1007/978-981-15-0860-8_37

piezoelements embedded on the structure [1], the vibration damping technique can be classified basically in three basic categories: passive control technique, active control technique and semi-active control technique. Passive control techniques described in [2] usually use four basic shunt circuits [3, 4] connecting with inductor, resistor and capacitor in parallel or in series to dissipate the mechanical energy in the form of electrical network. The external energy requirement and it can always get the damping effect, however, it is hard to get high damping effect and sometimes it suffers from less damping effect caused environmental variations. Although the active control technique [5] is global and effective, for generating a phase actuation to cancel the disturbance causing the vibration, an active actuators and sensors are required [6]. In that case, the external power is essential to supply both control and operative energy. For overcoming the drawbacks and limitations of passive control and active control to achieve the damping improvement, the semi-active vibration control techniques connecting the circuit to the piezoelectric element by implementing switches have been proposed. The strategy of the semi-active methods is to change the circuit operation conditions of the piezoelectric elements by a series of switching action which controls the circuit to perform energy conversion and dissipation [7]. For controlling the switches, just a little external energy is required. There are several methods have been developed on the semi-active vibration damping. In original research work [8], semi-active control method is proposed. In the method, the connection with piezoelectric element and circuit is synchronized by the structure vibration motioned to form a short or open circuit. The State-Switch Absorber technique was firstly proposed by Holdhusen et al. [9], the transducer stiffness is able to be changed smoothly through piezoelectric shunting while the strain becomes minimum.

More recently, Synchronized Switching Damping (SSD) technique developed by Richard and Guyomar [10, 11] indicates that the switch is closed intermittently when the piezoelectric voltage reaches an extreme. According to this SSD technique, the energy is dissipated because of a shift phase appearing between the piezoelectric voltage and resulting voltage. SSD techniques rely on the value of the strain and the voltage of piezoelectric elements and do not require any filtering device. Several typical methods have been developed, including Synchronized Switching Damping on Short circuit (SSDS), Synchronized Switching Damping on Inductor (SSDI) and Synchronized Switching Damping on Voltage source (SSDV) [12, 13]. The classical low power SSD technique can get good performance in mono-modal excitation case. However, because of difficultly defining the significant developing voltage using the appropriate phase, it is difficult to deal with the multimodal and complex vibrations case [14]. The multimodal control method, named SSDI was presented by [7, 15, 16]. The principles of this method is to get the maximal amplitude of the self contributed voltage by switching instantly on the basis of the given modal coordinate extremum, it reveals that the SSDI Max make use of a modal model permitting achieve the modal coordinate of vibration structure, it also uses on the switch to synchronized with extremum of the targeted modal [17, 18].

This present work develops a new SSD technique named Synchronized Switch Damping with Diode (SSDD) which improves the SSDI performance by means of transferring the energy and applying it on the vibrating structure by many piezoelectric elements. For improving the operative energy to enhance the damping of a given target mode, several piezoelectric elements are used to extract the vibration energy of the structure, and this novel method utilize a network of the piezoelectric elements connecting with the diodes, which have the characteristic of the unidirectional transmission to orientate the energy flow. Therefore, the advantage of the network applying on this energy transfer process is closely related to the energy level of structure itself, and to the energy allocated in the smart structure [14].

The theoretical models which present the principle of the basic SSDI and modal model strategy are explained in Sect. 2. Then, Sect. 3 declared the structure configuration and definition in simulation, and Sect. 4 introduce the SSDD technique concluding the principles of this method, the various networks of the configurations and the process the energy transfer. Finally, the simulation results reveal the performance of SSDD method with a plate and using enhanced electromechanical coupling coefficient are depict in Sect. 5.

2 The SSDI Technique

0.1 The Principles of Basic SSDI Method

A spring-mass model is presented in Fig. 1. It is composed of a structure mass symbolized as m, a damper with damping coefficient c and a spring with spring constant k^E. The F in the diagram denotes an external force, V is the voltage of the piezoelectric elements, u is the displacement of the structure and C_0 is the natural capacitance of piezoelectric element on the structure.

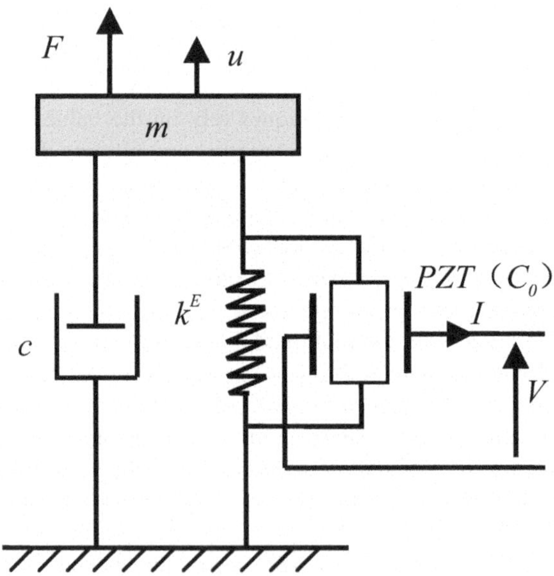

Fig. 1 The electromechanical model

Based on classical hypothesis of structure modeling, the mechanical equation and the electric equation are described by Eqs. (1) and (2):

$$m\ddot{u} + c\dot{u} + k^E u = -\alpha V + F \tag{1}$$

$$I = \alpha\dot{u} - C_0\dot{V} \tag{2}$$

where, α is the electromechanical coupling matrix, I is the electric current vector. The SSDI technique consists of the piezoelectric elements in parallel with the switch device, the SSDI circuit which is presented in Fig. 2a, L stands for an inductor, R presents a parasitic resistor and SW is a switching device. It shows that the capacitance of piezoelectric elements PZT (C_0), when the switch SW is turned on, the resistor R and the inductor L compose an *LCR* oscillator. It is trigged and turned on by a sensor which detects the extreme of the displacement, normally, the switch always keep open except a short instant when the extreme is detected.

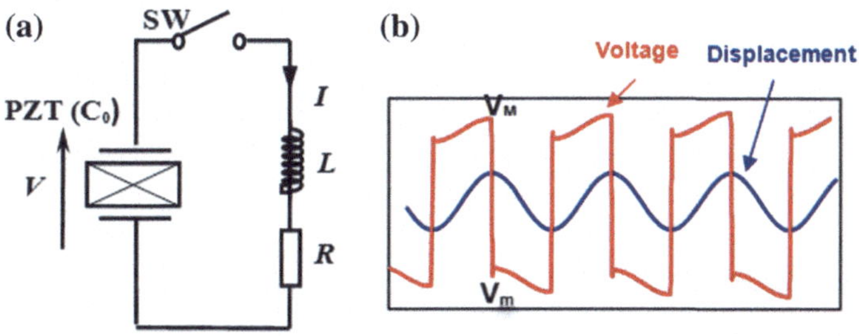

Fig. 2 **a** Classical SSDI circuit; **b** SSDI waveforms with a sinusoidal excitation

During these instants, the switch remains closed until the piezoelectric elements voltage reversed. The closing time is one half of the electric cycle T, and owing to the resistor R is very small, t_i is approximately expressed as Eq. (3), the smaller inductor can result in a shorter t_i. In this case, the switch can be closed on the maximum of the vibration cycle, the structural displacement and the voltage waveform can be drawn as Fig. 2b with a sinusoidal excitation. As indicated in the Fig. 2b, a voltage inversion caused by SW switching on happens at the maximum displacement of structure, after the switching, because of the energy loss in the circuit (switch and inductor), the absolute value of the voltage is smaller than that before the inversion, such loss can be modeled as an electrical quality factor Q_i which is expressed in Eq. (4), the relation between the before inversion voltage of piezoelectric element V_M and the after inversion voltage V_m can be expressed as Eq. (5),

$$t_i = \frac{T}{2} \approx \pi\sqrt{LC_0} \tag{3}$$

$$Q_i = \frac{1}{R}\sqrt{\frac{L}{C_0}} \tag{4}$$

$$\gamma = \frac{V_m}{V_M} = e^{-\frac{\pi}{2Q_i}} \tag{5}$$

$$k^2 = \frac{\omega_a^2 - \omega_0^2}{\omega_0^2} \tag{6}$$

k is the coupling coefficient of the piezoelectric element and can be computed from both open-circuit frequency ω_a and short-circuit resonance frequency ω_0, the energy dissipated by the switching E_S can be disintegrated as:

$$E_S = -\int_0^T \alpha V \dot{u} dt \tag{7}$$

$$E_S = 4\frac{\alpha^2}{C_0} U_m^2 \frac{1+\gamma}{1-\gamma} \tag{8}$$

The SSDI method increases the amplitude of the voltage, simultaneously, it also improve the operative energy which is described in Eq. (9):

$$W_{op} = \frac{1}{2} C_0 V^2 \tag{9}$$

The basic SSDI method has some limitations, there is no any energy transfer process occurred, the operative energy is extracted from the structure on the controlled mode by the driven piezoelectric element. The damping effect is only related to coupling coefficient of the piezoelectric elements for the target mode. The energy loss of the SSDI method is about $1 - \gamma^2$, usually it is half of the source energy.

0.2 The Modal SSDI Strategy

The classical SSDI method is performing well when the excitation is mono-modal or is located or focused on only one mode of the structure. To overcome this limitation and working on several modes with a wide bandwidth excitation, a modal SSDI technique was developed. It includes synchronizing switches at a given modal coordinate instead of voltage combinations so that the active control strategy and semi-active SSD damping have the advantages of both active control strategies and semi-active damping. When using classical hypothesis of finite element modelling, the dynamic behaviour of this kind of structure was modelled as electromechanical Eqs. (1) and (2), these two equations can be expressed in the modal basis ϕ after following change of variables:

$$u = \phi q \tag{10}$$

In the modal coordinate, the Eqs. (1) and (2) can be transformed to:

$$M\ddot{q} + C\dot{q} + K^E q = -\theta V + \beta F \tag{11}$$

$$I = \theta^T \dot{q} - C_0 \dot{V} \tag{12}$$

where, ϕ is the mode shape matrix related to n modes, q is the modal displacement; M is the structural mass, C is the structural damping and K^E is the generalized stiffness modal matrices. θ_i represents the modal electromechanical coupling coefficient with i piezoelectric elements:

$$\theta = \phi^T \alpha \tag{13}$$

The dimensions of θ are [n, i] matrix size, which is only related to modes of the structure. The strategy of the multimodal SSDI control is presented as Fig. 3. In this case, the smart structure is excited by an external force F and the controlled through the voltage of piezo-transductor $V_{transductor}$, the inversion of this voltage can be synchronized with modal displacement extremum q.

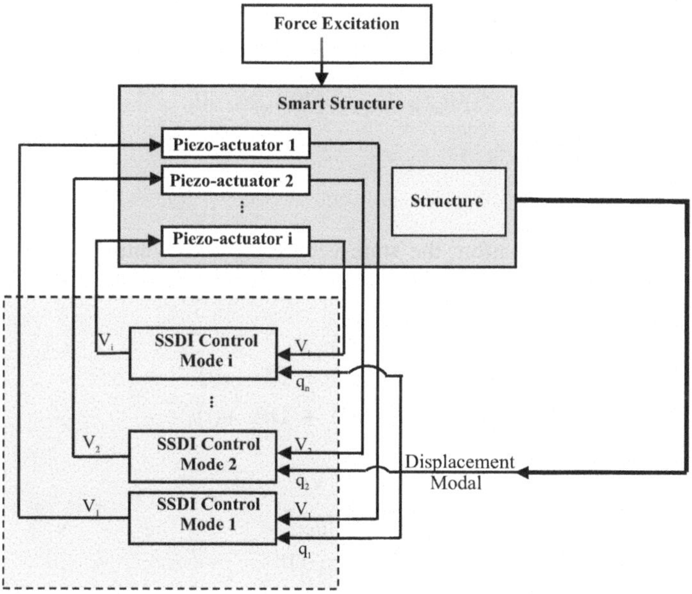

Fig. 3 The architecture of modal SSDI control

Under the multi-modes and multi-piezoelements, Eqs. (11) and (12) became:

$$\begin{bmatrix} [0]_{n\times n} & [m_n]_{n\times n} \\ [1]_{n\times n} & [0]_{n\times n} \end{bmatrix}_{2n\times 2n} \begin{bmatrix} \dot{q}_n \\ \ddot{q}_n \end{bmatrix}_{2n\times 1} = -\begin{bmatrix} [k_i^E]_{n\times n} & [c_n]_{n\times n} \\ [0]_{n\times n} & [-1]_{n\times n} \end{bmatrix}_{2n\times 2n} \begin{bmatrix} q_n \\ \dot{q}_n \end{bmatrix}_{2n\times 1} + \begin{bmatrix} [1]_{1\times i} & [-\theta]_{n\times i} \\ [0]_{1\times i} & [0]_{i\times i} \end{bmatrix}_{2i\times (i+1)} \begin{bmatrix} F \\ V_i \end{bmatrix}_{(i+1)\times 1} \tag{14}$$

$$[C_{0i}]_{i\times i}[V_i]_{i\times 1}= -\left[\frac{1}{R}\right]_{i\times i}[V]_{i\times 1}+[\theta]_{n\times i}[q_n]_{n\times 1} \tag{15}$$

The smart structure is modelled by Eqs. (11) and (12), which are also represented by a state space equation respectively. Under the mechanical equation, the state vector x(t), the control vector u(t) and the output vector y(t) are represented by Eq. (16), AM, BM, CM and DM are the state matrices.

$$\begin{aligned} x(t) &= \mathrm{AM}\cdot \mathrm{x(t)} + \mathrm{BM}\cdot \mathrm{u(t)} \\ y(t) &= CM\cdot \mathrm{x(t)} + \mathrm{DM}\cdot \mathrm{u(t)} \end{aligned}$$

$$\begin{aligned} \mathrm{x(t)} &= [\mathrm{q_n} \quad \dot{\mathrm{q}}_\mathrm{n}]^\mathrm{T} \\ u(t) &= [\mathrm{F} \quad V_i]^T \\ \mathrm{y(t)} &= [\mathrm{q_n} \quad \dot{\mathrm{q}}_\mathrm{n}]^\mathrm{T} \end{aligned} \tag{16}$$

$$\begin{aligned} \mathrm{AM} &= -\begin{bmatrix} [0]_{n\times n} & [m_n]_{n\times n} \\ [1]_{n\times n} & [0]_{n\times n} \end{bmatrix}^{-1}_{2n\times 2n} \begin{bmatrix} [k_i^E]_{n\times n} & [c_n]_{n\times n} \\ [0]_{n\times n} & [-1]_{n\times n} \end{bmatrix}_{2n\times 2n} \\ BM &= \begin{bmatrix} [0]_{n\times n} & [m_n]_{n\times n} \\ [1]_{n\times n} & [0]_{n\times n} \end{bmatrix}^{-1}_{2n\times 2n} \begin{bmatrix} [1]_{1\times i} & [-\alpha]_{i\times i} \\ [0]_{1\times i} & [0]_{i\times i} \end{bmatrix}_{2i\times (i+1)} \\ \mathrm{CM} &= [\mathrm{I}]_{2n\times 2n} \\ \mathrm{DM} &= 0 \end{aligned} \tag{17}$$

Under the electrical equation, the state vector x(t), the control vector u(t) and the output vector y(t) are represented by Eq. (18), AE, BE, CE and DE is the state matrices shown in Eq. (19):

$$\begin{aligned} x(t) &= \mathrm{AE}\cdot \mathrm{x(t)} + \mathrm{BE}\cdot \mathrm{u(t)} \\ y(t) &= CE\cdot \mathrm{x(t)} + \mathrm{DE}\cdot \mathrm{u(t)} \end{aligned}$$

$$\begin{aligned} \mathrm{x(t)} &= [\mathrm{V_i}]^\mathrm{T} \\ u(t) &= [\dot{q}_n]^T \\ \mathrm{y(t)} &= [V_i]^\mathrm{T} \end{aligned} \tag{18}$$

$$\begin{aligned} \mathrm{AE} &= -[C_{0i}]^{-1}_{i\times i}\left[\frac{1}{R}\right]_{i\times i} \\ BE &= [C_{0i}]^{-1}_{i\times i}[\alpha]_{i\times i} \\ \mathrm{CE} &= [\mathrm{I}]_{\mathrm{i\times i}} \\ \mathrm{DE} &= 0 \end{aligned} \tag{19}$$

3 Structure Configuration and Definition in Simulation

In simulation, the sketch of steel plate which is clamped on all four sides, bonded with four type P189 piezoelectric patches which is coupled with the four modes of the structure (Fig. 4). This structure has been designated as modal coordinate by means of the previously modal approach. The material properties of the steel plate and piezoelectric patches and the complete electromechanical coupling coefficient k with multi-modes are listed in Tables 1 and 2 respectively. The resonance frequencies of the first four modes, which are considered in the modal, are 96 Hz, 175 Hz, 204 Hz and 228 Hz respectively.

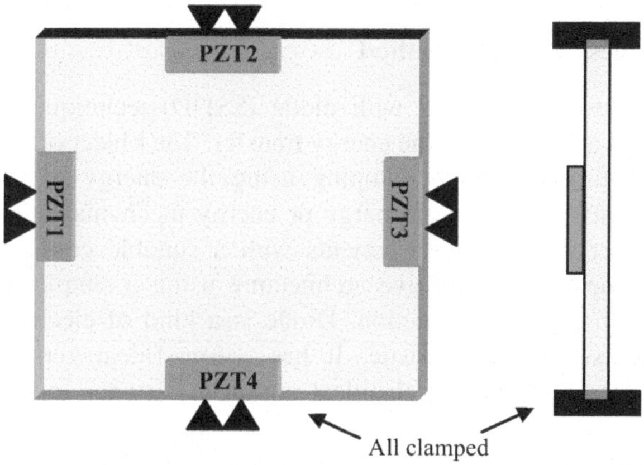

Fig. 4 The clamped plate bonded with piezoelectric elements

Table 1 The material properties

Steel plate	Length	0.6 m
	Width	0.4 m
	Thickness	1 mm
	Young modules	210 GPa
	Density	7500 kg/m^3
	Poisson ratio	0.345
Piezoelectric elements	Length	0.12 m
	Width	0.04 m
	Thickness	0.6 mm
	Density	7650 kg/m^3
	Elastic compliance constant (short circuit)	$S_{11}^{E} = 10.66\ \text{Pa}^{-1}$
	Permittivity	$\varepsilon_{33}^{T} = 10.17$ nF/m
	Piezoelectric coefficient	$d_{13} = -108$ pC/N

Table 2 The completed electromechanical coupling coefficient $k_{com-plated}$ ($\times 10^{-2}$)

Mode	PZT1	PZT2	PZT3	PZT4	Frequency (Hz)
1	6.43	8.06	0	0	96
2	0	0	11.42	9.1	175
3	11.96	8.54	0	0	204
4	7.64	3.82	6.38	3.96	228

4 The SSDD Technique

0.3 The Principles of SSDD Method

The synchronized switch damping with diode (SSDD) technique is base on SSDI method, which aims at optimizing the energy transfer. The object of this SSDD method is to enhance the targeted mode damping using the energy of a network of the piezoelectric elements and a proper energy or energy mechanism. It can increase the voltage of the target piezoelectric elements with a suitable configuration of piezoelectric elements; the specific elective architecture using a simple diode is allowing electricity to flow in only one direction. Diode is a kind of electronic element with asymmetric transmission characteristic. It has a low (ideal zero) forward current resistance in one direction and a high (ideal infinite) resistance in the other direction. The most common function of an ideal diode is to allow the current to pass in one direction (called the forward direction of the diode) while blocking the current in the opposite direction (the opposite direction). The circuit behaviour of the diode can be given by the voltage-current characteristic curve, shown in Fig. 5. This excellent characteristic is suitable for orienting current in only one way, the current will flow across the diode considering as a linear resistor, when the forward voltage exceeds its threshold voltage V_d equalled to 0.6, the zero and less forward voltage will result in zero current flowing.

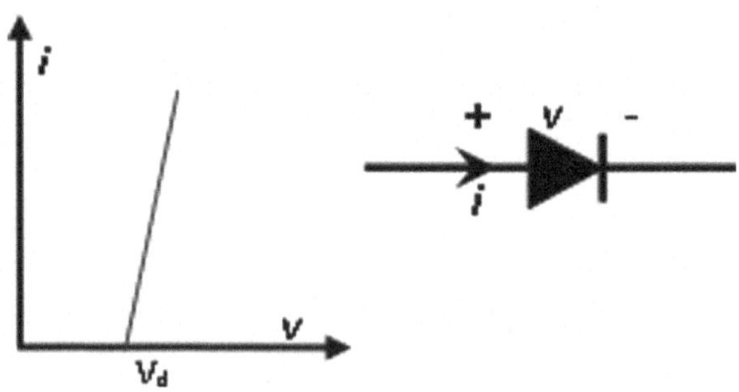

Fig. 5 I–V characteristics of an ideal diode

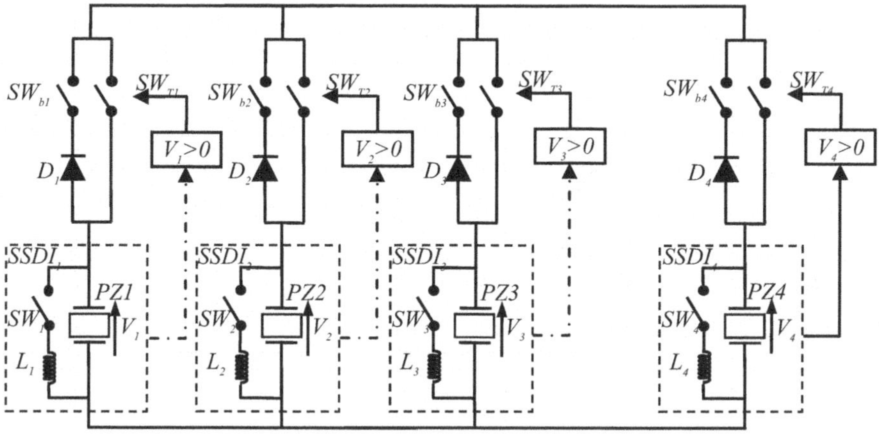

Fig. 6 The bidirectional SSDD network

We set up this piezoelectric elements network with a given plate structure presenting four decoupled modes, the bidirectional SSDD network is presented in Fig. 6, four parallel connected uniform branches form a bidirectional network which is adopted to achieve energy transfer and redistribution. Each pizeoelement could be an energy provider or receiver according to the specification of switch SWT. In each individual branch, such as covered by the gray box in the above figure, it consists of a basic SSDI component and two ideal diodes series connecting with two switches. To be specific, the ideal diode D8 is able to orientate the current into the branch when the switch SWt4 is on and SWb4 is off, which sets the PZT4 to be an energy receiver. Combining with closed SWb4 and open SWt4, the ideal diode D7 is capable of orientating the current out the branch, which determines PZT4 to work as an energy transmitter. With this flexibility of reconfiguration and by arranging properly the switches in the network, every pizeoelement is potentially a receiver which absorb energy from the network or a transmitter which provides energy to the other pizeoelements.

Established unidirectional SSDD network of this article is based on the one case of this bidirectional SSDD network, for improving the damping effect of a target pizeoelement on a target mode, an Optimized Oriented SSDD network which transfer and redistribute energy to a target pizeoelement PZT4 acting as an actuator to suppress the vibration on the target mode is configured as shown in Fig. 7. A target branch is configured to obtain energy from the other source branch, which aims at increasing the operative energy of target pizeoelement PZT4 and attains more damping improvement of the vibrational structure. The next discussion will introduce the evolution of the SSDD network.

0.4 Various Networks of the Configurations

For the sake of presenting the evolution of the SSDD network, in the simulation research, we set up three configurations of the network to show the different networks

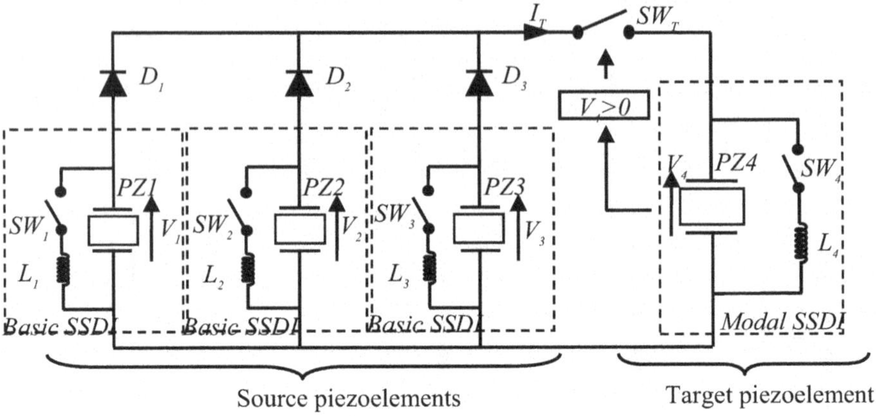

Fig. 7 The unidirectional SSDD network

about SSDD technique which are Basic SSDD method (Configuration 2), Oriented SSDD method (Configuration 3), and Optimized Oriented SSDD method (Configuration 4). Meanwhile, for comparison, two basic configurations which are basic SSDI method (Configuration 1) and SSDIF (Configuration 5) are taken into consideration. It should be emphasized that the piezoelements are classified as two parts, one part is the three source piezoelectric elements (PZT1, PZT2, PZT3) and another part is the one target piezoelement PZT4, each network of the configurations is set to target mode 4.

0.4.1 Configuration 1—SSDI Method

The network of the modal SSDI method is shown in Fig. 8.

In this configuration 1, apparently, there is no energy transfer between source piezoelements (PZT1, PZT2, PZT3) and target piezoelement PZT4 resulting in none the extra operative energy, the operative energy is only extracted from the structure on

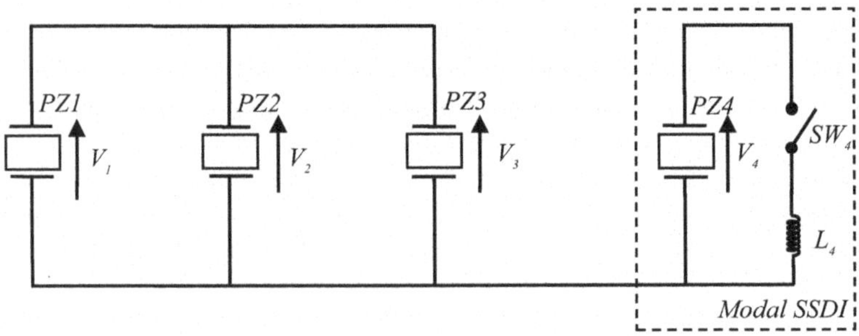

Fig. 8 The network of the basic SSDI method

the controlled mode 4 by the target piezoelement PZT4. The damping effect also only related to coupling coefficient in target mode 4. It always loses half of the source energy at each switching.

0.4.2 Configuration 2—Basic SSDD Method

The network of the basic SSDD method is shown in Fig. 9.

In configuration 2, named the basic SSDD (synchronized switch damping with diode) method, the network of piezoelements is used to realize the energy transfer, in each source piezoelements (PZT1 PZT2 PZT3)branch, the diode (D1 D2 D3) orient the

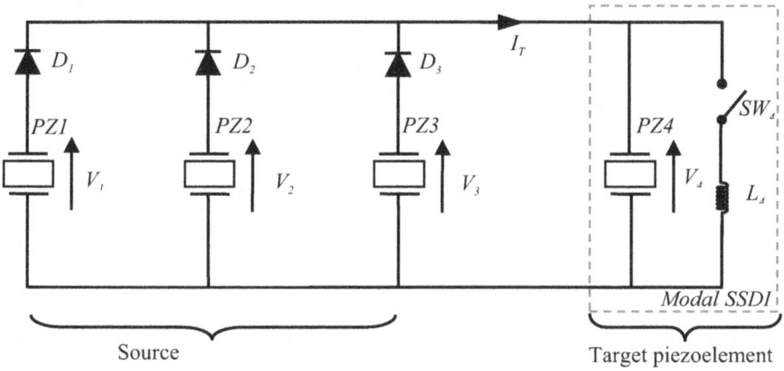

Fig. 9 The network of the basic SSDD method

current to the target piezoelement PZT4, which is controlled by modal SSDI targeting the mode 4, thanks to the diode's one way transmission characteristic, the source energy of the piezoelement (PZT1 PZT2 PZT3) can be transferred to target piezoelement PZT4, which results in improving the operative energy of target piezoelement PZT4 and achieve better damping effect than modal SSDI method in the target mode 4.

However, this basic SSDD method has some drawbacks and limitations. The energy transfer only happened in the beginning, which cannot improve the operative energy a lot. According to the diode model, I_T will be greater than zero when diodes work. For the power flow, the transferred energy P_T is equal to $V_4 \times I_t$, P_T will be negative if V_4 below zero, it means that the transferred energy will be lost and go back to network of the source piezoelectric elements. In this configuration, the diode can only orient the current flow, but cannot take care of the power flow and cannot orient the transfer energy. The performance of the Basic SSDD method is not so effective because of this drawback.

0.4.3 Configuration 3—Oriented SSDD Method

For overcoming the limitation of the basic SSDD method (Configuration 2), we proposed an oriented SSDD network shown in Fig. 10.

A switch named SWT is implemented between the source piezoelectric element (PZT1 PZT2 PZT3) and target piezoelectric element PZT4. The SW_T will be turned on only if the voltage V_4 of the PZT4 is positive. The switch SWT triggered by the voltage V_4, avoids effectively losing the transferred energy. Combing the diodes and switch SW_T, the power can be oriented to the targeted piezoelectric element PZT4 without going back to the source piezoelectric elements (PZT1 PZT2 PZT3). This oriented SSDD method can improve the operative energy using in the targeted mode 4. However, owing to the amplitude of the source piezoelements voltage (V_1 V_2 V_3) is not enough, the energy transfer only happen in sometimes, so limitations still exist in this configuration.

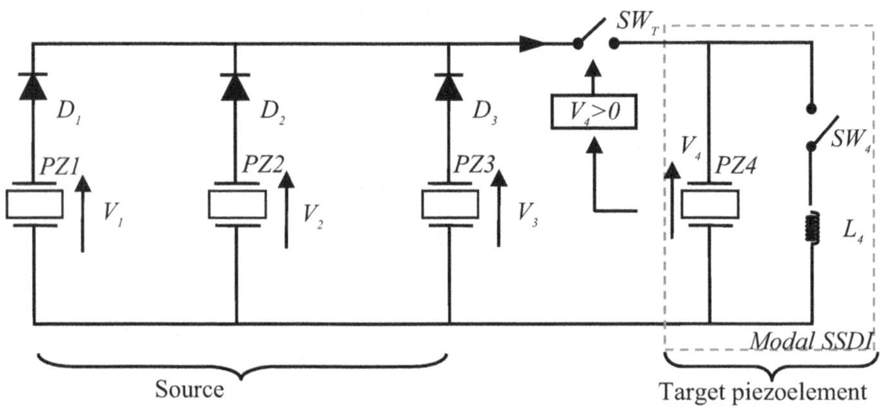

Fig. 10 The network of the oriented SSDD method

0.4.4 Configuration 4—Optimized Oriented SSDD Method

For continuing to improve the energy transfer and overcoming the limitation of the oriented SSDD method (Configuration 3), an optimized oriented SSDD network has been set up as shown in Fig. 11.

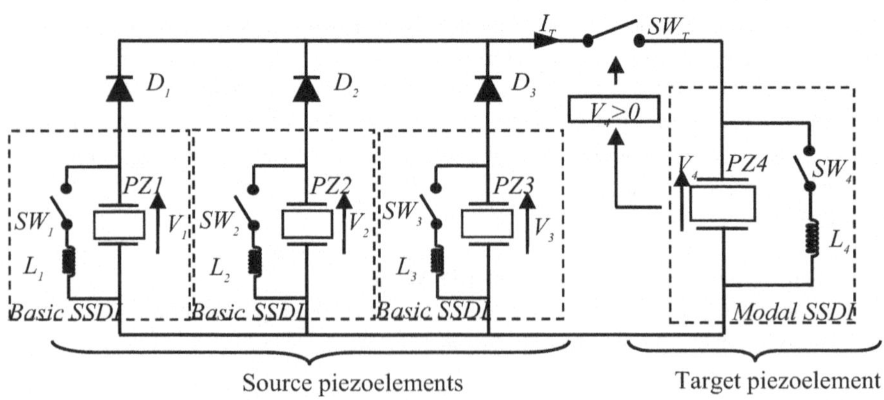

Fig. 11 The network of the optimized oriented SSDD method

In this configuration 4, the basic SSDI method is applied in each individual source piezoelement (PZT1 PZT2 PZT3). For improving its voltage, the energy extraction from the source piezoelements (PZT1 PZT2 PZT3) implemented through the basic SSDI process. The SW_1 SW_2 and SW_3 were triggered by the voltage of the source piezoelement (PZT1 PZT2 PZT3) respectively. Thanks to, the voltage in each source piezoelement (PZT1 PZT2 PZT3) has been enhanced thanks to the basic SSSI method, hence, more source energy of the piezoelements (PZT1 PZT2 PZT3) can be well transferred to target piezoelement PZT4, which results much more operative energy using in the targeted mode 4. The performance of damping effect using the optimized oriented SSDD network is much more effective than other configurations.

0.4.5 Configuration 5—SSDIF (Independent Piezoelectric Elements)

The configuration 5, called SSDIF (SSDI-full) is used for the comparison with the previous networks of the SSDD methods as shown in Fig. 12. Obviously, there is no network of piezoelectric elements at all and there should be no energy transfer between each piezoelements. The comparison will be presented at Sect. 5.

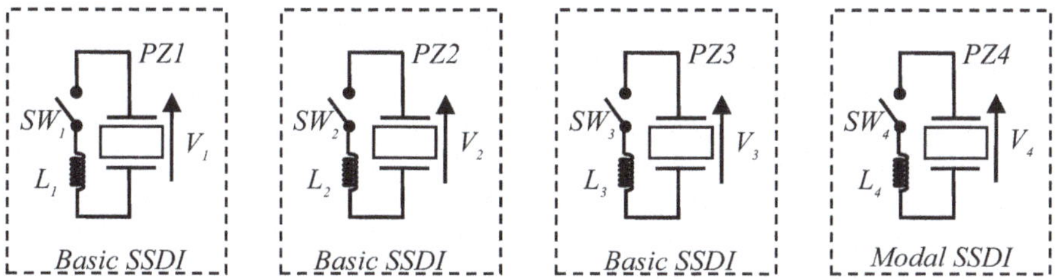

Fig. 12 The independent piezoelectric elements (SSDIF)

0.5 The Energy Transfer and Voltage Magnification Process

It is necessary to introduce the process of the energy transfer using SSDD network method. The configuration 1 (the modal SSDI method) and the configuration 4 (the optimized oriented SSDD method) as the example, Fig. 13 shows the value of target energy using SSDI method and the optimized oriented SSDD method in the time domain.

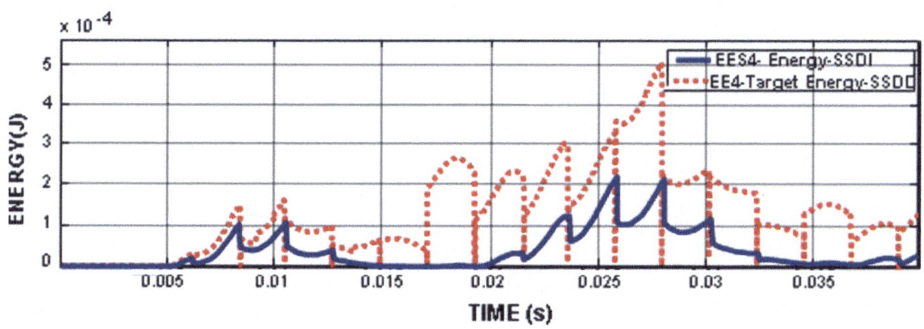

Fig. 13 The energy of target piezoelement PZT4 (operative energy) using the modal SSDI method and SSDD network method

Figure 13 illustrates the comparison about the energy of target piezoelement PZT4 (operative energy) using the modal method and SSDD network method. EE4 is the energy of target piezoelement PZT4 using SSDD network method; EES4 is the energy of the same piezoelement PZT4 using modal SSDI method, which can lose nearly half of operative energy. Contrarily, in the most of switch times, because the energy transfer happening in the SSDD network method, the energy of piezoelement PZT4 is not loosed even higher than before, it can improve the operative energy for the vibrational structure damping. Figure 14 can show the energy transfer process using SSDD network method.

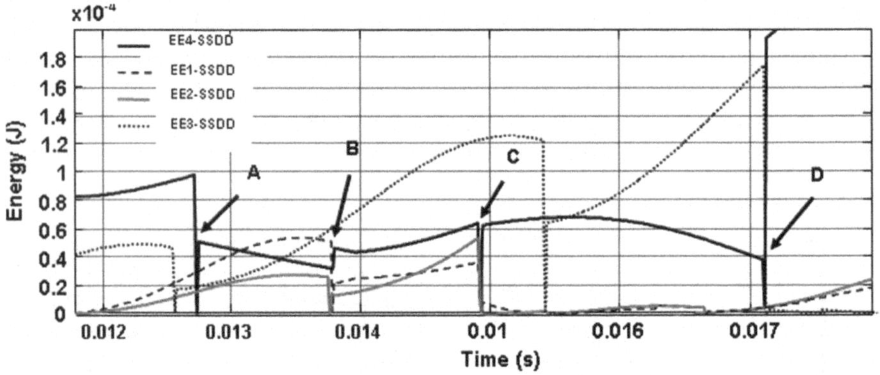

Fig. 14 The energy transfer process using SSDD network method

Table 3 Energy analysis

Time point	Source piezoelements			Target piezoelement
	PZT_1	PZT_2	PZT_3	PZT_4
A	No energy transfer	No energy transfer	No energy transfer	Modal $SSDI_1$` No energy
B	Energy transfer	Energy transfer	No energy transfer	Energy increase
C	Energy transfer	Energy transfer	No energy transfer	Energy increase
D	No energy transfer	No energy transfer	Energy transfer	Energy increase

For analysing the energy transfer behaviour in time domain, information of a series of time points such as A, B, C and D have been sampled. EE1 EE2 and EE3 are the energy of source piezoelectric elements (PZT1 PZT2 PZT3) using the SSDD network method. For simplifying the analysis procedure, we here just take the time point B for example. As the curve indicates in Fig. 14, the energy of target piezoelectric elements PZT4 has been augmented or increased whereas the energy of source piezoelectric elements PZT1 and PZT2 is decreased. The total decreased energy of PZT1 and PZT2 is mainly dissipated in the target piezoelement PZT4 and a small part of energy is distributed in the network, for example, the form of heat in the parasitic resistor of diode. After analyzing the energy waveform, Table 3 is summarized to indicate the energy transfer behaviour and characteristic of SSDD approach. We can also compare the voltage and displacement between modal SSDI method and SSDD network method shown in Fig. 15. The amplitude voltage of the target piezoelement PZT4 increase resulting from the transferred energy; it can improve the operative energy for damping in the target mode 4. It should be notable that the reference corresponding to the vibrational structure applying on the modal SSDI method has been well suppressed.

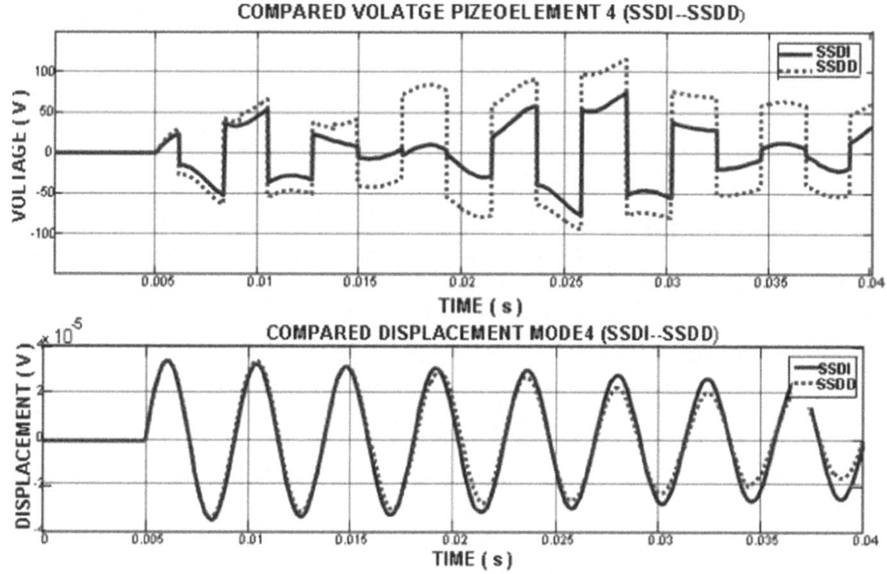

Fig. 15 The compared the voltage and displacement between modal SSDI and SSDD network method

5 The SSDD Implementation and Simulations

Simulations are made in the Matlab/SimulinkTM environment and using the SimscapeTM library. It is carried on a multi-modal model with a clamped plate including four piezoelectric elements fully coupled with the structure modes and the energy transfer network model takes into account realistic transfer losses. The multi-modal damping effect under pulse excitation is simulated.

From Figs. 16 and 17, the improvement of damping effect using the Optimized Oriented SSDD method (configuration 4) is clear illustrated. The network approach establishes a power transfer which modifies the damping of the targeted mode, the advantage of the network is closely interrelated with the energy level of the source piezoelements, and with the energy distribution in the smart structure. It is also strongly related to the coupling configuration of various PZT patches to the modes of the smart structure. At the same time, it is obviously that the damping effect of the SSDIF method (Configuration 5) is well, but just because the each piezoelement is well coupled with the structure in the each mode.

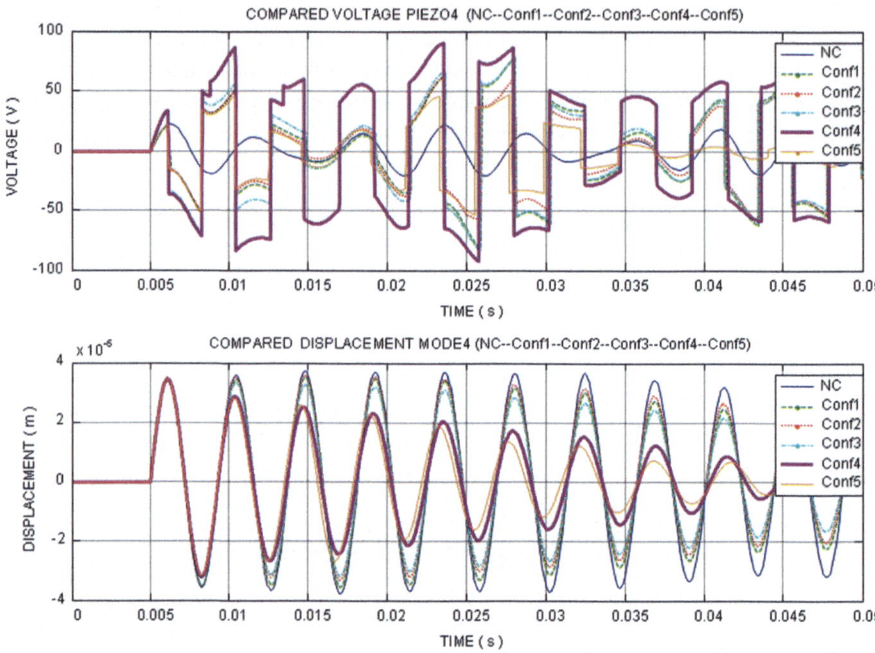

Fig. 16 The compared voltage of target piezoelement PZT4 and displacement of target mode 4 between different configurations

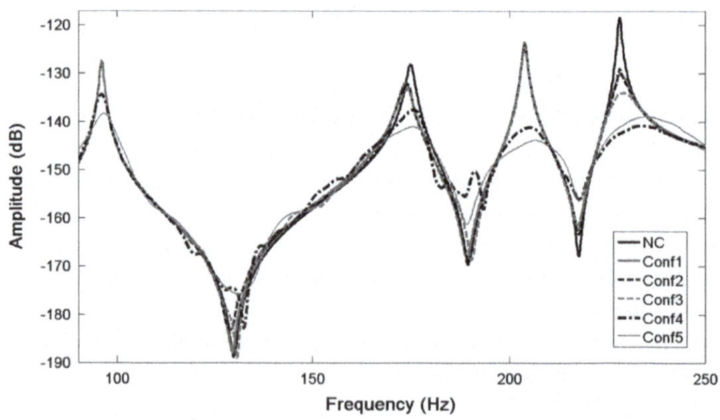

Fig. 17 The compared frequency between different configurations in the four modes of the structure

Owing to lack of transfer path between source and target piezoelectric elements, the energy cannot be transferred to the target piezoelectric element in the modal SSDI method (configuration 1) and the SSDIF method (configuration 5). The level of the cumulative transferred energy could estimate the efficiency of the energy transfer according to the power flow in the network; apparently, the more value of the cumulative transferred energy is expected and interested. Owing to the nice performance of the SSDD method with the transfer energy network, the Optimized Oriented SSDD method (configuration 4) can obtain more cumulative transferred energies (Fig. 18) acting on the structure vibration control in the target mode 4.

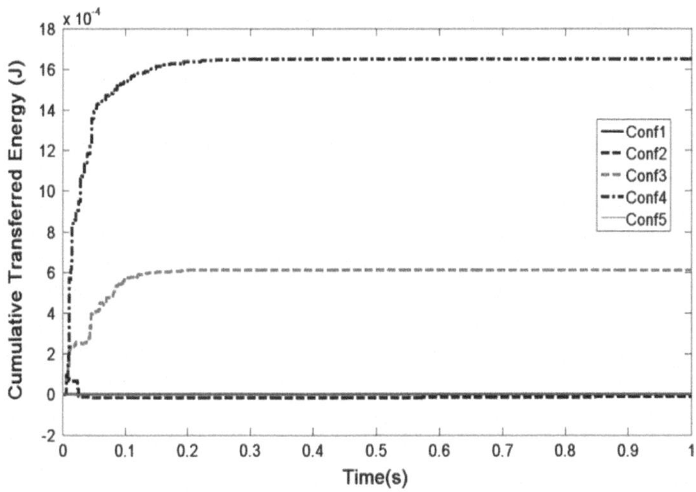

Fig. 18 The compared cumulative transferred energy between different configurations

6 Conclusion

A new network method of vibration control named the synchronized switch damping with diode (SSDD) technique was proposed and simulated. The simulation results indicate the relation between the damping performance on a given targeted mode and the ratio of the source and target energies. The Optimized Oriented SSDD technique has more excellent damping performance than SSDI method. Especially, when piezoelements weekly coupled with structure targeting given mode, the damping effect is still increased by 8 dB compared to classical SSDI at the pulse excitation. The result shows the ability of piezoelectric element network using energy redistribution; it also proves this new global SSDD method could improve vibration damping of smart structure.

References

1. Forward, R.L.: Electronic damping of vibrations in optical structures. Appl. Opt. **18**(5), 690–697 (1979)
2. Hagood, N.W., von Flotow, A.: Damping of structural vibrations with piezoelectric material and passive electric networks. J. Sound Vib. **146**, 243–268 (1991)
3. Lesieutre, G.A.: Vibration damping and control using shunted piezoelectric materials. Shock Vibr. Digest **30**, 187–195 (1998)
4. Sun, H., Yang, Z., Li, K., Li, B., Xie, J., Wu, D., Zhang, L.: Vibration suppression of a hard disk driver actuator arm using piezoelectric shunt damping with a topology-optimized PZT transducer. Smart Mater. Struct. **18**, 065010 (2009)
5. Snyder, S.D., Hansen, C.H.: Mechanisms of active noise control by vibration sources. J. Sound Vib. **147**(3), 519–525 (1991)
6. Guyomar, D., Richard, C., Mohammadi, S.: Damping behavior of semi-passive vibration control using shunted piezoelectric materials. J. Intell. Mater. Syst. Struct. **19**(8), 977–985 (2008)
7. Harari, S., Richard, C., Gaudiller, L.: Multimodal control of smart structures based on semi-passive techniques and modal observer. Motion Vib. Control **1**, 113–122 (2009)
8. Clark, W.W.: Vibration control with state-switching piezoelectric materials. J. Intell. Mater. Struct. **11**, 263–271 (2000)
9. Holdhusen, M.H., Wang, K.W., Cunefare, K.A.: Optimization of a state-switched absorber applied to a continuous vibrating system. In: Proceedings of SPIE Smart Structures and Materials, Damping and Isolation, **5386**, 319 (2004)
10. Richard, C., Guyomar, D., Audigier, D., Ching, G.: Semi-passive damping using continuous switching of a piezoelectric device. In: Proceedings of Smart Structures and Materials, Damping and Isolation, **3672**, 104–111 (1999)
11. Richard, C., Guyomar, D., Audigier, D., Bassaler, H.: Enhanced semi passive damping using continuous switching of a piezoelectric device on an inductor. In: Proceedings SPIE Smart Structures and Materials, Passive Damping and Isolation, **3989**, 288–299 (2000)
12. Lefeuvre, E., Badel, A., Petit, L., Richard, C., Guyomar, D.: Semi-passive piezoelectric structural damping by synchronized switching on voltage sources. J. Intell. Mater. Syst. Struct. **17**, 653–660 (2006)
13. Li, K., Gauthier, J.-Y., Guyomar, D.: Structural vibration control by synchronized switch damping energy transfer. J. Sound Vib. **330**(1), 49–60 (2011)
14. Wu, D., Richard, C.: A new global approach using a network of piezoelectric elements and energy redistribution for enhanced vibration damping of smart structure. In: Active, Passive Smart Structures and Integrated Systems, International Society for Optics and Photonics (2013)
15. Harari, S., Richard, C., Gaudiller, L.: Hybrid active/semi-active modal control of smart structures. In: Proceedings of SPIE Smart Structures and Materials (2009)
16. Harari, S., Richard, C., Gaudiller, L.: Enhanced piezoelectric voltage build-up for semi-active control of smart structures. In: Proceedings of SPIE Smart Structures and Materials (2009)
17. Cherif, A., Meddad. M., Eddiai. A., Zouhair, A., Zawadzka, A., Migalska-Zalas, A.: Multimodal vibration damping using energy transfer. Opt. Quantum Electron. **48**(5), 283 (2016)
18. Harari, S., Richard, C., Gaudiller, L.: New semi-active multi-modal vibration control using piezoceramic components. J. Intell. Mater. Syst. Struct. **20**(13), 1603–1613 (2009)

Laboratory Study of Solvent Assisted Steam Flooding in Thin Layer Offshore Heavy Oil Reservoir

Yong-chao Xue[1(✉)], Guan-yang Ding[2], Meng-ge Du[1], and Su-wei Wu[1]

[1] China University of Petroleum, Beijing, China
tesasydx@163.com, strjhl865@sohu.com, hdsjhfuisd@126.com
[2] China Petroleum Daily, Beijing, China
1602829387@qq.com

Abstract. There are a lot of offshore heavy oil resource in the world, which means it is necessary for us to study about the recovery of heavy oil. Nowadays, most of our heavy oil reservoirs are recovered by steam flooding. However, this technology has so many disadvantages, like low sweep efficiency and steam gravity overlap. To overcome that, we consider about using solvent assisted steam flooding to improve the recovery of heavy oil. Compare to traditional straight wells, horizontal wells have some unique advantages. Like contact more area, the wells that need to be drilled are less, which will save the cost and make the whole process more economic. As the technology of horizontal wells is growing rapidly, there has been many new technologies coming out. Such as steam assisted gravity drainage technology, solvent steam flooding technology, gas and chemical agents such as auxiliary level steam stimulation, and conventional horizontal well steam stimulation. In this paper, we do two test which are horizontal steam flooding and horizontal well solvent assisted steam flooding tests. After the tests, we get the temperature field, production speed, water content and so on. Using those figures, we find out that solvent assisted steam flooding can improve the impact of the effect, increase the sweep efficiency of the steam chamber, keep the water content at a small number, which will eventually enhance the heavy oil recovery. At last, the conclusion of the paper has some reference value for the analysis of the effect of using solvent assisted steam flooding in heavy oil reservoirs.

J. Lin (ed.), *Proceedings of the International Petroleum and Petrochemical Technology Conference 2019*, pp. 471–482, 2020.
https://doi.org/10.1007/978-981-15-0860-8_38

Keywords: Steam flooding · Horizontal well · Offshore heavy oil reservoir · Laboratory experiment

1 Introduction

Under reservoir conditions, we refer to crude oil with a viscosity greater than 50 mPa s and a relative density greater than 0.92 as heavy oil. The content of asphaltenes and gums in heavy oil is high, and the content of light ends is low. It has the characteristics of high starting pressure and poor fluidity. Although the viscosity of heavy oil is high, its viscosity is easily affected by temperature. The viscosity decreases rapidly with increasing temperature. The world's heavy oil resources are very rich, and its geological reserves are far more than conventional oil and gas. Steam flooding technology reduces the viscosity of heavy oil by injecting a certain amount of steam into the well. A large number of studies have found that the factors affecting the development of steam flooding mainly include reservoir parameters and operating conditions. The operating conditions mainly refer to the gas injection speed, the injection-production ratio and the dryness of the gas injection. The operating conditions of these two types have a greater impact on the development of steam flooding. Only when all operating conditions are met at the same time can a better development result be obtained. If you use a steam drive alone, the development will not be optimal. For better development results, the easiest way to do this is to inject solvent-assisted steam flooding. However, little research has been done on the effect of solvent-assisted steam flooding and the mechanism of action between solvent and steam.

This experiment simulates the effects and laws of solvent-assisted steam flooding in heavy oil reservoirs. At the same time, the sweep coefficient and mining degree of heavy oil reservoirs were evaluated. The experiment required thermal communication between the injection well and the production well before starting the solvent-assisted steam drive. In this experiment, the model was brought to the thermal communication state after the oil layer was swallowed by increasing the initial experimental temperature. Steam is then injected into the horizontal well to initiate steam flooding. According to the purpose of the experiment, a rectangular box-type closed sand-fillable model is used to simulate the actual formation. The horizontal well is used to test the row of wells and the arrangement of one production well in one injection well. The study will analyze the temperature field between the injection wells to describe the steam distribution in the horizontal wells to initially determine the extent of heavy oil reservoirs. At the same time, the results of oil recovery, water content and cumulative oil production were analyzed to evaluate the development and production dynamics of solvent-assisted steam flooding in horizontal wells.

2 Experimental Purpose and Devices

In this experiment, a two-dimensional horizontal well displacement physical model, a data collector, two incubators, a low-speed steam generator, a production liquid metering system, a precision pressure gauge, and the like were used. The specific information is as follows.

(1) Two-Dimensional Model

The model is 60 cm × 20 cm × 5 cm and has a layer of soft rubber on the bottom and around it. Inside the model are four injection well production well lines and 4 × 12 rows of equally spaced temperature points. At the same time the model is connected to the pressure temperature sensor.

(2) Incubator

As shown in Fig. 1, the oven was used primarily in high temperature saturated oil processes and solvent assisted steam flooding in this experiment. During the experiment, the equipment was placed in an incubator for preheating.

Fig. 1 Thermostat box

(3) ISCO Plunger Pump

As shown in Fig. 2, two plunger pumps are required in this experiment. The ISCO plunger pump is placed at the inlet and is used primarily in saturated oil, saturated water, and solvent assisted steam flooding. Another domestic pump is responsible for flooding 500 mL of solvent. The injection speed and the upper limit pressure can be set

Fig. 2 ISCO piston pump

at the plunger pump, and pressure changes, cumulative injection time, and cumulative injection amount can also be obtained.

(4) Steam Generator

As shown in Fig. 3, the device was used in a solvent-assisted steam flooding experiment to produce high temperature steam to displace the experimental liquid. Used in the experimental preparation stage to dehydrate the oil containing more water. During the dehydration process, the temperature is generally ninety degrees Celsius. When

Fig. 3 Schematic diagram of the steam generator

dehydration begins, the heating device should be turned on for a short period of time. When the fluidity of the oil is stable, the stirring device is turned on.

The oil sample used in the experiment was 3 L of dehydrated crude oil in an oil field, and the viscosity was 500 mPa s at 80 °C. The water sample used in the experiment is the formation water of an oil field. Glass beads of 120 mesh were used as the formation in the experiment. According to the phase of the solvent under the conditions of the reservoir and the ground conditions, the three solvents C_6, C_{10}, and C_{16} were selected in the experiment. Used as a light solvent, a medium quality solvent and a heavy solvent under laboratory conditions. During the experiment, we used C_6 to simulate propane undermine conditions. C_{10} was used to simulate C_6 undermine conditions, and C_{16} was used to simulate C_8 undermine conditions.

3 Experimental Design

As shown in Fig. 4, the first injection steam velocity is calculated based on the actual injection parameters of the reservoir, and the result is 2.0 ml/min. Adjust the temperature of the steam generator to the specified temperature, 150 °C. The experiment simulates solvent-assisted steam flooding under homogeneous heavy oil reservoir conditions. Finally, the data obtained from the experiment is analyzed and the development effect is obtained. By placing this rectangular two-dimensional physical model horizontally, it is possible to implement a solvent-assisted steam flooding experiment in a horizontal well under heavy oil reservoir conditions. The experiment used the outermost horizontal well to simulate the positively facing well pattern of a single injection unit. At the same time, the characteristics of different formation heterogeneity are simulated by setting the permeability distribution of the simulated formation. The

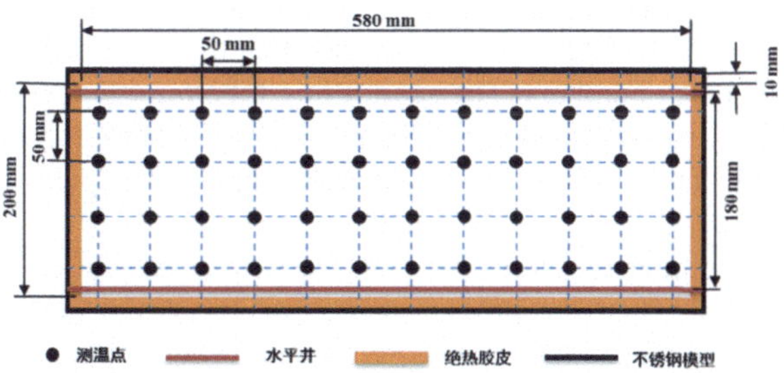

Fig. 4 Internal structure of 2D double well model

change in bottom hole pressure and temperature throughout the process is measured by pressure and temperature sensors installed at the injection and production finger tips and roots.

Saturate water and saturated oil before solvent-assisted steam flooding. The simulated formation pore volume is calculated according to the volume of saturated water, and then the recovery degree is calculated according to the saturated oil volume. Since the oven heating device is located on the side of the production well, the model cannot be uniformly heated during preheating and experimentation. This will result in a higher temperature near the injection well and a lower temperature near the production well. The measuring cylinder containing the produced liquid must be sealed with plastic wrap and then placed in an incubator at a temperature of 80° to allow the oil to hydrate.

Remove the model and first remove the screws that are fixed around with a screwdriver. Be careful to remove the middle screws first, then remove them in the diagonal and in the order of the two sides. This prevents the screw from being sheared and broken due to uneven force.

When the removal is complete, lift the lid open. Use a shovel to dispose of the sand inside into the trash can. Be careful with internal piping and temperature measurement points, and never destroy temperature sensors and piping. After removing the sand, add a small amount of kerosene to the model. Then rinse with 50 °C of clean water. Finally, wash with water repeatedly.

Clean the model cover and use a blade to scrape the grease and waste from the cover. Then wipe with kerosene and clean water. Check the soft rubber inside the model to see if it is close to the inner wall. If there is a loose slip, stick it with glue. Then use a blade to scrape off some of the rubber around them, leaving them slightly below the edge of the model and keeping them flush. Dry the model, ventilate and dry the washed model, and dry naturally.

Take 5 L of heavy oil sample in the laboratory's oil distribution room and place it in a 6 L large beaker. Allow the oil sample to stand for a while, and wait until it is stable before bringing the oil sample into the crude oil rapid dehydrator. Set the temperature to 95° and heat it for about 30 min. When the crude oil is heated to have a certain fluidity, the stirrer is turned on to start stirring and dehydrating. The speed must be moderate, if it is too fast, the oil will splash. Fill the sand filling tube with glass beads with multiple mesh numbers. The sand filling tube must be evenly tapped during the process so that the glass beads are closely packed. Check the air tightness after filling is completed. Then connect the line and connect the pressure sensor at the entrance.

The measuring cylinder measures the flow volume at the outlet. The pressure and flow data are recorded when the pressure sensor at the outlet and the flow rate are stable. The penetration rate is then calculated using Darcy's law. Each glass bead is subjected to three saturated water experiments and the average is taken as its permeability. The experimental simulated oil was prepared by mixing dehydrated crude oil and engine oil. The simulated oil viscosity temperature curve at 60–140 °C was measured according to the standard SY/T 7549-2000 operation.

4 Analysis of Development Effect of Solvent-Assisted Steam Flooding

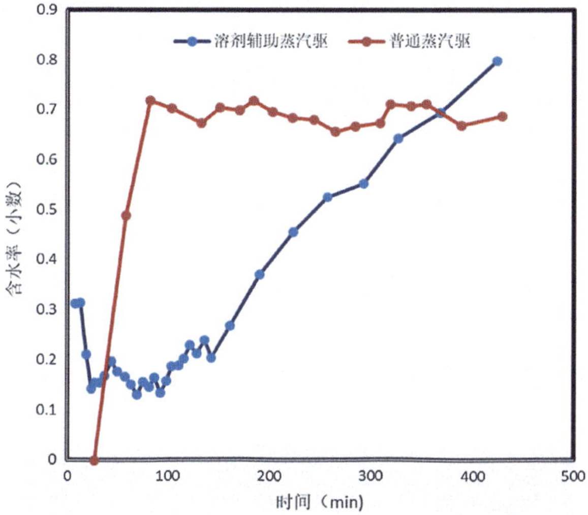

Fig. 5 Comparison of water content of oil sample

4.1 Development Indicator Evaluation

As shown in Fig. 5, solvent-assisted steam flooding has a longer low water recovery period compared to steam flooding. This shows that the solvent improves the expansion of the steam chamber and delays the steam breakthrough time. When the vapor chamber in the solvent-assisted steam flooding model breaks through, the water content increases. When the phenomenon of discouragement is serious, the displacement experiment is ended. At the beginning of the ordinary steam flooding, the steam

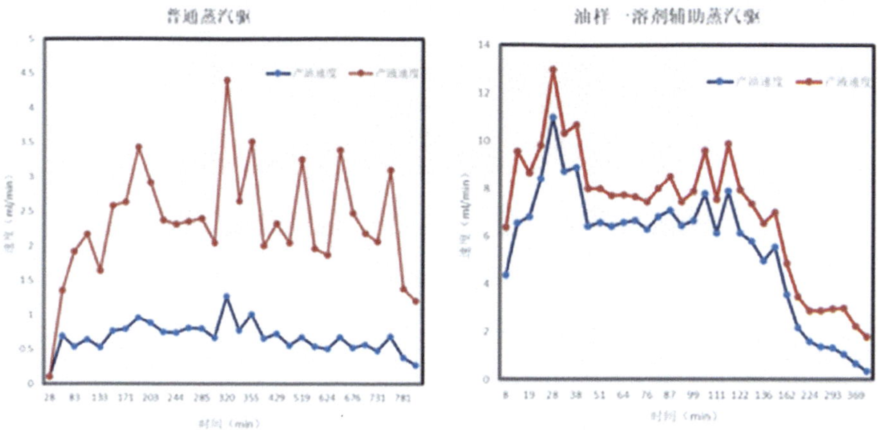

Fig. 6 Oil production rate by solvent assisted steam flooding and steam flooding

chamber broke through, and the gas enthalpy phenomenon soon appeared, so that the water content was at a very high level from the beginning.

As shown in Fig. 6, the solvent chamber of the solvent-assisted steam drive did not break before 200 min. At this time, the oil production rate is maintained at a very high level, and the liquid production rate is almost equal to the oil production rate. At this stage, the production rate of oil produced by ordinary steam flooding is low, indicating that the steam chamber has broken through and there is not enough time to expand. This proves that the solvent does improve the development of the vapor chamber. After 200 min, the steam chamber broke through, and the yield of solvent-assisted steam flooding decreased, gradually moving closer to the data of ordinary steam flooding.

4.2 Temperature Field Analysis

As shown in Fig. 7, in the early stage of ordinary steam flooding, the vapor chamber formed at a slower rate because of the greater viscosity and lower temperature of the

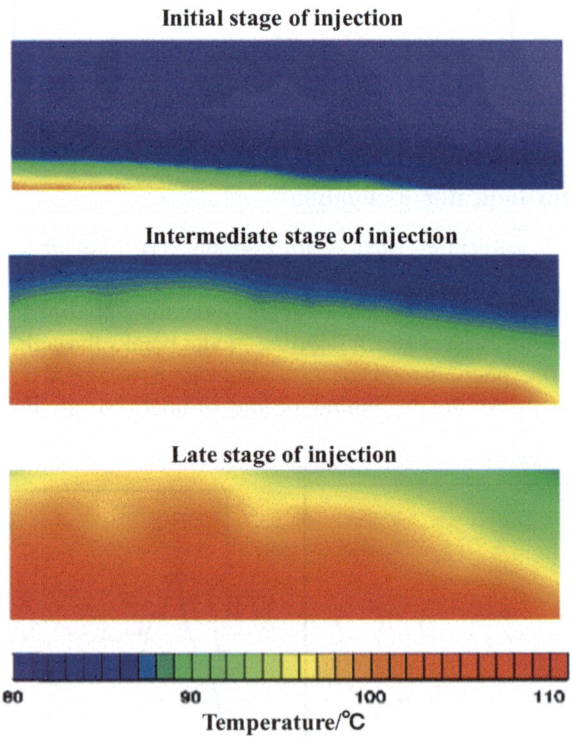

Fig. 7 The temperature profile of steam flooding

crude oil in the reservoir. The steam chamber first appears at the heel end of the injection well. In the mid-stage of production, the expansion of the steam chamber is not uniform. The steam chamber develops faster at the heel end of the injection well and the toe end is slower. In the later stages of production, the viscosity of the crude oil is reduced and the displacement speed is increased. The steam chamber develops too fast at the heel end of the injection well and develops slowly at the toe end of the injection well. This leaves part of the root end of the production well unaffected. At the end of the experiment, some parts of the injection toe end and the production well were not displaced.

As shown in Fig. 8, the formation temperature is lower in the early days of solvent-assisted steam flooding. Because the near-well zone is first exposed to steam and the viscosity of the crude oil is low, the vapor chamber is uniformly formed in the area near

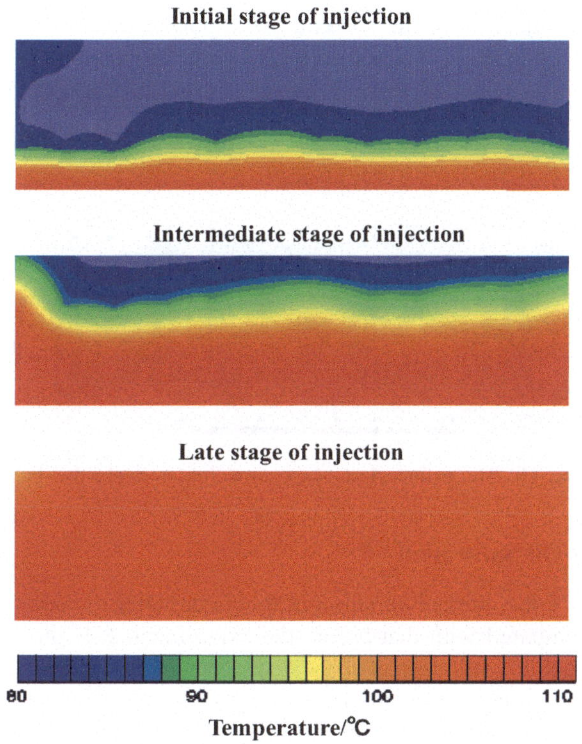

Fig. 8 The temperature profile of solvent assisted steam flooding

the injection well. In the mid-stage, the entire steam chamber begins to expand uniformly from the injection well to the production well due to the diffusion of steam heat and the action of the solvent. Due to the uneven distribution of the solvent in the model, the steam chamber is located at the toe end of the injection well. In the later stages, because the viscosity of the crude oil has decreased, and the steam chamber has also formed a certain scale. Therefore, the displacement speed will gradually increase, and the leading edge of the steam chamber will be uniformly pushed to the production well.

Fig. 9 The inside view of the model after steam flooding

After stopping the injection of the solvent, the vapor chamber at the root end of the injection well breaks through. Therefore, the slug phenomenon at the exit of the production well becomes more apparent. When the production speed drops to a certain level, the water content is very high.

Fig. 10 The inside view of the model after steam flooding

4.3 Remaining Oil Saturation

As shown in Fig. 9, the internal variation of the model after the end of the experiment was significant compared with that before the experiment. It is obvious in the model that the reservoir color is still deep after the displacement is completed. The color of the reservoir at the four corners of the model and near the production well is barely lighter. This shows that the amount of crude oil driven by ordinary steam flooding is very small and very uneven. Many parts of the reservoir have not been affected.

As shown in Fig. 10, the overall degree of recovery of the model is high compared to the experimental results of ordinary steam flooding. The color of the model is white, which proves that the displacement is good. This significantly improves the efficiency of displacement and the degree of recovery due to the addition of solvent. But in the four corners of the model, and in the area close to the production well, the color is still deep. There are still a lot of remaining oil in these areas that can be further mined.

5 Conclusions

1. Using a variety of solvent mixing assisted methods have better results than ordinary steam drive. This is because light solvents, medium solvents, and heavy solvents can each perform different functions at different stages, and can be more effectively displaced.
2. The solvent can significantly reduce the viscosity of the heavy oil in the heavy oil reservoir, so that the water content in the production process can be kept at a low level, and the yield will increase significantly. The oil production will also be shortened, which will improve the economics of production and improve the development of heavy oil.
3. Experimental results show that the addition of solvent makes the development of the steam chamber more complete. Moreover, the distribution of the steam chamber is more uniform, and the steam sweep efficiency is greatly improved. After the addition of the solvent, the vapor chamber does not have an untouched area like a normal steam drive.
4. Experimental results show that the solvent-assisted steam flooding is significantly higher than that of ordinary steam flooding. At the same time, after the addition of the solvent, the displacement is more uniform.

Acknowledgements. This paper has been supported by National Natural Science Foundation of China (No. 41672132).

References

1. Haitao, W., Zengmin, L., Chengyuan, L., et al.: Physical simulation experiment for the optimization of steam flooding horizontal well patterns in Pai 601 Block of Chunfeng Oilfield. Oil Drilling Prod. Technol. (2017)
2. Guilin, Z., Xuezhong, W.: Horizontal steam flooding pilot test for thin shallow super heavy oil of Chunfeng Oilfield in Junggar basin. Sci. Technol. Rev. (2014)
3. Luo, Y.Y., Cheng, L.S.: Physical simulation on temperature and pressure fields during steamflooding through horizontal wells in heterogeneous reservoir. Appl. Mech. Mater. **110–116**, 5 (2011)
4. Yang, Z., Dianfa, D., Yanwu, Z., et al.: Breakthrough laws of steam in heavy oil reservoirs stimulated with horizontal-well segmented steam flooding. Spec. Oil Gas Reservoirs (2017)
5. Huang, S., Chen, X., Liu, H., et al.: Experimental and numerical study of solvent optimization during horizontal-well solvent-enhanced steam flooding in thin heavy-oil reservoirs. Fuel **228**, 379–389 (2018)
6. Guanghuan, W.U., Zhengbin, W.U., Weizhong, L.I., et al.: Chemical performance evaluation and visualized experiment of thermal chemical-assisted steam flooding for horizontal well. Fault-Block Oil Gas Field (2016)
7. Chenghui, H., Shijun, H., Linsong, C., et al.: Experimental study on steam flooding in horizontal well in thin heterogeneous heavy oil reservoirs. Spec. Oil Gas Reservoirs (2015)
8. Zhongchen, B., Yuan, Z., Changhong, Z., et al.: Steam flooding optimization of vertical-horizontal well combination in super-heavy oil reservoir. Spec. Oil Gas Reservoirs (2017)

9. Yuanliang, Y.: Steam channeling sensitivity of horizontal well in thin-shallow super-heavy oil reservoir with steam flooding. Spec. Oil Gas Reservoirs (2016)
10. Dianfa, D., Yang, Z., Yanwu, Z., et al.: Horizontal well productivity equation of steam flooding with inverted nine-spot combination well pattern. Spec. Oil Gas Reservoirs (2016)
11. Dalei, J.: Similarity principle of physical simulation of steam flooding for horizontal well. Nat. Gas Explor. Dev. (2015)
12. Haitao, S., Shuo, W., Nan, L., et al.: Optimization of technical parameters for steam flooding in extra heavy oil horizontal wells. Spec. Oil Gas Reservoirs (2015)
13. Yanyan, L., Linsong, C., Shan, G., et al.: Experimental investigation of temperature and viscosity fields of steam flooding development of horizontal wells. Petrol. Drilling Tech. **40** (1), 74–77 (2012)
14. Jianmei, Y.: Study on adaptability of horizontal well pattern of steam flooding in shallow super-heavy oil reservoir. Petrochem. Ind. Technol. (2015)
15. Hongzhao, T., Ye, S.: Operation parameters optimization of steam flooding with the combination of vertical and horizontal wells. Lithologic Reservoirs **25**(3), 127–130 (2013)
16. Zhao, Y., Lu, S.T., Gao, H., et al.: Numerical simulation research on improving effect of steam flooding by using high permeability belt. Appl. Mech. Mater. **483**, 635–638 (2013)
17. Lu, H.S.: Single horizontal well conduction assisted steam drive process for removing viscous hydrocarbonaceous fluids (1993)
18. Jing, H.U., Xiao-Bo, L., Guo-Qiang, S.: Theory and application of the scaling laws for physical modeling for horizontal well steam assisted gravity flooding. Sci. Technol. Eng. (2014)
19. Hong, M., Zhenqi, C., Yuwen, C.: Numerical simulation study of steam injection development of super heavy oil with horizontal well. Spec. Oil Gas Reservoirs **48**(16), 3308–3318 (1998)
20. Chuan, L.U., Huiqing, L., Keqin, L.U., et al.: Flexibility study of horizontal well cyclic steam simulation assisted by combination gas and cleanup addictive for shallow-thin heavy oil reservoir. Oil Drilling Prod. Technol. **51**(6), 829–835 (2013)
21. Meng, W.: Numerical simulation research of the affections of geologic factors for horizontal steam flooding. Inner Mongolia Petrochem. Ind. (2013)

Dynamic Analysis of Steam Stimulation of Horizontal Wells in the Offshore Heavy Oil Reservoir with Bottom Water

Yong-chao Xue[1(✉)], Guan-yang Ding[2], Meng-ge Du[1], and Jun Liu[1]

[1] China University of Petroleum, Beijing, China
tesasydx@163.com, strjhl865@sohu.com,
hdsjhfuisd@126.com
[2] China Petroleum Daily, Beijing, China
1602829387@qq.com

Abstract. The horizontal well steam soaked in C heavy oil reservoir with bottom water located in Bohai has the problems of low sweep coefficient, unrecoverable large-area oil reservoirs and uneven steam injection profiles. Since steam recovery is used as a step-down development method, the reservoir energy will decrease as the throughput cycle increases. Under the influence of pressure difference, the bottom-bottom water continues to migrate to the production well, causing the bottom-bottom water to eventually invade the wellbore, affecting production. To this end, dynamic analysis of horizontal well development is needed to better understand the production characteristics of horizontal wells. This article mainly analyzes the typical oil production-water content characteristics, recovery water rate, oil-gas ratio, production model characteristics and other indicators, compares the production characteristics of different wells, divides different typical production modes, and analyzes the production effect. The main influencing factors, at the same time, aimed at the implementation of the two-nitrogen foam leveling wells on site, from the aspects of oil production, gas-oil ratio, etc., compared with the wells without measures, and analyzed the corresponding reasons. There are two factors that reduce the amount of oil produced during the cycle. First, since the density of steam is less than that of crude oil, under the influence of gravity, steam is caused to flow to the top of the oil layer. The leading edge of the steam is inclined and does not reach the lower middle oil layer. Second, since the viscosity of the steam is lower than that of the crude oil, the steam rushes. This limits the range of steam

J. Lin (ed.), *Proceedings of the International Petroleum and Petrochemical Technology Conference 2019*, pp. 483–493, 2020.
https://doi.org/10.1007/978-981-15-0860-8_39

heating and reduces the degree of mobilization in the longitudinal direction of the reservoir. The research results in this paper can better guide the development of such reservoirs.

Keywords: Offshore heavy oil reservoir · Steam stimulation · Dynamic analysis · Bottom water

1 Introduction

Heavy oil is widely distributed in the world, and heavy oil reservoirs are found in most oil-producing countries. Therefore, the development and utilization of heavy oil resources is of great practical significance. However, the development of heavy oil is also not easy. Higher viscosity and lower fluidity are real problems. Even under oil conditions, the general mining method cannot effectively extract heavy oil reservoirs. Not only is it inefficient, but the economic value is greatly reduced. Steam stimulation is a single well operation, and its construction difficulty is relatively lower than other methods, because it only needs to inject a good amount of hot steam into the well and close the oil well for the well, let the rock in the reservoir and the heavy oil and hot steam There is heat transfer between them to warm the reservoir, its effect can be quickly manifested, and it can be produced while testing, which has also won people's favour.

2 Manuscript Presentation

The C oil field is located in eastern Kazakhstan. The heavy oil reservoir is the main development block of the oil field, because the reservoir is not only shallowly buried, but also rich in abundance. The most important thing is that the reservoir has good physical properties. At the same time, the reservoir is a sandstone heavy oil reservoir affected by the bottom water. It also has the characteristics of high crude oil density, high viscosity of crude oil and small dissolved gas.

According to the first curve of Fig. 1, the irreducible water saturation is 0.2, the residual oil saturation is 0.29, and the isotonic point is 0.62, wherein the rock wettability is water wet, and the oil displacement efficiency is 0.6375.

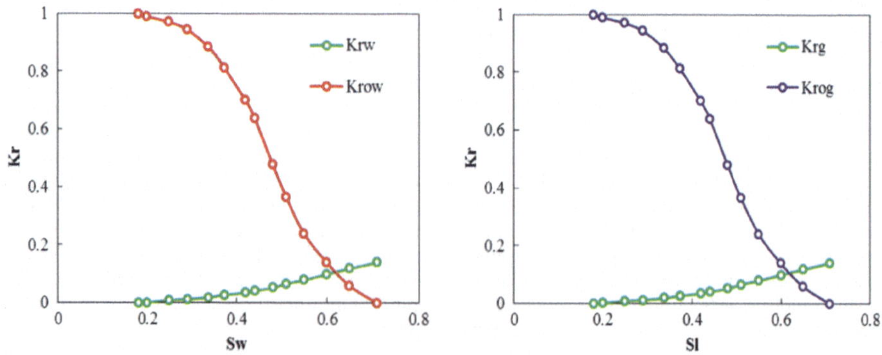

Fig. 1 Oil-water relative permeability curve

Through the dynamic analysis of horizontal wells, the cumulative steam injection amount, cumulative liquid production, cumulative oil production, and cumulative water production are counted. The statistical results are shown in Table 1.

The factors causing this phenomenon are various. The main factors are the characteristics of the reservoir itself, the injection and production parameters of production wells and injection wells, and the steam injection process of steam stimulation. In response to these factors, the reservoir began to test the nitrogen foam profile test well to improve the high-yield effect of horizontal well heavy oil steam stimulation.

Table 1 Thermal recovery well development history card table

Sweeping rounds	Cumulative steam injection/m^3	Cumulative oil production/t	Cumulative water production/t
First round	3855	5461.8	4321.03
Second round	2080	2932.8	8660.69
Third round	2394	2514.5	7330.41
Fourth round	5404	6727.5	16,608.37
Fifth round	4513	1348	3492.24

3 Horizontal Well Production Characteristic Curve and Analysis

3.1 Oil Production—Moisture Content Characteristics Analysis

According to the on-site production data, combined with the development method of each well, the two parameters of oil production and water content are used as indicators to calculate the production characteristic curve to facilitate the division of production mode.

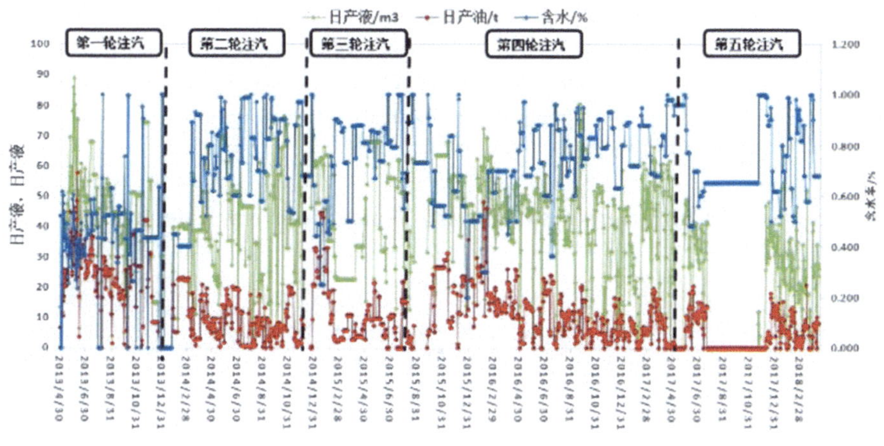

Fig. 2 Oil production—water content curve of H451

As shown in Fig. 2, this well H451 has 4–5 throughput cycles, and the number of cycles is more than other wells. During the first throughput cycle, the water content is lower and the oil content is high. However, from the second throughput cycle, the water content increases rapidly, and in the subsequent throughput cycle, the water content stabilizes within a certain range, and the oil content also decreases (Fig. 3).

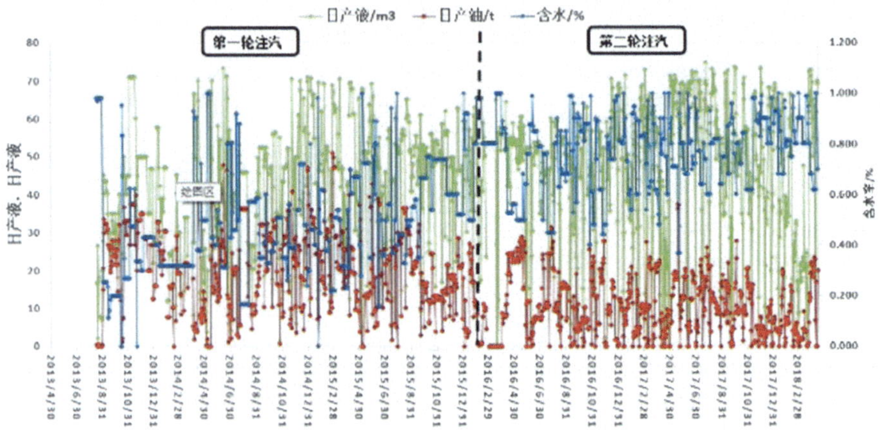

Fig. 3 Oil production—water content curve of H462

This well H462 has a low moisture content at the start of production. As time progresses, the water content increases slowly, and the oil content slowly decreases, roughly forming an "X"-shaped curve. This well is dominated by two throughput cycles. This also explains why a stable rising and falling production curve can be maintained. The moisture content of the first cycle of this well has been maintained at a relatively low level. This reflects the fact that the stratum itself is good at this well, and then the later mining operations are also very good. However, at the end of the second cycle or at the beginning of the third cycle, the oil production will become zero and the water content will be 100%, which is a common problem reflected in the C reservoir.

During the development process, the number of wells with a moisture content greater than 90% in the second and third cycles increased by 33.9% and 43.8%, respectively. In these two rounds, the number of shut-offs was 1.7% and 12.5%, respectively, because the water content reached 100%.

3.2 Analysis of Characteristics of Return Water Recovery Rate and Oil-Air Ratio

According to the statistics of on-site production data, combined with the development methods of each well, the two major parameters of the horizontal water wells and oil-gas ratio of each horizontal well are used as indicators to analyse the production status of different cycles.

As shown in Fig. 4, the oil-gas ratio in the first cycle of most horizontal wells is relatively high, while the recovery rate is relatively low. This shows that the mining method using steam stimulation in the heavy oil reservoir is effective.

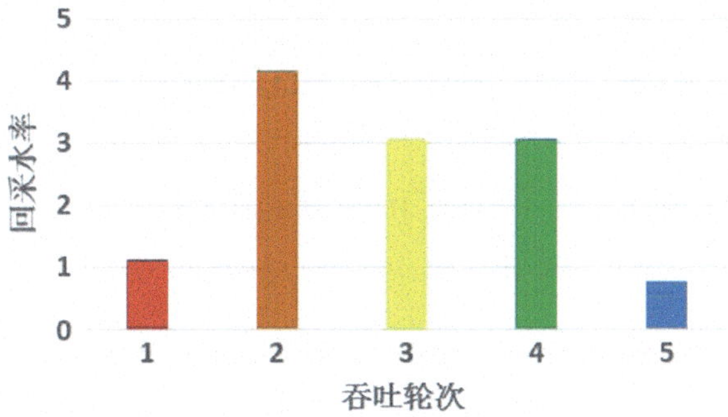

Fig. 4 Recovery water rate characteristic histogram

As shown in Fig. 5, the subsequent reduction of oil and gas ratio and the increase of recovery rate are also the processes that must be experienced in heavy oil steam stimulation. However, some wells vary greatly, and some wells have small variations.

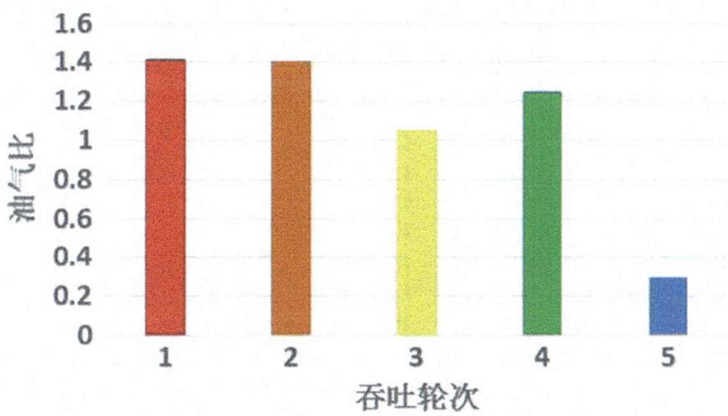

Fig. 5 Oil/gas ratio characteristic histogram

3.3 Horizontal Well Dynamic Mode Division

According to the production characteristic curve of the horizontal well, the production mode of the horizontal well is divided. The division method is mainly based on the oil content and water content. The results of the division show that the production dynamics of horizontal wells include four modes, as shown in Table 2.

From the summary table, most wells are low-water, high-yield models at the beginning of production. Many wells then enter the high-yield oil mode from the second throughput cycle and then continue to maintain high-yield wells. However, a small amount of well water content breaks through 90% and is close to 100%. The oil production is almost zero and can only be forced to shut down.

Table 2 Thermal recovery well development history card table

Sweeping rounds	High water content Low oil production	Low water content High-yield oil
First round	H497, H452, H451, H487	H451, H452, H462, H464, H467, H469, H471, H474, H476, H478, H481, H482
Second round	H471, H483, H497	H465, H422
Third round	H471, H476, H483	H478, H485, H412, H485

3.4 Analysis of the Monthly Variation of Horizontal Wells

In order to better and more clearly reflect the changing law of each well, the average daily production volume and average daily oil production of the export horizontal wells are statistically analysed in units of months.

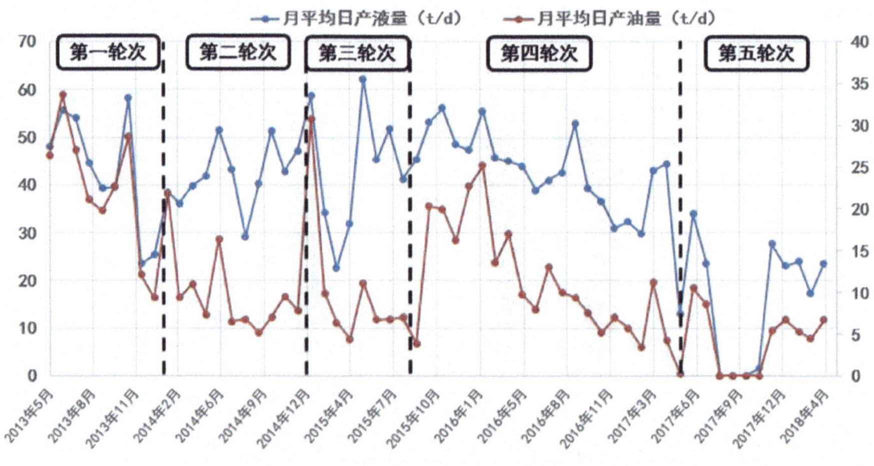

Fig. 6 Monthly average daily production of liquid and oil of H479

The table uses the average daily liquid production and the average daily oil production as the drawing ordinate, which can directly reflect the oil production of horizontal well steam stimulation. As shown in Fig. 6, the average daily oil production per well at the end of each throughput cycle has dropped to a low point. Then, starting the beginning of the next throughput cycle, the average daily oil production curve rises rapidly. This shows that for the mining of heavy oil reservoirs, the effect of steam stimulation is very obvious. Also obvious horizontal wells are H451, H462, H469,

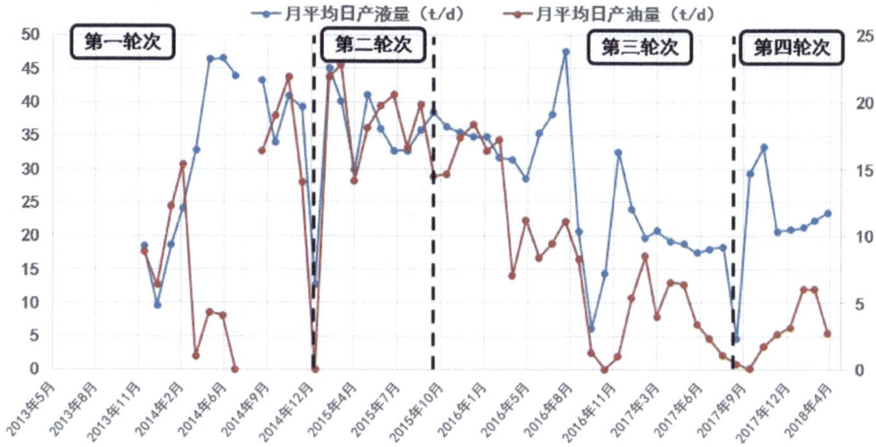

Fig. 7 Monthly average daily production of liquid and oil of H487

H471, H476 and H485. Although other horizontal wells do not have obvious oil production curves for these wells, they all reflect the effect of horizontal well steam stimulation.

As shown in Fig. 7, after experiencing the steam injection well stage at the beginning of each throughput cycle, the amount of oil produced increases as the amount of oil produced increases. Now, with each throughput cycle as the observation unit, it can be seen that the curves in more than 90% of the cycles are slowly decreasing after rising at the beginning of the cycle. The horizontal well then enters the next throughput cycle. In general, there is a step-down trend, but some wells have a downward trend and a large down-flow trend is small. The number of general cycles with a large downward trend is small, and the number of cycles with a small downward trend is relatively large.

As shown in Fig. 8, the amount of oil produced has a sudden drop in the second throughput cycle. Then, the oil production of the horizontal wells is maintained at a very low level and gradually decreases to zero. After entering the third cycle, the oil production of horizontal wells is still not going up. The main reason is that the horizontal well steam stimulation development method is a buck-type mining reservoir. There is more formation water around the C reservoir. Formation pressure will decrease as the number of throughput passes increases. Under the action of the pressure difference of the formation, the bottom and bottom water continuously move closer to the horizontal wellbore. This can easily lead to a decline in the production capacity of production wells.

In response to this phenomenon, according to the edge water intrusion in the horizontal well mining area, a nitrogen foam gel was used to profile the horizontal well to suppress the bottom water coning. This can avoid the invasion of the bottom water and improve the ultimate recovery.

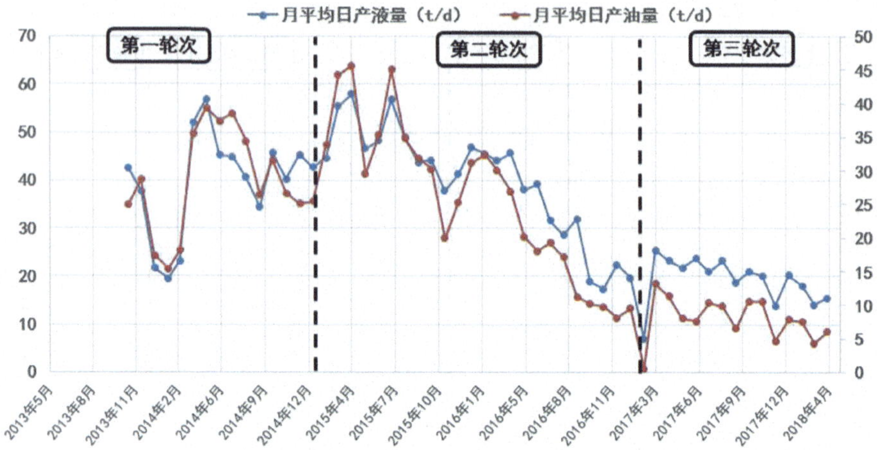

Fig. 8 Monthly average daily production of liquid and oil of H465

4 Nitrogen Foam Profile Control Thermal Recovery Well Dynamic Analysis

4.1 Steam and Nitrogen Foam System

In the production of heavy oil, thermal recovery is usually carried out in the form of steam injection into the well. The surfactant is flushed into the wet saturated steam with a certain degree of dryness, and slowly stirred until steam foam is generated. However, as the heat in the water vapor continues to be lost, the dryness of the water vapor begins to decrease until it condenses into water. The foam is foamed with certain instability, and the foam-assisted oil displacement effect is also minimal, so that it only shows the temporary plugging and shunting effect in the near-well zone.

In order to prevent the occurrence of steam foam reduction, the stability of the steam foam in the formation is effectively improved. In addition to injecting the foaming agent, an appropriate amount of nitrogen is mixed, so that even if the steam condenses into water, the gas phase still exists. This allows it to function as an oil displacement while also improving the foaming efficiency of the steam. In addition, with the increase of the nitrogen content in the gas phase, the foaming ability of the foaming agent is also effectively improved, and compared with those of the steam foam containing no nitrogen gas, the other characteristics are not greatly changed.

When the gas injection well does not require too much gas injection rate, and the injected steam is mixed with sufficient nitrogen, direct injection of saturated water can be performed. At this time, the foam needs to be called hot water foam, which not only satisfies the requirements of the gas phase flow rate reduction and the oil layer heating in the thermal recovery process, but also greatly reduces the steam injection cost.

4.2 Analysis of Nitrogen Foam Profile Control Well

In the C oil field, there are 12 well control wells. Only two horizontal wells were subjected to nitrogen foam gel profile control, and the rest were all vertical wells. This paper mainly studies the profile control effect of horizontal wells. The relevant production characteristics curves and data for the two wells H483 and H497 are shown in Figs. 9 and 10.

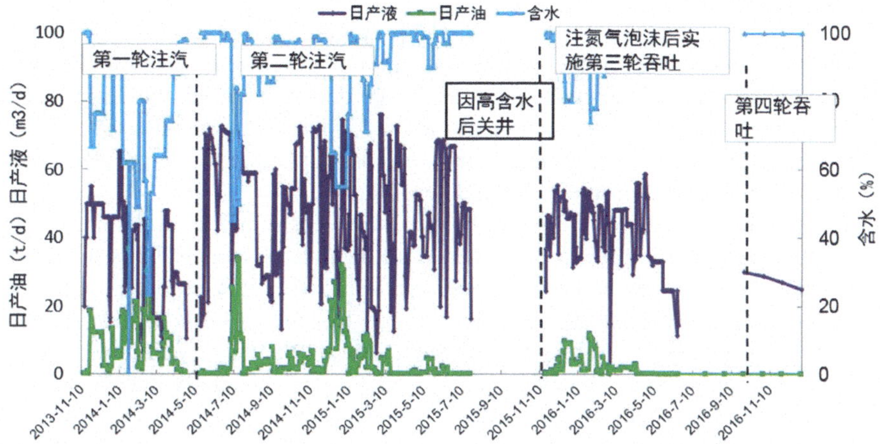

Fig. 9 Nitrogen foam gel profiled horizontal well production characteristics of H483

Both horizontal wells are closed because of the high water content caused by the immersion of the bottom water. H483 horizontal well was started in October 2015 (the end of the second throughput cycle) with nitrogen injection and gel for profile control. Once again, the production of the well is the third stage of the first throughput cycle. It can be seen that the water content is still high. There is not much change in oil production. Although the recovery rate has dropped a lot, the oil and gas ratio has not

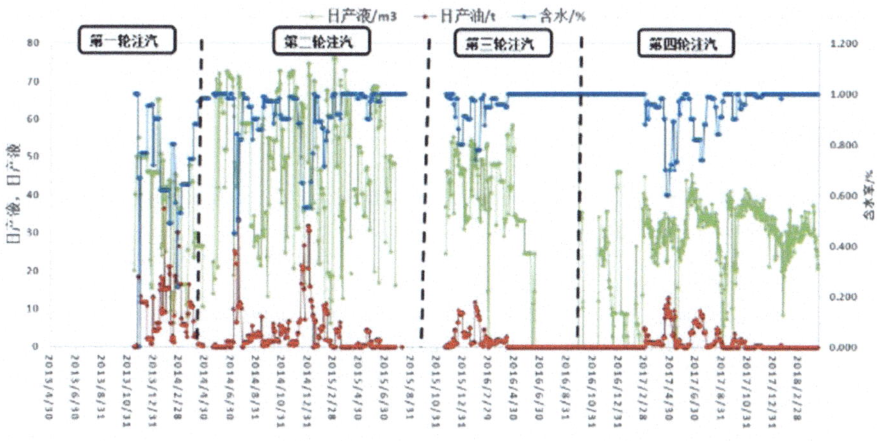

Fig. 10 Nitrogen foam gel profiled horizontal well production characteristics of H497

increased. It can also be seen from the monthly average daily oil production curve of H483 that the nitrogen foam profile control effect of the well is poor.

Another well H497 began to be profiled in June 2016 (at the end of the first round of the throughput cycle), and the well was re-opened for the second round of the throughput cycle. The oil increase was zero and the oil production was zero, indicating the well. Water intrusion is very serious, and profile control is difficult to achieve the desired results.

5 Conclusions

(1) There are two factors in the steam production and the decrease in oil production with the increase of the cycle. First, the steam density is small, the density of the crude oil is large, and the gravity tends to flow to the top of the oil layer due to the influence of gravity. The steam front is not perpendicular to the oil layer, but is inclined, and does not affect the middle oil layer.
(2) Second, the viscosity of the steam is low, the viscosity of the crude oil in the oil layer is high, and the steam protrudes in a finger shape, which limits the heating range of the steam and reduces the degree of use of the oil layer in the longitudinal direction. The combination of the two shows that the steam is heated longitudinally unevenly, and the remaining oil saturation in the middle and lower layers is high. The production of heavy oil in a single well gradually decreases as the number of throughput cycles increases. There is an urgent need to improve the profile of the gas injection steam to increase the volume of the sweep.
(3) Nitrogen has three main effects in assisting steam stimulation. First, by expanding the steam sweep volume, the steam is better heated to the crude oil in the middle and lower formations that were difficult to reach in the original steam stimulation, reducing the residual oil saturation in the middle and lower formations.
(4) Second, nitrogen is a poor conductor of heat, which insulates the top layer and prevents heat from escaping in the top layer. The heat that effectively acts on the crude oil is enhanced.
(5) Third, the gas drive of nitrogen can supplement a certain formation energy.

Acknowledgements. This paper has been supported by National Natural Science Foundation of China (No. 41672132).

References

1. Liu, Y., Fan, H.: The effect of hydrogen donor additive on the viscosity of heavy oil during steam stimulation. Energy Fuels **16**(4), 842–846 (2002)
2. Xu, A., Mu, L., Fan, Z., et al.: Mechanism of heavy oil recovery by cyclic superheated steam stimulation. J. Petrol. Sci. Eng. **111**(21), 197–207 (2013)
3. Sun, F., Li, C., Cheng, L., et al.: Production performance analysis of heavy oil recovery by cyclic superheated steam stimulation. Energy **121**, 356–371 (2017)
4. Yu-Xiao, Z., Lin, J.: Optimization of horizontal well production parameters for cyclic steam stimulation in ultra-heavy oil reservoir. Sci. Technol. Eng. (2013)

5. Jian, Y., Xiangfang, L., Zhangxing, C., et al.: A productivity prediction model for cyclic steam stimulation in consideration of non-Newtonian characteristics of heavy oil. Acta Petrolei Sin. (2017)
6. Bao-Shan, L., Company, L.O., CNPC.: Research and practice of air injection assisted steam stimulation mechanism in heavy oil reservoirs. Contemporary Chem. Ind. (2014)
7. Stewart, N.E.: Steam stimulation process for recovering heavy oil (1981)
8. Wang, X., Qiu, X.: Application of particle swarm optimization for enhanced cyclic steam stimulation in a offshore heavy oil reservoir (2013)
9. Huang, S.J., Hu, P., Li, Q.: Study on heating zone and producing zone of cyclic steam stimulation with horizontal well in heavy oil reservoir. Adv. Mater. Res. **594–597**, 4 (2012)
10. Chuanliang, Y., Yuanfang, C., Ji, T., et al.: The influence of steam stimulation on compression coefficient of heavy oil reservoirs. J. Petrol. Explor. Prod. Technol. (2017)
11. Wang, Y., Zhang, L., Deng, J., et al.: An innovative air assisted cyclic steam stimulation technique for enhanced heavy oil recovery. J. Petrol. Sci. Eng. **151**, 254–263 (2017)
12. Xuezhong, W.: Research on heavy oil nitrogen foam assisted steam stimulation in Chunfeng oilfield. Pet. Petrochemical Today (2017)
13. Hui, L.: Exploitation effects of heavy oil reservoir using steam stimulation and steam stimulation plus N_2 and CO_2. Tianjin Sci. Technol. (2016)
14. Liu, Y.T., Gu, W.H.: Optimal orthogonal design for herringbone wells injection and production parameters of steam stimulation in heavy oil reservoir. Adv. Mater. Res. **383–390**, 5 (2011)
15. Putra, E.M.F.: Continuous Improvement of Implementation of cyclic steam stimulation well design to optimize the heavy oil exploration program in IH field, south sumatra. Indonesian Petroleum Assoc. (2013)
16. Rui-Shan, L.I., Jianhua, H.E., Yin-Ming, T., et al.: Experiment on the mechanism of nitrogen-assisted steam stimulation in heavy oil reservoirs. J. Oil Gas Technol. (2006)
17. Rui-Lan, L., Lin-Song, C.: Feasibility study of CO_2 injection for heavy oil reservoir after cyclic steam stimulation. Xinjiang Petrol. Geol. (2004)
18. Rong, G.Y., Qi, L.S., Huang, S.D., et al.: Study on N_2 and solvent assisted steam stimulation in a super-heavy oil reservoir. Petrol. Explor. Devel. **30**(2), 73–75 (2003)
19. Dianfa, D.U., Qing, G., Qing, W., et al.: Feasibility analysis of nitrogen-assisted steam stimulation in super-heavy oil reservoir. Oil Drilling Prod. Technol. (2008)
20. Fumoto, E., Matsumura, A., Sato, S., et al.: Kinetic model for catalytic cracking of heavy oil with a zirconia-alumina-iron oxide catalyst in a steam atmosphere. Energy Fuels **23** (2009)
21. Zhan-Xiang, L.: Research on pressure drop interpretation in the stage of stewing in steam soak of heavy oil recovery. Pet. Geol. Oilfield Devel. Daqing (2006)
22. Farias, F.P.M, Buriti, C.J.O., de Barbosa, L.W.C.P., et al.: The effect of droplet diameter on the separation of heavy-oil from water using a hydrocyclone. Defect Diffus. Forum **303–304**, 131–137 (2010)
23. Farias, F.P.M, Buriti, C.J.O., de Barbosa, L.W.C.P., et al.: The effect of droplet diameter on the separation of heavy-oil from water using a hydrocyclone. Defect Diffus. Forum. (2010)

The manufacturer's authorised representative in the EU is Springer Nature Customer Service Centre GmbH, Europaplatz 3, 69115 Heidelberg, Germany. If you have any concerns regarding our products, please contact ProductSafety@springernature.com

Printed and bound by CPI Group (UK) Ltd, Croydon, CR0 4YY
25/09/2026
02222884-0013